MATHEMATICAL METHODS

for Engineers and Scientists

4th Edition

by

G. F. Fitz-Gerald

and

I. A. Peckham

with contributions from

G. T. Clarke

and

I. H. Grundy

Pearson SprintPrint is an imprint of Pearson Education Australia Pty Limited. It has been established to provide academics throughout Australia and New Zealand with fast and efficient access to the printing, warehousing and distribution services of Australia's leading educational publisher, ensuring a smooth supply channel to your campus bookseller.

Texts published under the **Pearson SprintPrint** banner do not undergo the full editorial production processes. The Author and Pearson Education Australia have made every effort, in the time available, to ensure an error-free book.

For more information about the **Pearson SprintPrint** service, contact the Editorial Department, Pearson Education Australia, Unit 4, Level 2, 14 Aquatic Drive, Frenchs Forest, New South Wales, 2086. Telephone: 02 9454 2200.

Acquisitions Editor: Frances Eden
Editorial Coordinator: Roisin Fitzgerald
Production Administrator: Barbara Honor

1 2 3 4 5 08 07 06 05 04

ISBN 978 0 7339 7512 7

Pearson Education Australia
Unit 4, Level 3,
14 Aquatic Drive
Frenchs Forest NSW 2086
www.pearsoned.com.au

An imprint of Pearson Education Australia

Contents

Preface

These notes are designed for part of a first year course in introductory calculus offered at first-year level in an Australian tertiary institution and assume the student has successfully completed a standard secondary school calculus course. They are *not* intended to be a "textbook" in the usual sense of the word but rather a set of working notes.

This set of notes has been developed by the authors after many years' experience teaching similar material both at RMIT and various other tertiary institutions within Australia. Many of the exercises included at the end of each chapter have appeared as questions in examination papers in such first year courses.

A feature of these notes is the inclusion of self-help exercises and a quick test in each chapter. These exercises and tests have a two-fold aim. They

- provide practice in the fundamental techniques underlying the theoretical material introduced in the chapter; and
- reinforce basic ideas that have been introduced in the chapter.

The self-help exercises are included at strategic points in the notes so that the student can continually assess how well fundamental concepts and ideas have been understood. These self-help exercises should be attempted *before* attempting any exercises at the end of each chapter. A poor response to these exercises and tests probably indicates that more effort needs to be put into thoroughly understanding the material contained in the corresponding chapter before attempting the miscellaneous exercises at the end of that chapter.

The quick tests are modifications of material that was first introduced by Ian Grundy of the Mathematics Department at RMIT. An example of the style adopted for these tests can be found on page xi. The material covered in this introductory test covers topics that you should already be familiar with as a result of your previous mathematical studies. Attempt this test now to check out your preparation for this course.

The authors thank Michael Liddell for his assistance in the preparation of many of the graphics images appearing throughout this text. His cheerfulness under the pressure of strict deadlines was of great benefit at a difficult time in the preparation of an earlier draft of these notes. They also thank Andy White for his excellent contribution to the conversion of all the graphics images to their current consistent state. He also managed to meet a tight deadline in spite of a period of several weeks deprivation from his beloved computer whilst it was undergoing urgent repairs.

The authors thank both Theo Doukakaros and Ian Grundy for their careful and diligent proofreading of these notes; particularly Ian Grundy's working of all the set exercises and the opportunity to include his "Quick Tests". They also thank Steve Gretton for his contribution in proofreading the entire document.

If you wish to contact the authors, you may send electronic mail to the address garyfitz@rmit.edu.au.

Sample Quick Test

Basic Mathematical Skills

Question	Selection
1. $2^a 2^b$ equals	(a) 2^{a+b} (b) 2^{ab} (c) 4^{ab} (d) 4^{a+b}
2. $2^2 3^3$ equals	(a) 6^6 (b) 5^5 (c) 6^5 (d) none of these
3. $\dfrac{1}{(x+y)^2} = (x+y)^{-2} = x^{-2} + y^{-2}$?	(a) True (b) False
4. $\sqrt{x^2 - 16} = x - 4$?	(a) True (b) False
5. $x = 1$ is a root (solution) of the equation $2x^3 - 4x^2 + 11x - 9 = 0$?	(a) True (b) False
6. $3x \sin \dfrac{x}{3} = \sin \dfrac{x}{3} 3x = \sin x^2$?	(a) True (b) False
7. $\dfrac{x}{x+1}$ equals	(a) 1 (b) $1 - \dfrac{1}{x+1}$ (c) $\dfrac{1}{2}$ (d) $\dfrac{1}{x}$ (e) 0
8. $2x + 2y = 9$ is the equation of a	(a) circle, $r = 3$ (b) circle, $r = 9$ (c) straight line
9. $\dfrac{1}{3x+2}$ equals	(a) $\dfrac{1}{3x} + \dfrac{1}{2}$ (b) $\dfrac{1}{3}x + \dfrac{1}{2}$ (c) neither
10. $(x+5)\,x - 1$ equals	(a) $x^2 + 4x - 5$ (b) $x^2 + 5x - 1$ (c) neither
11. $x + 2\,(x - 3)$ equals	(a) $3x - 6$ (b) $x^2 - x - 6$ (c) neither
12. $-2(x + y^2) = -2x + 2y^2$?	(a) True (b) False
13. The number of real roots of $x^2 - 5x + 8 = 0$ is	(a) 2 (b) 1 (c) 0
14. $(1 + \dfrac{4}{x^2})\,x^2 = 5$?	(a) True (b) False
15. $1 + x\,(1+x)^{-1/2} = \dfrac{1+x}{\sqrt{1+x}} = \sqrt{1+x}$?	(a) True (b) False
16. $\dfrac{(2x)^3}{3} - \dfrac{x^3}{3}$ equals	(a) $\dfrac{x^3}{3}$ (b) $\dfrac{7x^3}{3}$ (c) neither
17. $\dfrac{x^n}{x^{n-1}}$ equals	(a) $\dfrac{1}{x}$ (b) x^{2n-1} (c) x (d) $\dfrac{n}{n-1}$

Part I

First Semester

Topic 1

Vectors

1.1 Introduction

Physical quantities are either

- Scalars — completely specified by *magnitude* alone.

  Examples of scalar quantities include mass, temperature, volume, time, specific heat, conductivity, pressure, permeability etc.

or

- Vectors — specified by *magnitude and direction.*

  Examples of vector quantities include force, velocity, acceleration, momentum, torque, electric and magnctic ficld ctc.

1.2 Geometric Vectors

1.2.1 Definition

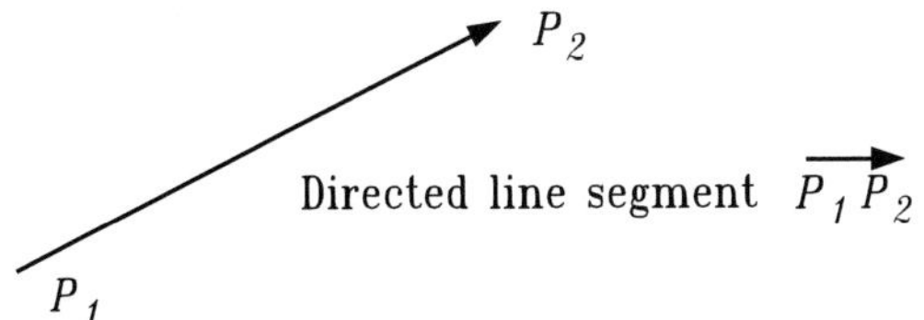

If P_1 and P_2 are two points in space, we denote by $\overrightarrow{P_1P_2}$ the directed line segment from P_1 to P_2. The length of the directed line segment is then the length of the line joining P_1 and P_2.

Definition 1.1 *A* geometric vector *is the set of* all *directed line segments having the same length and direction as a given directed line segment.*

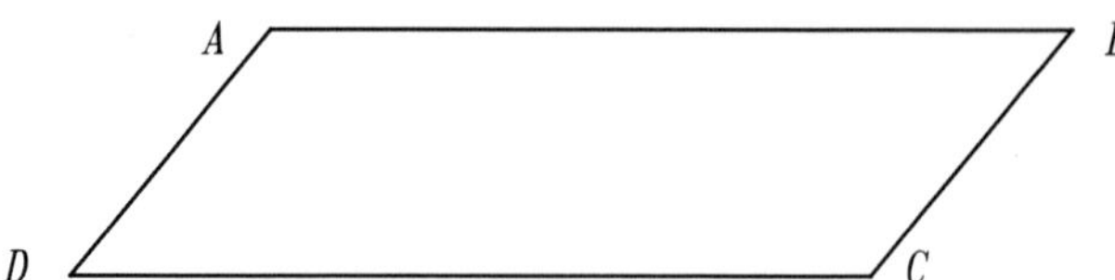

We will use the *same* symbol $\mathbf{a}$ to represent both the geometric vector and any representative directed line segment that can be used to define it. Thus in the parallelogram $ABCD$ pictured above, if we define the geometric vector $\mathbf{a}$ to consist of all the directed line segments with the same length and direction as $\overrightarrow{AB}$ and similarly the geometric vector $\mathbf{b}$ to consist of all directed line segments with the same length and direction as $\overrightarrow{AD}$, then we have

$$\overrightarrow{AB} = \overrightarrow{DC} = \mathbf{a} \text{ and } \overrightarrow{AD} = \overrightarrow{BC} = \mathbf{b}.$$

Also, we will use the term "vector" itself to mean geometric vector.

1.2.2 Multiplication by Scalars, Addition and Subtraction

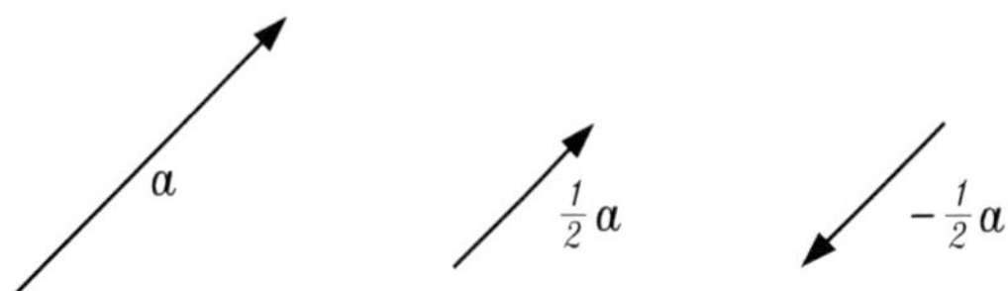

In this section we introduce the basic properties of adding and subtracting geometric vectors as well as the effect of multiplication of a geometric vector by a scalar quantity.

Multiplication by Scalars

Definition 1.2 *The vector* $r\,\mathbf{a}$ *is the vector of length* $|r|$ *times the length of* $\mathbf{a}$ *that points in the same direction as* $\mathbf{a}$ *if* $r > 0$ *and points in the opposite direction to* $\mathbf{a}$ *if* $r < 0$.

Definition 1.3 *Two vectors* $\mathbf{a}$ *and* $\mathbf{b}$ *are said to be* parallel *if and only if* $\mathbf{a} = k\,\mathbf{b}$ *for some real constant* k.

Addition of Vectors

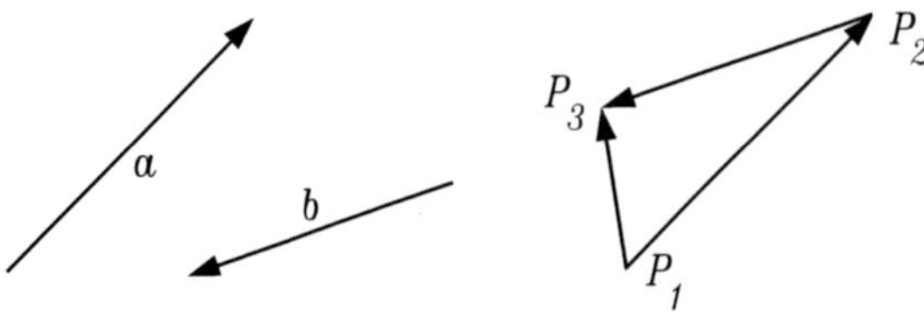

If the directed line segment $\overrightarrow{P_1P_2}$ represents the vector $\mathbf{a}$ and $\overrightarrow{P_2P_3}$ represents the vector $\mathbf{b}$ then the directed line segment $\overrightarrow{P_1P_3}$ represents a vector $\mathbf{c}$ which we call $\mathbf{a} + \mathbf{b}$.

The method of construction is referred to as the parallelogram or triangle rule of vector addition.

Subtraction of Vectors

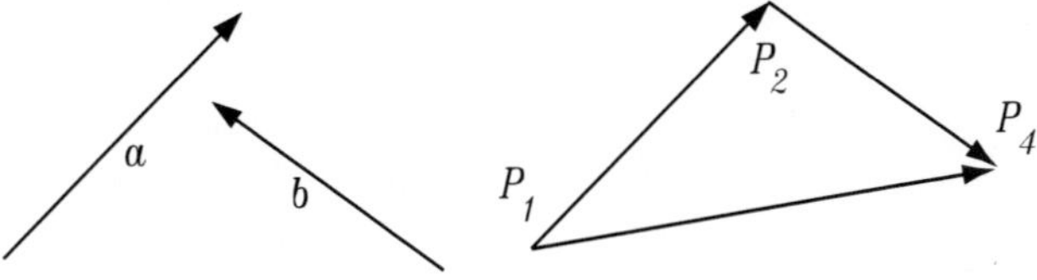

If the directed line segment $\overrightarrow{P_1P_2}$ represents the geometric vector **a** and $\overrightarrow{P_2P_4}$ represents the geometric vector $-\mathbf{b}$ then the directed line segment $\overrightarrow{P_1P_4}$ represents a geometric vector **c** which we call $\mathbf{a} - \mathbf{b}$ (that is $\mathbf{a} + (-\mathbf{b})$).

Worked Example 1.2.1 *Use vector arguments to show that the medians of a triangle intersect at the point of trisection of each.*

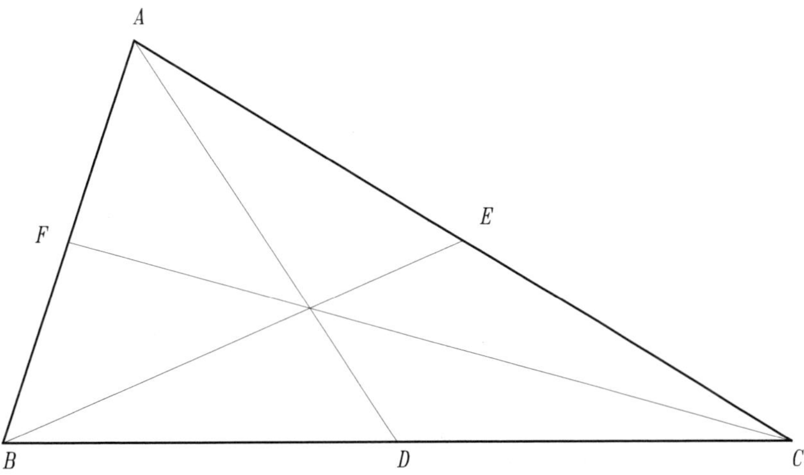

Figure 1.1: The triangle of Worked Example 1.2.1.

With reference to Figure 1.1, let $\overrightarrow{AB}$ represent the vector **a** and $\overrightarrow{BC}$ represent the vector **b**.
Then $\overrightarrow{AC} = \overrightarrow{AB} + \overrightarrow{BC}$ will represent the vector $\mathbf{a} + \mathbf{b}$.
Let G_1 be the point of trisection of the line segment AD.
Then $\overrightarrow{AG_1} = \frac{2}{3}\overrightarrow{AD}$ represents the vector $\frac{2}{3}(\mathbf{a} + \frac{1}{2}\mathbf{b}) = \frac{2}{3}\mathbf{a} + \frac{1}{3}\mathbf{b}$.
Let G_2 be the point of trisection of the line segment BE.
Then $\overrightarrow{AG_2} = \overrightarrow{AB} + \frac{2}{3}\overrightarrow{BE}$ represents the vector

$$\mathbf{a} + \frac{2}{3}(\mathbf{b} + \frac{1}{2}(-\mathbf{a} - \mathbf{b})) = \frac{2}{3}\mathbf{a} + \frac{1}{3}\mathbf{b}.$$

Therefore G_1 and G_2 are the same point.
Similarly, if G_3 is the point of trisection of the line segment CF then $\overrightarrow{AG_3}$ represents the same vector $\frac{2}{3}\mathbf{a} + \frac{1}{3}\mathbf{b}$. ■

Self-help exercises

Use vector methods to show that

1. the mid-points of the sides of a quadrilateral form the vertices of a parallelogram

2. the line segment joining the mid-points of two sides of a triangle is parallel to the third side and its length is half the length of this third side.

1.2.3 Components of a Vector

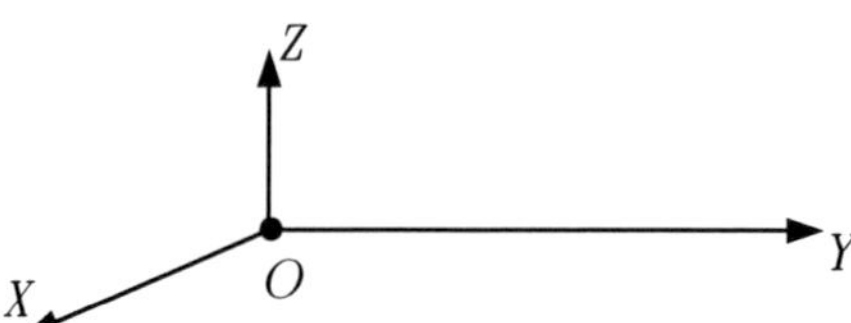

Consider the standard right-handed Cartesian co-ordinate system $OXYZ$ drawn at left. If $\mathbf{a}$ is a given vector then there will always be a point P with co-ordinates given by (x, y, z) with the property that the directed line segment $\overrightarrow{OP}$ can be used to represent this vector $\mathbf{a}$; that is, $\overrightarrow{OP} = \mathbf{a}$.

Definition 1.4 *The directed line segment* $\overrightarrow{OP}$ *is called the* position vector *of the point* P.

We write

$$\boxed{\overrightarrow{OP} = \mathbf{a} = (\mathrm{x},\ \mathrm{y},\ \mathrm{z})}$$

Also, defining addition of ordered triples and their multiplication by a scalar in the natural way; that is, by

$$(x_1, x_2, x_3) + (y_1, y_2, y_3) = (x_1 + y_1, x_2 + y_2, x_3 + y_3) \text{ and}$$
$$k(x, y, z) = (kx, ky, kz)$$

we can write

$$\begin{aligned} \overrightarrow{OP} = (x, y, z) &= (x, 0, 0) + (0, y, 0) + (0, 0, z) \\ &= x(1, 0, 0) + y(0, 1, 0) + z(0, 0, 1) \end{aligned}$$

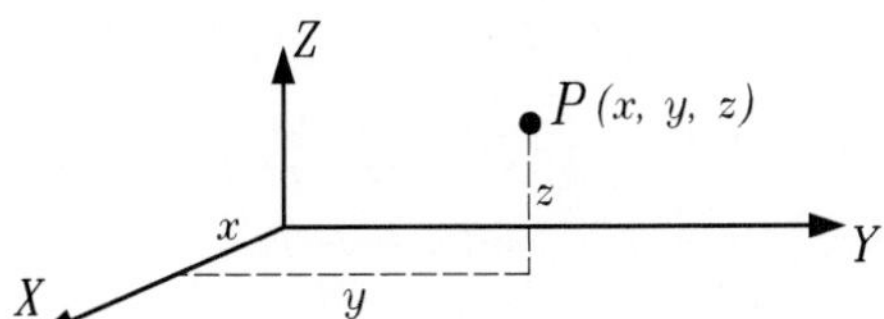

Here, $(1, 0, 0)$ is a vector of length 1 and is parallel to the X-axis which we call the unit vector $\mathbf{i}$. (We define a unit vector for general vectors later in this section — see Definition 1.7 on page 11.) Similarly, $(0, 1, 0)$ is a vector of length 1 that is parallel to the Y-axis which we denote by $\mathbf{j}$ and $(0, 0, 1)$ is a vector of length 1 parallel to the Z-axis which we denote by $\mathbf{k}$. Thus, with reference to the diagram above

$$\boxed{\overrightarrow{OP} = (\mathrm{x},\ \mathrm{y},\ \mathrm{z}) = \mathrm{x}\,\mathbf{i} + \mathrm{y}\,\mathbf{j} + \mathrm{z}\,\mathbf{k}.}$$

(The values of x, y and z are referred to as the *components* of the vector $\overrightarrow{OP}$.)

Convention

In these notes we make no distinction between the notation $a\,\mathbf{i} + b\,\mathbf{j} + c\,\mathbf{k}$ for a vector and the co-ordinates (a, b, c) of the point P for which $\overrightarrow{OP}$ is its position vector; the context in which the representation (a, b, c) is used will indicate whether it represents either a vector or the co-ordinates of a particular point in three-dimensional space.

▮**Example:** The position vector of the point $P(1, 2, 3)$ is the vector

$$\overrightarrow{OP} = (1, 2, 3) = \mathbf{i} + 2\,\mathbf{j} + 3\,\mathbf{k}.$$

▮**Example:** If A is the point $(1, -1, 2)$ and B is the point $(2, 1, -4)$ then

$$\overrightarrow{AB} = \overrightarrow{AO} + \overrightarrow{OB} = \overrightarrow{OB} - \overrightarrow{OA} = \mathbf{i} + 2\,\mathbf{j} - 6\,\mathbf{k} = (1, 2, -6).$$

That is, the vector $\overrightarrow{AB}$ is simply the co-ordinates of the point B minus the co-ordinates of the point A.

Self-help exercises

For the three points $A(1, -1, 4)$, $B(-2, 7, -2)$ and $C(3, 5, 11)$, determine the position vectors of

1. the mid-point P of the line segment AB $[(-\frac{1}{2}, 3, 1)]$
2. the point Q that divides the line segment CA in the ratio of 3:1 $[(\frac{3}{2}, \frac{1}{2}, \frac{23}{4})]$
3. the point D for which the four points $ABCD$ form a parallelogram. $[(6, -3, 17)]$

1.2.4 Dot (or Scalar) Product

Definition 1.5 *The* dot (or scalar) product $\mathbf{a} \cdot \mathbf{b}$ *of the two vectors* $\mathbf{a} = (a_1, a_2, a_3)$ *and* $\mathbf{b} = (b_1, b_2, b_3)$ *is the* real number (or scalar) *defined by*

$$\boxed{\mathbf{a} \cdot \mathbf{b} = a_1 b_1 + a_2 b_2 + a_3 b_3}$$

Note that the value of the dot product of two vectors does not depend on the order in which they appear in the formula, that is

$$\boxed{\mathbf{a} \cdot \mathbf{b} = \mathbf{b} \cdot \mathbf{a}}$$

▮**Example:** $(1, 2, 4) \cdot (-1, 3, 0) = -1 + 6 + 0 = 5.$

▮**Example:** $(2\,\mathbf{i} + 3\,\mathbf{j} - \mathbf{k}) \cdot (-\mathbf{i} - 2\,\mathbf{j} + 5\,\mathbf{k}) = -2 - 6 - 5 = -13.$

Self-help exercise

Using the vectors $\mathbf{a} = (-1, 2, 4)$, $\mathbf{b} = (2, 3, 1)$ and $\mathbf{c} = (5, -3, 2)$, show that

$$\boxed{\mathbf{a} \cdot (\mathbf{b} + \mathbf{c}) = \mathbf{a} \cdot \mathbf{b} + \mathbf{a} \cdot \mathbf{c}}$$

(That is, the dot product is distributive over vector addition.)

1.2.5 The Length of a Vector

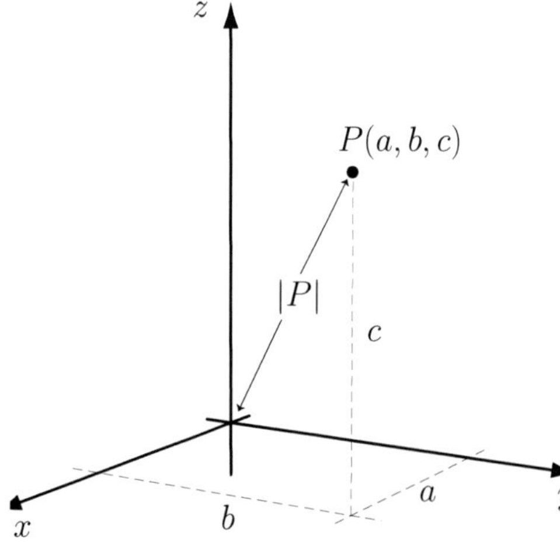

In this section we use the standard formula of analytic geometry that defines the distance between any two points lying in three-dimensional space to define the length of a vector. The result follows by two simple applications of Pythagoras' theorem.

Definition 1.6 *The* length *of a geometric vector* $\mathbf{a} = (a_1, a_2, a_3)$ *is denoted by* $|\mathbf{a}|$ *or just* a *and is given by*

$$\boxed{|\mathbf{a}| = \sqrt{a_1^2 + a_2^2 + a_3^2} = \sqrt{\mathbf{a} \cdot \mathbf{a}}}$$

▮**Example:** $|(1, -4, 7)| = \sqrt{1 + 16 + 49} = \sqrt{66}.$

Self-help exercises

For the vectors $\mathbf{a} = (1, -2, 4)$, $\mathbf{b} = (3, 5, -11)$ and $\mathbf{c} = (-1, 1, 6)$, determine the length of the vectors

1. $\mathbf{a}, \mathbf{b}$ and $\mathbf{c}$
2. $\mathbf{a} - 3\,\mathbf{c}$
3. $\mathbf{a} + 2\,\mathbf{b} - 4\,\mathbf{c}$.

[$\sqrt{21}$, $\sqrt{155}$, $\sqrt{38}$; $\sqrt{237}$; $\sqrt{1901}$]

1.2.6 The Angle Between Two Vectors

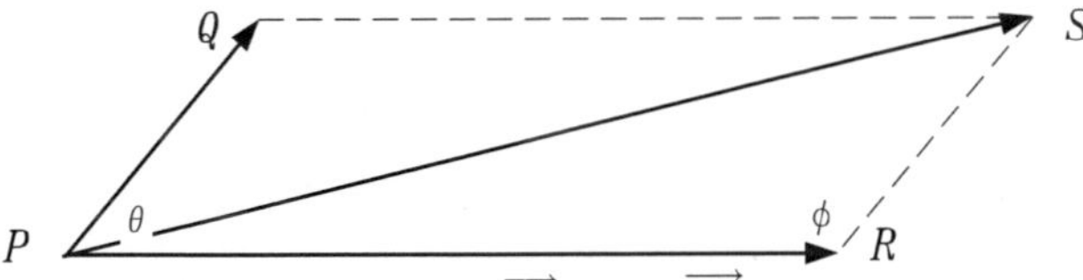

With reference to the diagram at left, let $\mathbf{a}$ and $\mathbf{b}$ be two given vectors represented by the directed line segments $\overrightarrow{PQ}$ and $\overrightarrow{PR}$ respectively. Let the angle between them be θ. (Using the triangle rule for vector addition, the directed line segment $\overrightarrow{PS}$ will represent the vector $\mathbf{a} + \mathbf{b}$.)

Now

$$\begin{aligned} |\mathbf{a}+\mathbf{b}|^2 &= |\mathbf{a}|^2 + |\mathbf{b}|^2 - 2|\mathbf{a}||\mathbf{b}|\cos\phi \\ &= |\mathbf{a}|^2 + |\mathbf{b}|^2 + 2|\mathbf{a}||\mathbf{b}|\cos\theta, \end{aligned}$$

since $\theta + \phi = 180°$. Also

$$\begin{aligned} |\mathbf{a}+\mathbf{b}|^2 &= (\mathbf{a}+\mathbf{b})\cdot(\mathbf{a}+\mathbf{b}) \\ &= \mathbf{a}\cdot\mathbf{a} + \mathbf{b}\cdot\mathbf{a} + \mathbf{a}\cdot\mathbf{b} + \mathbf{b}\cdot\mathbf{b} \\ &= |\mathbf{a}|^2 + |\mathbf{b}|^2 + 2\,\mathbf{a}\cdot\mathbf{b}. \end{aligned}$$

Thus, equating terms, we get

$$\boxed{\mathbf{a}\cdot\mathbf{b} = |\mathbf{a}||\mathbf{b}|\cos\theta}$$

(This last expression can be used as an alternative definition for the "dot product" of two vectors; but then, of course, the Cartesian representation needs to be *deduced* from it.)

The two formulae for the dot product of two vectors provide a method for determining the angle between two vectors. (Note that if $\mathbf{a}\cdot\mathbf{b} < 0$ then the angle between the vectors is obtuse.)

▮ **Worked Example 1.2.2** *Determine the angle between the two vectors*

$$\mathbf{a} = (2, -1, 4) \text{ and } \mathbf{b} = (-7, 11, 13).$$

Using the definition of dot product we have

$$\mathbf{a} \cdot \mathbf{b} = -14 - 11 + 52 = 27.$$

Also, by the alternative representation of the dot product, we have

$$\mathbf{a} \cdot \mathbf{b} = |\mathbf{a}||\mathbf{b}| \cos\theta = \sqrt{21}\sqrt{339} \cos\theta.$$

Therefore $\cos\theta = 27/\sqrt{7119}$ or $\theta \approx 71.34°$. ■

Note:

Two non-zero vectors $\mathbf{a}$ and $\mathbf{b}$ are perpendicular if and only if $\mathbf{a} \cdot \mathbf{b} = 0$. We say such vectors are *orthogonal.*

Example: The vectors $(1, -1, 4)$ and $(-2, 2, 1)$ are orthogonal because

$$(1, -1, 4) \cdot (-2, 2, 1) = 0.$$

Direction Cosines

If the angle between a vector $\mathbf{a}$ and the x-axis is denoted by α then

$$\mathbf{a} \cdot \mathbf{i} = |\mathbf{a}| \cos\alpha.$$

Thus

$$\boxed{\cos\alpha = \hat{\mathbf{a}} \cdot \mathbf{i}}$$

Similarly, if β and γ are the angles between the vector $\mathbf{a}$ and the y-axis and z-axis respectively, then

$$\boxed{\cos\beta = \hat{\mathbf{a}} \cdot \mathbf{j} \quad \text{and} \quad \cos\gamma = \hat{\mathbf{a}} \cdot \mathbf{k}}$$

The three angles α, β and γ are referred to as the *direction angles* of the vector $\mathbf{a}$, and their cosines the *direction cosines.*

Self-help exercises

1. For each of the following pairs of vectors, determine if they are orthogonal. If they are not orthogonal determine the cosine of the angle between them.

   (a) $(1, -2, 4)$ and $(-2, 4, 2)$ [$\cos\theta = -\frac{1}{3\sqrt{14}}$]

   (b) $(2, -1, 2)$ and $(1, -2, 3)$ [$\cos\theta = \frac{10}{3\sqrt{14}}$]

(c) $(1,-5,11)$ and $(-6,5,-1)$. $[\cos\theta = -\frac{42}{\sqrt{9114}}]$

2. Determine the value of α for which the vector $(1,-3,\alpha)$ and the vector $(2,-1,7)$ are orthogonal. For all other values of α determine the cosine of the angle between them. $[\alpha = -\frac{5}{7}]$

1.2.7 Projections — Geometric Interpretation of the Dot Product

Scalar Projection

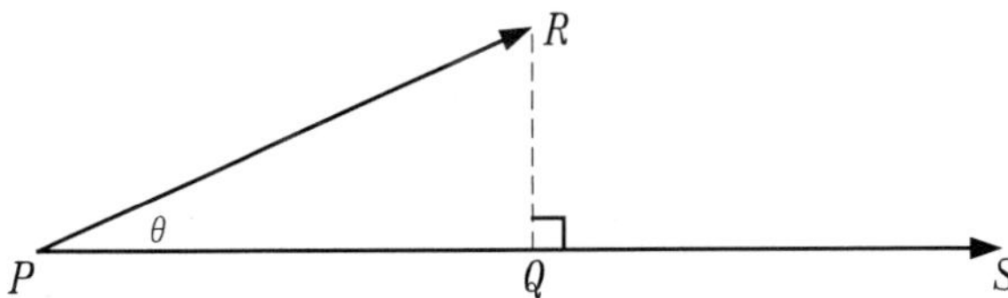

Projection simply means the component of a vector in a given direction. With reference to the diagram at left, let $\overrightarrow{PR}$ represent the geometric vector $\mathbf{a}$ and $\overrightarrow{PS}$ represent the geometric vector $\mathbf{b}$. The length of the straight line segment PQ is the (scalar) projection of the vector $\mathbf{a}$ onto the vector $\mathbf{b}$ (or in the direction of the vector $\mathbf{b}$). But

$$PQ \quad = \quad |\overrightarrow{PQ}| \quad = \quad |\mathbf{a}|\cos\theta \quad = \quad |\mathbf{a}|\cdot 1\cdot\cos\theta.$$

Therefore

$$PQ = |\mathbf{a}|\cdot\frac{|\mathbf{b}|}{|\mathbf{b}|}\cdot\cos\theta = \frac{|\mathbf{a}|\cdot|\mathbf{b}|\cdot\cos\theta}{|\mathbf{b}|} = \frac{\mathbf{a}\cdot\mathbf{b}}{|\mathbf{b}|} = \mathbf{a}\cdot\frac{\mathbf{b}}{|\mathbf{b}|}.$$

Note that $\mathbf{b}/|\mathbf{b}| = (1/|\mathbf{b}|)\mathbf{b}$ is a vector in the same direction as $\mathbf{b}$ with length $|\mathbf{b}|/|\mathbf{b}| = 1$.

Definition 1.7 *A vector* $\mathbf{a}$ *of unit length is called a* unit vector *and we use the symbol* $\hat{\mathbf{a}}$ *to denote such unit vectors.*

▌**Example:** A unit vector parallel to the vector

$$\mathbf{a} = (1,2,-1)$$

is the vector

$$\hat{\mathbf{a}} \;=\; (1/\sqrt{6}, 2/\sqrt{6}, -1/\sqrt{6}).$$

In terms of this notation, the (scalar) projection of the vector $\mathbf{a}$ onto the vector $\mathbf{b}$ is given by

$$\boxed{\mathbf{a}\cdot\hat{\mathbf{b}}}$$

Example: The (scalar) projection of the vector

$$\mathbf{c} = (1, 2, 3)$$

in the direction of the vector

$$\mathbf{a} = (1, -2, 5)$$

is given by

$$\mathbf{c} \cdot \hat{\mathbf{a}} = (1, 2, 3) \cdot \frac{(1, -2, 5)}{\sqrt{30}} = \frac{2}{5}\sqrt{30}.$$

Self-help exercises

Determine

1. the (scalar) projection of the vector $(-3, 5, 11)$ in the direction $(2, -7, 3)$ $[-\frac{8}{\sqrt{62}}]$

2. the work done by the force $\mathbf{F} = (5, -2, 3)$ in moving a particle of unit mass from the point $A(-3, 1, -7)$ to the point $B(2, 1, 5)$. [61]

Vector Projection

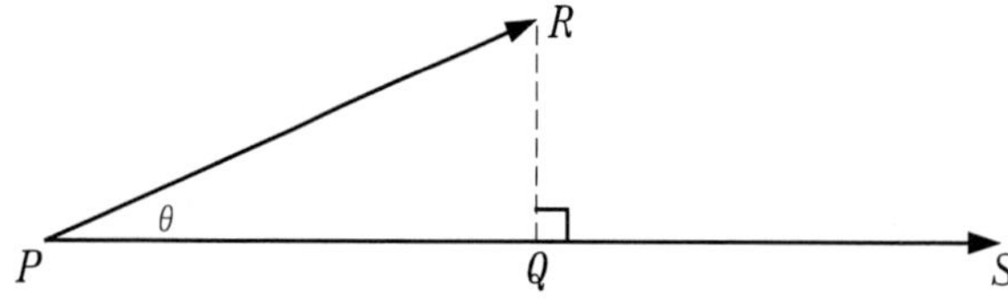

With reference to the diagram at left, let $\overrightarrow{PR}$ represent the vector **a** and $\overrightarrow{PS}$ represent the vector **b**. The vector $\overrightarrow{PQ}$ is the vector projection of **a** onto **b**. We have

$$\overrightarrow{PQ} = |\overrightarrow{PQ}|\,\hat{\mathbf{b}} = |\overrightarrow{PQ}|\frac{\mathbf{b}}{|\mathbf{b}|} = \left(\mathbf{a} \cdot \frac{\mathbf{b}}{|\mathbf{b}|}\right)\frac{\mathbf{b}}{|\mathbf{b}|}.$$

Thus the (vector) projection of the vector **a** onto the vector **b** is given by

$$\boxed{(\mathbf{a} \cdot \hat{\mathbf{b}})\,\hat{\mathbf{b}}}$$

Example: The (vector) projection of $(1, 2, 3)$ on $(1, 1, 1)$ is

$$2\sqrt{3}(1, 1, 1)/\sqrt{3} = (2, 2, 2).$$

Self-help exercise

Determine the vector projection of the vector $(-2,3,7)$ on the vector $(4,-1,2)$.

$[(\frac{4}{7},-\frac{1}{7},\frac{2}{7})]$

Resolution of vectors into mutually orthogonal components

The last worked example above provides us with a way to resolve any vector $\mathbf{a}$ ($\overrightarrow{PR}$ in the above diagram) into mutually orthogonal components — that is, to be able to express the vector $\mathbf{a}$ as the sum of two vectors (say $\mathbf{r}$ and $\mathbf{s}$), one of which may be parallel to a specified direction, in such a way that $\mathbf{r}\cdot\mathbf{s}=0$. If we treat $\mathbf{r}$ as the vector projection of $\mathbf{a}$ in the specified direction ($\overrightarrow{PQ}$ in the above diagram), then the vector $\mathbf{s}$ ($\overrightarrow{QR}$ in the above diagram) is simply the vector we add to this vector projection to return the original vector $\mathbf{a}$. That is, $\mathbf{s}=\mathbf{a}-\mathbf{r}$.

▮**Example:** Using the above worked example we have

$$(1,2,3)=(2,2,2)+(-1,0,1),$$

where $(2,2,2)$ is orthogonal to $(-1,0,1)$ (as it must by this method of construction).

▮ **Worked Example 1.2.3** *Resolve the vector* $(3,-2,5)$ *into the sum of two mutually orthogonal vectors, one of which is parallel to the direction* $(1,1,3)$.
The vector projection of $(3,-2,5)$ in the direction of $(1,1,3)$ is given by

$$(3,-2,5)\cdot\frac{(1,1,3)}{\sqrt{11}}\frac{(1,1,3)}{\sqrt{11}}=\frac{16}{11}(1,1,3).$$

Thus

$$(3,-2,5)=\frac{16}{11}(1,1,3)+\frac{1}{11}(17,-38,7),$$

and $(1,1,3)\cdot(17,-38,7)=0$, as required. ■

Self-help exercises

Resolve the vector $(7,1,-5)$ into the sum of two mutually orthogonal vectors, one of which is parallel to the

1. x-axis $[(7,0,0)+(0,1,-5)]$
2. y-axis $[(0,1,0)+(7,0,-5)]$
3. vector $(1,1,1)$. $[(1,1,1)+(6,0,-6)]$

1.3 Lines and Planes

1.3.1 Lines

To uniquely define a line we need to know a point on the line and the line's direction. (Note that this information can easily be obtained from any other specification of the line; for example, if we are given two distinct points lying on it.)

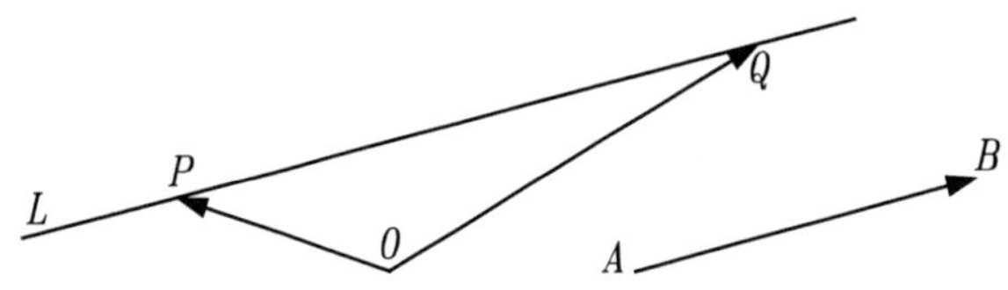

With reference to the diagram at left, we require the equation of the straight line (denoted by L) that passes through the point P and is parallel to the directed line segment $\overrightarrow{AB}$ (that represents the vector •).

Let the co-ordinates of the point P be (a_1, a_2, a_3). Then the position vector of P is given by

$$\overrightarrow{OP} = (a_1, a_2, a_3) = \mathbf{a}.$$

A point Q also lies on this line if and only if $\overrightarrow{PQ}$ is parallel to the line's given direction • ; that is

$$\overrightarrow{PQ} = t\bullet\ , \text{ for some real number } t. \tag{1.1}$$

Let the co-ordinates of the general point Q on L be (x, y, z). Then the position vector of Q is given by

$$\overrightarrow{OQ} = (x, y, z) = \mathbf{r}.$$

Thus

$$\overrightarrow{PQ} = (x, y, z) - (a_1, a_2, a_3) = \mathbf{r} - \mathbf{a},$$

so that Equation 1.1 becomes

$$\mathbf{r} - \mathbf{a} = t\bullet \text{ for some real } t.$$

That is

$$\mathbf{r} = \mathbf{r}(t) = \mathbf{a} + t\bullet \text{ for some real } t. \tag{1.2}$$

Note that each value of $t = t_0$ defines a unique point P_0 on L.

Equation 1.2 is called the *vector equation* of the line L. (In words, the vector equation of the line is simply "point on the line plus a scalar multiple of the line's direction".)

▌**Worked Example 1.3.1** *Determine the vector equation of the line passing through the two points $A(1,2,5)$ and $B(-2,1,7)$.*

The direction of the line is parallel to $\overrightarrow{AB} = (-3,-1,2)$.
Take A as the point on the line.
Therefore, the vector equation of the line is

$$\mathbf{r}(t) = (x,y,z) = (1,2,5) + t(-3,-1,2) \text{ for all real } t.$$

(You should show that if the point B were selected instead, we would nevertheless obtain the same set of points.) ■

Note that the vector equation of the line in the last worked example can be written in the alternative form

$$\mathbf{r}(t) = (x,y,z) = (1,2,5) + t(-3,-1,2) = (1-3t, 2-t, 5+2t).$$

Therefore, equating components of the vectors on each side of this equation, we get

$$\left.\begin{array}{rcl} x & = & 1-3t \\ y & = & 2-t \\ z & = & 5+2t \end{array}\right\} \text{ for all real } t.$$

These three equations also define the line L and are called the *parametric equations* of the line L. (Note that the first column on the right-hand side of these equations represents the co-ordinates of the given point on L and the coefficients of t represent the components of the direction of the line L.)

If we eliminate the parameter t between these three equations we get

$$\frac{x-1}{-3} = \frac{y-2}{-1} = \frac{z-5}{2} \; (=t)$$

These TWO equations also define L and are called the *Cartesian equations* of the line L. (Sometimes also called the *freedom form* or *symmetric form* of the equations of the line.) Note the coefficients of x, y and z are each unity and (because of this) the denominators of each fraction represent the components of the direction of the line L; the co-ordinates of the given point on L correspond to the value $t = 0$.

▌**Example:** The Cartesian equation of the line passing through the point $(1,0,7)$ and parallel to the direction $(2,1,6)$ is given by

$$\frac{x-1}{2} = \frac{y-0}{1} = \frac{z-7}{6}.$$

The parametric form of the same line is

$$\left.\begin{array}{rcl} x &=& 1+2t \\ y &=& t \\ z &=& 7+6t \end{array}\right\} \text{ for all real } t.$$

The vector form of the equation of the line is

$$\mathbf{r}(t) = (x, y, z) = (1, 0, 7) + t(2, 1, 6) \text{ for all real } t.$$

Example: The line defined by the Cartesian equations

$$\frac{x+3}{-2} = \frac{4-2y}{2} = \frac{1-z}{0}$$

passes through the point[1] $(-3, 2, 1)$ and is parallel to the direction $(-2, -1, 0)$. (Why not $(-2, 2, 0)$ or $(-2, 1, 0)$?)

Worked Example 1.3.2 *A line L contains the points $(1, 2, 3)$ and $(-1, 5, 6)$.*
Which of the points $(-3, 8, 9)$ and $(3, -1, 1)$ also lie on this line?
The vector equation of the given line is $\mathbf{r}(t) = (1, 2, 3) + t(-2, 3, 3)$ for all real t.
The point $(-3, 8, 9)$ lies on L if there exists t such that

$$(-3, 8, 9) = (1, 2, 3) + t(-2, 3, 3).$$

Clearly $t = 2$, so this point $(-3, 8, 9)$ lies on L.
The point $(3, -1, 1)$ lies on L if there exists t such that

$$(3, -1, 1) = (1, 2, 3) + t(-2, 3, 3).$$

Clearly no such t exists. Therefore the point $(3, -1, 1)$ does not lie on L. ∎

Worked Example 1.3.3 *Which of the following lines are parallel?*
If they are not parallel do they intersect?

1. $L_1 : \left\{\begin{array}{rcl} x &=& -1+t \\ y &=& -2t \\ z &=& 4-t \end{array}\right. \qquad L_2 : \left\{\begin{array}{rcl} x &=& 7-2t \\ y &=& -5+4t \\ z &=& 2t \end{array}\right.$

---

[1] Note that this equation of the line is meaningless unless $z \equiv 1$.

2. *L_1: Passes through the point $P(1,1,1)$ and parallel to the direction $(1,2,3)$*

   L_2: Passes through the point $Q(2,1,0)$ and parallel to the direction $(3,8,13)$

3. *L_1: Passes through the points $P_1(2,2,1)$ and $P_2(-1,3,0)$*

   L_2: Passes through the points $Q_1(0,5,-1)$ and $Q_2(1,3,-4)$.

1. Line L_1 is parallel to the direction $(1,-2,-1)$ and line L_2 is parallel to the direction $(-2,4,2)$.

   Since $(1,-2,-1)$ is a scalar multiple of $(-2,4,2)$, L_1 and L_2 are parallel.

   (Are they in fact the *same* line?)

2. L_1 is parallel to $(1,2,3)$ and L_2 is parallel to $(3,8,13)$.

   Therefore L_1 is not parallel to L_2.

   An arbitrary point on L_1 has co-ordinates $(1+t, 1+2t, 1+3t)$.

   An arbitrary point on L_2 has co-ordinates $(2+3s, 1+8s, 13s)$.

   Such a point also lies on both L_1 and L_2 if

   $$(1+t, 1+2t, 1+3t) = (2+3s, 1+8s, 13s).$$

   That is, if $\begin{cases} 1+t &= 2+3s \\ 1+2t &= 1+8s \\ 1+3t &= 13s. \end{cases}$

   The first two equations imply that

   $$\begin{cases} t-3s &= 1 \\ 2t-8s &= 0 \end{cases} \Rightarrow s=1, t = 4.$$

   But then $1+3t = 13$ and $13s = 13$ so that the last equation is also satisfied.

   Thus L_1 and L_2 intersect at the point $(5,9,13)$.

3. Line L_1 is parallel to $(-3,1,-1)$ and line L_2 is parallel to $(1,-2,-3)$.

   Thus L_1 is not parallel to L_2.

   An arbitrary point on L_1 has co-ordinates $(2-3t, 2+t, 1-t)$.

An arbitrary point on L_2 has co-ordinates $(s, 5-2s, -1-3s)$.
Such a point also lies on both L_1 and L_2 if

$$(2-3t, 2+t, 1-t) = (s, 5-2s, -1-3s).$$

That is, if $\begin{cases} 2-3t &= s \\ 2+t &= 5-2s \\ 1-t &= -1-3s. \end{cases}$

The first two of these equations imply that $t = 1/5$ and $s = 7/5$.

But then $1-t = 4/5$ and $-1-3s = -26/5$ so that the last equation is *not* satisfied.

Thus L_1 and L_2 do not intersect. Such lines are said to be *skew*.

■

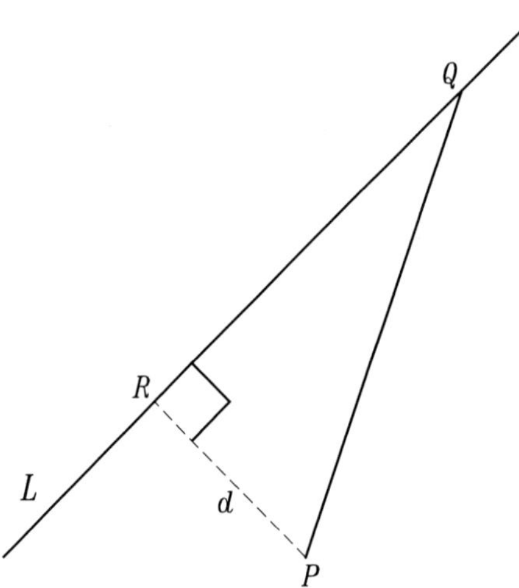

Figure 1.2: Calculation of the shortest distance in Worked Example 1.3.4.

Worked Example 1.3.4 *Find the distance of the point P* $(3, 5, -11)$ *from the straight line*

$$\mathbf{r}(t) = (x, y, z) = (5, -7, 1) + t(-2, 4, 13).$$

The given line is parallel to the direction $(-2, 4, 13)$ and a point on it is $Q(5, -7, 1)$.

With reference to Figure 1.2, the required distance d is the length of the line segment PR where R is the point on the line such that PR is perpendicular to RQ.

We obtain this distance by determining the length RQ as the (scalar) projection of PQ onto the line's direction $(-2, 4, 13)$ and then applying Pythagoras' theorem.

Thus $RQ = \overrightarrow{PQ} \cdot \widehat{\overrightarrow{RQ}} = (2, -12, 12) \cdot \frac{1}{\sqrt{189}}(-2, 4, 13) = 104/\sqrt{189}$.

Hence, by Pythagoras' theorem $d^2 = PQ^2 - RQ^2 = 292 - 10816/189 = 44372/189$.

That is the required distance $d = \sqrt{44372/189} \approx 15.32$. ■

Self-help exercises

1. Determine both the direction and a point on the line

$$\frac{2x-4}{3} = \frac{1-3y}{9} = \frac{z+4}{-3}.$$

[$\frac{3}{2}\,\mathbf{i} - 3\,\mathbf{j} - 3\,\mathbf{k}$; $(2, \frac{1}{3}, -4)$]

2. Determine the parametric equation of the straight line that passes through both points $A(4, -1, 5)$ and $B(-2, 3, -11)$.

[$x = 4 - 6t, y = -1 + 4t, z = 5 - 16t$]

3. Determine the point of intersection of the two lines

$$\begin{aligned} x &= -3 - 2t \\ y &= 5 + 3t \\ z &= -13 - 7t \end{aligned} \qquad \text{and} \qquad \begin{aligned} x &= s \\ y &= 4 - 5s \\ z &= -5 + 6s. \end{aligned}$$

[$(1, -1, 1)$]

1.3.2 Planes

Vector Form

A plane Π is determined once we know three non-collinear points A, B and C on it.

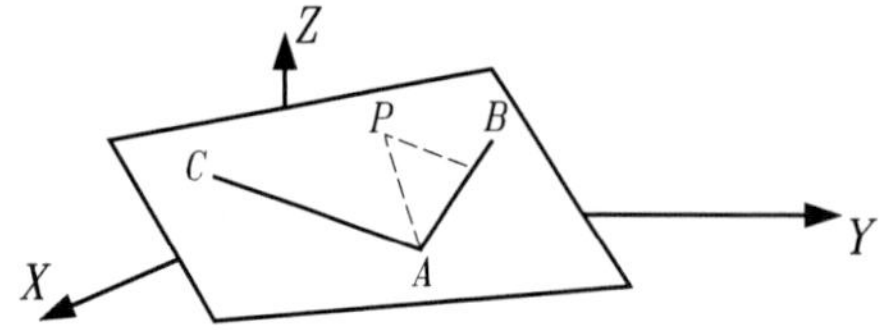

With reference to the diagram at left, the directed line segments $\overrightarrow{AB}$ and $\overrightarrow{AC}$ will represent non-parallel vectors in Π. An arbitrary point P lies on Π if and only if

$$\overrightarrow{AP} = c_1\overrightarrow{AB} + c_2\overrightarrow{AC},$$

where c_1 and c_2 are two real scalar quantities. Using the triangle rule we have

$$\overrightarrow{OP} = \overrightarrow{OA} + \overrightarrow{AP},$$

so that Π consists of all points P such that

$$\overrightarrow{OP} = \overrightarrow{OA} + c_1\overrightarrow{AB} + c_2\overrightarrow{AC}, \tag{1.3}$$

where c_1 and c_2 are any real scalar quantities. Equation 1.3 is called the *vector equation* of the plane Π defined by the three points A, B and C.

▌**Worked Example 1.3.5** *Determine the vector equation of the plane defined by the three points $A(1,0,1)$, $B(0,1,2)$ and $C(1,3,2)$.*
An arbitrary point $P(x,y,z)$ lies on this plane if and only if

$$\overrightarrow{OP} = \overrightarrow{OA} + c_1\overrightarrow{AB} + c_2\overrightarrow{AC},$$

where c_1 and c_2 are any real scalar quantities.
Thus the plane is defined by the set of points (x,y,z) that satisfy

$$(x,y,z) = (1,0,1) + c_1(-1,1,1) + c_2(0,3,1),$$

for all real values of c_1 and c_2. ■

We could proceed to eliminate c_1 and c_2 from the *three* equations that define x, y and z in the same way we eliminated the parameter t from the parametric equation of the line to derive its Cartesian equation. However, it is easier in this case to use the vector cross product; see Section 1.3.3.

Cartesian Form

A plane Π is also completely specified once we know a point A on it and a direction *normal* to it.

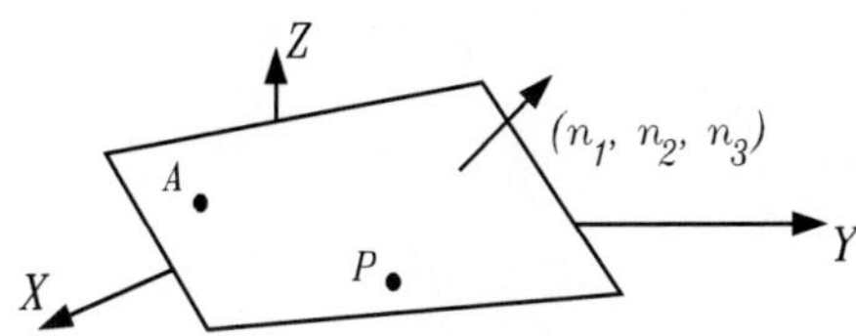

With reference to the diagram at left, let $\mathbf{n} = (n_1, n_2, n_3)$ be a given vector normal to the plane and A be the point (a_1, a_2, a_3). An arbitrary point $P(x, y, z)$ also lies on this plane if and only if the vectors $\overrightarrow{AP}$ and $\mathbf{n}$ are orthogonal; that is, $\overrightarrow{AP} \cdot \mathbf{n} = 0$. Hence

$$(x - a_1, y - a_2, z - a_3) \cdot (n_1, n_2, n_3) = 0.$$

That is

$$n_1x + n_2y + n_3z = n_1a_1 + n_2a_2 + n_3a_3 = d, \text{ a known constant.}$$

Thus the general *Cartesian form* of the equation of the plane is

$$\boxed{ax + by + cz = d}$$

where the vector (a, b, c) is a vector normal to the plane and d is determined from a known point on the plane.

Example: The plane with its normal parallel to the direction $(4, 5, 6)$ and containing the point $A(-1, 2, 1)$ has the Cartesian equation

$$4x + 5y + 6z = d = 4(-1) + 5(2) + 6(1) = 12.$$

Definition 1.8 *Two planes* Π_1 *and* Π_2 *are said to be* parallel *if their normal dirrections are parallel.*

Example: The two planes

$$\begin{aligned} \Pi_1 : \quad 2x - y + z &= 3 \text{ and} \\ \Pi_2 : \quad -4x + 2y - 2z &= 7 \end{aligned}$$

have normals parallel to the vectors $(2, -1, 1)$ and $(-4, 2, -2)$. Since $(2, -1, 1) = -\frac{1}{2}(-4, 2, -2)$, Π_1 is parallel to Π_2.

Worked Example 1.3.6 *Consider the straight line* L *and the plane* Π *defined by*

$$L : \quad \frac{x-1}{3} = \frac{y-2}{-1} = z + 4 \text{ and}$$

$$\Pi : \quad 2x + 2y - 4z = 7.$$

1. *Show that the line L is parallel to the plane Π.*
2. *Determine the distance d of the line L from the plane Π.*

1. The line L is parallel to the vector $(3, -1, 1)$ and the normal $\mathbf{n}$ to the plane is parallel to the vector $(2, 2, -4)$.

   But $(3, -1, 1) \cdot (2, 2, -4) = 0$. Therefore, the line L is parallel to the plane Π.

2. The required distance is the (scalar) projection onto the plane's normal direction of a vector joining *any* point on the line L to *any* point on the plane Π. (Why?)

   The point $A(1, 2, -4)$ is one point on the line L.

   The x-intercept of the plane Π is the point $B(\frac{7}{2}, 0, 0)$.

   The distance d of L from Π is the (scalar) projection of $\overrightarrow{AB}$ onto $\mathbf{n}$. That is

$$d = \overrightarrow{AB} \cdot \hat{\mathbf{n}} = \frac{15}{\sqrt{24}}.$$

■

Self-help exercises

1. With reference to the above Worked Example 1.3.6, by selecting a different point on the line L and a different point on the plane Π, show that the same distance d is nevertheless obtained.

2. Determine the Cartesian equation of the plane passing through the point $P(1, 7, 13)$ and with a normal direction parallel to the vector $(4, -2, -1)$. $[4x - 2y - z + 23 = 0]$

3. Find the distance of the point $P(1, -7, 9)$ from the plane $7x - 2y - 3z = -21$. $[\frac{15}{\sqrt{62}}]$

1.3.3 Cross Product

If A, B and C are three non-collinear points then there is a unique plane Π containing them. One method of determining the normal $\mathbf{n}$ is as follows. Let

$$\overrightarrow{AB} \text{ be the vector } (b_1, b_2, b_3) \text{ and}$$

$$\overrightarrow{AC} \text{ be the vector } (c_1, c_2, c_3).$$

Then the normal $\mathbf{n}$ is orthogonal to both $\overrightarrow{AB}$ and $\overrightarrow{AC}$. That is

$$\mathbf{n} \cdot \overrightarrow{AB} = \mathbf{n} \cdot \overrightarrow{AC} = 0,$$

or

$$\begin{cases} n_1 b_1 + n_2 b_2 + n_3 b_3 &= 0 \\ n_1 c_1 + n_2 c_2 + n_3 c_3 &= 0. \end{cases}$$

These two last equations are TWO equations for the THREE unknowns n_1, n_2 and n_3. One possible solution is

$$\mathbf{n} = (n_1, n_2, n_3) = (b_2c_3 - b_3c_2, b_3c_1 - b_1c_3, b_1c_2 - b_2c_1).$$

(You should verify that this is the case.)
All other solutions of these two equations will be scalar multiples of this solution. (Why?)

Definition 1.9 *The* cross product *of two vectors* $\mathbf{b}$ *and* $\mathbf{c}$ *is the vector* $\mathbf{b}\times\mathbf{c}$ *given by (see Appendix A for a brief introduction to the determinants being used here)*

$$\begin{aligned} \mathbf{b}\times\mathbf{c} &= (b_2c_3 - b_3c_2, b_3c_1 - b_1c_3, b_1c_2 - b_2c_1) \\ &= (b_2c_3 - b_3c_2)\,\mathbf{i} + (b_3c_1 - b_1c_3)\,\mathbf{j} + (b_1c_2 - b_2c_1)\,\mathbf{k} \\ &= \mathbf{i}\begin{vmatrix} b_2 & b_3 \\ c_2 & c_3 \end{vmatrix} - \mathbf{j}\begin{vmatrix} b_1 & b_3 \\ c_1 & c_3 \end{vmatrix} + \mathbf{k}\begin{vmatrix} b_1 & b_2 \\ c_1 & c_2 \end{vmatrix} \\ &= \begin{vmatrix} \mathbf{i} & \mathbf{j} & \mathbf{k} \\ b_1 & b_2 & b_3 \\ c_1 & c_2 & c_3 \end{vmatrix}. \end{aligned}$$

▮**Example:** $(1,-3,-1)\times(-2,1,5) = \begin{vmatrix} \mathbf{i} & \mathbf{j} & \mathbf{k} \\ 1 & -3 & -1 \\ -2 & 1 & 5 \end{vmatrix}$

$$= \mathbf{i}(-14) - \mathbf{j}(3) + \mathbf{k}(-5)$$

$$= (-14, -3, -5).$$

Note:

1. The cross product $\mathbf{a}\times\mathbf{b}$ is a *vector* quantity — whereas $\mathbf{a}\cdot\mathbf{b}$ is a scalar quantity.
2. The cross product $\mathbf{a}\times\mathbf{b}$ is perpendicular to *both* $\mathbf{a}$ and $\mathbf{b}$. (That is $\mathbf{a}\cdot\mathbf{a}\times\mathbf{b} = 0$ and $\mathbf{b}\cdot\mathbf{a}\times\mathbf{b} = 0$.)
3. $\mathbf{a}\times\mathbf{b} = -\mathbf{b}\times\mathbf{a}$. (That is, the cross product is *not* commutative.)
4. The cross product is *not* associative.
   (For example, $\mathbf{i}\times(\mathbf{i}\times\mathbf{j}) = -\mathbf{j}$ whereas $(\mathbf{i}\times\mathbf{i})\times\mathbf{j} = \mathbf{0}$.)

We know that if $\mathbf{a}$ and $\mathbf{b}$ are vectors and θ is the angle between them then by the alternative definition of dot product

$$\mathbf{a}\cdot\mathbf{b} = |\mathbf{a}||\mathbf{b}|\cos\theta.$$

A similar expression holds for the length (or modulus) of the cross product $\mathbf{a}\times\mathbf{b}$. We have

$$\begin{aligned}
|\mathbf{a}\times\mathbf{b}|^2 + |\mathbf{a}\cdot\mathbf{b}|^2 &= (\mathbf{a}\times\mathbf{b})\cdot(\mathbf{a}\times\mathbf{b}) + (\mathbf{a}\cdot\mathbf{b})\cdot(\mathbf{a}\cdot\mathbf{b}) \\
&= (a_2b_3 - a_3b_2, \ldots)\cdot(a_2b_3 - a_3b_2, \ldots) + (a_1b_1 + \cdots)^2 \\
&\vdots \\
&= (\mathbf{a}\cdot\mathbf{a})(\mathbf{b}\cdot\mathbf{b}) \\
&= |\mathbf{a}|^2|\mathbf{b}|^2.
\end{aligned}$$

(You should complete all the steps outlined in the above derivation.)

Therefore $|\mathbf{a}\times\mathbf{b}|^2 = |\mathbf{a}|^2|\mathbf{b}|^2 - |\mathbf{a}|^2|\mathbf{b}|^2\cos^2\theta = |\mathbf{a}|^2|\mathbf{b}|^2\sin^2\theta$.
That is

$$\boxed{|\mathbf{a}\times\mathbf{b}| = |\mathbf{a}||\mathbf{b}|\sin\theta}$$

(Note that geometrically this formula for the length of the vector cross product $|\mathbf{a}\times\mathbf{b}|$ represents the area of the parallelogram — or twice the area of the triangle — defined by the vectors $\mathbf{a}$ and $\mathbf{b}$.)

Thus $\mathbf{a}\times\mathbf{b}$ is a vector

- of length $|\mathbf{a}||\mathbf{b}|\sin\theta$
- which is perpendicular to both $\mathbf{a}$ and $\mathbf{b}$.

The sense in which the vector $\mathbf{a}\times\mathbf{b}$ is directed is that in which a right-handed screw would move if rotated from $\mathbf{a}$ to $\mathbf{b}$. (This sense is obtained by considering a special case for $\mathbf{a}$ and $\mathbf{b}$, such as $\mathbf{i}\times\mathbf{j} = \mathbf{k}$.)

▌**Worked Example 1.3.7** *Find the Cartesian equation of the plane containing the three points* $A(1,1,1)$*,* $B(0,0,2)$ *and* $C(7,2,-9)$*.*
The plane's normal $\mathbf{n}$ is parallel to $\overrightarrow{AB}\times\overrightarrow{AC} = (9,-4,5)$.
Thus the equation of the plane is $9x - 4y + 5z = d = 10$. ■

▌**Worked Example 1.3.8** *Determine the equation of the line passing through the point* $P(1,-1,2)$ *and parallel to the line of intersection of the planes*

$$\begin{aligned} 2x - y + z &= 1 \textit{ and} \\ 3x + 3y - z &= 5. \end{aligned}$$

The normal to the first plane is parallel to the vector $\mathbf{n}_1 = (2,-1,1)$ and the second to $\mathbf{n}_2 = (3,3,-1)$.
Therefore the line of intersection is parallel to the direction $\mathbf{n}_1\times\mathbf{n}_2 = (-2,5,9)$. (The line of intersection lies in each of the two planes and thus is perpendicular to each of the two normals $\mathbf{n}_1$ and $\mathbf{n}_2$.)
Thus the equation of the line is

$$L: \quad \frac{x-1}{-2} = \frac{y+1}{5} = \frac{z-2}{9}.$$

■

▌**Worked Example 1.3.9** *Find the area of the triangle whose vertices are the points* $A(1,-1,0)$*,* $B(2,1,-1)$ *and* $C(-1,1,2)$*.*
Here $\overrightarrow{AB} = (1,2,-1)$ and $\overrightarrow{AC} = (-2,2,2)$.
Thus the area of the triangle is given by

$$\frac{1}{2}|\overrightarrow{AB}||\overrightarrow{AC}|\sin\theta = \frac{1}{2}|\overrightarrow{AB}\times\overrightarrow{AC}| = \frac{1}{2}|(6,0,6)| = 3\sqrt{2}.$$

■

Self-help exercises

1. Determine the cross product of $\mathbf{a} = (-1,2,-3)$ and $\mathbf{b} = (2,-6,9)$.
[$(0,3,2)$]

2. Find the Cartesian equation of the plane that contains the three points $A(3,4,9)$, $B(2,1,-7)$ and $C(-2,1,4)$. [$-11x + 25y - 4z = 31$]

3. Determine the Cartesian equation of the plane containing the lines
$$\frac{3x-2}{3}=\frac{y+2}{4}=\frac{z-3}{-2} \text{ and } \frac{x}{-5}=\frac{2y+8}{5}=\frac{z-7}{-3}.$$
[Lines are skew!]

4. Find the Cartesian equation of the plane that is perpendicular to the line
$$\mathbf{r}(t)=(x,y,z)=(2,-3,5)+t(-1,4,7)$$
and passes through the point $P(1,0,5)$. $[-x+4y+7z=34]$

1.4 Scalar and Vector Triple Product

In this section we examine two different combinations of any three vectors. The first leads to a scalar result and the second to a vector result.

1.4.1 Scalar Triple Product

Refers to combinations such as $\mathbf{a}\cdot\mathbf{b}\times\mathbf{c}$. There is only *one* possible interpretation of this combination (why?), so no brackets are required to indicate the order in which the cross and dot product should be performed.

With reference to Figure 1.3, if the directed line segment $\overrightarrow{PA}$ represents the vector $\mathbf{a}$, $\overrightarrow{PB}$ represents $\mathbf{b}$ and $\overrightarrow{PC}$ represents $\mathbf{c}$, then

$$\begin{aligned}
\mathbf{a}\cdot\mathbf{b}\times\mathbf{c} &= |\mathbf{a}||\mathbf{b}\times\mathbf{c}|\cos\varphi \\
&= |\mathbf{b}\times\mathbf{c}|\cdot|\mathbf{a}|\cos\varphi \\
&= (\text{area of base } \times \text{ height}) \text{ of parallelepiped} \\
&= \text{volume of parallelepiped} \\
&= \mathbf{b}\cdot\mathbf{c}\times\mathbf{a} \\
&= \mathbf{c}\cdot\mathbf{a}\times\mathbf{b}; \text{ Why? (See Figure 1.3.) Note the cyclic order!} \\
&= \mathbf{a}\times\mathbf{b}\cdot\mathbf{c}; \text{ since } \cdot \text{ is commutative.}
\end{aligned}$$

(In Figure 1.3, AM is perpendicular to PQ, the directed line segment $\overrightarrow{PQ}$ represents the vector $\mathbf{b}\times\mathbf{c}$ and the segment $PM=|\mathbf{a}|\cos\varphi$ is the height of the parallelepiped.)

An immediate consequence of the above is that "dot and cross are interchangeable" in the evaluation of a scalar triple product and "vectors making up the product can be replaced cyclically without altering the value of the product".

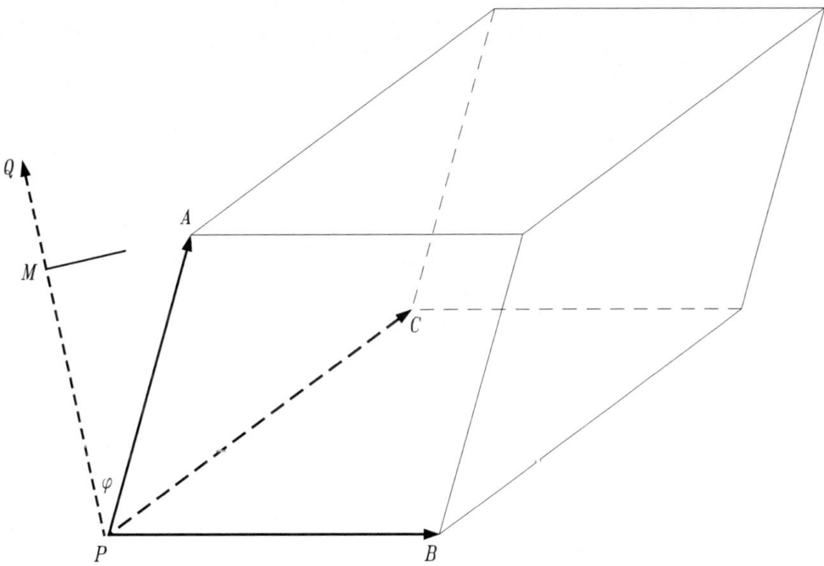

Figure 1.3: Geometric interpretation of the scalar triple product.

Note:

Three non-zero vectors **a**, **b** and **c** are *co-planar* if and only if

$$\mathbf{a} \cdot \mathbf{b} \times \mathbf{c} = 0.$$

In such a case any one vector can be expressed (linearly) in terms of one or more of the other two. We say that such vectors are *linearly dependent*.

In evaluating scalar triple products it is convenient to recognize that

$$\begin{aligned}
\mathbf{a} \cdot \mathbf{b} \times \mathbf{c} &= \mathbf{a} \cdot \begin{vmatrix} \mathbf{i} & \mathbf{j} & \mathbf{k} \\ b_1 & b_2 & b_3 \\ c_1 & c_2 & c_3 \end{vmatrix} \\
&= \mathbf{a} \cdot \left((b_2 c_3 - b_3 c_2)\,\mathbf{i} - (b_1 c_3 - b_3 c_1)\,\mathbf{j} + (b_1 c_2 - b_2 c_1)\,\mathbf{k} \right) \\
&= a_1 \begin{vmatrix} b_2 & b_3 \\ c_2 & c_3 \end{vmatrix} - a_2 \begin{vmatrix} b_1 & b_3 \\ c_1 & c_3 \end{vmatrix} + a_3 \begin{vmatrix} b_1 & b_2 \\ c_1 & c_2 \end{vmatrix} \\
&= \begin{vmatrix} a_1 & a_2 & a_3 \\ b_1 & b_2 & b_3 \\ c_1 & c_2 & c_3 \end{vmatrix}
\end{aligned}$$

▮Example: $(-1,2,5)\cdot(3,-2,7)\times(6,0,-9)$

$$= \begin{vmatrix} -1 & 2 & 5 \\ 3 & -2 & 7 \\ 6 & 0 & -9 \end{vmatrix}$$

$$= (-1)(18) - 2(-69) + 5(12) = 180$$

where we have evaluated the determinant by expanding along the first row.

Self-help exercises

1. Determine the scalar triple product $(-2,4,6)\cdot(1,-4,-7)\times(3,11,-3)$. [−112]

2. Test for co-planarity the vectors $(1,4,-5)$, $(-3,2,8)$ and $(-9,-8,31)$. [Co-planar]

1.4.2 Vector Triple Product

Refers to combinations such as $\mathbf{a}\times(\mathbf{b}\times\mathbf{c})$. Because of the possible ambiguity in such triple products *brackets are essential*.

The vector triple product $\mathbf{a}\times(\mathbf{b}\times\mathbf{c})$ is perpendicular to $\mathbf{b}\times\mathbf{c}$ and hence is parallel to the plane determined by the vectors $\mathbf{b}$ and $\mathbf{c}$. Therefore, there exists scalars d_1 and d_2 such that

$$\mathbf{a}\times(\mathbf{b}\times\mathbf{c}) = d_1\,\mathbf{b} + d_2\,\mathbf{c}.$$

Thus

$$0 = \mathbf{a}\cdot\mathbf{a}\times(\mathbf{b}\times\mathbf{c}) = d_1\,\mathbf{a}\cdot\mathbf{b} + d_2\,\mathbf{a}\cdot\mathbf{c}.$$

Hence $d_2 = -d_1\,\mathbf{a}\cdot\mathbf{b}/\mathbf{a}\cdot\mathbf{c}$ (assuming $\mathbf{a}\cdot\mathbf{c}\neq 0$; why is this justified?) and therefore

$$\begin{aligned} \mathbf{a}\times(\mathbf{b}\times\mathbf{c}) &= d_1\,\mathbf{b} - d_1\frac{\mathbf{a}\cdot\mathbf{b}}{\mathbf{a}\cdot\mathbf{c}}\mathbf{c} \\ &= \frac{d_1}{\mathbf{a}\cdot\mathbf{c}}((\mathbf{a}\cdot\mathbf{c})\mathbf{b} - (\mathbf{a}\cdot\mathbf{b})\mathbf{c}) \qquad (1.4) \\ &= d_3((\mathbf{a}\cdot\mathbf{c})\mathbf{b} - (\mathbf{a}\cdot\mathbf{b})\mathbf{c}). \end{aligned}$$

Since this equation is true for all vectors $\mathbf{a}$, $\mathbf{b}$ and $\mathbf{c}$, we can determine the constant d_3 by choosing them suitably. For example, if we choose

$$\mathbf{a} = \mathbf{i}, \mathbf{b} = \mathbf{j} \text{ and } \mathbf{c} = \mathbf{i},$$

then

$$\mathbf{a}\times(\mathbf{b}\times\mathbf{c}) = \mathbf{i}\times(-\mathbf{k}) = \mathbf{j} \text{ and}$$

$$d_3((\mathbf{a}\cdot\mathbf{c})\mathbf{b} - (\mathbf{a}\cdot\mathbf{b})\mathbf{c}) = d_3(\mathbf{j} - \mathbf{0}) = d_3\mathbf{j}.$$

Thus in Equation 1.4, the scalar d_3 takes the value $d_3 = 1$.

Hence

$$\boxed{\mathbf{a}\times(\mathbf{b}\times\mathbf{c}) = (\mathbf{a}\cdot\mathbf{c})\mathbf{b} - (\mathbf{a}\cdot\mathbf{b})\mathbf{c}}$$

This last formula is often written as

$$\mathbf{a}\times(\mathbf{b}\times\mathbf{c}) = (\mathbf{c}\cdot\mathbf{a})\mathbf{b} - (\mathbf{b}\cdot\mathbf{a})\mathbf{c},$$

using the commutativity of the dot product. In this form, the expansion of the vector triple product on the left-hand side into the difference of two dot products is referred to as the "cab minus bac rule".

Example:

$$\begin{aligned}(1,2,4)\times((-2,3,1)\times(2,-4,6)) &= (18)(-2,3,1) - (8)(2,-4,6)\\ &= (-52,86,-30).\end{aligned}$$

Worked Example 1.4.1 *Simplify (by expanding) the vector triple product*

$$(\mathbf{a}\times\mathbf{b})\times\mathbf{c}.$$

We first rearrange the given vector triple product into the required form to apply the above expansion formula. Thus

$$\begin{aligned}(\mathbf{a}\times\mathbf{b})\times\mathbf{c} &= -\mathbf{c}\times(\mathbf{a}\times\mathbf{b})\\ &= -((\mathbf{c}\cdot\mathbf{b})\mathbf{a} - (\mathbf{c}\cdot\mathbf{a})\mathbf{b})\\ &= (\mathbf{a}\cdot\mathbf{c})\mathbf{b} - (\mathbf{b}\cdot\mathbf{c})\mathbf{a}.\end{aligned}$$

■

In this topic we have

- Reviewed basic vector algebra (including the definitions of scalar and vector products) and explained scalar and vector projections.
- Used vector algebra techniques to define lines and planes.
- Introduced scalar and vector triple products.

1.5 Quick Test Number 1

Basic Mathematical Skills

Question **Selection**

1. $\frac{1}{1/a+1/b}$ is equal to (a) $a+b$ (b) $\frac{ab}{a+b}$ (c) $\frac{a+b}{ab}$
2. $\sqrt{x^2+16x+49} = x+4\sqrt{x}+7$? (a) True (b) False
3. $\dfrac{1}{x^2+16x+49} = \dfrac{1}{x^2}+\dfrac{1}{16x}+\dfrac{1}{49}$? (a) True (b) False
4. $x\cos 2x = \cos 2x^2$? (a) True (b) False
5. 2^{ab} equals (a) 2^a2^b (b) $(2^a)^b$ (c) 2^a+2^b

Vectors

Question **Selection**

1. If $P_1(0,1,3)$ and $P_2(2,4,-5)$ are points in 3D, the vector $\overrightarrow{P_1P_2}$ equals (a) $2\mathbf{i}+5\mathbf{j}-2\mathbf{k}$ (b) $-2\mathbf{i}-3\mathbf{j}+8\mathbf{k}$ (c) $2\mathbf{i}+3\mathbf{j}-8\mathbf{k}$

In Questions 2–8, use $\mathbf{a}=\mathbf{i}-2\mathbf{j}+4\mathbf{k}$, $\mathbf{b}=2\mathbf{i}+3\mathbf{j}-5\mathbf{k}$ and $\mathbf{c}=-\mathbf{i}-\mathbf{j}+3\mathbf{k}$.

2. $\mathbf{a}\cdot\mathbf{b}$ is (a) a scalar (b) a vector
3. $\mathbf{a}\cdot\mathbf{b}$ equals (a) $2\mathbf{i}-6\mathbf{j}-20\mathbf{k}$ (b) -24 (c) $-\mathbf{i}-5\mathbf{j}+9\mathbf{k}$
4. $\mathbf{b}\times\mathbf{c}$ is (a) a scalar (b) a vector
5. $\mathbf{b}\times\mathbf{c}$ equals (a) $-2\mathbf{i}-3\mathbf{j}-15\mathbf{k}$ (b) -20 (c) $4\mathbf{i}-\mathbf{j}+\mathbf{k}$
6. $\mathbf{a}\cdot\mathbf{b}\times\mathbf{c}$ is (a) a scalar (b) a vector (c) impossible
7. $\mathbf{a}\cdot\mathbf{b}\times\mathbf{c}$ equals (a) $24\mathbf{i}+24\mathbf{j}-72\mathbf{k}$ (b) 10 (c) $-2\mathbf{i}+6\mathbf{j}-60\mathbf{k}$
8. $|\mathbf{a}|$ is (a) 7 (b) $\sqrt{13}$ (c) $\sqrt{21}$
9. $2\mathbf{i}-14\mathbf{j}-4\mathbf{k}=\mathbf{i}-7\mathbf{j}-2\mathbf{k}$? (a) True (b) False

10. If two vectors are perpendicular, their dot product is

(a) zero (b) nothing special

11. If two vectors are perpendicular, their cross product is

(a) zero (b) nothing special

12. If two vectors are parallel, their dot product is

(a) zero (b) nothing special

13. If two vectors are parallel, their cross product is

(a) zero (b) nothing special

14. A vector of four times the length and parallel to $2\mathbf{i} - 3\mathbf{j} + 6\mathbf{k}$ is:

(a) $8\mathbf{i} - 12\mathbf{j} + 24\mathbf{k}$ (b) $\frac{1}{2}\mathbf{i} - \frac{3}{4}\mathbf{j} + \frac{3}{2}\mathbf{k}$

(c) $\frac{24}{7}\mathbf{i} - \frac{36}{7}\mathbf{j} + \frac{72}{7}\mathbf{k}$ (d) $-8\mathbf{i} + 12\mathbf{j} - 24\mathbf{k}$

15. A vector of length three units parallel to $2\mathbf{i} - 3\mathbf{j} + 6\mathbf{k}$ is:

(a) $6\mathbf{i} - 9\mathbf{j} + 18\mathbf{k}$ (b) $\frac{2}{7}\mathbf{i} - \frac{3}{7}\mathbf{j} + \frac{6}{7}\mathbf{k}$

(c) $\frac{6}{7}\mathbf{i} - \frac{9}{7}\mathbf{j} + \frac{18}{7}\mathbf{k}$ (d) $\frac{6}{\sqrt{31}}\mathbf{i} - \frac{9}{\sqrt{31}}\mathbf{j} + \frac{18}{\sqrt{31}}\mathbf{k}$

Lines and Planes

Question **Selection**

---

1. A straight line is uniquely determined by a point on the line and a direction vector parallel to the line.

(a) True (b) False (c) I'm definitely not sure

2. The equation of the line through the points $(1, 1, 0)$ and $(2, -1, 2)$ is

(a) $(x, y, z) = (1, 1, 0) + t(2, -1, 2)$ (b) $(x, y, z) = (2, -1, 2) + t(1, 1, 0)$
(c) $(x, y, z) = (1, 1, 0) + t(1, -2, 2)$ (d) $(x, y, z) = (1, -2, 2) + t(1, 1, 0)$

3. $(0, -1, 1)$ lies on the line $(x, y, z) = (-1, -3, 5) + t(-1, -2, 2)$

(a) True (b) False

4. The lines $(x, y, z) = (1, 2, 1) + t(1, 0, 1)$ and $(x, y, z) = (2, -1, 5) + t(1, 1, 0)$ are

(a) parallel (b) orthogonal

(c) neither (a) nor (b) (d) both (a) and (b)

5. A plane is uniquely defined by a point in the plane and a vector parallel to the plane. (a) True (b) False (c) My brain hurts

6. The equation of the plane through the point $(1, 0, 1)$ which is normal to the vector $\mathbf{n} = 2\mathbf{i} + 3\mathbf{j} + 5\mathbf{k}$ is

   (a) $2\mathbf{i} + 3\mathbf{j} + 5\mathbf{k} = 7$ (b) $(x, y, z) = (1, 0, 1) + t(2, 3, 5)$

   (c) $\dfrac{x-1}{2} = \dfrac{y}{3} = \dfrac{z-1}{5}$ (d) $2x + 3y + 5z = 7$

7. Which of the points $(1, 2, 3)$, $(1, 1, 2)$ and $(-1, 3, 4)$ lie in the plane $x + 5y - 3z = 2$?

   (a) $(1, 2, 3)$ (b) $(1, 1, 2)$ (c) $(-1, 3, 4)$ (d) none of them

8. Is the vector $2\mathbf{i} - \mathbf{j} - \mathbf{k}$ normal or parallel to the plane $x + 5y - 3z = 2$, or neither of these? (a) normal (b) parallel (c) neither

9. A unit vector normal to the plane $x + 5y - 3z = 2$ is

   (a) $2\mathbf{i} - \mathbf{j} - \mathbf{k}$ (b) $\dfrac{2}{\sqrt{6}}\mathbf{i} - \dfrac{1}{\sqrt{6}}\mathbf{j} - \dfrac{1}{\sqrt{6}}\mathbf{k}$

   (c) $\mathbf{i} + 5\mathbf{j} - 3\mathbf{k}$ (d) $\dfrac{1}{\sqrt{35}}\mathbf{i} + \dfrac{5}{\sqrt{35}}\mathbf{j} - \dfrac{3}{\sqrt{35}}\mathbf{k}$

10. The planes $x + 3y + 7z = 1$ and $-2x - 6y - 14z = -2$ are

   (a) normal (b) parallel (c) the same (d) different

1.6 Exercises

Before attempting any of these miscellaneous exercises, make sure that you have successfully answered all the self-help exercises appearing throughout the chapter.

1. Let $\mathbf{a} = 2\mathbf{i} + 2\mathbf{j} + \mathbf{k}$ and $\mathbf{b} = 4\mathbf{i} + 4\mathbf{j} + 7\mathbf{k}$. Determine
   (a) $3\mathbf{a} + \mathbf{b}$
   (b) $|\mathbf{b}|$
   (c) $\hat{\mathbf{b}}$
   (d) a vector parallel to **b** and five time as long
   (e) a vector of length 5 units parallel to **b**
   (f) the scalar component of **a** in the direction of **b**.

2. Let $\mathbf{a} = \mathbf{i} - 2\mathbf{j} - 2\mathbf{k}$, $\mathbf{b} = 6\mathbf{i} + 3\mathbf{j}$ and $\mathbf{c} = 7\mathbf{k}$. Determine
   (a) $\mathbf{a} - \mathbf{b} + \mathbf{c}$
   (b) $|\mathbf{a}|$
   (c) $\hat{\mathbf{a}}$
   (d) the scalar component of **c** in the direction of **a**
   (e) a vector of length 6 units parallel to **a**
   (f) a vector parallel to **a** and six times as long.

3. Let $\mathbf{a} = 2\mathbf{i} - \mathbf{j} + \mathbf{k}$, $\mathbf{b} = 3\mathbf{i} + \mathbf{j} - 5\mathbf{k}$ and $\mathbf{c} = 2\mathbf{i} + \mathbf{j} + 2\mathbf{k}$.
   (a) Find
      i. $2\mathbf{a} - \mathbf{b} + 3\mathbf{c}$
      ii. $\mathbf{a} \cdot \mathbf{b}$; is **a** perpendicular to **b**?
      iii. $\mathbf{b} \cdot (\mathbf{a} - 2\mathbf{c})$.
   (b) Determine the angle between **b** and **c**.
   (c) Find the scalar projection of **a** in the direction of **c**.
   (d) Determine the vector projection of **b** in the direction of **a**.

4. If $\mathbf{a} = \mathbf{i} + 2\mathbf{j} - 3\mathbf{k}$ and $\mathbf{b} = 2\mathbf{i} - 3\mathbf{j} + m\mathbf{k}$, determine m such that **a** and **b** are perpendicular.

5. ABC is a triangle with $\overrightarrow{AB} = \mathbf{i} + 2\mathbf{j} + 2\mathbf{k}$ and $\overrightarrow{AC} = 2\mathbf{i} - \mathbf{j} - \mathbf{k}$. Determine the *largest* angle in the triangle.

6. Let $\mathbf{a} = \mathbf{i} - 2\mathbf{j} + 3\mathbf{k}$, $\mathbf{b} = 2\mathbf{i} - \mathbf{j} + 2\mathbf{k}$ and $\mathbf{c} = 3\mathbf{i} + \mathbf{j} + \mathbf{k}$.

   (a) Find

      i. $\mathbf{a} - 2\mathbf{b} + \mathbf{c}$

      ii. $\mathbf{a} \cdot \mathbf{b}$ and $\mathbf{a} \times \mathbf{b}$

      iii. the length of $\mathbf{b}$ and hence $\hat{\mathbf{b}}$

      iv. $\mathbf{a} \cdot \mathbf{b} \times \mathbf{c}$.

   (b) Verify that $\mathbf{a} \times (\mathbf{b} \times \mathbf{c}) = (\mathbf{a} \cdot \mathbf{c})\mathbf{b} - (\mathbf{a} \cdot \mathbf{b})\mathbf{c}$.

7. If $\mathbf{a} = \mathbf{i} - \mathbf{j} + 2\mathbf{k}$ and $\mathbf{b} = 2\mathbf{i} + 3\mathbf{j} - \mathbf{k}$, find

   (a) $\mathbf{a} \cdot \mathbf{b}$

   (b) $\mathbf{a} \times \mathbf{b}$

   (c) $\mathbf{b} \cdot \mathbf{b}$

   (d) $\mathbf{a} \times \mathbf{b} \cdot \mathbf{b}$

   (e) a unit vector perpendicular to both $\mathbf{a}$ and $\mathbf{b}$.

8. Given the vectors $\mathbf{a} = \mathbf{i} + 2\mathbf{j} + 3\mathbf{k}$, $\mathbf{b} = -2\mathbf{i} + \mathbf{j} + 2\mathbf{k}$ and $\mathbf{c} = \mathbf{i} + \mathbf{k}$,

   (a) verify that $\mathbf{a} \times (\mathbf{b} \times \mathbf{c}) = (\mathbf{a} \cdot \mathbf{c})\mathbf{b} - (\mathbf{a} \cdot \mathbf{b})\mathbf{c}$

   (b) find the scalar component of $\mathbf{a}$ in the direction of $\mathbf{b}$

   (c) find the vector component of $\mathbf{a}$ in the direction of $\mathbf{b}$.

9. If $\mathbf{a} = 2\mathbf{i} + 3\mathbf{k}$ and $\mathbf{b} = \mathbf{i} - 2\mathbf{j} + \mathbf{k}$, find

   (a) $\mathbf{a} \cdot \mathbf{b}$

   (b) $\mathbf{a} \times \mathbf{b}$

   (c) $\mathbf{a} \times \mathbf{b} \cdot \mathbf{a}$

   (d) the scalar component of $\mathbf{a}$ in the direction of $\mathbf{b}$

   (e) the area of the triangle whose vertices are at the points $(0, 0, 0)$, $(2, 0, 3)$ and $(1, -2, 1)$.

10. Given the vectors $\mathbf{a} = 2\mathbf{i} + 2\mathbf{j} - \mathbf{k}$, $\mathbf{b} = \mathbf{i} + 5\mathbf{j} + 2\mathbf{k}$ and $\mathbf{c} = 3\mathbf{j}$, find

    (a) $\mathbf{a} \times \mathbf{b} \cdot \mathbf{c}$

    (b) the vector projection of $\mathbf{b}$ in the direction of $\mathbf{c}$.

11. Let $\mathbf{a} = \mathbf{i} + \mathbf{j} + \mathbf{k}$, $\mathbf{b} = 2\mathbf{i} + 2\mathbf{j} - \mathbf{k}$ and $\mathbf{c} = \mathbf{i} - 2\mathbf{j} + 2\mathbf{k}$.

    (a) Find the angle between $\mathbf{b}$ and $\mathbf{c}$.

    (b) Determine

       i. $\mathbf{a} \times \mathbf{b}$

       ii. $\mathbf{a} \cdot \mathbf{b} \times \mathbf{c}$.

    (c) Find a unit vector perpendicular to both $\mathbf{a}$ and $\mathbf{b}$.

    (d) Calculate the scalar component of $\mathbf{a}$ in the direction of $\mathbf{b}$.

12. Given $\mathbf{u} = (1, 0, 1)$, $\mathbf{v} = (2, -1, 2)$ and $\mathbf{w} = (0, 2, -1)$, find

    (a) the angle between the vectors $\mathbf{u}$ and $\mathbf{w}$

    (b) the length of the vectors $\mathbf{u}$ and $2\,\mathbf{v} - 3\,\mathbf{w}$

    (c) the value of $\mathbf{u} \cdot \mathbf{v}$, $\mathbf{v} \times \mathbf{w}$ and $\mathbf{u} \cdot \mathbf{v} \times \mathbf{w}$

    (d) a unit vector perpendicular to both $\mathbf{u}$ and $\mathbf{w}$

    (e) the scalar component of $\mathbf{u}$ in the direction of $\mathbf{v}$.

13. Determine the component of the force vector $\mathbf{F} = F_0(\mathbf{i} - 3\,\mathbf{j} + \mathbf{k})$ along the direction $2\,\mathbf{i} + \mathbf{j} + 2\,\mathbf{k}$.

14. Resolve the vector $(3, -4, 2)$ into the sum of two mutually orthogonal vectors, one of which is parallel to the vector $(1, 2, 3)$.

15. Let P be the point $(2, 1, 3)$ and Q be the point $(4, 1, 2)$.

    (a) Find the parametric equations of the line L passing through P and Q.

    (b) Determine the point R on the line L such that $\overrightarrow{OR}$ is perpendicular to L.

    (c) Calculate the distance of the origin from the line L.

16. Given the vectors $\mathbf{a} = \mathbf{i} - \mathbf{j} + 2\,\mathbf{k}$ and $\mathbf{b} = 2\,\mathbf{i} + 3\,\mathbf{j} - \mathbf{k}$, determine

    (a) the equation of the plane Π that contains the vectors $\mathbf{a}$ and $\mathbf{b}$ and the point $P(2, 1, -1)$

    (b) the shortest distance from the point $Q(2, 3, 4)$ to the plane Π.

17. A line L is parallel to the vector $3\,\mathbf{i} - 2\,\mathbf{j} - 2\,\mathbf{k}$ and passes through the point $P(1, 0, -\frac{1}{2})$.

    Find

    (a) the parametric equations of the line L

    (b) the point of intersection Q of the line L with the plane Π

    $$x + y + z = 2$$

    (c) the angle the line L makes with the plane Π at the intersection point Q.

18. Determine the area of the triangle whose vertices are at the points $(0, 0, 0)$, $(2, 0, 3)$ and $(1, -2, 1)$.

19. The line L is defined by the intersection of the planes

$$\begin{aligned} \Pi_1 : \quad 2x + y - z &= 4 \\ \Pi_2 : \quad 3x - 2y + z &= 6. \end{aligned}$$

Determine

(a) the parametric equations of the line L

(b) the point of intersection of the line L and the plane

$$\Pi_3 : \quad 2x - 3y + 2z = 6.$$

20. Find the equation of the plane that passes through the points $P(2, 1, -3)$ and $Q(1, -2, 2)$ and is perpendicular to the plane

$$x + y + z = 3.$$

21. A plane Π intersects the co-ordinate axes OX, OY and OZ at the points $(1, 0, 0)$, $(0, 2, 0)$ and $(0, 0, 3)$.

(a) Determine a normal vector $\mathbf{n}$ to the plane Π and hence obtain the Cartesian equation of the plane Π.

(b) Find the distance from the origin to the plane Π.

(c) Determine the parametric equations of the *normal* line L to the plane Π that passes through the origin O.
Hence find the co-ordinates of the point P on the plane Π that is closest to the origin O. Deduce the distance d of the plane Π from the origin O.

22. Given the vectors

$$\begin{aligned} \mathbf{a} &= -2\mathbf{i} + \mathbf{j} \\ \mathbf{b} &= \mathbf{i} + \mathbf{j} + 2\mathbf{k} \\ \mathbf{c} &= \lambda\mathbf{i} + \mathbf{j} + 2\mathbf{k}, \end{aligned}$$

(a) determine a unit vector normal to both $\mathbf{a}$ and $\mathbf{b}$

(b) find values of λ so that the vector $\mathbf{c}$ is parallel to the family of planes defined by the vectors $\mathbf{a}$ and $\mathbf{b}$.

23. The two planes

$$\begin{aligned} \Pi_1 : \quad 3x - y + 4z &= 3 \\ \Pi_2 : \quad 4x - 2y + 7z &= 8 \end{aligned}$$

intersect in a line L.

(a) Determine the angle α between the two planes Π_1 and Π_2.

(b) Find parametric equations for the line L.

24. A parallelogram $ABCD$ (taken cyclically) in three-dimensional space is defined by the position vectors

$$\overrightarrow{OA} = (2, -3, 1),\ \overrightarrow{OB} = (1, 2, -5) \text{ and } \overrightarrow{OD} = (-2, 0, 3).$$

(a) Find the area of the parallelogram $ABCD$.

(b) Determine the equation of the plane Π that contains the parallelogram.

(c) Find the shortest distance of the plane Π from the origin O.

25. Two straight lines L_1 and L_2 are defined by the Cartesian equations

$$L_1 : \quad \frac{x-1}{2} = \frac{y-1}{-4} = \frac{z-5}{-1}$$

$$L_2 : \quad \frac{x-4}{-2} = \frac{y+1}{6} = \frac{z-4}{1}.$$

Do these lines L_1 and L_2 intersect?

If they do intersect, find

(a) their point of intersection

(b) the Cartesian equation of the plane that contains L_1 and L_2.

If they do not intersect, find the shortest distance between the two lines L_1 and L_2.

26. The vectors $\mathbf{u}$, $\mathbf{v}$ and $\mathbf{w}$ are defined by

$$\begin{aligned} \mathbf{u} &= 2\,\mathbf{i} - \mathbf{j} + \mathbf{k} \\ \mathbf{v} &= \mathbf{i} - \mathbf{j} + 3\,\mathbf{k} \\ \mathbf{w} &= 3\,\mathbf{i} - \mathbf{j} + \alpha\,\mathbf{k} \end{aligned}$$

where α is a constant.

(a) Find the value of the constant α so that the vectors $\mathbf{u}$, $\mathbf{v}$ and $\mathbf{w}$ are co-planar.

For this value of α, determine the Cartesian equation of the plane Π containing the three vectors $\mathbf{u}$, $\mathbf{v}$ and $\mathbf{w}$ passing through the point $(-2, 3, 7)$.

(b) Determine the Cartesian equation of the straight line L that passes through the point $(1, 2, 3)$ and is parallel to the vector $\mathbf{u}$.

(c) If the line L intersects the plane Π, determine the point P of intersection. Otherwise, find the (shortest) distance of the line L from the plane Π.

(d) Verify the identity

$$(\mathbf{u}\cdot\mathbf{v})^2 + |\mathbf{u}\times\mathbf{v}|^2 = |\mathbf{u}|^2|\mathbf{v}|^2.$$

Can you *prove* this result for *general* vectors $\mathbf{u}$ and $\mathbf{v}$?

27. The equations for two intersecting lines are

$$x = y - 2 = 1 - z \text{ and } \frac{x-2}{-3} = \frac{y+2}{3} = z - 1.$$

(a) Determine the point A of intersection of these two lines.

(b) Find the equation of the plane Π that contains these two lines.

(c) Determine the equation of the line perpendicular to these two lines and which passes through their point of intersection A.

28. Two lines are defined by

$$\frac{x-2}{1} = \frac{y-4}{3} = \frac{z-2}{1} \text{ and } \frac{x-3}{1} = \frac{y-9}{4} = \frac{z-5}{2}.$$

(a) Show that these lines intersect.

(b) Determine the equation of the plane determined by these lines.

29. (a) i. Determine the distance from the point $P(1, 1, 1)$ to the plane

$$\Pi : \quad 2x + 3y - 6z = 1.$$

ii. Are P and the origin on the same "side" of the plane Π?

(b) Find a vector parallel to the line of intersection of the plane Π and the plane

$$x - 2y + 2z = -1.$$

30. Determine whether or not the straight lines

$$\frac{x-1}{1} = \frac{y-2}{2} = \frac{z-3}{3}$$

and

$$\frac{x+3}{3} = \frac{y+2}{2} = \frac{z+1}{1}$$

intersect.

If they intersect, determine the point of intersection.

31. Determine the plane Π perpendicular to the planes

$$x + y + z = 13 \text{ and } 2x - y - 3z = 11$$

and passing through the point $(1, 1, 0)$.

How far is the plane Π from the origin?

32. Determine the equation of the plane Π that passes through the point $P(-1, 2, 0)$ and is perpendicular to the line L

$$L: \quad \frac{x-2}{3} = 1 - y = \frac{z+2}{3}.$$

33. Given the points $A(0, 1, 0)$, $B(1, 2, 1)$ and $C(2, 2, -1)$, find

    (a) the cosine of the angle $\angle BAC$ formed in the triangle ABC

    (b) the area of the triangle ABC

    (c) the equation of the plane that passes through the points A, B and C.

34. Show that

    (a) $(\mathbf{a}\times\mathbf{b})\times(\mathbf{c}\times\mathbf{d}) = (\mathbf{b}\cdot\mathbf{d}\times\mathbf{c})\mathbf{a} + (\mathbf{a}\cdot\mathbf{c}\times\mathbf{d})\mathbf{b}$

    (b) $(\mathbf{a}\times\mathbf{b})\cdot(\mathbf{c}\times\mathbf{d}) = (\mathbf{a}\cdot\mathbf{c})\,(\mathbf{b}\cdot\mathbf{d}) - (\mathbf{a}\cdot\mathbf{d})\,(\mathbf{b}\cdot\mathbf{c})$.

35. Determine the points of intersection of the line

$$L: \quad \frac{x-1}{4} = y - 5 = z$$

and the sphere $x^2 + y^2 + z^2 = 62$.

Topic 2

Complex Numbers

2.1 Introduction

Quadratic equations such as $x^2 + 1 = 0$ and $x^2 + 4x + 13 = 0$ which have negative discriminants have no solutions in the real number system.

However if we were to proceed purely formally to attempt to solve $x^2 + 4x + 13 = 0$ using the formula for a quadratic, we would obtain

$$\begin{aligned} x &= \frac{-4 \pm \sqrt{16 - 52}}{2} \\ &= \frac{-4 \pm \sqrt{-36}}{2} \\ &= \frac{-4 \pm 6\sqrt{-1}}{2} \\ &= -2 \pm 3\sqrt{-1}. \end{aligned}$$

In the 18th century the symbol i was introduced to denote $\sqrt{-1}$, to provide solutions to equations such as $x^2 + 4x + 13 = 0$. The solutions would then have been denoted by $-2 \pm 3i$.

Numbers such as $a + b\sqrt{-1}$ or $a + bi$ were called *complex* numbers and were manipulated in the same manner as ordinary real numbers with $i^2 = -1$. For example:

$(3 + 2i) + (-4 + 5i) = -1 + 7i$;

$2(1 - 3i) - 3(2 - 4i) = 2 - 6i - 6 + 12i = -4 + 6i$; and

$(3 + 2i)(-4 + 5i) = -12 + 15i - 8i + 10i^2 = -22 + 7i$ since $i^2 = -1$.

Note that a complex number $a + ib$ is just an ordered pair (a, b) of real numbers.

Definition 2.1 *A complex number is a number of the form*

$$\boxed{z = x + iy}$$

(or $x + yi$) where $i^2 = -1$ and x, y are real numbers.

- *x is called the real part of z and is denoted by $\Re e\ z$.*
- *y is called the imaginary part of z and is denoted by $\Im m\ z$.*

  (Note that the imaginary part of a complex number is real.)
- *Numbers such as $3i$ which have zero real part are said to be purely imaginary.*
- *Numbers having zero imaginary part are simply real numbers.*

For example:

$$\Re\text{e }(3+2i) = 3,\ \Im\text{m }(3+2i) = 2,\ \Re\text{e }(3i-2) = -2 \text{ and } \Im\text{m }(3i-2) = 3.$$

Equality

Two complex numbers are equal if and only if they have the same real part and the same imaginary part.

Addition

If $z_1 = x_1 + iy_1$ and $z_2 = x_2 + iy_2$, we *define*

$$z_1 + z_2 = (x_1 + x_2) + i(y_1 + y_2).$$

For example, $(-7 + 3i) + (4 + 8i) = -3 + 11i$.

It follows that

- $\Re\text{e }(z_1 + z_2) = \Re\text{e }z_1 + \Re\text{e }z_2$
- $\Im\text{m }(z_1 + z_2) = \Im\text{m }z_1 + \Im\text{m }z_2$.

Multiplication

If $z_1 = x_1 + iy_1$ and $z_2 = x_2 + iy_2$, we *define*

$$z_1 z_2 = (x_1 + iy_1)(x_2 + iy_2) = (x_1x_2 - y_1y_2) + i(x_1y_2 + x_2y_1).$$

Note that this is equivalent to formally multiplying $x_1 + iy_1$ by $x_2 + iy_2$ and using $i^2 = -1$.
For example:

$$(3 - 2i)(4 + 5i) = 12 + 15i - 8i - 10i^2 = 22 + 7i.$$

Furthermore, since $i^2 = -1$ it follows that $i^3 = -i$, $i^4 = 1$ and so on.

Complex Conjugate

The conjugate of the complex number $z = x + iy$ is denoted by $\overline{z}$ and *defined* by $\overline{z} = x - iy$.
For example, if $z = 2 - 3i$ then $\overline{z} = 2 + 3i$. Note also that $z\overline{z} = 4 + 9 = 13$.

Division

The quotient z_1/z_2 is calculated by multiplying the numerator and denominator by the conjugate $\overline{z_2}$ of the denominator. (Recall that expressions such as $1/(a + b\sqrt{2})$ can be rationalized by multiplying numerator and denominator by the *conjugate surd* $a - b\sqrt{2}$. Division of complex numbers is carried out in a similar manner.)
For example:

$$\begin{aligned}\frac{2+3i}{1-3i} &= \frac{(2+3i)(1+3i)}{(1-3i)(1+3i)} \\ &= \frac{2+6i+3i+9i^2}{1-9i^2} = \frac{-7+9i}{10} = -0.7 + 0.9i.\end{aligned}$$

You should verify that

1. $\overline{z_1 + z_2} = \overline{z_1} + \overline{z_2}$
2. $\overline{z_1 z_2} = \overline{z_1}\,\overline{z_2}$
3. $\overline{\left(\dfrac{z_1}{z_2}\right)} = \dfrac{\overline{z_1}}{\overline{z_2}}.$

Worked Example 2.1.1 *If* $z_1 = 2 - 3i$, $z_2 = 2 + i$ *and* $z_3 = 3 - i$, *find*

1. $\Re e\ 1/z_1$ *2.* $z_1\overline{z_2}/z_3$ *3.* $\left(\dfrac{1}{z_2} + \dfrac{1}{z_3}\right)^2.$

1. $$\frac{1}{z_1} = \frac{1}{2-3i} = \frac{2+3i}{(2-3i)(2+3i)} = \frac{2+3i}{4-9i^2} = \frac{2+3i}{13}.$$
   Therefore $\Re e\ 1/z_1 = 2/13$.

2. $$\frac{z_1\overline{z_2}}{z_3} = \frac{(2-3i)(2-i)}{3-i} = \frac{4-8i+3i^2}{3-i} = \frac{1-8i}{3-i} \times \frac{3+i}{3+i}$$
$$= \frac{11-23i}{10}.$$

3. $$\left(\frac{1}{z_2}+\frac{1}{z_3}\right)^2 = \left(\frac{1}{2+i}+\frac{1}{3-i}\right)^2 = \left(\frac{2-i}{5}+\frac{3+i}{10}\right)^2$$
$$= \left(\frac{7-i}{10}\right)^2 = \frac{49-14i+i^2}{100} = \frac{48-14i}{100}$$
$$= \frac{24-7i}{50}.$$

■

Self-help exercises

If $z = -3+7i$ and $w = 3-2i$, determine

1. $\Re\text{e}\ (2z-3w), \Im\text{m}\ (z+w^2)$ and $\bar{w}$ [−15; −5; 3 + 2i]
2. $\bar{z}w^2$ and $2/w$. [−99 + i; (6 + 4i)/13]

2.2 Solution of Equations

Solving equations involving complex numbers is similar to solving the corresponding equations with real numbers.
For example:

- The method of solving the equation
$$\frac{z+2+i}{z+i} = \frac{z-1}{z+2i}$$
is similar to the method used to solve
$$\frac{x+2}{x+1} = \frac{x-1}{x+2}$$
where x is real.
- The "formula" for a quadratic can be used to solve equations such as
$$z^2 - (2+2i)z + (1+2i) = 0.$$

Worked Example 2.2.1 *Solve*

1. $\dfrac{z+2+i}{z+i} = \dfrac{z-1}{z+2i}$ *2.* $z^2-(2+2i)z+(1+2i)=0.$

1. Cross multiplication yields

$$\begin{aligned}
&(z+2i)(z+2+i) = (z-1)(z+i)\\
\Rightarrow\ & z^2+(2+3i)z+4i-2 = z^2+(-1+i)z-i\\
\Rightarrow\ & (3+2i)z = 2-5i.
\end{aligned}$$

Therefore

$$z = \frac{2-5i}{3+2i} = \frac{(2-5i)(3-2i)}{(3+2i)(3-2i)} = \frac{-4-19i}{13}.$$

2. $$\begin{aligned}
z &= \frac{2+2i \pm \sqrt{4+8i+4i^2-4-8i}}{2}\\
&= \frac{2+2i\pm\sqrt{-4}}{2} = \frac{2+2i\pm 2i}{2} = 1+2i, 1.
\end{aligned}$$

■

If $p(z)=0$ is a polynomial equation whose *coefficients are real* its roots occur in conjugate pairs; that is, if $\alpha+i\beta$ is a root then $\alpha-i\beta$ is also a root. This follows since

$$p(\alpha+i\beta)=0 \Rightarrow \overline{p(\alpha+i\beta)}=\overline{0}.$$

Therefore $p(\overline{\alpha+i\beta}) = 0$ since $\overline{z_1+z_2} = \overline{z_1}+\overline{z_2}$ and $\overline{z_1 z_2} = \overline{z_1}\,\overline{z_2}$. Thus $p(\alpha-i\beta)=0$.

For example, since $z=2-3i$ is a root of

$$p(z) = z^4-4z^3+9z^2+16z-52 = 0 \text{ (verify this!)},$$

it follows that $2+3i$ is also a root. Thus

$$(z-(2-3i))(z-(2+3i)) = z^2-4z+13$$

is a factor of $p(z)$.

2.3 Geometrical Interpretation

Since a complex number $z=x+iy$ is simply an ordered pair (x,y) of real numbers it can be represented geometrically by a point (x,y) in the plane or by a vector from the origin to the point (x,y). This plane is called the

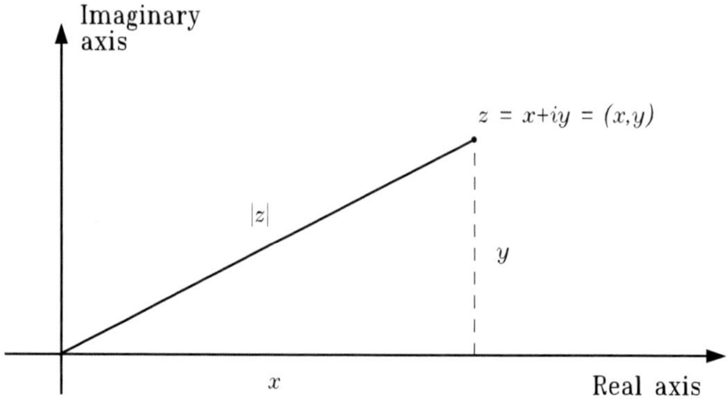

Figure 2.1: Cartesian representation.

complex plane or *z-plane* or the *Argand diagram*; the x-axis is referred to as the *real axis* and the y-axis as the *imaginary axis* — see Figure 2.1.

Since

$$z_1 + z_2 = (x_1 + iy_1) + (x_2 + iy_2) = (x_1 + x_2) + i(y_1 + y_2),$$

complex numbers are added in a similar manner to vectors — see Figure 2.2.

Although real numbers can be ordered (that is, we can determine if $a > b$, $a < b$ or $a = b$ for any two real numbers a and b), no such ordering is possible for complex numbers. For example, statements such as $56+212i > 1+i$ and $z^2 \geq 0$ (consider $z = i$) are meaningless and totally incorrect. However, we can compare the moduli (absolute values) of complex numbers. If $z = x+iy$ we define the modulus of z (or absolute value of z), denoted by $|z|$, by

$$\boxed{|z| = |x + iy| = \sqrt{x^2 + y^2}}$$

Note that, geometrically, $|z|$ measures the distance of the point representing z from the origin in the complex plane. (See Figure 2.1.)
For example:

$$\begin{aligned} |2 + 3i| &= \sqrt{2^2 + 3^2} &= \sqrt{13} \\ |2 - 3i| &= \sqrt{2^2 + (-3)^2} &= \sqrt{13} \\ |-2i| &= 2. \end{aligned}$$

Properties of the Modulus

1. $\boxed{|z|^2 = z\bar{z}}$

2. $\boxed{|\overline{z}| = |z|}$

3. $\boxed{|z_1 + z_2| \leq |z_1| + |z_2|}$ (Triangle inequality; see Figure 2.2.)

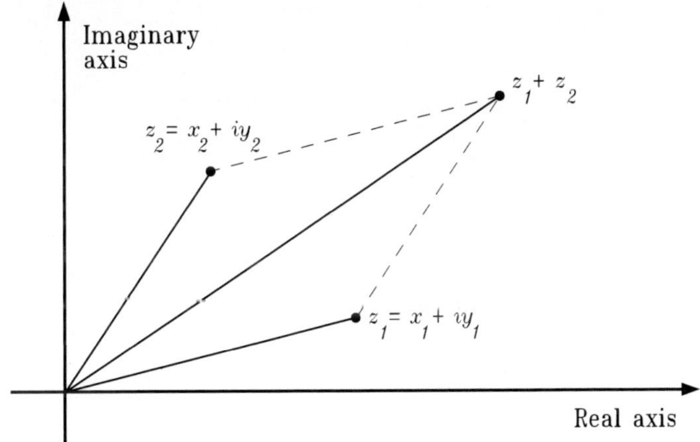

Figure 2.2: Addition of complex numbers.

4. $\boxed{|z_1 z_2| = |z_1||z_2|}$

$$\begin{aligned}
(\text{For } |z_1 z_2|^2 &= (z_1 z_2)\overline{(z_1 z_2)} \text{ by Property 1} \\
&= z_1 z_2 \overline{z_1}\,\overline{z_2} \\
&= (z_1 \overline{z_1})(z_2 \overline{z_2}) \\
&= |z_1|^2 |z_2|^2 \text{ by Property 1.)}
\end{aligned}$$

Thus, *the modulus of a product equals the product of the individual moduli*.

5. $\boxed{\left|\dfrac{z_1}{z_2}\right| = \dfrac{|z_1|}{|z_2|} \text{ if } z_2 \neq 0}$

Thus, *the modulus of a quotient equals the quotient of the moduli*. This is established in a similar way to the result in 4.

These properties are useful in simplifying expressions involving complex numbers.

For example:

$$\left|\frac{(2-3i)(1+2i)}{2-i}\right| = \frac{|2-3i||1+2i|}{|2-i|} = \frac{\sqrt{13}\sqrt{5}}{\sqrt{5}} = \sqrt{13}$$

$$|(-1+i)^4| = |-1+i|^4 = (\sqrt{2})^4 = 4.$$

Worked Example 2.3.1 *Find the modulus of*

$$\frac{(-1+i)^{10}(\sqrt{3}-i)^4}{(1+i\sqrt{3})^8}.$$

We have

$$\left|\frac{(-1+i)^{10}(\sqrt{3}-i)^4}{(1+i\sqrt{3})^8}\right|$$

$$= \frac{|-1+i|^{10}|\sqrt{3}-i|^4}{|1+i\sqrt{3}|^8} = \frac{(\sqrt{2})^{10}2^4}{2^8} = 2.$$

■

Self-help exercises

1. If $z = -1 - i$, write down $|z|$, $|1/z|$, $|z^4|$, $|2z|$ and $|iz|$. $[\sqrt{2}, 1/\sqrt{2}, 4, 2\sqrt{2}, \sqrt{2}]$
2. If $z = 1 - \sqrt{3}i$ and $w = \sqrt{3} + i$, determine $|z + 2w|$, $|z| + |2w|$, $|z/w|$ and $|w/z|$. $[2\sqrt{5}, 6, 1, 1]$

2.4 Polar Form/Exponential Form

If $z = x + iy$ ($z \neq 0$) we can express x, y in polar form $x = r\cos\theta$, $y = r\sin\theta$ where $r = |z|$ and θ is the polar angle called the *argument of* z and is denoted by Arg z. Note that Arg z is determined only up to multiples of 2π. The value of θ is not unique, but if we restrict θ to lie in the interval $(-\pi, \pi]$, θ is called the *principal argument* of z and is denoted by $\arg z$.

In order to calculate $\arg z$, it is strongly recommended that you draw a diagram.

For example, with reference to Figure 2.3:

$\arg(-1 + \sqrt{3}i) = \pi - \pi/3 = 2\pi/3$ (or $120°$),

$\arg(\sqrt{3} - i) = -\pi/6$ (or $-30°$).

When r and θ have been found, z can be expressed in polar form as

$$\boxed{z = x + iy = r\cos\theta + ir\sin\theta = r(\cos\theta + i\sin\theta)}$$

— see Figure 2.4. This polar co-ordinate representation of z is usually abbreviated as $z = r\operatorname{cis}\theta$. (In electrical work, $r\operatorname{cis}\theta$ is often written as $r\angle\theta$.)

Note that if $z = r\operatorname{cis}\theta$ then $\overline{z} = r\operatorname{cis}(-\theta)$. In polar form two complex numbers are equal if they have the same moduli and their arguments differ by a multiple of 2π.

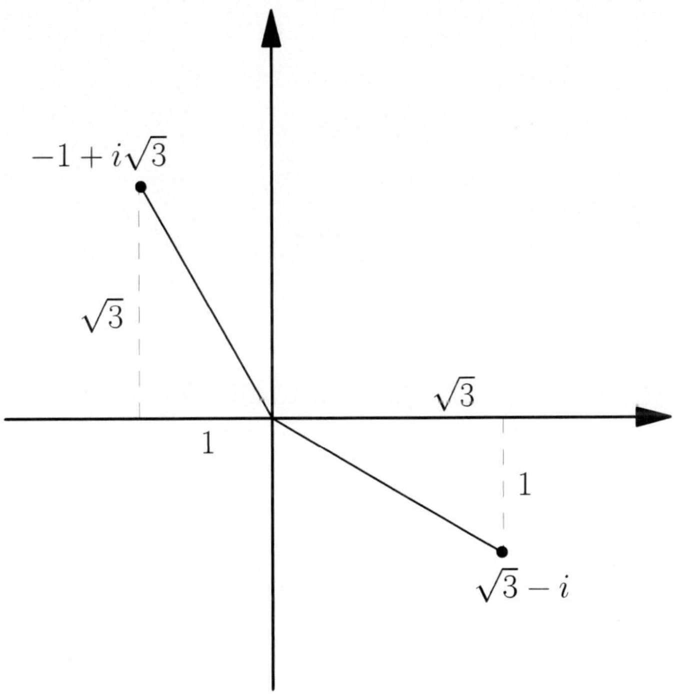

Figure 2.3: Calculation of arg z.

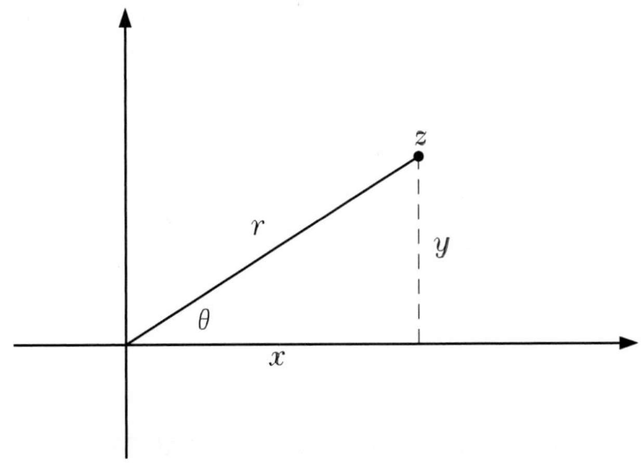

Figure 2.4: Polar representation of z.

Alternatively, z can be expressed in the exponential form $z = re^{i\theta}$ where θ, the polar angle, *must be expressed in radians*. For, using the standard series expansions for the exponential, sin and cos functions (given on Page 489 in Appendix B)

$$\begin{aligned} e^{i\theta} &= 1 + i\theta + \tfrac{(i\theta)^{\bullet}}{2!} + \tfrac{(i\theta)^{\bullet}}{3!} + \tfrac{(i\theta)^{\bullet}}{4!} + \cdots \\ &= 1 + i\theta - \tfrac{\theta^{\bullet}}{2!} - i\tfrac{\theta^{\bullet}}{3!} + \tfrac{\theta^{\bullet}}{4!} + \cdots \\ &= (1 - \tfrac{\theta^{\bullet}}{2!} + \tfrac{\theta^{\bullet}}{4!} + \cdots) + i(\theta - \tfrac{\theta^{\bullet}}{3!} + \tfrac{\theta^{\bullet}}{5!} + \cdots) \\ &= \cos\theta + i\sin\theta \end{aligned}$$

This last identity

$$\boxed{e^{i\theta} = \cos\theta + i\sin\theta}$$

is known as Euler's formula.

Thus, a complex number can be expressed in either of three ways:

1. Cartesian form $a + ib$
2. Polar form $r\,\mathrm{cis}\,\theta$ where $r = |z|$ and $\theta = \mathrm{Arg}\ z$
3. Exponential form $re^{i\theta}$ where $r = |z|$ and $\theta = \mathrm{Arg}\ z$ (expressed in radians) and $e^{i\theta} = \mathrm{cis}\,\theta = \cos\theta + i\sin\theta$.

▌Example:

1. Express $2e^{i\pi/3}$ in the form $a + ib$.
2. Express $3\,\mathrm{cis}\,120^\circ = 3\angle 120^\circ$ in the form $a + ib$.
3. Express $-2 + 3i$ in both polar and exponential forms.

Solution:

1. $2e^{i\pi/3} = 2(\cos\pi/3 + i\sin\pi/3) = 2(1/2 + i\sqrt{3}/2) = 1 + i\sqrt{3}$.
2. $3\,\mathrm{cis}\,120^\circ = 3(\cos 120^\circ + i\sin 120^\circ) = 3(-1/2 + i\sqrt{3}/2) = -3/2 + 3\sqrt{3}i/2$.
3. With reference to Figure 2.5

   $r = |z| = |-2 + 3i| = \sqrt{13}$,

   $\arg z = \pi - \alpha = \pi - \arctan(3/2) \approx 123.7^\circ$;

   or 2.159 (in radians)

   Thus $-2 + 3i \approx \sqrt{13}\,\mathrm{cis}\,123.7^\circ$ or $\sqrt{13}\,e^{2.159i}$

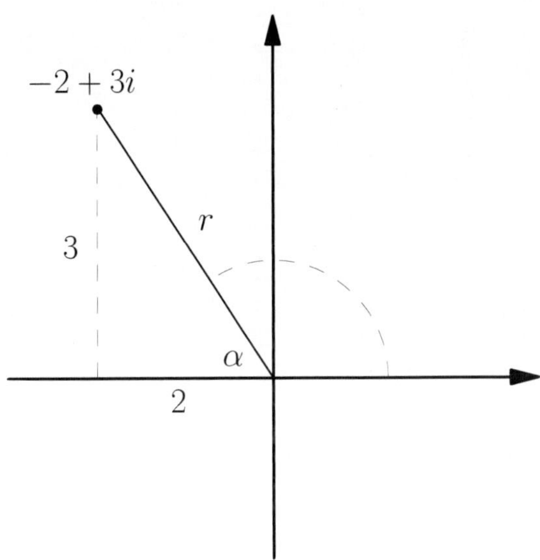

Figure 2.5: Illustration of calculation of arg z.

2.5 Multiplication and Division in Polar (Exponential) Form

If $z_1 = r_1 e^{i\theta_{\cdot}} = r_1 \operatorname{cis}\theta_1$ and $z_2 = r_2 e^{i\theta_{\cdot}} = r_2 \operatorname{cis}\theta_2$ then

$$z_1 z_2 = (r_1 e^{i\theta_{\cdot}})(r_2 e^{i\theta_{\cdot}}) = r_1 r_2 e^{i(\theta_{\cdot}+\theta_{\cdot})}$$

by the law of indices. (See Figure 2.6.)

Alternatively, in polar form

$$\begin{aligned}
z_1 z_2 &= (r_1 \operatorname{cis}\theta_1)(r_2 \operatorname{cis}\theta_2) \\
&= r_1 r_2 (\cos\theta_1 + i\sin\theta_1)(\cos\theta_2 + i\sin\theta_2) \\
&= r_1 r_2 [(\cos\theta_1\cos\theta_2 - \sin\theta_1\sin\theta_2) + i(\sin\theta_1\cos\theta_2 + \sin\theta_2\cos\theta_1)] \\
&= r_1 r_2 [\cos(\theta_1+\theta_2) + i\sin(\theta_1+\theta_2)] \\
&= r_1 r_2 \operatorname{cis}(\theta_1+\theta_2)
\end{aligned}$$

Thus:

1. $|z_1 z_2| = |z_1||z_2|$
2. $\arg(z_1 z_2) = \arg z_1 + \arg z_2 + 2n\pi$.

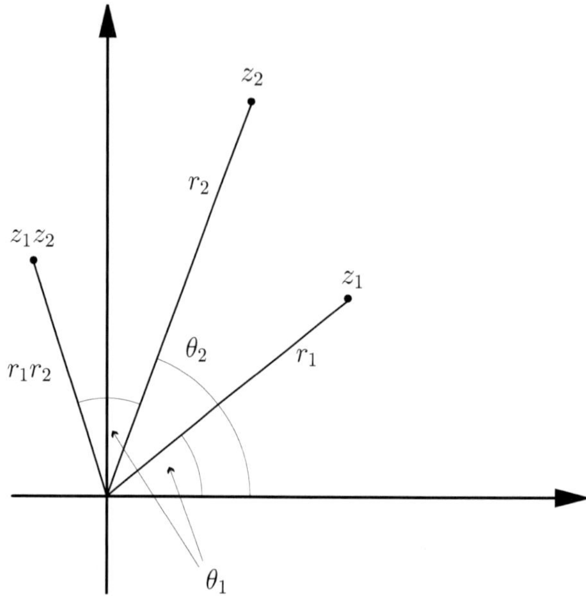

Figure 2.6: Geometric interpretation of multiplication.

The integer n is chosen so that the value of $\arg(z_1z_2)$ lies between $-\pi$ and π.
For example:

$$(2\,\text{cis}\,160°)(3\,\text{cis}\,140°) = 6\,\text{cis}\,300° = 6\,\text{cis}\,(-60°)$$

$$(2e^{i2\pi/3})(3e^{i5\pi/6}) = 6e^{i9\pi/6} = 6e^{i3\pi/2} = 6e^{-i\pi/2}$$

Geometrically, multiplication by i has the effect of an anticlockwise rotation through 90° since if $z = re^{i\theta}$ then

$$iz = e^{i\pi/2}re^{i\theta} = re^{i(\theta+\pi/2)}.$$

(See Figure 2.7.)

Similarly it can be shown that

$$\frac{z_1}{z_2} = \frac{r_1\,\text{cis}\,\theta_1}{r_2\,\text{cis}\,\theta_2} = \frac{r_1}{r_2}\,\text{cis}\,(\theta_1 - \theta_2)$$

or equivalently

$$\frac{r_1e^{i\theta_{\cdot}}}{r_2e^{i\theta_{\cdot}}} = \frac{r_1}{r_2}e^{i(\theta_{\cdot} - \theta_{\cdot})}.$$

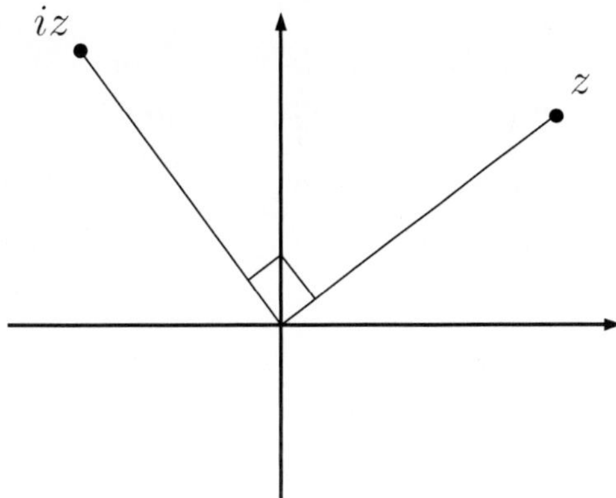

Figure 2.7: Geometrical interpretation of multiplication by i.

Thus:

1. When multiplying complex numbers the moduli are multiplied and the arguments (angles) are added.
2. When dividing complex numbers the moduli are divided and the arguments (angles) are subtracted.

You should note, however, that the polar form of complex numbers is not really appropriate for addition. For example, to calculate

$$2\,\mathrm{cis}\,30^\circ + 4\,\mathrm{cis}\,60^\circ = 2e^{i\pi/6} + 4e^{i\pi/3}$$

we must first express each number in the form $a + ib$.

Thus:

$$\begin{aligned} 2\,\mathrm{cis}\,30^\circ + 4\,\mathrm{cis}\,60^\circ &= 2(\cos 30^\circ + i \sin 30^\circ) + 4(\cos 60^\circ + i \sin 60) \\ &= (\sqrt{3} + i) + (2 + 2\sqrt{3}i) \\ &= (\sqrt{3} + 2) + (1 + 2\sqrt{3})i. \end{aligned}$$

Self-help exercises

1. Express each of the following complex numbers in polar form:

   (a) $-2 + 2\sqrt{3}i$ $[4\,\mathrm{cis}\,\frac{2\pi}{3}]$ (c) $-\sqrt{3} - i$ $[2\,\mathrm{cis}\,\frac{-5\pi}{6}]$

   (b) $1 - i$ $[\sqrt{2}\,\mathrm{cis}\,\frac{-\pi}{4}]$

2. Express each of the following complex numbers in the form $a + ib$:

(a) $3e^{-i\pi/2}$ $[-3i]$ (b) $4\,\text{cis}\,120°$ $[-2+2\sqrt{3}i]$

3. Express each of the following complex numbers in the form $a+ib$:

   (a) $(4\,\text{cis}\,120°)(2\,\text{cis}\,30°)$ $[-4\sqrt{3}+4i]$

   (b) $(4\,\text{cis}\,120°)/(2\,\text{cis}\,30°)$ $[2i]$

2.6 Powers of a Complex Number

From the previous section it follows that integral powers of complex numbers are easily calculated if polar or exponential forms are used.

Note carefully the following special result. If $z = r\,\text{cis}\,\theta = re^{i\theta}$ where $r = |z|$ and $\theta = \arg z$ then

$$\begin{aligned} z^2 &= (r\,\text{cis}\,\theta)(r\,\text{cis}\,\theta) & \text{OR} \quad z^2 &= (re^{i\theta})(re^{i\theta}) \\ &= r^2\,\text{cis}\,2\theta & &= r^2e^{2i\theta} \\ z^3 &= z^2 z & & \\ &= (r^2\,\text{cis}\,2\theta)(r\,\text{cis}\,\theta) & \text{OR} \quad &= (r^2e^{2i\theta})(re^{i\theta}) \\ &= r^3\,\text{cis}\,3\theta & &= r^3e^{3i\theta} \end{aligned}$$

and, by induction

$$z^n = r^n\,\text{cis}\,n\theta \qquad \text{OR} \qquad = r^n e^{ni\theta}$$

for each positive integer n.

This last result is known as de Moivre's theorem:

$$(r\,\text{cis}\,\theta)^n = r^n\,\text{cis}\,n\theta$$
$$(re^{i\theta})^n = r^n e^{ni\theta}$$

It readily follows (upon replacing n by $-m$) that this theorem also holds for all negative integers.

Note also that:

1. $|z^n| = |z|^n$
2. $\text{Arg}\,z^n = n\arg z$.

For example, if $z = -\sqrt{3} + i$ so that $|z| = 2$ and $\arg z = 5\pi/6$, then

$$|z^6| = 2^6 = 64 \text{ and}$$

$$\operatorname{Arg} z^6 = 6(5\pi/6) = 5\pi \Rightarrow \arg z^6 = \pi.$$

Worked Example 2.6.1 *Calculate*

1. $(-2+2i)^4$

2. $\dfrac{(-2+2i)^4(1-i\sqrt{3})^6}{(\sqrt{3}+i)^{12}}$.

1. $-2+2i = \sqrt{8}\,\text{cis}\,135^\circ$

Therefore

$$\begin{aligned}(-2+2i)^4 &= (\sqrt{8}\,\text{cis}\,135^\circ)^4\\ &= 64\,\text{cis}\,540^\circ\\ &= 64\,\text{cis}\,180^\circ\\ &= -64\end{aligned}$$

2. $\sqrt{3}+i = 2\,\text{cis}\,30^\circ$

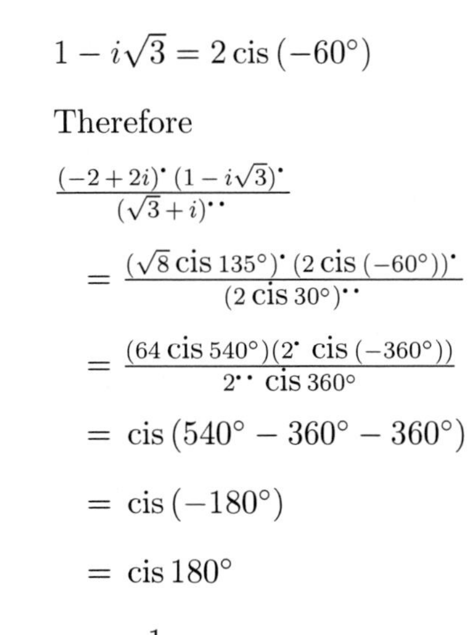

$1 - i\sqrt{3} = 2\,\text{cis}\,(-60^\circ)$

Therefore

$$\frac{(-2+2i)^{\cdot}(1-i\sqrt{3})^{\cdot}}{(\sqrt{3}+i)^{\cdot\cdot}}$$

$$= \frac{(\sqrt{8}\,\text{cis}\,135^\circ)^{\cdot}(2\,\text{cis}\,(-60^\circ))^{\cdot}}{(2\,\text{cis}\,30^\circ)^{\cdot\cdot}}$$

$$= \frac{(64\,\text{cis}\,540^\circ)(2^{\cdot}\,\text{cis}\,(-360^\circ))}{2^{\cdot\cdot}\,\text{cis}\,360^\circ}$$

$$= \text{cis}\,(540^\circ - 360^\circ - 360^\circ)$$

$$= \text{cis}\,(-180^\circ)$$

$$= \text{cis}\,180^\circ$$

$$= -1$$

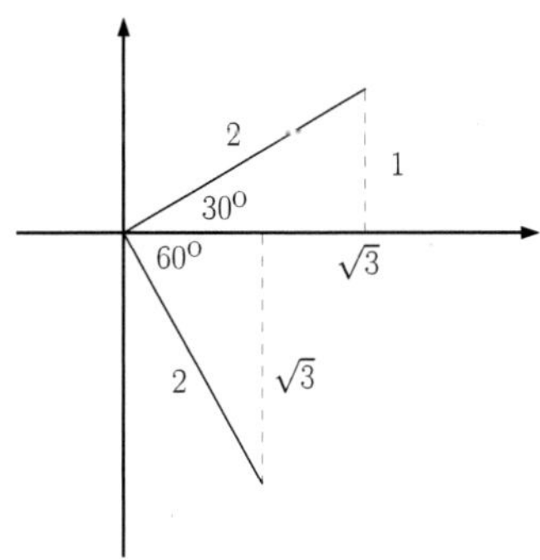

■

2.7 Roots of Complex Numbers — Fractional Powers

We now consider the problem of determining the square roots, cube roots and, in general, the nth roots of a non-zero complex number. It is instructive to consider a particular example:

Example: Find the cube roots of -1.
Solution: Note that z is a cube root of -1 if $z^3 = -1$.
If we let $z = r \operatorname{cis} \theta = re^{i\theta}$ we have

$$(r \operatorname{cis} \theta)^3 = -1 \qquad \text{or} \quad (re^{i\theta})^3 = -1.$$

$$\text{Therefore} \quad r^3 \operatorname{cis} 3\theta = 1 \operatorname{cis} \pi \quad \text{or} \quad r^3 e^{3i\theta} = 1e^{i\pi}.$$

Clearly $r = 1$ and $3\theta = \pi + 2n\pi$ or $\theta = (2n+1)\pi/3$. The values of θ corresponding to $n = 0, 1$ and 2 give the three cube roots

$$\begin{aligned} z &= 1 \operatorname{cis} \pi/3, 1 \operatorname{cis} \pi, 1 \operatorname{cis} 5\pi/3 \\ &= (1 + i\sqrt{3})/2, -1, (1 - i\sqrt{3})/2. \end{aligned}$$

(You should verify that $z^3 = -1$ for each of these values of z.)
Note that these three roots all lie on a circle of radius 1 centered on the origin, and are equally spaced around this circle.
Other values of θ corresponding to $n = 3, 4, 5, \ldots$ give repetitions of the three cube roots obtained above.

We now consider the general problem of finding the n distinct n^{th} roots of the complex number c ($\neq 0$); that is, we find all z which satisfy $z^n = c$.

Suppose $c = \rho \operatorname{cis} \phi$ and $z = r \operatorname{cis} \theta$.
Then

$$\begin{aligned} z^n = c \quad &\Rightarrow \quad (r \operatorname{cis} \theta)^n = \rho \operatorname{cis} \phi \\ &\Rightarrow \quad r^n \operatorname{cis} n\theta = \rho \operatorname{cis} \phi \end{aligned}$$

It follows that $r^n = \rho \Rightarrow r = \rho^{1/n}$

and $n\theta = \phi + 2k\pi$

i.e. $\theta = \phi/n + 2k\pi/n$

The values of θ corresponding to $k = 0, 1, 2, \ldots n-1$ give the n distinct n^{th} roots

$$z = \rho^{1/n} \operatorname{cis} (\phi/n + 2k\pi/n).$$

Note that this is equivalent to expressing c in n equivalent forms

$$\rho \operatorname{cis} \phi, \rho \operatorname{cis} (\phi + 2\pi), \ldots, \rho \operatorname{cis} (\phi + 2(n-1)\pi)$$

and then using

$$(\rho \operatorname{cis} \alpha)^{1/n} = \rho^{1/n} \operatorname{cis} (\alpha/n).$$

Note also that these n roots all lie on a circle of radius ρ centered on the origin, and are equally spaced around this circle.

For example,

$$-8 = 8 \operatorname{cis} \pi, 8 \operatorname{cis} 3\pi, 8 \operatorname{cis} 5\pi,$$

so that the three cube roots of -8 are

$$\begin{aligned}&(8 \operatorname{cis} \pi)^{1/3}, (8 \operatorname{cis} 3\pi)^{1/3}, (8 \operatorname{cis} 5\pi)^{1/3}\\&= 2 \operatorname{cis} \pi/3, 2 \operatorname{cis} \pi, 2 \operatorname{cis} 5\pi/3\\&= 1 + i\sqrt{3}, -2, 1 - i\sqrt{3}.\end{aligned}$$

Note also that these three roots all lie on a circle of radius 2 centered on the origin, and are equally spaced around this circle.

Worked Example 2.7.1 *Find the fourth roots of i.*
Note that four equivalent ways of writing i are:

$$i = 1 \operatorname{cis} \pi/2, 1 \operatorname{cis} 5\pi/2, 1 \operatorname{cis} 9\pi/2, 1 \operatorname{cis} 13\pi/2$$

Therefore the fourth roots of i are

$$\begin{aligned}&(1 \operatorname{cis} \pi/2)^{1/4}, (1 \operatorname{cis} 5\pi/2)^{1/4},\\&\quad (1 \operatorname{cis} 9\pi/2)^{1/4}, (1 \operatorname{cis} 13\pi/2)^{1/4}\\&= \operatorname{cis} \pi/8, \operatorname{cis} 5\pi/8, \operatorname{cis} 9\pi/8, \operatorname{cis} 13\pi/8.\end{aligned}$$

■

Worked Example 2.7.2 *Find all values of* $(1+i\sqrt{3})^{4/3}$.

Note that $(1+i\sqrt{3})^{4/3} = ((1+i\sqrt{3})^4)^{1/3}$
But $1+i\sqrt{3} = 2\,\text{cis}\,60°$ or $2e^{i\pi/3}$.

Therefore

$$(1+i\sqrt{3})^4$$

$$= 16\,\text{cis}\,240°, 16\,\text{cis}\,600°, 16\,\text{cis}\,960°.$$

(Three equivalent representations are taken since we need this expression's cube roots.)

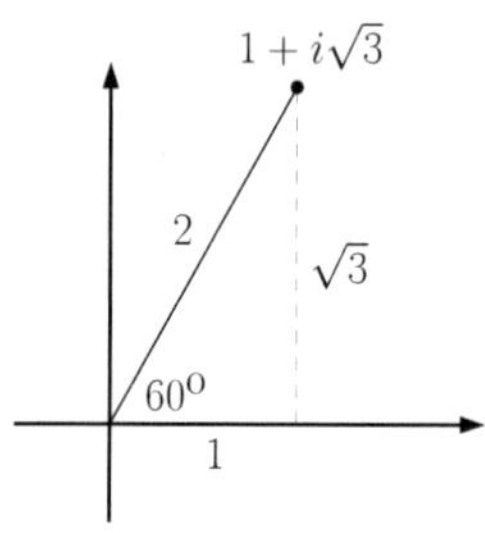

Therefore $(1+i\sqrt{3})^{4/3}$

$$= 16^{1/3}\,\text{cis}\,80°, 16^{1/3}\,\text{cis}\,200°, 16^{1/3}\,\text{cis}\,320°$$

$$\text{or } 16^{1/3}e^{4\pi i/9}, 16^{1/3}e^{10\pi i/9}, 16^{1/3}e^{16\pi i/9}.$$

■

Self-help exercises

Determine

1. $(2\,\text{cis}\,2\pi/3)^3$ $[8\,\text{cis}\,2\pi = 8]$
2. $(\sqrt{3}e^{-i\pi/4})^4$ $[-9]$
3. $(4e^{i\pi/3})^{1/2}$. $[2e^{i\pi/6} = \sqrt{3}+i]$

2.8 Regions in the Complex Plane

Note that $|z-c|$ represents the distance in the complex plane between the points z and c (see Exercise 16). In particular, $|z|$ represents the distance of z from the origin.

Therefore, for real values of r, the equation $|z-c| = r$ represents the set of points on a circle with centre at $z = c$ and of radius r. For example $|z-1+2i| = 4$ is the equation of a circle with centre at the point $1-2i$ and of radius 4. Equivalently, $|z-3i| \leq 2$ represents the region inside and on a circle with centre at the point $3i$ and of radius 2.

Worked Example 2.8.1 *Describe the set of points in the complex plane which satisfy the inequality*

$$|z-1| \geq |z-2+i|.$$

Algebraic Approach

Put $z = x + iy$ then

$$|z-1| \geq |z-2+i|$$

$$\Rightarrow |(x-1)+iy| \geq |(x-2)+i(y+1)|$$

$$\Rightarrow \sqrt{(x-1)^2+y^2} \geq \sqrt{(x-2)^2+(y+1)^2}$$

$$\Rightarrow (x-1)^2+y^2 \geq (x-2)^2+(y+1)^2$$

$$\Rightarrow -2x+1 \geq -4x+2y+5$$

$$\Rightarrow 2x-4 \geq 2y$$

$$\Rightarrow y \leq x-2$$

Thus the region in the complex plane defined by $|z-1| \geq |z-2+i|$ consists of all points on and below the straight line $y = x - 2$.

Geometric Approach

We know that $|z-1|$ represents the distance of z from 1 and $|z-2+i| = |z-(2-i)|$ represents the distance of z from $2-i$.

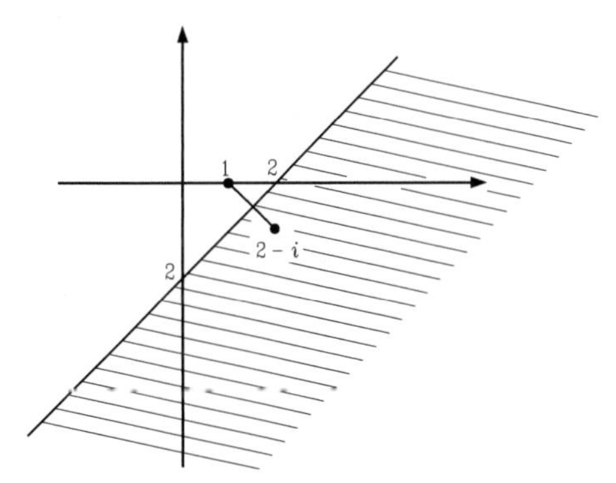

Thus the equation

$$|z-1| = |z-2+i|$$

represents all points equidistant from both the points 1 and $2-i$ and hence lying on the perpendicular bisector of the line joining the points 1 and $2-i$ (i.e. the line $y = x-2$).

Thus $|z-1| \geq |z-2+i|$ is satisfied by all points z lying on and below this line. (That is, points closer to $2-i$ than to 1.) ■

Worked Example 2.8.2 *Describe the set of points in the complex plane which satisfies*

$$\left|\frac{z-1}{z+i}\right| \geq 2.$$

Since $|z_1/z_2| = |z_1|/|z_2|$ we have $|z-1| \geq 2|z+i|$.

Putting $z = x + iy$ gives

$$|(x-1)+iy| \geq 2|x+i(y+1)|$$

$$\Rightarrow \sqrt{(x-1)^2+y^2} \geq 2\sqrt{x^2+(y+1)^2}$$

$$\Rightarrow (x-1)^2+y^2 \geq 4x^2+4(y+1)^2$$

$$\Rightarrow 3x^2+2x+3y^2+8y \leq -3$$

$$\Rightarrow x^2+2x/3+y^2+8y/3 \leq -1.$$

Completing the square gives

$$(x+1/3)^2+(y+4/3)^2 \leq 8/9,$$

which is satisfied by all points inside and on the circle with centre located at the point $(-1/3, -4/3) = -1/3 - 4/3i$ and of radius $2\sqrt{2}/3$. ■

Self-help exercise

If $z = x + iy$ show that $|z + 1 - 2i| = \sqrt{(x+1)^2 + (y-2)^2}$.

2.9 Functions of a Complex Variable

The expressions $w = 3iz+2-3i$, $w = (2z-3i)/(iz+1)$, $w = 1/z$, $w = z^2$ and $w = \exp(z)$ are all examples of *functions of a complex variable* $z = x + iy$. Geometrically, functions such as $w = f(z)$ can be interpreted as representing a *mapping* or *transformation* from one complex plane (the z-plane) to a second complex plane (the w-plane).

To illustrate this idea we shall consider the function $w = z^2$ in some detail. It is clear that to each point $z = x + iy$ there corresponds a point $w = u + iv$ according to the rule $w = z^2$.
For example,

$z = 2 - 3i$ corresponds to $w = -5 - 12i$

$z = -2$ corresponds to $w = 4$, and

$z = 3i$ corresponds to $w = -9$, and so on.

This correspondence between points in the z-plane and points in the w-plane can be thought of as a mapping from the z-plane to the w-plane. In such a representation points in the z-plane map into points in the w-plane according to the rule $w = z^2$. It follows that points lying on a curve in the z-plane map into points lying on a corresponding curve in the w-plane; that is, curves in the z-plane map into curves in the w-plane.

Example: Let us determine the image of the line $y = x$ under the mapping $w = z^2$.
Solution: If $w = u + iv$ then we can write $w = z^2$ in the form $u + iv = (x + iy)^2$ so that $u = x^2 - y^2$ and $v = 2xy$.
Thus each point along the line $y = x$ must map into a point $u + iv$ in the w-plane which satisfies $u = 0, v = 2x^2$. These points define the non-negative imaginary axis in the w-plane.
The effect of the mapping $w = z^2$ could perhaps have been more easily seen if we had expressed z in the polar form $z = re^{i\theta}$.
Then $w = (re^{i\theta})^2 = r^2 e^{2i\theta}$ so that $|w| = r^2$ and $\arg w = 2 \arg z$.
Thus the mapping has the effect of squaring distances from the origin and doubling angles. Consequently the sector defined by $z : |z| \leq 2$ and $0 \leq \arg z \leq \pi/4$ would map into the sector $w : |w| \leq 4$ and $0 \leq \arg w \leq \pi/2$ under the mapping defined by $w = z^2$. (See Figure 2.8.)

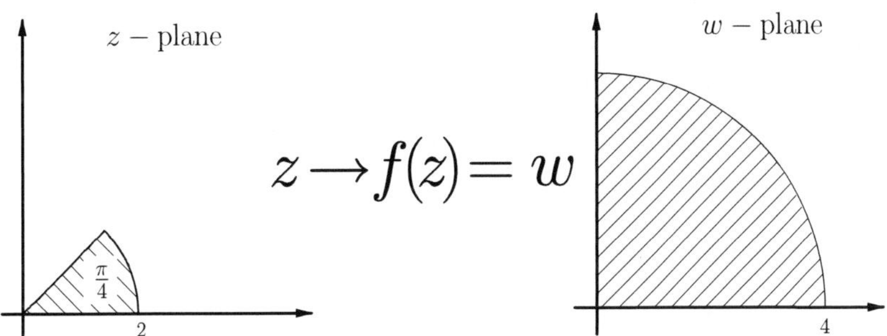

Figure 2.8: Mapping $w = z^2$.

What effect would the mapping $w = z^3$ have on this sector?

In general, under a mapping defined by $w = f(z)$:

1. points map onto points
2. curves map onto curves
3. regions map onto regions.

Worked Example 2.9.1 *Consider the mapping defined by the function*

$$w = (z + i)/(z - i).$$

1. *What are the images of the points:*

   (a) $z = -i$, *(b)* $z = 1 + i$ *and (c)* $z = 1$?

2. *If* $z = x + iy$ *and* $w = u + iv$, *express* u *and* v *in terms of* x *and* y. *Hence show that under this mapping*

   (a) the line $y = 0$ *(i.e. the real axis) maps into the unit circle* $u^2 + v^2 = 1$.

   (b) the unit circle $x^2 + y^2 = 1$ *maps into the line* $u = 0$.

3. *What is the image of the region* $|z| \leq 1$?

1. (a) $z = -i$ maps into $w = 0$

   (b) $z = 1 + i$ maps into $w = (1 + 2i)/(1 + i - i) = 1 + 2i$

   (c) $z = 1$ maps into $w = (1 + i)/(1 - i) = (1 + i)^2/2 = i$.

2. $u + iv = (x + iy + i)/(x + iy - i) = \dfrac{x^2 + y^2 - 1 + 2ix}{x^2 + (y-1)^2}$ so that $u = \dfrac{x^2 + y^2 - 1}{x^2 + (y-1)^2}$ and $v = \dfrac{2x}{x^2 + (y-1)^2}$

   (a) Each point on the line $y = 0$ maps onto a point satisfying $u = \dfrac{x^2 - 1}{x^2 + 1}$ and $v = \dfrac{2x}{x^2 + 1}$; that is, points on the circle $u^2 + v^2 = 1$. (You should verify this; eliminate x from the above parametric representation of this curve.)

   (b) Points that lie on the circle $x^2 + y^2 = 1$ map to the points on the curve $u = 0, v = \dfrac{2x}{x^2 + (y-1)^2}$; that is, the unit circle in the z-plane maps onto the imaginary axis in the w-plane.

3. Since the unit circle $|z| = 1$ maps onto the imaginary axis $u = 0$ the inside of this unit circle maps onto either the right half-plane $u > 0$ or the left half-plane $u < 0$. To check which is the corresponding half-plane, consider any point inside the unit circle. The simplest point to test is $z = 0$ and this maps onto -1 which is in the half-plane $u < 0$. Thus the inside of the unit circle maps onto the half-plane $u < 0$. (See Figure 2.9.)

■

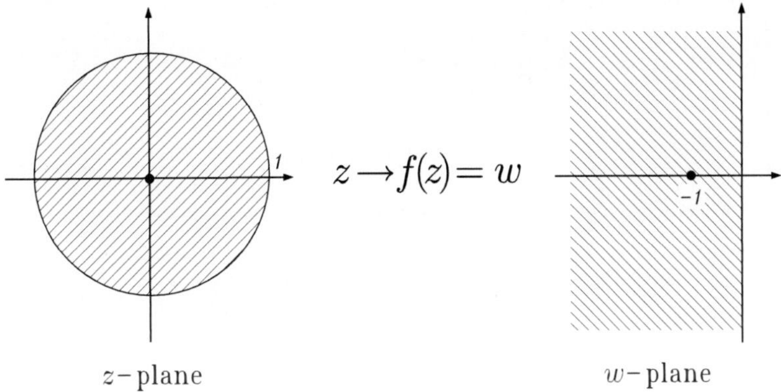

Figure 2.9: Mapping $w = \dfrac{z+i}{z-i}$.

Self-help exercises

1. Under the mapping
$$w = \frac{z-2i}{z+1+i}$$
determine the images of the points $2i, 1+i$ and -1. $[\,0, -i/2, -2+i]$

2. Consider the mapping defined by $w = z^2 + 2z - 3i$.
If $w = u + iv$ and $z = x + iy$, express u and v in terms of x and y.
$[u = x^2 - y^2 + 2x; v = 2xy + 2y - 3]$

In this topic we have

- Introduced the basic algebra of complex numbers.
- Described the geometrical interpretation of complex numbers.
- Defined Euler's formula.
- Manipulated powers and determined roots of complex numbers.
- Examined regions in the complex plane.
- Discussed functions of a complex variable.

2.10 Quick Test Number 2

Basic Mathematical Skills

Question	**Selection**
1. $1 - \dfrac{x}{1-x}$ equals	(a) 1 (b) $\dfrac{1}{1-x}$ (c) $\dfrac{1-2x}{1-x}$ (d) 2
2. $x^2 + y^2$ is	(a) never zero (b) sometimes negative (c) always non-negative
3. $\sin 2x$ equals	(a) $2\sin x$ (b) $2\sin x \cos x$ (c) $2\cos x$
4. $\dfrac{1}{1-\sqrt{3}}$ equals	(a) $-\dfrac{1}{2}(1+\sqrt{3})$ (b) $1 - \dfrac{1}{\sqrt{3}}$ (c) neither
5. If $1 = x - y + z$, then $y = x + z$.	(a) True (b) False

Complex Numbers

Question	**Selection**
1. $(2+5i)(1-3i)$ equals	(a) $2-15i$ (b) $17-i$ (c) $-13-i$
2. $(1+i)^2$ equals	(a) $2i$ (b) $2+2i$ (c) 0
3. $\Im\mathrm{m}\left\{\dfrac{1}{1+3i}\right\}$ equals	(a) $\dfrac{1}{3}$ (b) 3 (c) $-\dfrac{3}{10}$
4. $(4-3i)-(2-i)$ equals	(a) $2-2i$ (b) $2-4i$ (c) neither
5. If $z = 3i - 2$ then $\overline{z}$ equals	(a) $3i+2$ (b) $\frac{1}{3i-2}$ (c) $-3i-2$
6. If $z = 4-2i$, then $\lvert z\rvert$ equals	(a) 4 (b) $\sqrt{20}$ (c) $\sqrt{12}$ (d) 20
7. If $z = 2+3i$ and $w = 3-i$, then $\lvert z+w\rvert$ equals	(a) 5 (b) $\sqrt{29}$ (c) $\sqrt{13}+\sqrt{10}$ (d) $\sqrt{21}$ (e) $5-2i$
8. If $z = x+yi$, then $\lvert z-1\rvert$ equals	(a) $\sqrt{x^2-y^2+1}$ (b) $\sqrt{(x-1)^2-y^2}$ (c) $\sqrt{(x-1)^2+y^2}$ (d) $\sqrt{x^2+(y-1)^2}$

9. If $\arg z = -\dfrac{\pi}{6}$, then $\arg z^5$ equals (a) $-\dfrac{5\pi}{6}$ (b) $\left(-\dfrac{\pi}{6}\right)^5$ (c) $\dfrac{5\pi}{6}$

10. $2 + 2i = \sqrt{8}e^{i\pi/4}$? (a) True (b) False

11. $\sqrt{8}e^{i\pi/4}$ equals (a) $\sqrt{8}e^{i9\pi/4}$ (b) $\sqrt{8}e^{i5\pi/4}$ (c) $\sqrt{8}e^{i17\pi/4}$ (d) $\sqrt{8}e^{-i3\pi/4}$

12. If $|z| = \sqrt{8}$, then $|z^{1/3}|$ equals (a) $\sqrt[3]{8}$ (b) $\sqrt{8}$ (c) $\sqrt{2}$

13. If $z = 3 + 2i$, then $|z|$ equals (a) $\sqrt{5}$ (b) $\sqrt{13}$ (c) neither

---

2.11 EXERCISES

Before attempting any of these miscellaneous exercises, make sure that you have successfully answered all the self-help exercises appearing throughout the chapter.

1. Express each of the following in the form $a + bi$:

   (a) i^{12}
   (b) i^{10}
   (c) i^{53}
   (d) $1 + i + i^2 + i^3$
   (e) $(2 + 3i) - 2(3 - 5i)$
   (f) $i(1 + 2i)$
   (g) $(-1 + i)^2$
   (h) $(-1 + i)^4$
   (i) $(2 - 7i)(3 + 4i)$
   (j) $(1 - 2i)^2(3 + 2i)$
   (k) $(1 + 2i)(3 - 4i)(7 + 5i)$
   (l) $(1 + i)^3 + (1 - i)^3$.

2. If $z = 5i - 1$ and $w = 2 - 3i$, find

   (a) $\Re\text{e } w$ (b) $\Im\text{m } \overline{w}$ (c) $\Re\text{e } z$ (d) $\overline{z}$
   (e) $\Re\text{e } (1/z)$ (f) $\Re\text{e } z^2$ (g) $(\Re\text{e } z)^2$.

3. Express each of the following in the form $a + bi$:

   (a) $\dfrac{1}{1 + 2i}$
   (b) $(3 - 2i)^{-1}$
   (c) $\dfrac{3 + 2i}{i}$
   (d) $\dfrac{2 + 6i}{3 - i}$
   (e) $\dfrac{2 + 3i}{3 + 4i} + \dfrac{2 - 3i}{3 - 4i}$
   (f) $\dfrac{(2 + i)^2}{1 + 2i}$
   (g) $\dfrac{1}{1 + 3i} + \dfrac{1}{3 - i}$.

4. If $z = -2 + 5i$ and $w = 1 - 2i$, express each of the following in the form $a + bi$:

   (a) $2z - 3w$ (b) $z\overline{z}$ (c) $1/z$ (d) $(z + 1/z)^2$ (e) $zw^2/\overline{w}$.

5. If $z = x + iy$ find

   (a) $\Re\text{e } z^2$ (b) $\Re\text{e } (2z + \overline{z} + 4 + 2i)$.

6. Find $z = x + iy$ if $(z - \overline{z})^2 = 2z + 6i$.

7. If $z = x + iy$ find x and y when $z\overline{z} + 2z - 2\overline{z} = 5 + 8i$.

8. Explain why there is no complex number z which satisfies

$$z\bar{z} + 2z + 2\bar{z} = 2i.$$

9. If $z = x + iy$, express each of the following functions in the form $u(x, y) + iv(x, y)$:

(a) $1/z$ (b) $z^2 - iz + 1$ (c) $\dfrac{1}{z - i}$ (d) $\dfrac{z - i}{z + i}$.

10. Solve the following equations for z:

(a) $\dfrac{z}{z + i} = 1 + i$ (b) $\dfrac{z + i}{z - i} = \dfrac{z + 1}{z - 3}$ (c) $\dfrac{z + 2 + i}{z - 3 - i} = \dfrac{z - 1 + i}{z + 1 - i}$.

11. Verify that $z = 3 + i$ is a root of $z^3 - 7z^2 + 16z - 10 = 0$. Hence find all the roots of this equation.

12. Solve each of the following equations for z:

(a) $z^2 - 2z + 2 = 0$ (b) $z^2 - (3 - 2i)z + (1 - 3i) = 0$.

13. Find the modulus of each of the following complex numbers:

(a) $2 + 3i$
(b) $3 - 4i$
(c) $-2i$
(d) -4
(e) $6 + i$
(f) $1/(1 + 3i)$
(g) $\dfrac{2 + 3i}{3 - 2i}$
(h) $3(1 + i) - 2(2 - i)$
(i) $(3 + 2i)(1 - i)$
(j) $(1 + 3i)^2$.

14. Find thc modulus of
(a) $\dfrac{(3 - i)^4(-1 + i)^8}{(1 - 3i)^6}$ (b) $\dfrac{(2 - 2\sqrt{3}i)^6(\sqrt{3} + i)^4}{(1 + i)^{12}}$.

15. If $z = x + iy$, determine

(a) $|z + i|$ (b) $|z - 1|$ (c) $|z + 2 - i|$.

16. If $z_1 = x_1 + iy_1$ and $z_2 = x_2 + iy_2$, give a geometrical interpretation of $|z_1 - z_2|$.

17. Describe the set of points in the complex plane which satisfy

(a) $|z| = 2$ (b) $|z| \le 2$ (c) $|z| > 2$ (d) $1 \le |z| \le 2$

(e) $|z - 2| = 3$ (f) $|z + 3i| \le 3$ (g) $|z - 2 + i| \ge 4$.

(Hint: Recall that $|z|$ is just the distance of z from the origin in the complex plane.)

18. If $z = j\omega L + R + 1/(j\omega C)$ is the complex impedance of a particular electrical circuit and $j = \sqrt{-1}$, express z in the form $a + jb$ and hence find $|z|$. For what value of ω is $|z| = R$?

    (Note that in electrical circuit theory i is usually denoted by j to avoid confusion with the symbol representing current.)

19. The complex equation

$$z\overline{z} + (1+i)z + (1-i)\overline{z} - 7 = 0$$

    represents a circle in the complex plane.

    By putting $z = x + iy$, find the Cartesian equation of this circle.

    Determine the centre and radius of the circle.

20. Express each of the following in the form $a + ib$:

    (a) $3\operatorname{cis}\pi/4$ (b) $2\operatorname{cis}120°$ (c) $4\operatorname{cis}1000°$

    (d) $3e^{i\pi/6}$ (e) $4e^{-i\pi/3}$ (f) $3e^{1.4i}$

    (g) $2\angle 120°$ (h) $5\angle 300°$ (i) $2e^{2+3i}$.

21. Express each of the following complex numbers in polar and exponential forms:

    (a) $2 + 2\sqrt{3}i$ (b) -8 (c) $2i$ (d) $-\sqrt{3} + i$

    (e) $1 - i$ (f) $3 + 4i$ (g) $-4 - 3i$.

22. Express each of the following in the form $a + ib$:

    (a) $(2\operatorname{cis}\pi/3)(3\operatorname{cis}3\pi/4)$

    (b) $\dfrac{4\operatorname{cis}\pi/3}{2\operatorname{cis}3\pi/4}$

    (c) $\dfrac{(3\operatorname{cis}2\pi/3)(4\operatorname{cis}\pi/2)}{6\operatorname{cis}3\pi/4}$

    (d) $(2e^{i\pi/6})(3e^{i\pi/3})$

    (e) $\dfrac{(2e^{i\pi})(4e^{i\pi/3})}{(e^{i\pi/6})(8e^{-i\pi/4})}$

    (f) $(\sqrt{3}e^{i2\pi/3})^2$.

23. Sketch the set of points in the complex plane which satisfy

    (a) $-\pi/3 \le \arg z \le \pi/4$ (b) $-\pi/3 \le \arg z \le \pi/4$ and $|z| \le 1$.

24. Solve for real x the equation: $2e^{ix} = -1 + i\sqrt{3}$.

25. (a) Show that $e^{2x}\sin 3x = \Im\text{m}\; e^{(2+3i)x}$.

    Integrate $e^{(2+3i)x}$ and, by finding the imaginary part of your result, show that

$$\int e^{2x}\sin 3x\,dx = \frac{1}{13}e^{2x}(2\sin 3x - 3\cos 3x).$$

(b) Using the method outlined in the above, find $\int e^{2x} \cos 3x \, dx$.

26. Evaluate and express in the form $a + ib$:

(a) $(-1+i)^6$ (b) $(-1-i\sqrt{3})^7$ (c) $(-\sqrt{3}+i)^8$

(d) $\dfrac{(1-i\sqrt{3})^7}{(1+i)^{10}}$ (e) $\dfrac{(\sqrt{3}-i)^4(2+2\sqrt{3}i)^6}{(-2+2i)^5}$ (f) $1/(-1+\sqrt{3}i)^5$.

27. By considering the real and imaginary parts of the equation

$$(\operatorname{cis}\theta)^3 = \operatorname{cis} 3\theta$$

show that

$$\cos 3\theta = \cos^3\theta - 3\cos\theta\sin^2\theta$$

$$\sin 3\theta = 3\cos^2\theta\sin\theta - \sin^3\theta.$$

28. (a) If $z = \cos\theta + i\sin\theta$, show that

i. $z + 1/z = 2\cos\theta$

ii. $z^n + 1/z^n = 2\cos n\theta$

iii. $z - 1/z = 2i\sin\theta$

iv. $z^n - 1/z^n = 2i\sin n\theta$.

(b) By expanding $(z+1/z)^5$, show that

$$\cos^5\theta = \frac{1}{16}\left(\cos 5\theta + 5\cos 3\theta + 10\cos\theta\right).$$

(c) Obtain a similar expression for $\sin^5\theta$ by expanding $(z-1/z)^5$.

29. Determine the square roots of

(a) i (b) -9 (c) $-1+i\sqrt{3}$.

30. Determine the cube roots of

(a) $2-2\sqrt{3}i$ (b) 1 (c) $-2+2i$.

31. Determine all values of each of the following in the form $a+ib$:

(a) $(2+2i)^{4/3}$ (b) $(-2+3i)^{2/3}$ (c) $(1-3i)^{3/2}$.

32. Find all roots of each of the following equations:

(a) $z^3 = 8i$ (b) $z^2 - 2z + i = 0$ (c) $z^4 + 4iz^2 - 1 = 0$.

33. Describe the region in the complex plane defined by

(a) $|z - i| \geq 2$ (b) $|2z - i| \geq 4$ (c) $1 < |z - 2i| < 3$

(d) $|z| \leq |z + 2i|$ (e) $|z - 1 + i| \geq |z - 2i|$ (f) $\left|\dfrac{z - 1 + i}{z + 2}\right| \leq 3.$

34. Show that

$$az\overline{z} + \overline{b}z + b\overline{z} + c = 0,$$

where a, c are real and $b\overline{b} > ac$ represents a circle in the complex plane if $a \neq 0$.

What does this equation represent if $a = 0$?

35. For each of the following mappings, determine the images of the points

$$1, i, 1 - i, 2 + i, 3i, -1 + 2i$$

(a) $w = 2z + i$ (c) $w = \exp(z)$

(b) $w = (3z - i)/(z + 2i)$ (d) $w = z^3$.

36. For the mappings (a) and (c) in Question 35, determine the images of the lines

(a) $x = 0$, (b) $x = 2$, (c) $y = 0$ and (d) $y = 2$.

37. Consider the function $w = \exp(z)$. If $z = x + iy$ and $w = u + iv$, express u and v in terms of x and y.

Hence show that lines $x = c$ in the z-plane map into circles in the w-plane.

What are the images of the lines $y = d$?

38. Under the mapping defined by the transformation $w = z^3$, determine the image of the sector defined by $z : |z| < 2$ and $0 < \arg z < \pi/3$.

39. Under the mapping $w = (z + 1)/(z - 2i)$, determine the images of the curves

(a) $x = 0$, (b) $y = 0$ and (c) the unit circle $|z| = 1$.

Topic 3

Functions and their Derivatives — Part 1

3.1 Introduction

We begin by reviewing and extending some of the ideas relating to the derivative of a function. Some of these ideas, such as derivatives of very simple functions, the product rule, the quotient and the chain rule, will already be familiar to you from your previous studies.

Throughout this chapter we only consider continuous functions.

3.1.1 The Derivative

A continuous function f is said to be differentiable at the point x if

$$\lim_{h \to 0} \frac{f(x+h) - f(x)}{h}$$

exists. The value of this limit (if it exists) is called the **derivative** of f at x and is denoted by $f'(x)$ or $\frac{dy}{dx}$ if we write $y = f(x)$.

We can give a geometric interpretation of the derivative as follows. Suppose $P(x, y)$ and $Q(x+h, y+k)$ are neighbouring points on the graph of the function defined by $y = f(x)$; see Figure 3.1. The chord PQ which joins the points P and Q has gradient $\frac{f(x+h)-f(x)}{h}$.

As h tends to 0 (that is, as the point Q approaches the point P), the line through P and Q tends to the line which is tangential to the curve at P. That is, as $h \to 0$, the expression $\frac{f(x+h)-f(x)}{h}$ tends to the gradient of the tangent at P. In other words, we have

The gradient of the tangent at P is given by $\lim_{h \to 0} \frac{f(x+h) - f(x)}{h}$

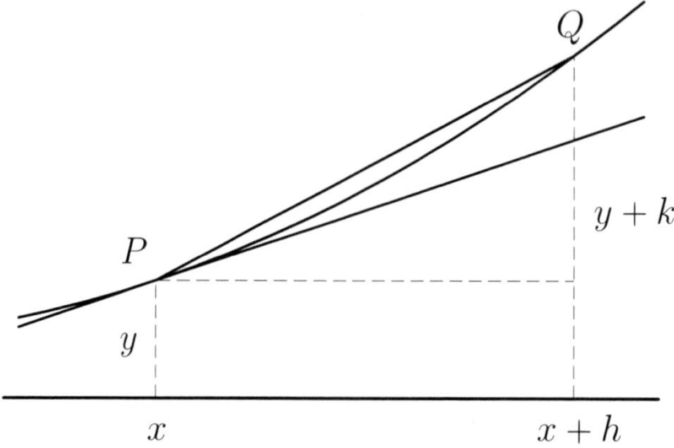

Figure 3.1: Geometric interpretation of the derivative.

That is

$$\boxed{f'(x) = \lim_{h \to 0} \frac{f(x+h) - f(x)}{h}}$$

is the gradient (or slope) of the curve at P. It follows that if a function is not continuous at a point then it certainly cannot be differentiable at that point; however, continuity is not sufficient to guarantee differentiability.

Roughly speaking, a function is differentiable if its graph is smooth; that is, if its graph has no "sharp corners". More precisely, a function is differentiable at a point if its graph has a unique tangent at that point. For example, it can be shown (see Exercise 1) that the function defined by $f(x) = |x|$ is not differentiable at $x = 0$. This is not unexpected when one looks at the graph of the function (see Figure 3.2); it is clear that the graph does not have a unique tangent at the origin. Another simple illustration of a function which is continuous but not differentiable at a point is provided in the following worked example.

▌ **Worked Example 3.1.1** *Consider the function f defined by* $f(x) = x^{2/3}$.
Is f continuous at $x = 0$?
Is f differentiable at $x = 0$?
These questions can perhaps be answered by looking at the graph of f; see Figure 3.3.
From the graph it appears that f is continuous at $x = 0$ but not differentiable there.
More formally, since

$$\lim_{x \to 0} f(x) = \lim_{x \to 0} x^{2/3} = 0 = f(0)$$

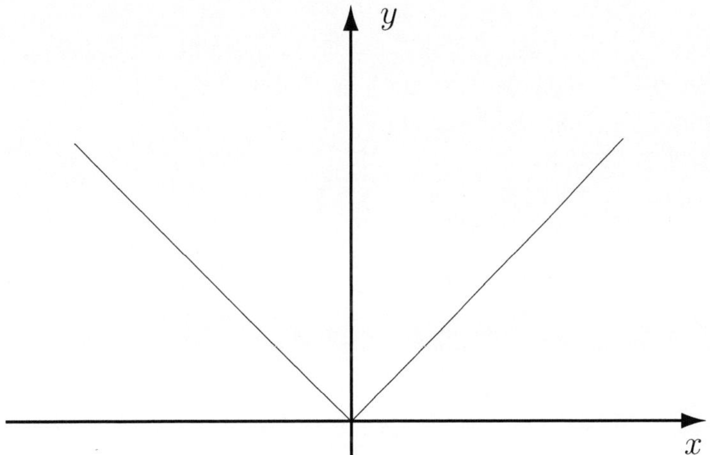

Figure 3.2: Graph of the function $y = |x|$.

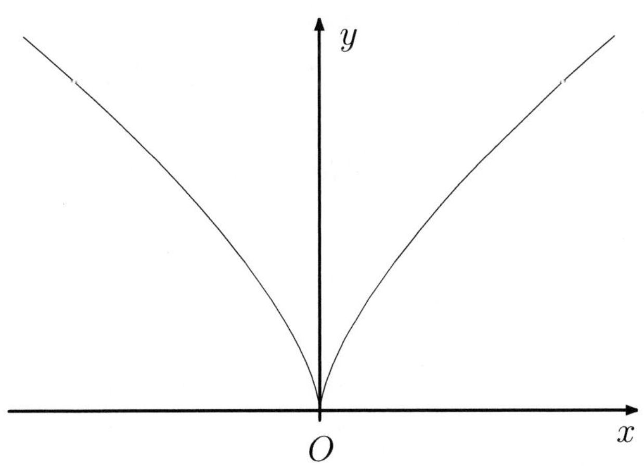

Figure 3.3: A continuous but not differentiable function.

f is certainly continuous at the origin.
However

$$\lim_{h\to 0}\frac{f(0+h)-f(0)}{h}=\lim_{h\to 0}\frac{h^{2/3}-0}{h}=\lim_{h\to 0}\frac{1}{h^{1/3}}$$

and this limit does not exist, so f is not differentiable at the origin. From Figure 3.3 it is clear that the graph does not have a unique tangent at the origin.

■

The derivatives of the following common functions should already be familiar to you

1. $\dfrac{d}{dx}x^n = nx^{n-1}$
2. $\dfrac{d}{dx}\sin x = \cos x$
3. $\dfrac{d}{dx}\cos x = -\sin x.$

If not, you should refer to Exercises 2–4.

The linearity property

$$\frac{d}{dx}\{af(x)+bg(x)\}=a\frac{d}{dx}f(x)+b\frac{d}{dx}g(x)=af'(x)+bg'(x)$$

where a and b are constants also extends naturally to a linear combination of any finite number of differentiable functions. For example:

$$\frac{d}{dx}\left(3x^2+5+3\sqrt{x}-\frac{2}{x}+2\sin x-4\cos x\right)=6x+\frac{3}{2}x^{-1/2}+\frac{2}{x^2}+2\cos x+4\sin x.$$

In order to obtain derivatives of more complicated functions such as $x^2\sin x$, $(x^3+2x)\cos x$, $\frac{x^{\cdot}+2x-1}{x\sin x}$, $\sqrt{1+x^2}$ and $\sin^3 4x$, we need rules for differentiation of products and quotients of functions, and also for composite functions.

The Product and Quotient Rules

The product rule:

$$\frac{d}{dx}\{u(x)v(x)\}=u(x)\frac{d}{dx}v(x)+v(x)\frac{d}{dx}u(x)=u(x)v'(x)+u'(x)v(x).$$

The quotient rule:

$$\frac{d}{dx}\left\{\frac{u(x)}{v(x)}\right\}=\frac{v(x)\frac{d}{dx}u(x)-u(x)\frac{d}{dx}v(x)}{(v(x))^2}=\frac{v(x)u'(x)-u(x)v'(x)}{(v(x))^2}.$$

These are often simplified to

$$\boxed{\frac{d}{dx}(uv)=uv'+u'v \quad\text{and}\quad \frac{d}{dx}\left(\frac{u}{v}\right)=\frac{vu'-uv'}{v^2}}$$

For example:

1. $\dfrac{d}{dx}\left(x^2 \sin x\right) = x^2 \cos x + 2x \sin x$

2. $\dfrac{d}{dx}(4x \cos x) = 4 \cos x - 4x \sin x$

3. $\dfrac{d}{dx}(\tan x) = \dfrac{d}{dx}\left(\dfrac{\sin x}{\cos x}\right) = \dfrac{\cos x \cos x + \sin x \sin x}{\cos^2 x} = \sec^2 x$

4. $\dfrac{d}{dx}\left(\dfrac{3+2x}{1-x}\right) = \dfrac{(1-x)2-(-1)(3+2x)}{(1-x)^2} = \dfrac{5}{(1-x)^2}.$

The result obtained in (3) is so important that it should be memorized along with the three basic results for the derivatives of x^n, $\sin x$ and $\cos x$.

▌**Worked Example 3.1.2** *If $y = x(1+3x^2)\sin x$, determine y'.*
Since $x(1+3x^2) = (x+3x^3)$, we can write $y = (x+3x^3)\sin x$. Using the product rule, we have

$$y' = (1+9x^2)\sin x + (x+3x^3)\cos x.$$

Alternatively we can use the product rule extended to a product of three functions as follows:

$$\begin{aligned}&\frac{d}{dx}(u(x)v(x)w(x))\\&\quad = u'(x)v(x)w(x) + u(x)v'(x)w(x) + u(x)v(x)w'(x)\end{aligned}$$

or more briefly

$$(uvw)' = u'vw + uv'w + uvw'.$$

Applying this to $y = x(1+3x^2)\sin x$, we obtain

$$y' = 1(1+3x^2)\sin x + x \cdot 6x \sin x + x(1+3x^2)\cos x.$$

You should verify that this agrees with the result obtained above.

■

Self-help exercises

1. Write down the derivatives of

   (a) $3x^3 - 2x + 6$ $[9x^2 - 2]$

   (b) $3/x + 4/x^2$ $[-3/x^2 - 8/x^3]$

(c) $3\sin x$ $[3\cos x]$

(d) $2\sin x - 3\cos x$ $[2\cos x + 3\sin x]$

(e) $2\tan x$ $[2\sec^2 x]$

(f) $4\cos x + 3x - \sqrt{x} + \frac{1}{x}$ $[-4\sin x + 3 - \frac{1}{2\sqrt{x}} - \frac{1}{x^2}]$

(g) $x\tan x$ $[\tan x + x\sec^2 x]$

2. Determine the derivatives of:

(a) $x\sin x$ $[x\cos x + \sin x]$

(b) $x(1+x)^2$ $[(1+x)(1+3x)]$

(c) $\frac{1}{1-x}$ $[\frac{1}{(1-x)^2}]$

(d) $x^3\cos x$ $[x^2(3\cos x - x\sin x)]$

The Chain Rule

If $y = f(g(x))$ then $\frac{dy}{dx} = f'(g(x))g'(x)$ or equivalently

$$\frac{dy}{dx} = \frac{dy}{du}\frac{du}{dx} \text{ where } y = f(u) \text{ with } u = g(x).$$

The chain rule can be used to derive a rule for differentiating functions of the form $\{f(x)\}^n$; that is, for differentiating **a power of a function** as follows:

We first put $u = f(x)$ so that $y = \{f(x)\}^n = u^n$.
On differentiating with respect to x and using the chain rule, we obtain

$$\frac{dy}{dx} = \frac{dy}{du}\frac{du}{dx} = nu^{n-1}f'(x).$$

This leads directly to the important result

$$\boxed{\frac{d}{dx}\{f(x)\}^n = n\{f(x)\}^{n-1} \cdot f'(x)}$$

With practice you will find it straightforward to write down the derivative of any power of a differentiable function as illustrated in the following worked example.

Worked Example 3.1.3 *Determine the derivative of each of the following functions:*

1. $(1 + 2x - 3x^2)^7$

2. $\sqrt{1+2x^2}$

3. $\sin^3 x$

4. $3x^2\cos^2 x$

5. $\frac{3}{1+4x^2}$

6. $\frac{1}{\sqrt{1+4x^2}}$.

1. $\frac{d}{dx}(1+2x-3x^2)^7 = 7(1+2x-3x^2)^6(2-6x)$

2. $\frac{d}{dx}\sqrt{1+2x^2}$

$$= \frac{d}{dx}\left(1+2x^2\right)^{1/2} = \frac{1}{2}(1+2x^2)^{-1/2}(4x) = \frac{2x}{\sqrt{1+2x^2}}$$

3. $\frac{d}{dx}\left(\sin^3 x\right)$

$$= \frac{d}{dx}(\sin x)^3 = 3(\sin x)^2 \cos x = 3\sin^2 x \cos x$$

4. Using the product rule as well as the chain rule, we obtain

$$\begin{aligned}&\frac{d}{dx}\left(3x^2\cos^2 x\right)\\ &= 6x\cos^2 x + 3x^2 2\cos x(-\sin x)\\ &= 6x\cos x(\cos x - x\sin x)\end{aligned}$$

5. $\frac{d}{dx}\left(\frac{3}{1+4x^2}\right)$

$$= \frac{d}{dx}3(1+4x^2)^{-1} = -3(1+4x^2)^{-2}8x = -\frac{24x}{(1+4x^2)^2}$$

6. $\frac{d}{dx}\frac{1}{\sqrt{1+4x^2}}$

$$\begin{aligned}&= \frac{d}{dx}(1+4x^2)^{1/2} = -\frac{1}{2}(1+4x^2)^{-3/2}8x\\ &= -\frac{4x}{(1+4x^2)^{3/2}}\end{aligned}$$

■

The chain rule can also be used to obtain the following generalizations to the other three basic results discussed previously. You should endeavour to thoroughly understand how the chain rule applies in these cases; then you will be able to simply write down the derivatives of most functions without having to go through the formal change of variable.

The derivations of these general results are left as exercises (see Exercise 6) but are very similar to the manner in which we extended the result for x^n to the derivative of $\{f(x)\}^n$. For the sake of completeness we also include these results in the following table:

Function	Derivative
$\{f(x)\}^n$	$n\{f(x)\}^{n-1} \cdot f'(x)$
$\sin f(x)$	$\cos f(x) \cdot f'(x)$
$\cos f(x)$	$-\sin f(x) \cdot f'(x)$
$\tan f(x)$	$\sec^2 f(x) \cdot f'(x)$

Worked Example 3.1.4 *Determine the derivative of each of the following functions:*

1. $\cos 2x$

2. $\sin x^2$

3. $\sin^3 4x$

4. $3x \tan 3x$.

Using the results from the above table, we obtain:

1. $\frac{d}{dx}(\cos 2x) = -\sin 2x \cdot 2 = -2\sin 2x$

2. $\frac{d}{dx}\left(\sin x^2\right) = \cos x^2 \cdot 2x = 2x\cos x^2$

3. $$\begin{aligned}\frac{d}{dx}\left(\sin^3 4x\right) &= \frac{d}{dx}(\sin 4x)^3 \\ &= 3(\sin 4x)^2 \cos 4x \cdot 4 = 12\sin^2 4x \cos 4x\end{aligned}$$

4. $$\begin{aligned}&\frac{d}{dx}(3x\tan 3x) \\ &= 3\tan 3x + 3x\sec^2 3x \cdot 3 = 3\tan 3x + 9x\sec^2 3x.\end{aligned}$$

■

Self-help exercises

1. Write down the derivatives of

(a) $3\sin 4x$ [$12\cos 4x$]

(b) $2\cos 4x$ [$-8\sin 4x$]

(c) $\tan 2x$ [$2\sec^2 2x$]

(d) $\cos(1-x^2)$ [$2x\sin(1-x^2)$]

(e) $\sqrt{1-x^2}$ [$\frac{-x}{\sqrt{1-x^2}}$]

2. Determine the derivative of

(a) $(1-2x)^8$	$[-16(1-2x)^7]$	(e) $\sin^2 x$	$[2\sin x\cos x]$
(b) $(1+2x^2)^3$	$[12x(1+2x^2)^2]$	(f) $\sin^2 3x$	$[6\sin 3x\cos 3x]$
(c) $\tan x^2$	$[2x\sec^2 x^2]$		
(d) $\cos^3 x$	$[-3\cos^2 x\sin x]$	(g) $x\sqrt{1-x^2}$	$[\frac{1-2x^{\bullet}}{\sqrt{1-x^{\bullet}}}]$

3.1.2 Higher Derivatives

We define the second derivative $\frac{d^{\bullet} y}{dx^{\bullet}}$ to be the derivative of the first derivative $\frac{dy}{dx}$. It is often denoted by y''; that is,

$$y'' = \frac{d^2y}{dx^2} = \frac{d}{dx}\left(\frac{dy}{dx}\right)$$

Just as the first derivative $\frac{dy}{dx}$ measures the rate at which y changes with respect to x, the second derivative $\frac{d^{\bullet} y}{dx^{\bullet}} = \frac{d}{dx}(\frac{dy}{dx})$ measures the rate at which y' changes with respect to x. Consequently, it measures the rate at which the slope or gradient of the graph of $y = f(x)$ changes. In particular:

1. If the graph is **concave upwards** (see Figure 3.4) the slope increases as x increases and $y'' > 0$.

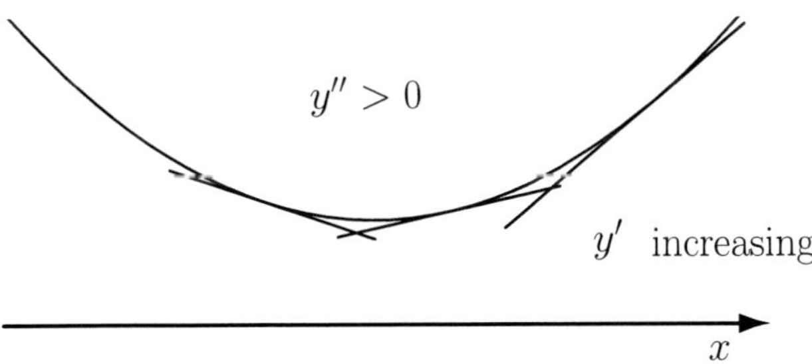

Figure 3.4: A function that is concave up.

2. If the graph is **concave downwards** (see Figure 3.5) the slope decreases as x increases and $y'' < 0$.

3. If the slope of the graph is constant, $y'' = 0$.

In summary, the graph of a differentiable function defined by $y = f(x)$

- has a positive gradient wherever $f'(x) > 0$; that is, the function is increasing wherever $f'(x) > 0$

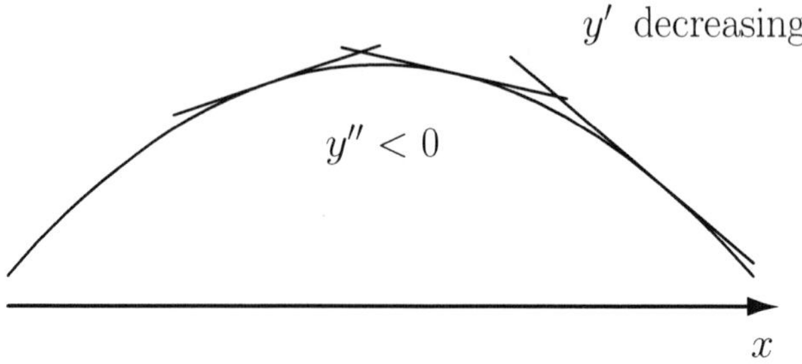

Figure 3.5: A function that is concave down.

- has a negative gradient wherever $f'(x) < 0$; that is, the function is decreasing wherever $f'(x) < 0$
- is concave up wherever $f''(x) > 0$
- is concave down wherever $f''(x) < 0$.

3.2 Implicit Differentiation

Until now we have been concerned with differentiating functions which are given by an explicit formula such as $y = f(x)$. In applications, however, a relation such as $f(x, y) = 0$ may arise which cannot be solved for y explicitly in terms of x. Nevertheless this implicit representation **may** imply that y is a function of x.

When the implicit equation $f(x, y) = 0$ does imply that y is a function of x, the derivatives $\frac{dy}{dx}$ and $\frac{d^2 y}{dx^2}$ can be found without knowing y explicitly as a function of x. For example, the relation $x^3 - y^3 = 8$ does define y as a function of x. To obtain the derivative $\frac{dy}{dx}$ we could first solve for y in terms of x to obtain $y = \sqrt[3]{x^3 - 8}$ and then differentiate both sides with respect to x to give

$$y' = \frac{1}{3}(x^3 - 8)^{-2/3} 3x^2 = x^2 (x^3 - 8)^{-2/3}.$$

Unfortunately this is not always possible; for example, can you solve

$$\sin(x + 2y) = 1 - 3x^2 - y^2$$

for y in terms of x? In such cases we can nevertheless determine $\frac{dy}{dx}$ by formally differentiating both sides of the equation with respect to x.

Before considering implicit differentiation in general, it is instructive to consider how to differentiate with respect to x terms like y^2, y^3, xy, $3xy^2$ and $\sin(x + 2y)$ which often occur in such implicitly defined functions. Here we need to be careful to note that y is to be treated as a function of x. The differentiation of terms such as

1. y^2, y^3, $\sin y$ and $\cos y$ requires the use of the chain rule. For example
   - $\dfrac{d}{dx}\left(y^2\right) = \dfrac{d}{dy}\left(y^2\right)\dfrac{dy}{dx} = 2y\,y'$
   - $\dfrac{d}{dx}\left(y^3\right) = \dfrac{d}{dy}\left(y^3\right)\dfrac{dy}{dx} = 3y^2\,y'$
   - $\dfrac{d}{dx}(\sin y) = \dfrac{d}{dy}(\sin y)\dfrac{dy}{dx} = \cos y\ y'$
   - $\dfrac{d}{dx}\left(\cos y^2\right) = \dfrac{d}{dy}\left(\cos y^2\right)\dfrac{dy}{dx} = -2y\sin y^2\ y'$
   - $\dfrac{d}{dx}\sin(x+2y) = \cos(x+2y)\,(1+2\dfrac{dy}{dx}) = (1+2y')\cos(x+2y).$
2. $2xy$ and yy' requires the use of the product rule. We have
   - $\dfrac{d}{dx}(2xy) = 2x\dfrac{d}{dx}(y) + 2\cdot y = 2x\,y' + 2y$
   - $\dfrac{d}{dx}(yy') = \dfrac{d}{dx}(y)\,y' + y\dfrac{d}{dx}(y') = y'^2 + y\,y''.$

Returning to the previous relation $x^3 - y^3 = 8$ we have, on differentiating both sides with respect to x, $3x^2 - 3y^2y' = 0$. This implies $y' = \frac{x^2}{y^2}$, agreeing with the result obtained earlier. (You should verify this!)

Worked Example 3.2.1 *A curve has equation*

$$x^2 + 2xy - y^2 + 3y = 7.$$

1. *Determine y'.*
2. *Find the equations of the tangent and normal to the curve at the point $P(1,2)$.*
3. *Determine the value of $\frac{d^2y}{dx^2}$ at the point P.*

1. Differentiating both sides with respect to x gives

   $$2x + 2y + 2xy' - 2y\,y' + 3y' = 0. \qquad (3.1)$$

   Thus $(2x - 2y + 3)y' = -2x - 2y$.

   Hence $y' = \dfrac{2x+2y}{2y-2x-3}$.
2. At the point $P(1,2)$, $y' = \frac{2+4}{4-2-3}$ (from part (1)) so that the tangent to the curve at P has gradient -6 and the normal[1] has slope $1/6$.

[1] Recall that if two lines are perpendicular their slopes m_1 and m_2 satisfy $m_1 m_2 = -1$.

Thus, the equation of the tangent at P has the form $y = -6x + c$ where c can be found using the fact that P lies on the tangent. That is, when $x = 1$, y must equal 2. Hence $c = 8$.

Therefore the equation of the tangent at P is $y = -6x + 8$.

In a similar manner the normal to the curve at P has the equation

$$y = \frac{1}{6}x + c.$$

As P also lies on the normal we have $2 = \frac{1}{6} + c$ so $c = 11/6$, giving the equation $6y - x = 11$.

3. To obtain the second derivative we simply differentiate Equation 3.1 with respect to x to obtain

$$2 + 2y' + 2y' + 2xy'' - (2y'\,y' + 2y\,y'') + 3y'' = 0.$$

Thus $(2x - 2y + 3)y'' = -2 - 4y' + 2y'^2$.
But $y' = -6$ at P so that finally

$$(2 - 4 + 3)y'' = -2 + 24 + 72 \Rightarrow y'' = 94 \text{ at } P.$$

■

Self-help exercises

1. In each of the following cases determine y'':

(a) $y = x^4 + 2x + 12$ $[12x^2]$ (c) $y = \frac{1}{1+x}$ $[\frac{2}{(1+x)^3}]$
(b) $y = \sin 2x + \cos 2x$ $[-4y]$

2. If $x^2 + y^2 = 4$, determine y' at the point $(\sqrt{3}, 1)$ using implicit differentiation. $[-\sqrt{3}]$

3.3 Applications of Differentiation

3.3.1 Curve Sketching

Here only a brief review of the approach to curve sketching is given.

When sketching the graph of $y = f(x)$, some or all of the following points should be considered.

1. **The domain of the function**

   Determine (if not already specified) the domain of the function (that is, the set of values of the independent variable x for which the function is defined).

   For example:

   (a) $f(x) = \frac{1}{x^2 - 2x}$ is defined for all x except 0 and 2

   (b) $f(x) = \sqrt{2 - x}$ is defined for $x \leq 2$ so that it has domain $(-\infty, 2]$.

2. **Intercepts**

   Determine where the graph cuts the axes.

   To find where the graph cuts the x-axis put $y = 0$, and to find where the graph cuts the y-axis put $x = 0$.

3. **Symmetry**

   If $f(-x) = f(x)$ (which is the case for the functions defined by $f(x) = x^2, x^4, \frac{1}{x^2}, \cos x$ and $\sin^2 x$) the graph of $y = f(x)$ is symmetrical about the y-axis. Such functions are called **even** functions. (See Figure 3.6.)

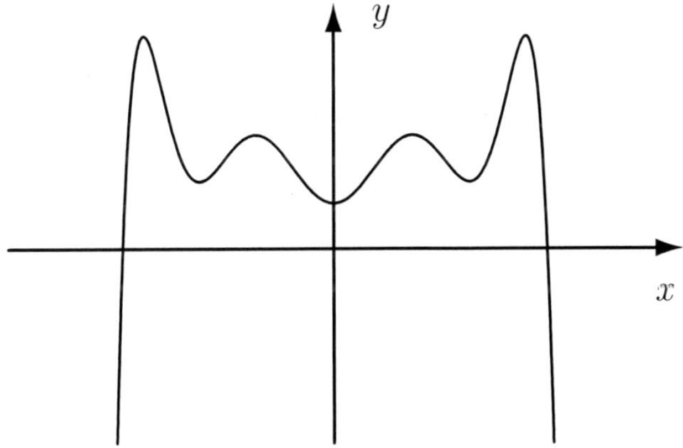

Figure 3.6: Even function.

If $f(-x) = -f(x)$ (which is the case for the functions defined by $f(x) = x, x^3, \frac{1}{x}$ and $\sin x$) the graph of $y = f(x)$ is symmetrical about the origin. Such functions are called **odd** functions. (See Figure 3.7.)

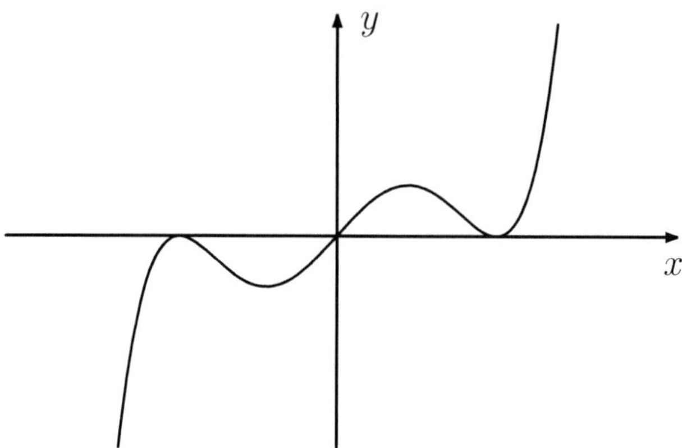

Figure 3.7: Odd function.

4. **Asymptotes**

- **Vertical asymptotes**

  The line $x = a$ is called a vertical asymptote of the curve $y = f(x)$ if $f(x)$ is unbounded at $x = a$; that is, if $f(x) \to \pm\infty$ as $x \to a$ from either side (that is, $x > a$ and $x < a$).

  For example, the graph of the function $f(x) = \frac{1}{x^2 - x - 2} = \frac{1}{(x-2)(x+1)}$ has vertical asymptotes at $x = 2$ and $x = -1$. (See Figure 3.8.)

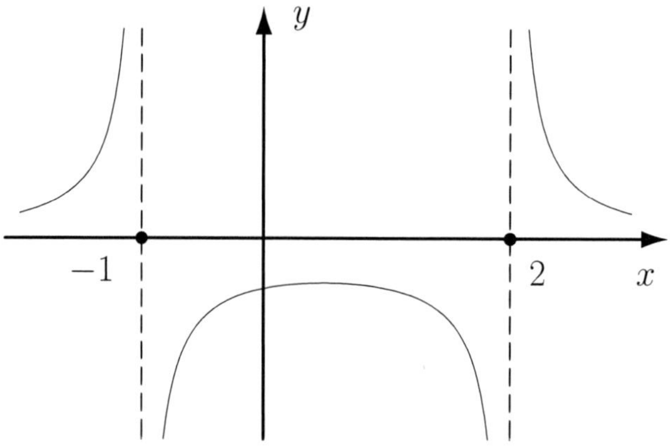

Figure 3.8: Graph of the function $y = \frac{1}{x^2 - x - 2}$.

- **Horizontal or oblique asymptotes**

  The line $y = b$ is called a horizontal asymptote of the curve $y = f(x)$ if

$$\lim_{x \to \infty} f(x) = b.$$

For example, $y = 2$ is a horizontal asymptote of the curve $y = \frac{2x+1}{x-2} = 2 + \frac{5}{x-2}$. (Note that as $x \to \pm\infty$, the term $\frac{5}{x-2} \to 0$.) (See Figure 3.9.)

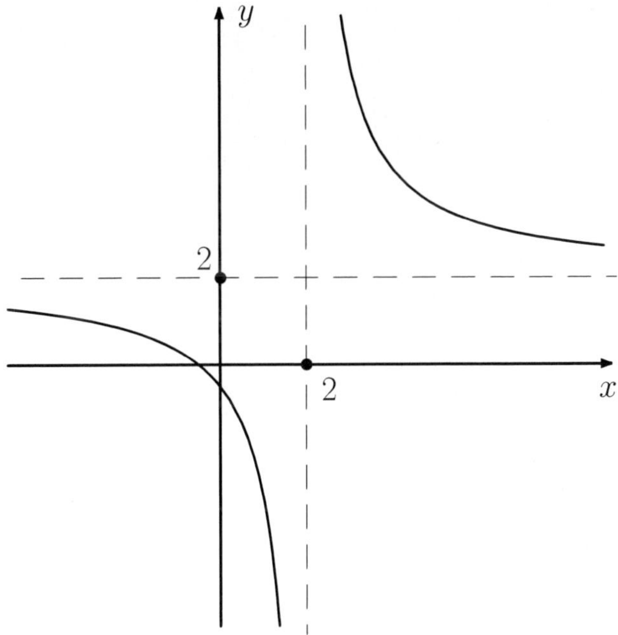

Figure 3.9: Graph of the function $y = \frac{2x+1}{x-2}$.

The curve $y = g(x)$ is called an oblique asymptote of the curve $y = f(x)$ if the difference

$$f(x) - g(x) \to 0 \text{ as } x \to \pm\infty.$$

For example, $y = x$ is an (oblique) asymptote of the curve $y = x + \frac{2}{x}$. (See Figure 3.10.)

5. **The Sign of the Gradient — Turning Points**

   Investigation of the sign of the derivative $f'(x)$ will indicate where the function is increasing, decreasing or has turning points.

6. **Concavity**

   Investigation of the sign of the second derivative $f''(x)$ will indicate where the graph of the function is concave up or concave down.

Furthermore, the sketching of graphs can often be simplified if a composite graph is built up by adding ordinates, translating, reflecting and taking reciprocals of known elementary curves such as straight lines, quadratics (in

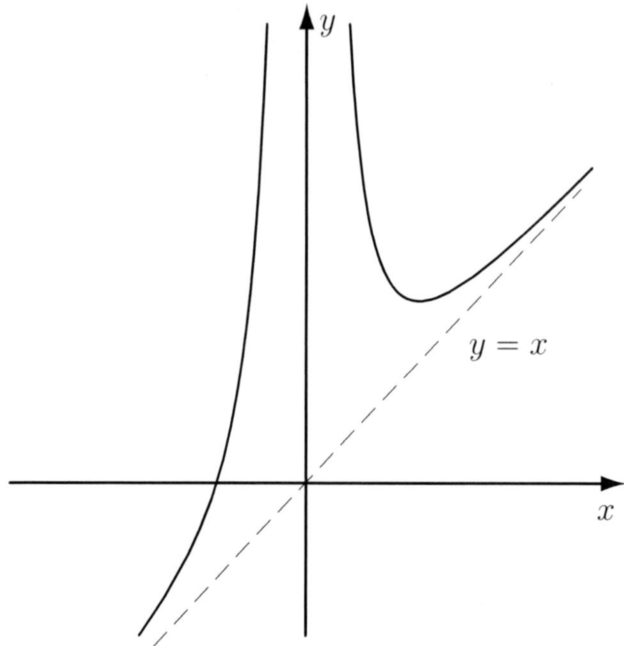

Figure 3.10: Graph of the function $y = x + \frac{2}{x^2}$.

particular x^2), reciprocals such as $\frac{1}{x}$ and $\frac{1}{x^2}$, and the circular functions $\sin x$ and $\cos x$.

For example:

- The graph of $y = (x-2)^2 + 1$ can be obtained from that of $y = x^2$ with a translation of two units to the right (in the positive x-direction) followed by a translation of one unit up (in the positive y-direction). (See Figure 3.11.)

- The graph of $y = x + \frac{3}{x+1}$ can be obtained from those of $y = x$ and $y = \frac{3}{x+1}$ by addition of ordinates. (The graph of $y = \frac{3}{x+1}$ is obtained from that of $\frac{1}{x}$ with a translation of one unit to the left (in the negative x-direction).) (See Figure 3.12.)

- The graph of $y = 4 + 3\sin 2(x-1)$ can be obtained as follows. Translate the graph of $y = 3\sin 2x$ (what is the amplitude and period of this?) one unit to the right and then translate four units in the positive y-direction.

  Has the amplitude and period remained the same? (See Figure 3.13.)

- The graph of $y = \frac{x^2}{x-1}$ can be obtained by first dividing out to obtain $y = x + 1 + \frac{1}{x-1}$, sketching the graphs of $y_1 = x + 1$, $y_2 = \frac{1}{x-1}$ and then

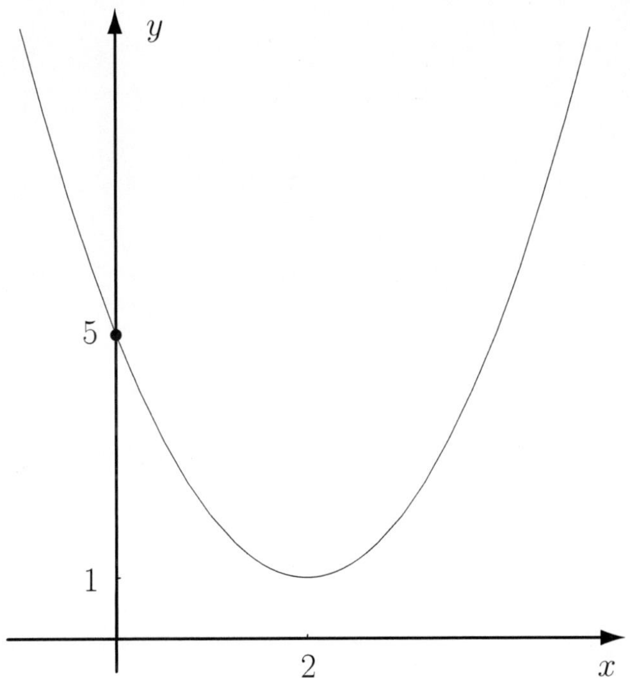

Figure 3.11: Graph of the function $y = (x-2)^2 + 1$.

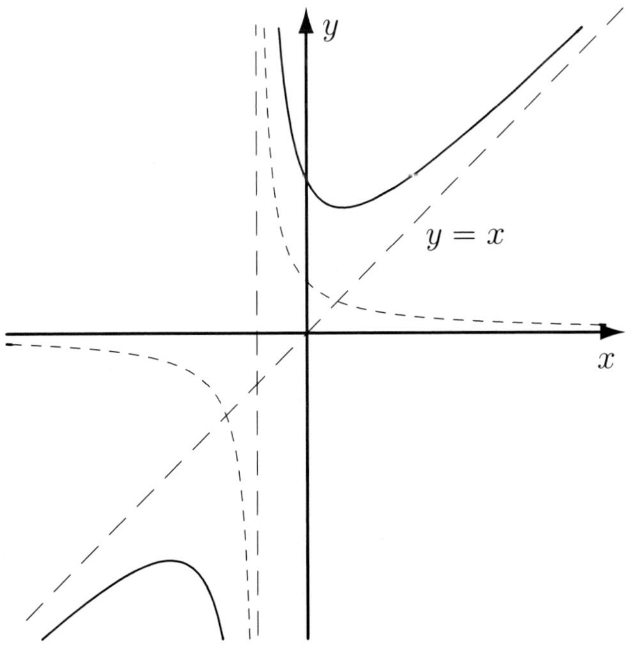

Figure 3.12: Graph of the function $y = x + \frac{3}{x+1}$.

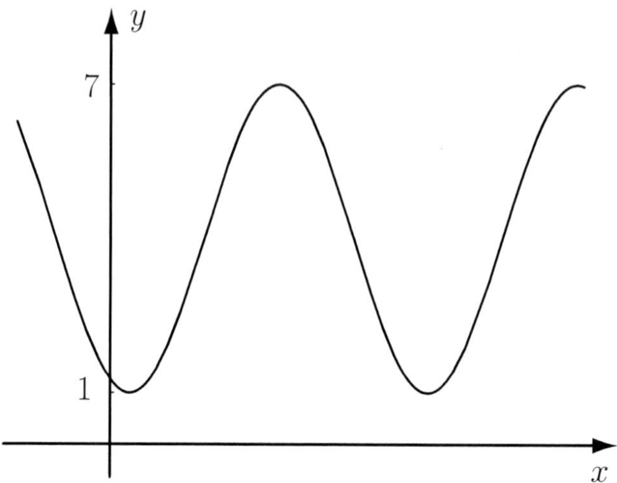

Figure 3.13: Graph of the function $y = 4 + 3\sin 2(x - 1)$.

adding ordinates. (See Figure 3.14.)

3.3.2 Rates of Change

You should already be familiar with the fact that, if $y = f(x)$, then $\frac{dy}{dx}$ measures the rate at which y changes with respect to x. (This is precisely what is being measured by the slope of the graph of $y = f(x)$.)

For example:

- If $s = s(t)$ is the distance travelled in time t of a body moving in a straight line, then $\frac{ds}{dt}$ measures the rate of change of distance travelled with respect to time; that is, the velocity.

- If $p = p(V)$ is the pressure of a gas as a function of its volume V, then $\frac{dp}{dV}$ measures the rate of change of the pressure of the gas with respect to volume; that is, the volume rate of change of pressure.

- If $T = 30 + 40e^{-t/10}$ is the temperature of a body at time t, then $\frac{dT}{dt}$ measures the rate of change of its temperature with respect to time.

The concept of a derivative being a measure of a rate of change is further developed in the following examples.

Worked Example 3.3.1 *A plane which is flying horizontally at an altitude of 2000 metres passes directly over an observation tower.*

If the plane flies with a constant speed of 150 km/hr, how fast is its distance from the base of the observation tower increasing one minute later? (See Figure 3.15.)

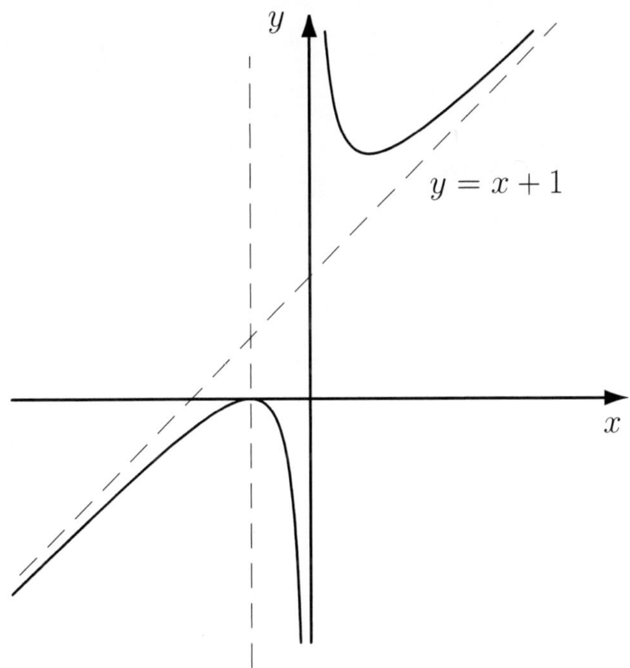

Figure 3.14: Graph of the function $y = \frac{x^{\bullet}}{x-1}$.

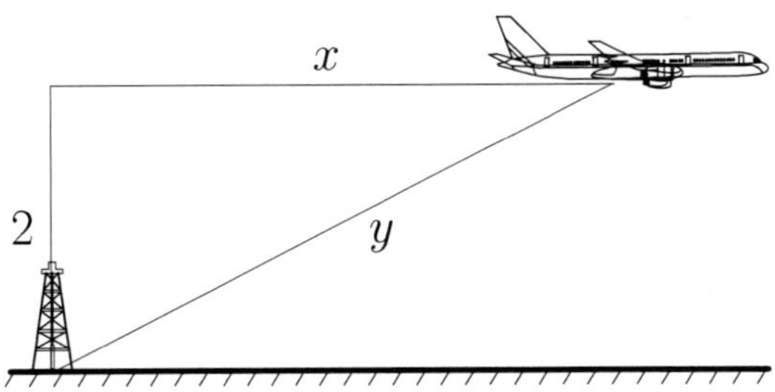

Figure 3.15: Sketch in support of Worked Example 3.3.1.

Let y km denote the distance from the plane to the base of the observation tower at time t minutes after passing over the tower and let x km be the horizontal distance travelled by the plane in that time.
The rate at which x increases is just the speed of the plane; that is

$$\frac{dx}{dt} = 150/60 = 2.5$$

if we use *minute* as the unit of time.
We need to find $\frac{dy}{dt}$. To do this we need a geometric relationship between x and y; implicit differentiation will then yield $\frac{dy}{dt}$.
From Figure 3.15, using Pythagoras' theorem we have

$$x^2 + 4 = y^2.$$

If we differentiate both sides with respect to t we get:

$$2x\frac{dx}{dt} = 2y\frac{dy}{dt} \text{ which implies } \frac{dy}{dt} = \frac{5x}{2y}.$$

After one minute, $x = 2.5$ and hence $y = \sqrt{4 + 6.25} = \sqrt{10.25}$. Thus $\frac{dy}{dt} = \frac{12.5}{2\sqrt{10.25}} \approx 1.95$; that is, after one minute the distance from the observation tower is increasing at the rate of 1.95 km/min.
(Note that an alternative approach would have been to use the chain rule to determine $\frac{dy}{dt}$ using $\frac{dy}{dt} = \frac{dy}{dx}\frac{dx}{dt}$ where $\frac{dy}{dx}$ is obtained using $y = \sqrt{x^2 + 4}$.) ■

Worked Example 3.3.2 *Water flows into a large hemispherical tank at a constant rate of 2 m^3/hour. If the tank has a radius of 4 metres, determine the rate at which the depth of water increases.*
Let V m^3 be the volume and h m the depth of the water in the tank at time t. (See Figure 3.16.)
As water enters at a constant rate of 2 m^3/hour, we have $\frac{dV}{dt} = 2$.
We need to find the rate at which the depth increases; that is, $\frac{dh}{dt}$.
By the chain rule

$$\frac{dh}{dt} = \frac{dh}{dV}\frac{dV}{dt} = 2\frac{dh}{dV}.$$

It can be shown (using the techniques of Topic 5) that the volume V corresponding to the depth h is given by

$$V = \frac{\pi h^2}{3}(12 - h) = \frac{\pi}{3}\left(12h^2 - h^3\right).$$

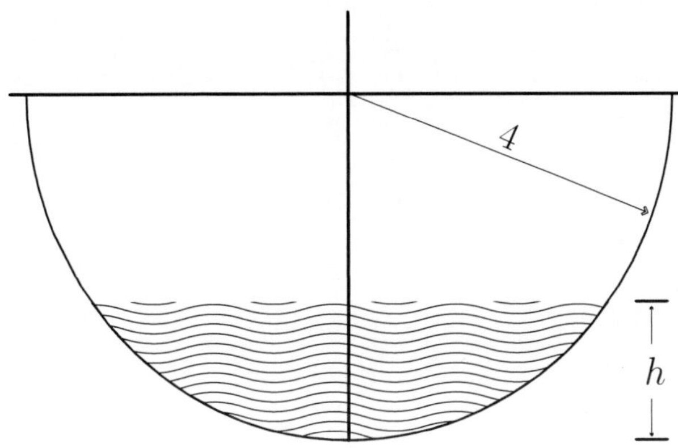

Figure 3.16: Sketch in support of Worked Example 3.3.2.

Thus

$$\frac{dV}{dh} = \frac{\pi}{3}(24h - 3h^2) = \pi(8h - h^2).$$

Therefore

$$\frac{dh}{dt} = \frac{dh}{dV}\frac{dV}{dt} = 2\frac{dh}{dV} = \frac{2}{\pi(8h - h^2)}.$$

As is to be expected, the rate at which the depth increases depends on the depth.

■

3.3.3 Maxima and Minima

Optimization plays an important role in the application of mathematics to real life. For example, business managers are concerned with maximizing profits and minimizing costs (such problems are studied in the discipline of operations research) whereas in experimental work we are often concerned with fitting the best model (in the sense that errors are minimized) to experimental data. In other situations we are often confronted with the problem of optimizing physical quantities such as energy expended, time taken, capacity of a container, etc.

Here we shall only be concerned with the mathematical problem of minimizing and maximizing (differentiable) functions defined on a specified closed interval $[a, b]$. On most occasions we need to find the **absolute** (or global) extrema of the function on the interval $[a, b]$ rather than just the relative extrema. Fortunately, the problem of determining the absolute maximum and minimum is rather easy once the relative extrema have been found. It turns out that the absolute extrema of a function defined on a closed interval $[a, b]$ occur at either a relative extremum or at one of the endpoints $x = a$ and $x = b$.

The following figures indicate some of the situations that may occur. In Figure 3.17, the absolute extrema occur at turning points; the absolute maximum occurs at $x = c$ and the absolute minimum at $x = d$. In Figure 3.18, the absolute maximum occurs at the turning point at $x = c$ but the absolute minimum occurs at the endpoint $x = b$. In Figure 3.19, the absolute maximum and minimum occur at the endpoints $x = b$ and $x = a$ respectively.

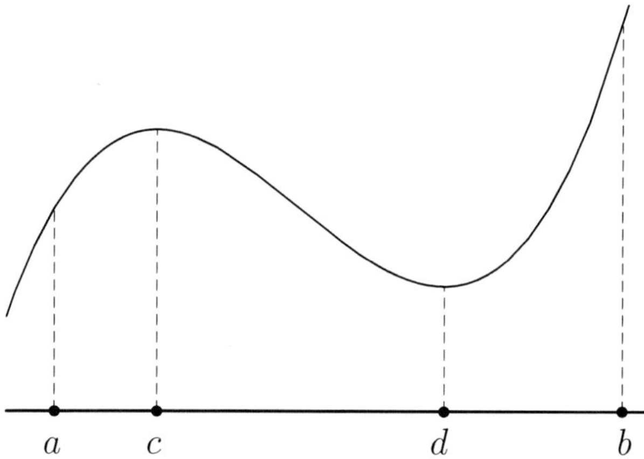

Figure 3.17: Absolute maximum and minimum at turning points.

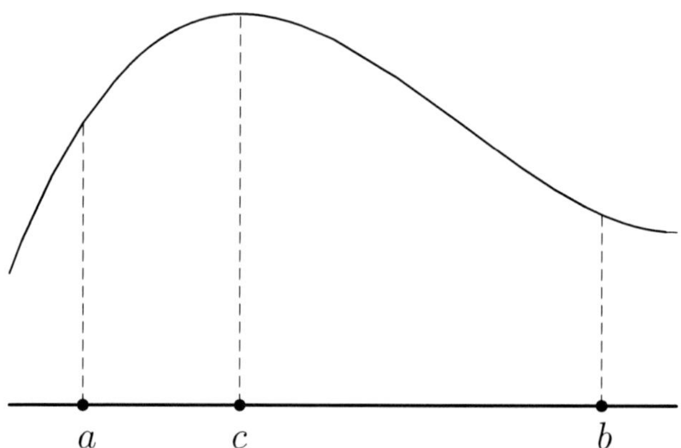

Figure 3.18: Absolute maximum at a turning point and minimum at an endpoint.

Thus to determine the extreme values of a given function defined by $f(x)$ for $x \in [a, b]$ we proceed as follows.

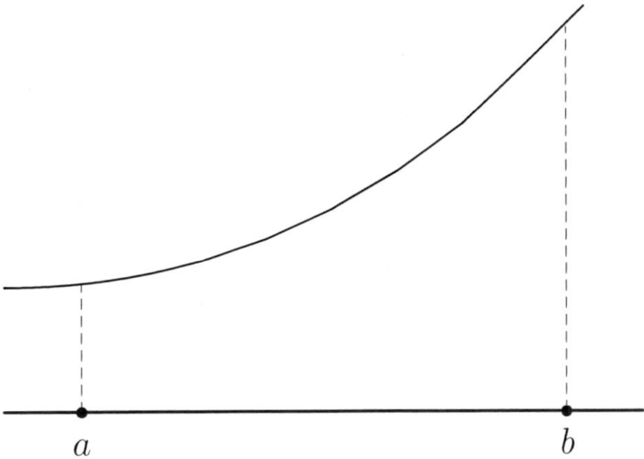

Figure 3.19: Absolute maximum and minimum at endpoints.

1. **Find the critical points (stationary or turning points, or relative extrema) of $f(x)$; that is, the points where $f'(x) = 0$.**

   To distinguish between relative (or local) extrema we can either utilize the second derivative test or examine the slope (or gradient) of the curve on either side of the turning points.

   If at a turning point $x = c$ (that is, where $f'(c) = 0$) we also have

   - $f''(c) > 0$, then the graph is concave up at $x = c$ and so $f(x)$ has a local minimum at c
   - $f''(c) < 0$, then the graph is concave down at $x = c$ and so $f(x)$ has a local maximum at c
   - $f''(c) = 0$, then the second derivative test is inconclusive.

   Alternatively if we examine the sign of $f'(x)$ on either side of (and near to) $x = c$ we obtain the three cases illustrated in Figures 3.20, 3.21 and 3.22.

2. **Determine the values of $f(a)$ and $f(b)$.**

Then compare the values taken by the function at each of the turning points and the values $f(a)$ and $f(b)$ respectively.

Worked Example 3.3.3 *Metallic spheres having radii of 10 cm are turned down into cylinders. Determine the dimensions of the cylinders so that there is minimum wastage of metal. (See Figure 3.23.)*
Suppose the cylinders have height h and radii r.
Let V denote the volume of each cylinder.

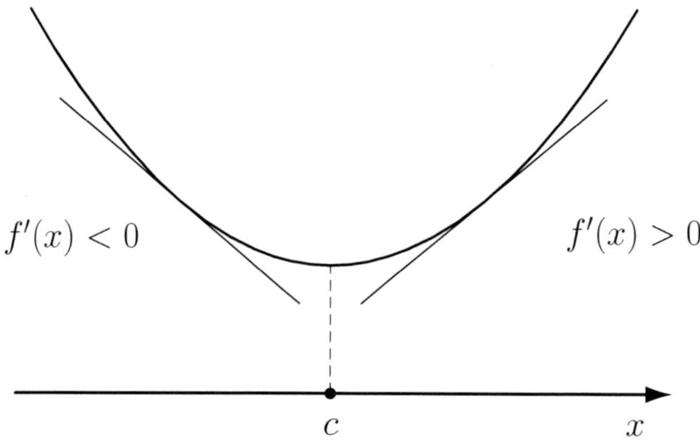

Figure 3.20: A local minimum.

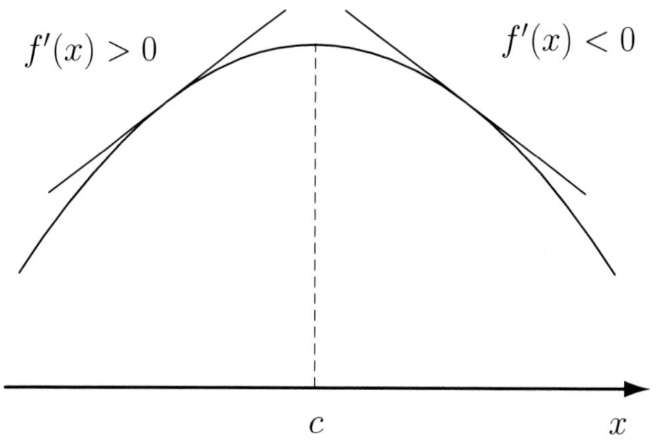

Figure 3.21: A local maximum.

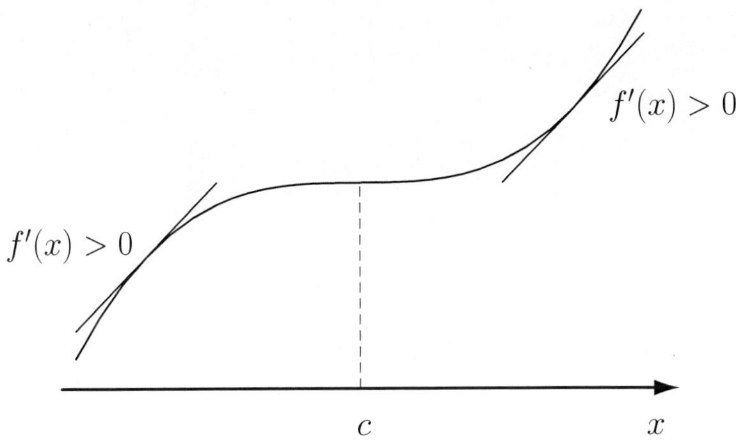

Figure 3.22: Neither maximum nor minimum.

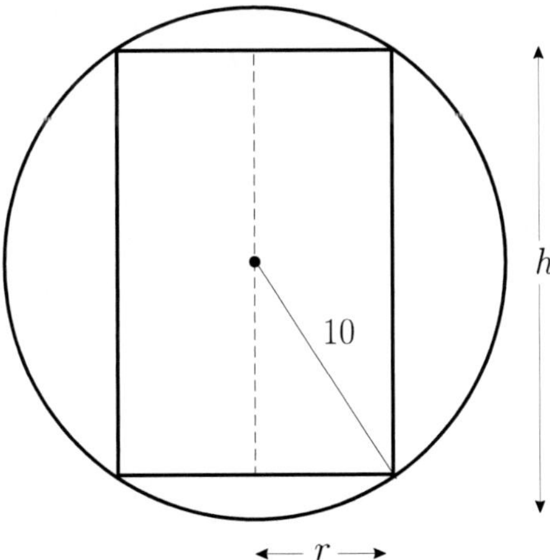

Figure 3.23: Graph in support of Worked Example 3.3.3.

In order to minimize wastage of metal it is clear that we need to maximize

$$V = \pi r^2 h.$$

From the geometry of the problem (see Figure 3.23) we see that $r^2 + \frac{h^2}{4} = 100$ so that

$$V = \pi h\left(100 - \frac{h^2}{4}\right) = \pi\left(100h - \frac{h^3}{4}\right) \text{ for } 0 \le h \le 20.$$

V will be greatest (why?) when $\frac{dV}{dh} = 0$; that is, when

$$\pi\left(100 - \frac{3h^2}{4}\right) = 0 \Rightarrow h^2 = 400/3.$$

Thus, wastage is minimized if the cylinders have radii $r = \sqrt{\frac{200}{3}}$ cm and height $h = \frac{20}{\sqrt{3}}$ cm.

■

Worked Example 3.3.4 *Determine the point on the curve* $y = x^2$ *which is nearest the point* $(3, 0)$. *(See Figure 3.24.)*

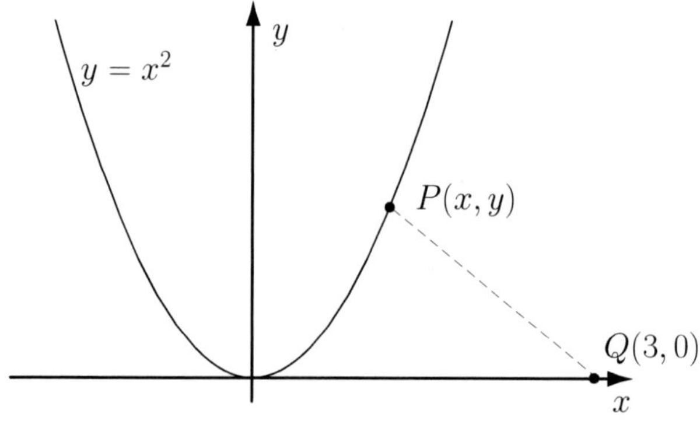

Figure 3.24: Graph in support of Worked Example 3.3.4.

If $P(x, y)$ is any point on the parabola and Q is the point $(3, 0)$ we need to minimize the distance PQ.
Now

$$PQ = \sqrt{(x-3)^2 + (y-0)^2} = \sqrt{(x-3)^2 + x^4} \text{ since } y = x^2.$$

The square root is a little awkward to handle when determining the local extrema. However, PQ will clearly be least whenever $(PQ)^2$ is least, and so it is sufficient to minimize $(x-3)^2 + x^4$.

Let $f(x) = (x-3)^2 + x^4$.
Then $f'(x) = 2(x-3) + 4x^3 = 0$ when $2x^3 + x - 3 = 0$ and this equation has only one real solution, namely $x = 1$. On geometrical grounds this must minimize PQ. (Why? Hint: What is the greatest distance from Q to the parabola?)
Thus the point $(1, 1)$ is the closest point on the parabola to the given point Q. ■

Worked Example 3.3.5 *An oil company ULTRA has an oil-rig situated in Bass Strait 2 km offshore from a section of a straight coastline. ULTRA wishes to construct a pipeline from the rig to its refinery which is situated on the coastline and 8 km along the coast from the point directly opposite the oil rig. (See Figure 3.25.)*
It costs $\$C_1$ *per km to construct the pipeline under water and* $\$C_2$ *per km on land.*
Design the most economic pipeline from the oil rig to the refinery if $C_1 = kC_2$ *where* $k > 1$ *is a known value.*
Explain what happens if $k = 1$.

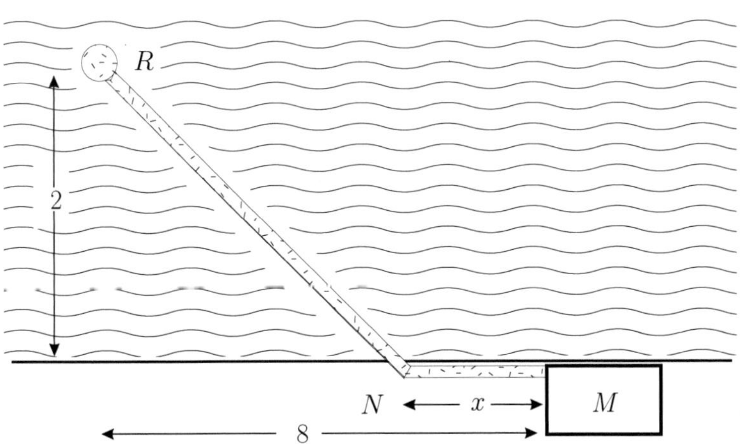

Figure 3.25: Graph in support of Worked Example 3.3.5.

To obtain the most general formula for the cost of construction, suppose the pipeline is laid in a straight line from the rig at R to the coastline at N and then along the coast to the refinery at M.
Let $NM = x$. It is clear that $0 \leq x \leq 8$.
The required length of pipeline under water is

$$RN = \sqrt{4 + (8-x)^2}$$

so its corresponding cost is $C_1\sqrt{4+(8-x)^2}$; the cost for the pipeline on land is obviously C_2x.
Therefore the total cost of construction C is given by

$$C = C_1\sqrt{4+(8-x)^2} + C_2x.$$

We need to minimize C as a function of x and this will occur either where $\frac{dC}{dx} = 0$ or where $x = 0$ or $x = 8$.
Now

$$\frac{dC}{dx} = C_2 - C_1\frac{(8-x)}{\sqrt{4+(8-x)^2}}.$$

This is zero when

$$C_1(8-x) = C_2\sqrt{4+(8-x)^2}.$$

That is

$$C_1^2(8-x)^2 = 4C_2^2 + C_2^2(8-x)^2$$

which implies

$$x = 8 - \frac{2C_2}{\sqrt{C_1^2 - C_2^2}}.$$

When $C_1 = kC_2$, this gives $x = 8 - \frac{2}{\sqrt{k^2-1}}$.
For example, if $k = 4$, then C is minimized when $x \approx 7.48$.
However when $k = 1$,

$$\frac{dC}{dx} = C_2\left\{1 - \frac{(8-x)}{\sqrt{4+(8-x)^2}}\right\} > 0$$

so that C is increasing on $[0, 8]$. In this case the least cost occurs when $x = 0$ and this corresponds to constructing the pipeline directly from the rig to the refinery. (Here we are simply minimizing the total distance from R to M via N which is the hypotenuse of the triangle RNM.) ■

3.3.4 Errors and Small Approximations

The total surface area A of a closed circular cylinder having radius r, height h and fixed volume $64\pi cm^3$ is given by

$$A = 2\pi r^2 + \frac{128\pi}{r}.$$

Consider the following two related questions in connection with changes in the value of A as a result of changes in the value of r:

- If the radius of the cylinder is increased by 5% what is the consequential percentage change in the surface area?

- If the radius of the cylinder is measured to be 4.0 ± 0.2 what is the corresponding maximum error in the surface area?

Exact answers to these questions can certainly be obtained by direct substitution. For example, for the latter question we could substitute $r = 3.8$ and $r = 4.2$ into the formula for A to obtain an estimate for the maximum error in the surface area.

Approximate answers (which are usually sufficient) can often be obtained more easily using a calculus method as described in what follows. Consider a function defined by $y = f(x)$. If x is increased by a small amount Δx to $x + \Delta x$ then what is the corresponding change Δy in y? From Figure 3.26 we see that, for small Δx, the gradient of the chord PQ is approximately the same as the gradient of the tangent at P. That is,

$$\frac{\Delta y}{\Delta x} \approx \frac{dy}{dx} \text{ which implies } \Delta y \approx \frac{dy}{dx}\Delta x.$$

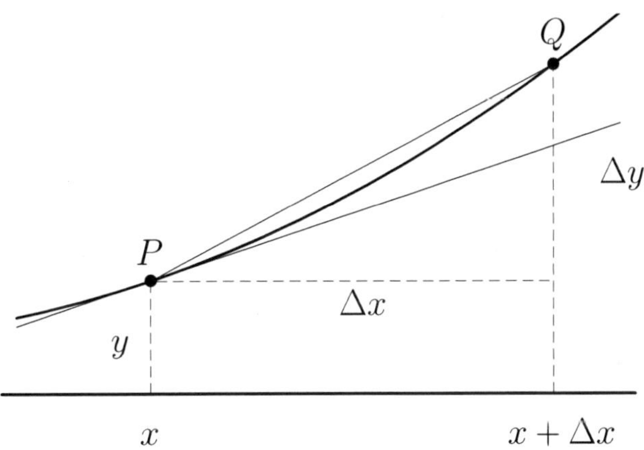

Figure 3.26: Approximation of gradient using the chord.

Therefore

$$\boxed{\Delta y \approx f'(x)\Delta x}$$

Note that in the above discussion, Δx can be interpreted as the maximum error in measuring the quantity x and Δy the corresponding approximate maximum error in y.

Worked Example 3.3.6 *The surface area A of a circular cylinder having radius r, height h and fixed volume 64π cm^3 is given by*

$$A = 2\pi r^2 + \frac{128\pi}{r}.$$

1. *If the radius of the cylinder is increased by 5%, find an expression for the percentage change in the surface area.*
2. *If the radius of the cylinder is measured to be 4.0± 0.2, what is the corresponding maximum error in the surface area?*

1. If r is increased by 5% then $\Delta r = \frac{5r}{100}$ so that

$$\Delta A \approx \frac{dA}{dr}\Delta r = \left(4\pi r - \frac{128\pi}{r^2}\right)\Delta r = \left(4\pi r - \frac{128\pi}{r^2}\right)\frac{5r}{100}.$$

The percentage error in A is then

$$\frac{\Delta A}{A} \times 100 \approx \frac{\left(4\pi r - \frac{128\pi}{r^2}\right)}{\left(2\pi r^2 + \frac{128\pi}{r}\right)} 5r = \frac{10\left(r^3 - 32\right)}{r^3 + 64}.$$

The percentage change in A clearly depends on the value of r.

When $r = 4$, the percentage change in A is approximately 2.5%.

2. If r is measured as 4.0 ± 0.2 then $\Delta r = \pm 0.2$.

Thus

$$\begin{aligned} \Delta A &\approx \frac{dA}{dr}\Delta r = \left(4\pi r - \frac{128\pi}{r^2}\right)\Delta r \\ &= \left(16\pi - \frac{128\pi}{16}\right)(\pm 0.2) \text{ since } r = 4 \\ &= \pm 1.6\pi. \end{aligned}$$

Since $A = 64\pi$ when $r = 4$ the corresponding percentage error in A is at most $\frac{1.6\pi}{64\pi} \times 100 = 2.5\%$.

■

3.3.5 Newton's Method

Newton's method (or the Newton-Raphson method as it is often referred to) is a common method of determining roots of non-linear equations such as $2\sin x = x$ and $x^3 + 4x - 6 = 0$ which are not easily solved by elementary methods.

To assist in the derivation of the formula, consider Figure 3.27. We want to find an approximation to x^*; that is, where the graph of $y = f(x)$ cuts the x-axis. If x_0 is an initial approximation (usually a guess) to the required solution, the point where the tangent to the curve at P cuts the x-axis will usually (provided the initial approximation x_0 is sufficiently close to x^*) give

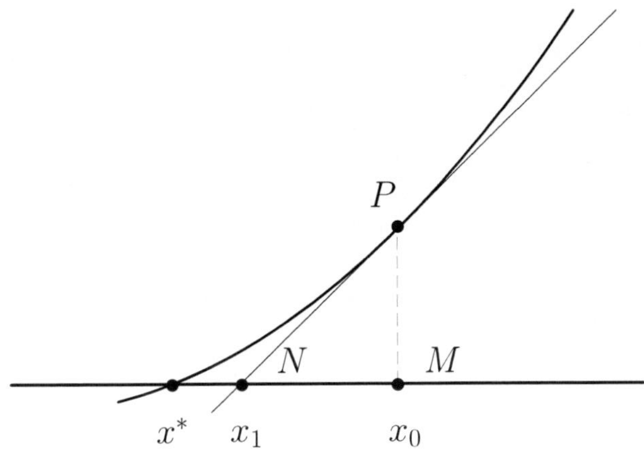

Figure 3.27: Derivation of Newton's method.

a better approximation to the required root; that is, x_1 will usually give a better approximation to x^*.

From the diagram, we have that the gradient of the tangent at P is given by

$$\frac{PM}{MN} = \frac{f(x_0)}{x_0 - x_1}.$$

But the gradient of the tangent to the curve at $x = x_0$ is also $f'(x_0)$ so that

$$f'(x_{0)} = \frac{f(x_0)}{x_0 - x_1} \Rightarrow x_1 = x_0 - \frac{f(x_0)}{f'(x_0)} \text{ provided } f'(x_0) \neq 0.$$

We can obtain an even better approximation by repeating this process, this time with x_0 replaced by x_1.

This iterative process can be continued until an approximation to x^* is obtained to the required accuracy. The entire process can be expressed recursively as follows:

$$\boxed{x_{n+1} = x_n - \frac{f(x_n)}{f'(x_n)} \text{ for } n = 0, 1, 2, 3, \ldots}$$

The main advantage of Newton's method over other root-finding techniques is that convergence is extremely rapid provided the initial approximation is sufficiently close to x^*; problems may occur if, for example, the initial approximation is at or near a turning point of the graph of $f(x)$.

▌**Worked Example 3.3.7** *Show that the cubic equation $x^3 + 4x - 6 = 0$ has a solution between $x = 1$ and $x = 2$.*
Use Newton's method to find an approximation to this root that is correct to three decimal places.

Let $f(x) = x^3 + 4x - 6$.
We have $f(1) = -1$ and $f(2) = 10$ so that $f(x)$ must equal 0 somewhere between $x = 1$ and $x = 2$. (Why?)
Now $f'(x) = 3x^2 + 4$ so that Newton's method gives

$$x_{n+1} = x_n - \frac{f(x_n)}{f'(x_n)} = x_n - \frac{x_n^3 + 4x_n - 6}{3x_n^2 + 4}.$$

Starting with $x_0 = 1$ we obtain

$$x_1 = 1 - \frac{-1}{7} \approx 1.1429.$$

Repeated application of the Newton algorithm above yields the following table.

n	x_n	$f(x_n)$	$f'(x_n)$
0	1	−1	7
1	1.1429	0.0645	7.9187
2	1.1348	0.0006	7.8633
3	1.1347	−0.0002	7.8626
4	1.1347		

Thus we feel justified in claiming the required solution is 1.135 correct to three decimal places. ■

Worked Example 3.3.8 *Show that the only positive root of the non-linear equation* $2\sin x = x$ *lies between* $x = 1$ *and* $x = 2$. *Use Newton's method to find an approximation to this root that is correct to three decimal places.*
From Figure 3.28 it is evident that the graphs of $y = 2\sin x$ and $y = x$ have only one point of intersection for $x > 0$ and this point lies between $x = \pi/2$ and $x = \pi$. Thus the equation $2\sin x = x$ has only one positive root.
Alternatively, if $f(x) = 2\sin x - x$ then

$$f(1) = 2\sin 1 - 1 \approx 0.6829 > 0$$

and

$$f(2) = 2\sin 2 - 2 \approx -0.1814 < 0$$

so that (by continuity) $f(x) = 0$ for some x between 1 and 2.

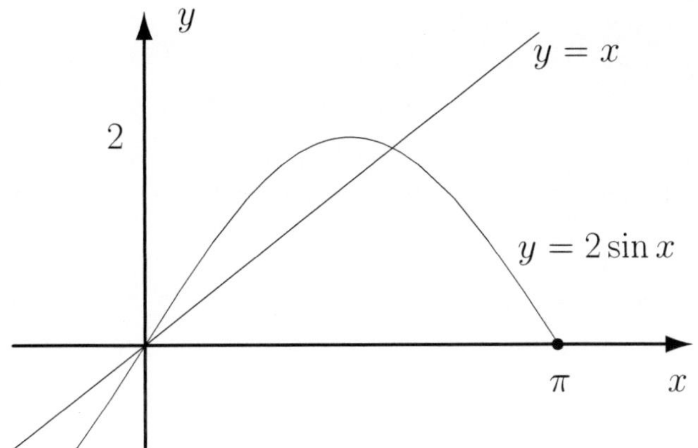

Figure 3.28: Graph to support Worked Example 3.3.8.

Newton's algorithm yields:

$$x_{n+1} = x_n - \frac{f(x_n)}{f'(x_n)} = x_n - \frac{2\sin x_n - x_n}{2\cos x_n - 1}.$$

If we take $x = 2$ as our first approximation (that is, $x_0 = 2$) we obtain recursively the following table. (Note: When evaluating $\sin x$ and $\cos x$ it is important to remember that you must use the "radian mode" in your calculator.)

n	x_n	$f(x_n)$	$f'(x_n)$
0	2	−0.1814	−1.8323
1	1.9010	−0.0090	−1.6485
2	1.8955	9.4×10^{-6}	

Thus we feel justified in claiming the required solution is $x = 1.896$ correct to three decimal places. ■

In this topic we have

- Reviewed differentiation including product, quotient and chain rules.
- Examined implicit functions and their derivatives.
- Applied differentiation to curve sketching; solved rates of change and maxima and minima problems; estimated errors; and generated solutions of non-linear equations using Newton's method.

3.4 Quick Test Number 3

Basic Mathematical Skills

Question **Selection**

1. If $\arg z = \frac{\pi}{3}$, then $\arg z^4$ equals (a) $\frac{4\pi}{3}$ (b) $\left(\frac{\pi}{3}\right)^4$ (c) neither

2. $\frac{d}{dx}(\cos x)$ equals (a) $\sin x$ (b) $-\cos x$ (c) $-\sin x$ (d) $\cos x$

3. $\frac{d}{dx}(\sin x)$ equals (a) $\sin x$ (b) $-\cos x$ (c) $-\sin x$ (d) $\cos x$

Differentiation

Question **Selection**

1. $\frac{d}{dx}\left(x^3 \cos x\right)$ equals
   (a) $-3x^2 \sin x$ (b) $3x^2 \cos x - x^3 \sin x$ (c) $3x^2 - \sin x$

2. $\frac{d}{dx}\left(\tan x^2\right)$ equals
   (a) $\sec^2 x^2 + \tan 2x$ (b) $2x \sec^2 x^2$ (c) $\sec^2 x^2$ (d) $\sec^2 2x$

3. $\frac{d}{dx}(-2xy)$ equals (a) $-2y'$ (b) $-2xy' + 2y$ (c) $-2xy' - 2y$

4. $\frac{d}{dx}(yy')$ equals (a) $y'y''$ (b) $y' + yy''$ (c) $(y')^2 + yy''$

5. $\frac{d}{dx}\left(\cos x^2\right)$ equals
   (a) $-\sin 2x$ (b) $-\sin x^2 + \cos 2x$ (c) $-\sin x^2$ (d) $-2x \sin x^2$

Newton's method for solving non-linear equations of the form $f(x) = 0$ uses the formula

$$x_{n+1} = x_n - \frac{f(x_n)}{f'(x_n)}, \qquad n = 0, 1, 2, \ldots$$

6. In this method, x_0 is:
   (a) an initial guess (b) an initial guess of $x = 0$
   (c) the exact solution of $f(x) = 0$

7. Applying Newton's method once, $x_0 - \frac{f(x_\bullet)}{f'(x_\bullet)}$ is calculated to give 0.23. We now have the:

   (a) next iterate, $x_1 = 0.23$ (b) next iterate, $x_{0.23}$
   (c) exact solution of $f(x) = 0$

8. To solve the equation $\cos x = x$ using Newton's method, we should set

   (a) $f(x) = \cos x - x$ (b) $f(x) = \cos x$ (c) $f(x) = x$

9. $\cos(0.7391)$ equals (a) 0.9999 (b) 0.7391 (c) 0.4568

10. $\sin 1$ equals (a) 0.0175 (b) 0.8415 (c) 1.5708

Chain rule: For the following functions write down $\frac{dy}{dx}$.

11. $y = x^2 + 1$
12. $y = (x^2 + 1)^5$
13. $y = \sqrt{x^2 + 1}$
14. $y = \frac{1}{x^{\cdot} + 1}$
15. $y = \cos(x^2 + 1)$
16. $y = \sin(x^2 + 1)$
17. $y = \tan(x^2 + 1)$

3.5 Exercises

Before attempting any of these miscellaneous exercises, make sure that you have successfully answered all the self-help exercises appearing throughout the chapter.

1. Consider $f(x) = |x|$ at $x = 0$.

   (a) Show that

   $$\lim_{h\to 0}\frac{f(h)-f(0)}{h} = \lim_{h\to 0}\frac{|h|}{h}.$$

   (b) By considering separately the cases $h > 0$ and $h < 0$ show that this limit does not exist.

   What does this say about the differentiability of $|x|$ at the origin?

   Could you have deduced this from the graph of $y = |x|$?

2. (a) Let $f(x) = x^2$. By considering

   $$\lim_{h\to 0}\frac{f(x+h)-f(x)}{h}$$

   show that $f'(x) = 2x$.

   (b) Let $f(x) = 3x^2 - 2x + 5$. By considering

   $$\lim_{h\to 0}\frac{f(x+h)-f(x)}{h}$$

   show that $f'(x) = 6x - 2$.

3. By considering

   $$\lim_{h\to 0}\frac{f(x+h)-f(x)}{h}$$

   where $f(x) = \sin x$ show that

   $$f'(x) = \cos x.$$

   (You may assume that

   $$\lim_{k\to 0}\frac{\sin k}{k} = 1$$

   and

   $$\lim_{k\to 0}\frac{1-\cos k}{k} = 0.)$$

4. Repeat Exercise 3 with

   $$f(x) = \cos x$$

   to show that

   $$f'(x) = -\sin x.$$

5. In each of the following cases determine the first and second derivatives

   (a) $4x^3 + 7x^2 + 4$

   (b) $x^2 + 4/x - \sqrt{x}$

   (c) $2x^2 + 4\sin x - \dfrac{3}{x} + 12\cos x + (1+2x)^2$

   (d) $t^2 \sin t$

   (e) $\dfrac{x}{1+x}$

   (f) $x^3 \cos x$

   (g) $5\theta \sin\theta$

   (h) $\dfrac{1+x}{1-x}$

   (i) $x^2 \tan x$

(j) $(\sin x + \cos x)^2$.

6. Use the chain rule to show that

A $\dfrac{d}{dx}\sin f(x)$
$= \cos f(x) \cdot f'(x)$

B $\dfrac{d}{dx}\cos f(x)$
$= -\sin f(x) \cdot f'(x)$

C $\dfrac{d}{dx}\tan f(x)$
$= \sec^2 f(x) \cdot f'(x)$.

Hence write down the derivatives of:

(a) $\sin 4x$
(b) $\cos 7x$
(c) $\tan 3x$
(d) $\sin x^3$
(e) $\cos(3x+1)$
(f) $4\sin 2x$
(g) $2\sin 3x - \cos 6x$
(h) $\cos(1+x^2)$
(i) $\tan(2x+1)$.

7. Determine the derivatives of each of the following functions.

(a) $(1+x)^3$
(b) $\sqrt{1-2x^2}$
(c) $\sin^2 x$
(d) $4x^2 \tan 3x$
(e) $\dfrac{3}{1+4x}$
(f) $\dfrac{1}{\sqrt{1+2x+4x^2}}$
(g) $(1+x-x^4)^4$
(h) $\sqrt{1+3x^2}$
(i) $\sin^2 3x$
(j) $6x\sin^2 x$
(k) $\dfrac{2}{(1-3x)^2}$
(l) $\sqrt{1+x}$
(m) $\sqrt{\frac{1+x}{1-x}}$
(n) $(x+\dfrac{1}{x})^3$
(o) $\cos^3 4x$
(p) $\tan^2 2x$.

8. Determine the derivatives of each of the following functions:

(a) $\sqrt{\sin 2x}$
(b) $\sqrt{1+4x+x^2}$
(c) $\sin^2 4x$
(d) $x\tan^2 x$
(e) $\sec x = \dfrac{1}{\cos x}$
(f) $\dfrac{1+2x}{1-x}$
(g) $\sin^2(2x-3)$
(h) $\cos^2 3x$
(i) $\tan^3 4x$
(j) $x\tan(1+x^2)$
(k) $\dfrac{\cos x}{\sin x + \cos x}$
(l) $\dfrac{1+\sin x}{1-\sin x}$
(m) $x\sqrt{1+x^2}$.

9. Determine the equation of the tangent and the normal to the curve

$$y = x^3 + 2x + 3 + \sin 2x$$

at the point $(0,3)$.

10. A curve has equation

$$x^2 - 2xy + 2y^2 + 5 = 10x.$$

(a) Find the equation of the tangent to the curve at the point $P(1, 2)$.

(b) Determine y'' at P.

11. For the curve whose equation is

$$x^2 + 2xy - y^2 - 4x + 3y = 2$$

(a) show that $y' = 0$ at the point $(0, 2)$

(b) determine y'' at the point $(0, 2)$.

12. For the curve whose parametric equations are

$$x = 1 + 4\sin\theta, y = \sqrt{3} - 2\cos\theta$$

for $0 \le \theta \le 2\pi$

(a) determine $\frac{dx}{d\theta}$ and $\frac{dy}{d\theta}$

(b) hence find $\frac{dy}{dx}$

(c) find the equation of the tangent to the curve at the point $(3, 0)$

(d) find $\frac{d}{d\theta}\left(\frac{dy}{dx}\right)$

(e) deduce that

$$\frac{d^2y}{dx^2} = \frac{1}{8}\sec^3\theta$$

(f) identify the curve.

13. A curve is specified in parametric form by

$$x = t\sin 2t$$

and

$$y = t\cos 2t$$

for $0 \le t \le \pi$.

(a) Find $\frac{dx}{dt}$ and $\frac{dy}{dt}$.

(b) Hence determine $\frac{dy}{dx}$.

(c) Find $\frac{d^2x}{dt^2}$ and $\frac{d^2y}{dt^2}$.

(d) Determine $\frac{d^2y}{dx^2}$.

14. In each of the following cases determine $\frac{dy}{dx}$.

(a) $xy^3\cos x + x^2y = 1$

(b) $xy^2\sin x + 2x^2y = 4$

15. Sketch the graphs of

(a) $y = (x-1)^2 + 3$

(b) $y = 1 + \dfrac{1}{x-1}$

(c) $y = x + \dfrac{2}{x}$

(d) $y = 2\sin x + 1$

(e) $y = \dfrac{2x}{x+1}$

(f) $y = \sec x = \dfrac{1}{\cos x}$

(g) $y = \sqrt{x-1}$

(h) $y = 1 + \sqrt{2-x}$

(i) $y = \sqrt{1-x^2}$

(j) $y = \dfrac{1}{\sqrt{x-4}}$

(k) $y = x + \dfrac{2}{x^2}$

(l) $y = 3\sin 2x - 1$

(m) $y = 1 - 2\cos(x - \pi/4)$

(n) $y = \dfrac{1}{x^2 - 3x + 2}$

(o) $y = \dfrac{1}{x^2 + 4}$

(p) $y = \dfrac{1}{x^2 + 2x + 5}$.

16. Determine the extreme values of

$$f(x) = \sin x + \cos x$$

where $0 \le x \le 2\pi$.

Use this information to help you sketch the graph of f.

17. When a large but thin circular metal disk is being heated its radius is increasing at the rate of 0.4 mm/minute.

    How fast is its area increasing when the radius of the disk is 20 cm?

18. Air is blown into a balloon at the rate of 8 cm^3/second.

    Assuming that the balloon remains spherical while being inflated, determine how fast its

    (a) radius
    (b) surface area

    are increasing when its radius is 10 cm.

19. Water is poured into a conical vat at the rate of 2 m^3/minute.

    If the radius of the vat is 3 metres and its height is 6 metres, how fast is the water level rising when

    (a) the depth of water is 3 metres
    (b) the vat is half full?

20. A plane flying north at 800 km/hr passes over a small Pacific island at midnight. Twenty minutes later another aircraft flying west at 900 km/hr passes over the island at the same altitude.

    Assuming that both planes maintain the same courses, the same speeds and the same altitudes for at least another hour, determine how quickly the distance between the two planes is increasing at 12.30 am.

21. ABC is a triangle in which $\overline{AB} = 20$ cm, $\overline{AC} = 32$ cm and $\angle BAC = \theta$. If θ is increasing at the rate of 2° per minute, determine the rate at which the triangle's **area** is changing when $\theta = 120°$.

22. Determine the point on the line

    $$2x + y = 4$$

    which is closest to the origin.

23. Given a set of numbers $\{a_1, a_2, \ldots a_n\}$ show that

    $$S(x) = (a_1 - x)^2 + (a_2 - x)^2 + \cdots + (a_n - x)^2$$

    is a minimum when x is the mean (average) of the numbers.

24. P and Q are two hot bodies 8 metres apart with P emitting ten times as much heat as Q. Assuming that the intensity of heat from a hot body at any point varies inversely as the square of the distance of that point from the body, find the point on the line joining P and Q which receives the least amount of heat.

25. A retailer wishes to minimize the costs incurred in ordering and storing stock for which there is a known demand of n items per week.

    Each time she places an order she incurs an ordering cost (invoicing, delivery, etc.) of $\$C_1$.

Stock holding costs amount to $\$C_2$ per item per week.

If it is decided to order Q items every T weeks, the total of these costs is

$$C = C_1 + C_2 \frac{Q}{2} T$$

where it is assumed that delivery occurs immediately.

Given $T = \frac{1}{n} Q$, deduce that the average cost per week is

$$\overline{C}(Q) = \frac{C_1 n}{Q} + \frac{C_2 Q}{2}.$$

(a) Sketch the graph of $\overline{C}(Q)$.

If $C_1 = 100, C_2 = 0.5$ and $n = 100$

(b) Determine the most economic order quantity Q.

(c) How often should orders be placed?

26. A steel storage tank for propane gas is to be constructed in the shape of a right circular cylinder with a hemisphere at each end. Suppose the cylinder has length h and radius r metres.

    (a) If the desired capacity of the tank is $10\pi\ m^3$, show that

    $$3r^2 h = 30 - 4r^3.$$

    (b) The cost per square metre of constructing the end pieces is twice that of the cylindrical piece.

    Determine the dimensions r and h which minimizes the cost of construction.

27. Cardboard boxes are to be manufactured from rectangular sheets of cardboard having dimensions 30×20 cm by cutting squares from each corner and then turning up the edges.

    Determine the size of the squares to be cut from each corner in order that the resulting boxes have maximum capacity.

28. Show that the volume of the largest circular cylinder that can be inscribed in a given right circular cone is $4/9$ the volume of the cone.

29. A stained glass window is to be constructed consisting of a rectangular sheet of clear glass surmounted by a semicircular sheet of coloured glass. Coloured glass admits only half as much light as clear glass.

    Assuming that the amount of light entering through a window is proportional to the area of the window, determine the dimensions of the window having a perimeter of 6 metres which admits the most light.

30. The costs of running a train comprise

    - overhead costs, including wages, which are \$400 per hour
    - running costs which vary with the cube of the train's average speed and are \$150 per hour at an average speed of 50 km/hr.

Determine the most economical average speed for the train over a long journey.

31. A metal of density ρ is to be cast into cylindrical blocks having mass M and **semi-circular** cross-section.

    The rate at which a block cools is proportional to the surface area of the block.

    Determine the dimensions of the blocks so that they cool as slowly as possible.

32. A sector is to be cut from a thin circular metal sheet and the two straight edges soldered together to form a hollow cone.

    Determine the angle of the sector remaining so that the cone has maximum volume.

33. Light travels from point P to point Q with velocity c after being reflected at point N from a surface (see Figure 3.29).

    The distances $\overline{PA}$, $\overline{AB}$, $\overline{NB}$ and $\overline{BQ}$ are respectively a, d, x and b where a, b and d are constants.

    (a) Show that the time taken T for the light to travel from P to Q is

    $$T = \frac{1}{c}\left(\sqrt{a^2 + (d-x)^2} + \sqrt{b^2 + x^2}\right)$$

    for $0 < x < d$.

    (b) Light travels between two points in such a way so as to minimize the time taken.

    Show that T is a minimum when

    $$x = \frac{bd}{a+b}.$$

    (c) Deduce the law of reflection for light; that is, deduce that $\alpha = \beta$.

34. Use Newton's method *once*, with an initial guess of $x_0 = 1.0$, to find an approximation to the solution of $\cos x = x$.

35. (a) Show that the equation $x^3 - 2x = 5$ has a root between $x = 2$ and $x = 3$.

    (b) Taking $x = 2.0$ as an initial approximation to this root, use Newton's method to obtain the solution correct to two decimal places.

36. (a) Apply Newton's method to the equation $x^2 = 2$ to obtain the formula

    $$x_{n+1} = \frac{1}{2}\left(x_n + \frac{2}{x_n}\right).$$

    Hence find $\sqrt{2}$ correct to four decimal places.

    (b) Generalize the result from (a) to determine an algorithm to find an approximation to $\sqrt{N}$ for any positive integer N.

37. Use Newton's method to locate the root of $x^2 \log x = 1$ correct to three decimal places.

38. (a) Show that the equation $x^3 + 3x = 5$ has a root between $x = 1$ and $x = 2$.

(b) Taking $x = 1.2$ as an initial approximation to this root, use Newton's method to obtain the solution correct to two decimal places.

39. The kinetic energy K of a body of mass m moving with speed v is given by

$$K = \frac{1}{2}mv^2.$$

If a body's speed is increased by 2%, what is the approximate percentage change in its kinetic energy?

40. The period T of a simple pendulum of length L is given by

$$T = 2\pi\sqrt{\frac{L}{g}}$$

where g is the acceleration due to gravity.

If L is increased by 5%, what is the approximate percentage increase in T?

41. The efficiency E of a petrol engine is given by

$$E = 100\left(1 - \frac{1}{R^{0.25}}\right)$$

where R is the expansion ratio.

If R is increased by 1%, show that the approximate increase in E is $\frac{1}{4}(1 - \frac{E}{100})$.

42. A charged circular disc of radius a has a uniform charge density σ per unit area on one side of the disc. The electric field strength E at a point on the axis of the disc at a distance h from its centre is

$$E = \frac{\sigma}{2\epsilon_0}\left(1 - \frac{h}{\sqrt{a^2 + h^2}}\right)$$

where ϵ_0 is a constant.

If $h = 4$ and $a = 3$ find the approximate percentage change in E if h is increased to 4.04.

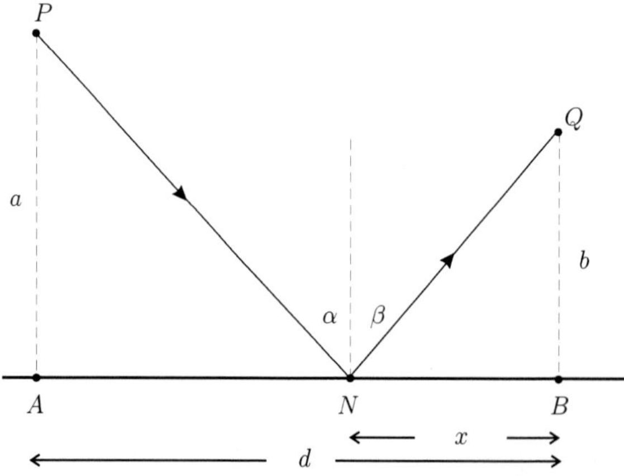

Figure 3.29: Light travel through a reflective medium.

Topic 4

Functions and their Derivatives — Part 2

4.1 Exponential Functions

4.1.1 Introduction

Functions such as 2^x, 3^x and $\left(\frac{1}{2}\right)^x$ are members of a family of functions called exponential functions.

They are defined as follows:

Definition 4.1 *An exponential function is any function of the form*

$$f : \mathbf{R} \to \mathbf{R} \text{ where } f(x) = a^x \text{ for some positive constant } a \neq 1.$$

Note that the range of f is $\mathbf{R}^+$ since $a^x > 0$ for all $x \in \mathbf{R}$.

For such functions we *define*

$$a^0 = 1 \text{ and } a^{-x} = \frac{1}{a^x}.$$

Exponential functions also

1. satisfy the following basic index laws:

   - $a^x a^y = a^{x+y}$
   - $\dfrac{a^x}{a^y} = a^{x-y}$
   - $(a^x)^y = a^{xy}$
   - $(ab)^x = a^x b^x$.

2. have the limiting properties

   If $a > 1$ then $a^x \to \infty$ as $x \to \infty$ and

   $a^x \to 0$ as $x \to -\infty$

whereas if $0 < a < 1$ then $a^x \to 0$ as $x \to \infty$ and

$a^x \to \infty$ as $x \to -\infty$.

All of the above properties enable us to sketch the graph appearing as Figure 4.1.

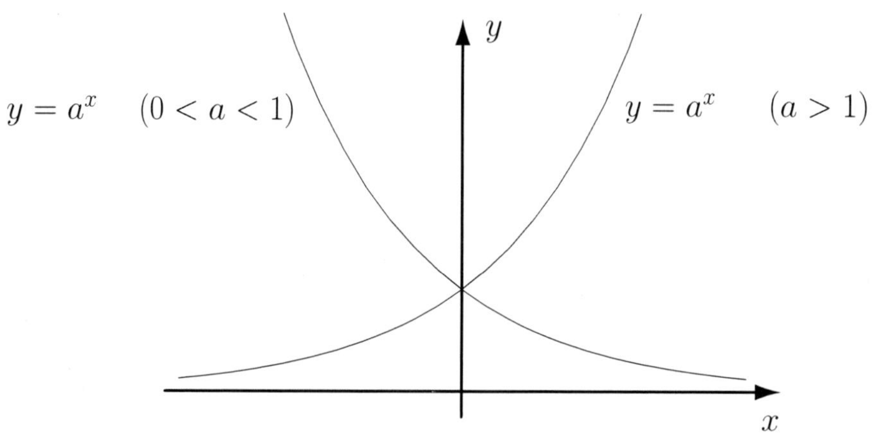

Figure 4.1: Exponential functions.

4.1.2 The Exponential Function

If $f(x) = a^x$ then

$$\begin{aligned} f'(x) &= \lim_{h\to 0} \frac{a^{x+h} - a^x}{h} \\ &= a^x \left(\lim_{h\to 0} \frac{a^h - 1}{h} \right) \\ &= k.a^x \end{aligned}$$

where $k = \lim_{h\to 0} \frac{a^{\cdot} - 1}{h}$ is a constant. Note that the value of this constant k actually depends on a.

In particular,

- when $a = 2$, $k = \lim_{h\to 0} \dfrac{a^h - 1}{h} \approx 0.6931$ so that

$$\frac{d}{dx}(2^x) = k \cdot 2^x \text{ where } k \approx 0.6931.$$

- when $a = 3$, $k = \lim_{h\to 0} \dfrac{a^h - 1}{h} \approx 1.0986$ so that

$$\frac{d}{dx}(3^x) = k \cdot 3^x \text{ where } k \approx 1.0986.$$

Of particular importance is the special case where the constant k in the derivative above is 1; that is, when

$$\lim_{h\to 0} \frac{a^h - 1}{h} = 1.$$

The value of a for which this is true (which, from the above, lies between 2 and 3) is approximately 2.7182818284... and is denoted by e. The corresponding function e^x is usually referred to as **the** exponential function. It has the remarkable property that

$$\boxed{\frac{d}{dx}e^x = e^x}$$

It is the *only* function (up to a multiplicative constant) which exhibits this property. The exponential function arises in the modelling of many natural phenomena such as population growth, radioactive decay, the cooling of a heated body etc.

It follows, by an application of the chain rule, that the derivative of $e^{f(x)}$ is $e^{f(x)}f'(x)$; that is

$$\boxed{\frac{d}{dx}e^{f(x)} \quad = \quad e^{f(x)} \cdot f'(x)}$$

Worked Example 4.1.1 *Determine the derivatives of*

1. e^{x^2-7x+2} *2.* $e^{-2x}\sin 3x$ *3.* $\sqrt{x^4 + e^{2x^2}}$

1. By the chain rule

$$\begin{aligned}\frac{d}{dx}e^{x^2-7x+2} &= e^{x^2-7x+2}\frac{d}{dx}(x^2 - 7x + 2)\\ &= (2x-7)e^{x^2-7x+2}.\end{aligned}$$

2. By the product rule

$$\begin{aligned}\frac{d}{dx}e^{-2x}\sin 3x &= e^{-2x}3\cos 3x + e^{-2x}(-2)\sin 3x\\ &= e^{-2x}(3\cos 3x - 2\sin 3x).\end{aligned}$$

3. Writing $\sqrt{x^4 + e^{2x^2}}$ as $(x^4 + e^{2x^2})^{1/2}$ and using the chain rule we obtain

$$\frac{d}{dx}\sqrt{x^4 + e^{2x^2}} = \frac{1}{2}(x^4 + e^{2x^2})^{-1/2}(4x^3 + e^{2x^2}\,4x)$$
$$= \frac{2x^3 + 2xe^{2x^2}}{\sqrt{x^4 + e^{2x^2}}}.$$

■

Worked Example 4.1.2 *Sketch the graphs of*

1. $y = 2(e^{2x} - 1)$ *2.* $y = 3(1 - e^{-3x})$

1. Probably the simplest way of sketching the graph is to use addition of ordinates. First, sketch the graph of $y = 2e^{2x}$ (see Figure 4.2) and then subtract 2 from each ordinate; that is, shift the graph two units vertically down. An al-

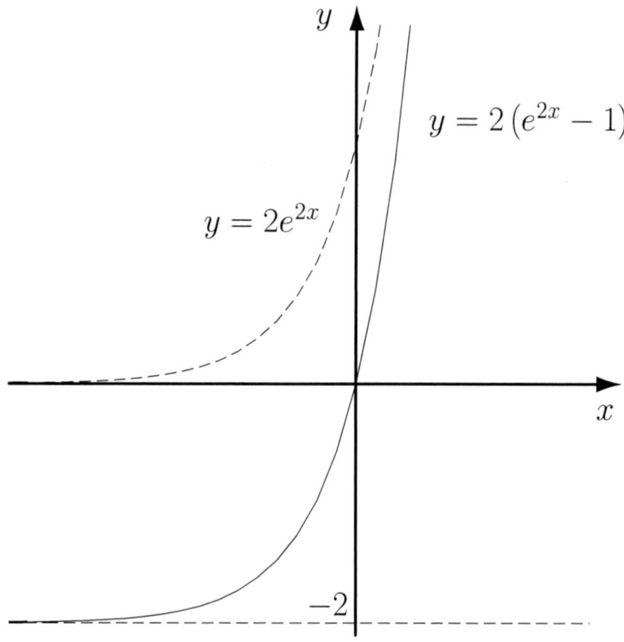

Figure 4.2: Shifting of a graph.

ternative approach is to use some or all of the following features to sketch the graph:

- If $x = 0$, then $y = 0$. That is, the graph passes through the origin.

- As $x \to -\infty$, then $y \to -2$ since $e^{2x} \to 0$ as $x \to -\infty$. Thus, the line $y = -2$ is a horizontal asymptote.
- As $x \to \infty$, then $y \to \infty$ since $e^{2x} \to \infty$ as $x \to \infty$.
- Since $y' = 4e^{2x} > 0$ for all x, the graph is monotonically increasing.
- Since $y'' = 8e^{2x} > 0$ for all x, the graph is concave upwards.

2. The graph can be sketched in stages as follows:
   (a) Sketch the graph of $y = 3e^{-3x}$.
   (b) Sketch the graph of $y = -3e^{-3x}$ by reflecting the graph from (a) in the x-axis.
   (c) Move the graph from (b) vertically upwards three units.

   This gives the required graph as illustrated in Figure 4.3.

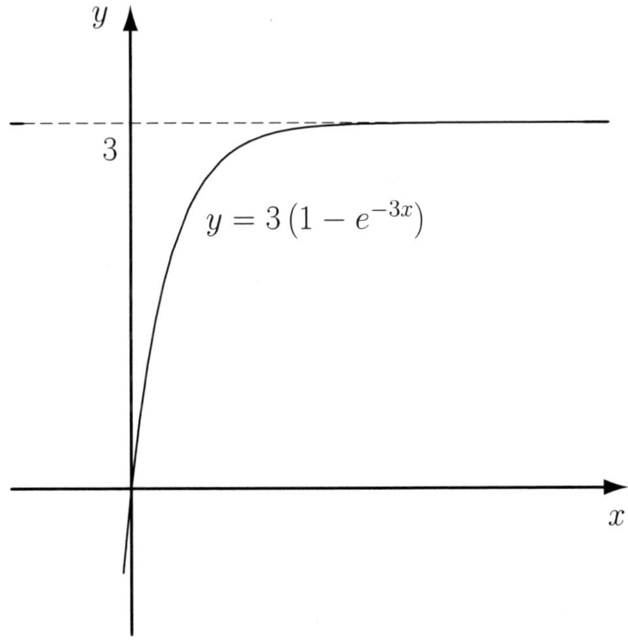

Figure 4.3: Graph of the function $y = 3(1 - e^{-3x})$.

■

Worked Example 4.1.3 *Sketch the graph of the function*

$$f(x) = 2xe^{-x}$$

after determining where the (graph of the) function is monotonic increasing, monotonic decreasing, concave upwards and concave downwards.

First note (with $y = f(x)$) that

- if $x = 0$, then $y = 0$ so that the graph passes through the origin.
- if $x \to \infty$, then $y \to 0$. Thus $x = 0$ is a horizontal asymptote.

  (Note that as $x \to \infty$, $2x \to \infty$ and $e^{-2x} \to 0$; however, the exponential "dominates" the polynomial $2x$.)
- if $x \to -\infty$ then $y \to -\infty$.

Furthermore

$$y' = 2e^{-x} + 2xe^{-x}(-1) = (2 - 2x)e^{-x}$$

so that

- $y' > 0$ for $2 - 2x > 0$; that is, for $x < 1$. (Note that $e^{-x} > 0$ for all x.)
- $y' < 0$ for $2 - 2x < 0$; that is, for $x > 1$.

Thus, the graph is increasing for $x < 1$, decreasing for $x > 1$ and consequently has a maximum at $(1, 2e^{-1})$.

Also $y'' = (2x - 4)e^{-x}$ so that $y'' > 0$ for $x > 2$. That is, the graph is concave up for $x > 2$. Similarly $y'' < 0$ for $x < 2$. That is, the graph is concave down for $x < 2$.

At $x = 2$ the graph has a point of inflexion.

Using this information we can now sketch the graph; see Figure 4.4. ■

Self-help exercises

1. Write down the derivatives of

   (a) e^{-3x} $[-3e^{-3x}]$ (c) $e^x + e^{-x}$ $[e^x - e^{-x}]$

   (b) e^{-3x^2} $[-6xe^{-3x^2}]$ (d) xe^{-x} $[e^{-x} - xe^{-x}]$

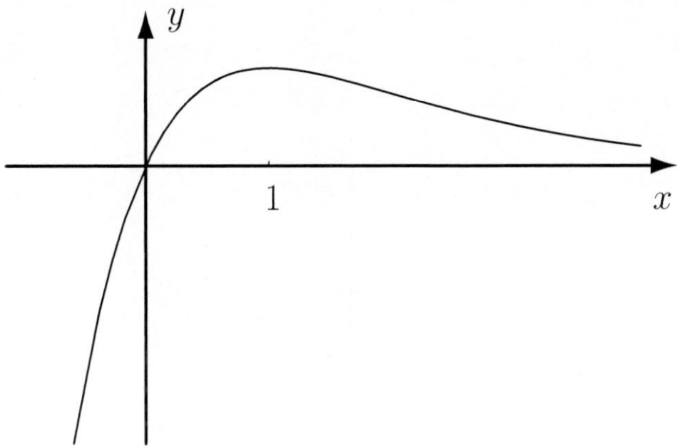

Figure 4.4: Graph of the function $y = 2xe^{-x}$.

2. Write down

(a) $\lim_{x\to\infty}(2 - e^{-x})$ [2]

(b) $\lim_{x\to\infty}(3 + 7e^{-x})$ [3]

(c) $\lim_{x\to-\infty}(2 + 3e^{x})$ [2]

4.2 The Hyperbolic Functions

Another class of functions involving the exponential function which has applications in science and engineering is the family of **hyperbolic** functions.

4.2.1 Definitions and Properties

The basic hyperbolic functions are the hyperbolic sine function, which is denoted by sinh (and usually pronounced "shine") and the hyperbolic cosine function which is denoted by cosh (and is pronounced phonetically).

They are defined as follows:

$$\sinh x = \tfrac{1}{2}(e^{x} - e^{-x}) \quad \text{and} \quad \cosh x = \tfrac{1}{2}(e^{x} + e^{-x})$$

From the graphs of e^x and e^{-x} it is a simple matter to construct the graphs for $\cosh x$ and $\sinh x$; see Figure 4.5. The shape of the graph of $y = \cosh x$ is similar to that of a "hanging chain" or catenary; that is, the shape assumed by a heavy string hanging under its own weight between two points which are not in the same vertical line.

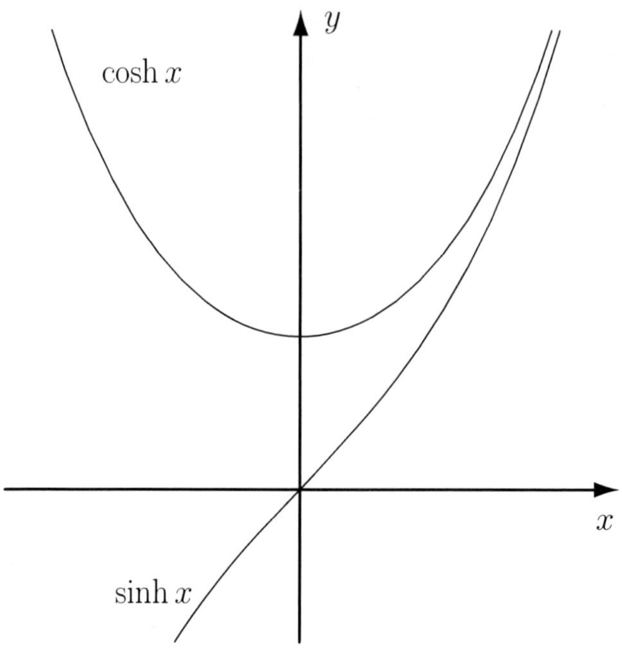

Figure 4.5: Graph of the functions $y = \cosh x$ and $y = \sinh x$.

Note that

- $\cosh 0 = 1$ and $\sinh 0 = 0$.
- $\cosh x \geq 1$ for all x.
- $\sinh x < \cosh x$ for all x.
- $\cosh x$ is an even function; that is, $\cosh(-x) = \cosh x$ for all x and consequently the graph of $y = \cosh x$ is symmetric about the y-axis.
- $\sinh x$ is an odd function; that is, $\sinh(-x) = -\sinh x$ for all x.
- for large positive x, both $\cosh x$ and $\sinh x$ behave like $\frac{1}{2}e^x$.
- as $x \to \infty$, $\dfrac{\sinh x}{\cosh x} \to 1$.

We also have, directly from the definitions of sinh and cosh, that

$$\boxed{\cosh x + \sinh x = e^x \text{ and } \cosh x - \sinh x = e^{-x}}$$

The other hyperbolic functions are defined — by analogy with the trigonometric (circular) functions — as follows:

$$\tanh x = \frac{\sinh x}{\cosh x} = \frac{e^x - e^{-x}}{e^x + e^{-x}}$$

$$\operatorname{cosech} x = \frac{1}{\sinh x} = \frac{2}{e^x - e^{-x}}$$

$$\operatorname{sech} x = \frac{1}{\cosh x} = \frac{2}{e^x + e^{-x}}$$

$$\coth x = \frac{1}{\tanh x} = \frac{\cosh x}{\sinh x} = \frac{e^x + e^{-x}}{e^x - e^{-x}}$$

[**A note on pronunciation:** tanh is read as "than" where the "th" is pronounced as in thousand; sech is read as "sheck", cosech as "co-sheck" and coth (which is a contraction of "cotanh") is read phonetically.]

The graph of $y = \tanh x$ may be obtained from the graphs of $y = \cosh x$ and $y = \sinh x$ by division of ordinates; however, it is not immediately clear by inspection what happens as $x \to \infty$ and $x \to -\infty$. In fact

$$\begin{aligned} \lim_{x\to\infty} \tanh x &= \lim_{x\to\infty} \frac{e^x - e^{-x}}{e^x + e^{-x}} \\ &= \lim_{x\to\infty} \frac{1 - e^{-2x}}{1 + e^{-2x}} \\ &= 1 \text{ since } e^{-2x} \to 0 \text{ as } x \to \infty. \end{aligned}$$

Similarly, $\lim_{x\to-\infty} \tanh x = -1$. (See Exercise 11.)

Thus the lines $y = 1$ and $y = -1$ are horizontal asymptotes and

$$-1 < \tanh x < 1 \text{ for all } x.$$

(See Figure 4.6.)

The graphs of the other hyperbolic functions are easily sketched using the reciprocal of ordinates. (See Figures 4.7, 4.8 and 4.9.) The hyperbolic functions satisfy identities very similar to those satisfied by the trigonometric functions as illustrated in the following worked example.

Worked Example 4.2.1 *Show that*

1. $\cosh^2 x - \sinh^2 x = 1$

2. $\cosh(x + y) = \cosh x \cosh y + \sinh x \sinh y$

3. $\cosh^2 x + \sinh^2 x = \cosh 2x$.

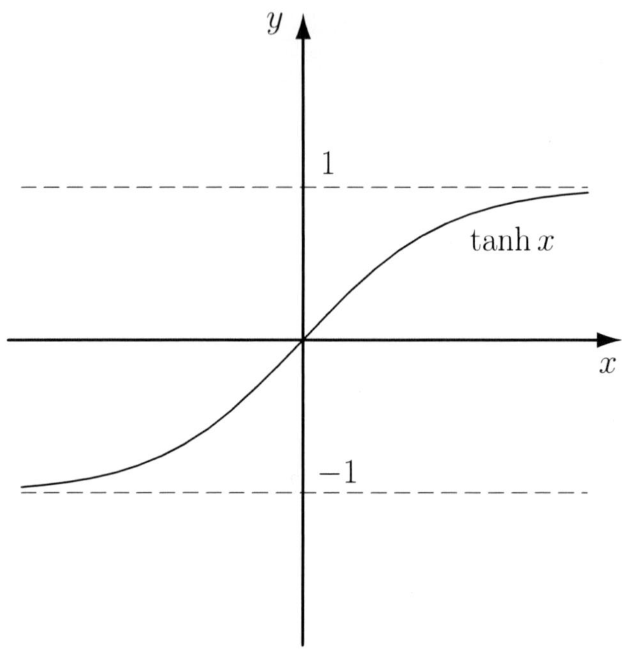

Figure 4.6: Graph of the function $y = \tanh x$.

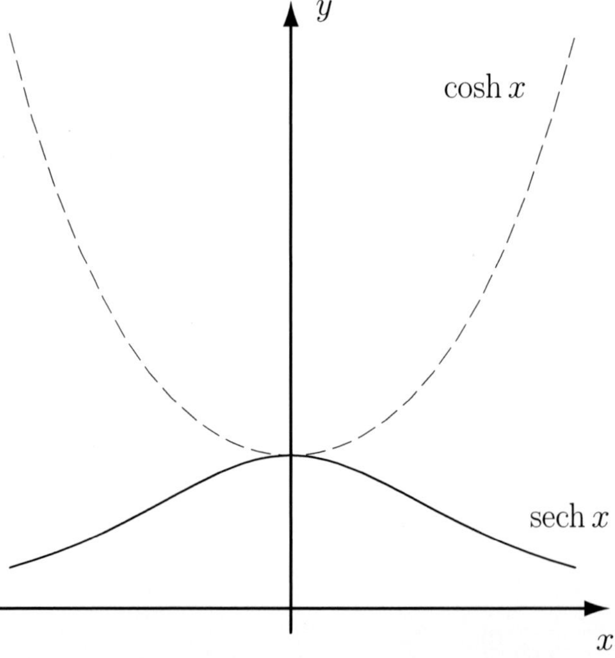

Figure 4.7: Graph of the function $y = \operatorname{sech} x$.

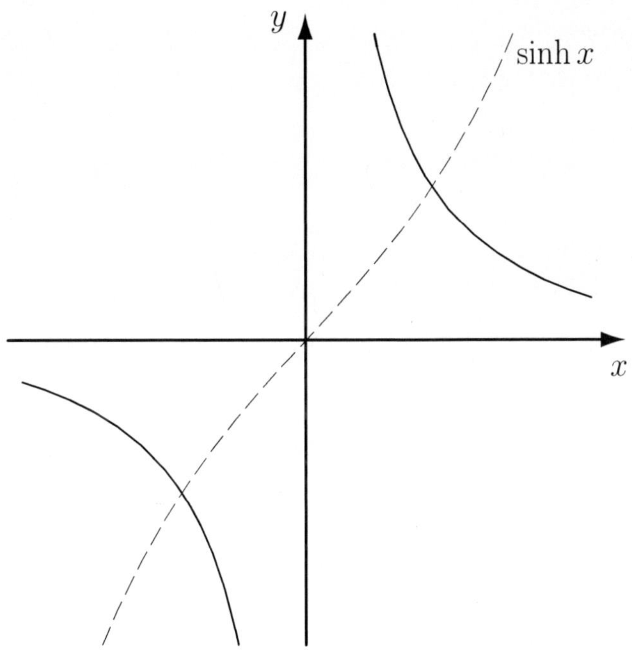

Figure 4.8: Graph of $y = \operatorname{cosech} x$.

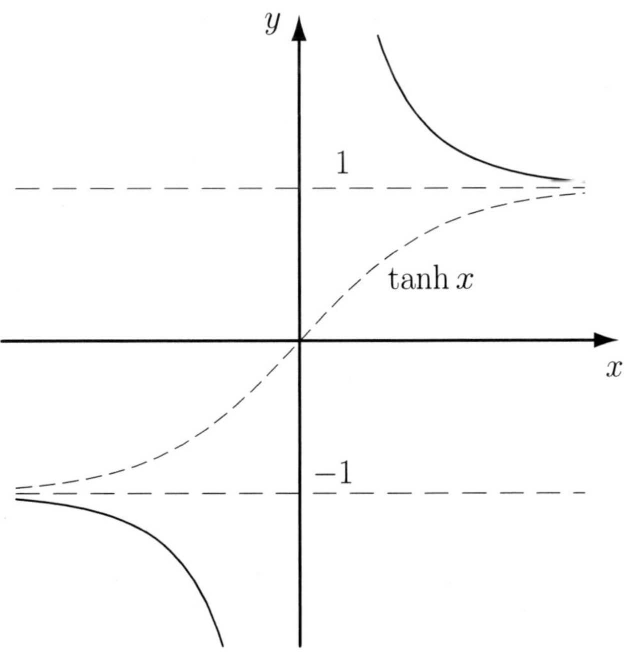

Figure 4.9: Graph of $y = \coth x$.

1. $\cosh^2 x - \sinh^2 x$

$$= \left(\frac{e^x + e^{-x}}{2}\right)^2 - \left(\frac{e^x - e^{-x}}{2}\right)^2$$

$$= \frac{1}{4}\left\{(e^{2x} + 2 + e^{-2x}) - (e^{2x} - 2 + e^{-2x})\right\}$$

$$= 1.$$

Alternatively

$\cosh^2 x - \sinh^2 x$

$$= (\cosh x + \sinh x)(\cosh x - \sinh x)$$

$$= e^x \cdot e^{-x} = e^0 = 1.$$

2. $\cosh x \cosh y + \sinh x \sinh y$

$$= \left(\frac{e^x + e^{-x}}{2}\right)\left(\frac{e^y + e^{-y}}{2}\right) + \left(\frac{e^x - e^{-x}}{2}\right)\left(\frac{e^y - e^{-y}}{2}\right)$$

$$= \frac{1}{4}\{(e^{x+y} + e^{x-y} + e^{-x+y} + e^{-x-y})$$

$$+ (e^{x+y} - e^{x-y} - e^{-x+y} + e^{-x-y})\}$$

$$= \frac{1}{2}(e^{x+y} + e^{-x-y})$$

$$= \cosh(x + y).$$

3. This can be readily shown by putting $y = x$ in the result from (2) or can be derived directly as follows:

$\cosh^2 x + \sinh^2 x$

$$= \left(\frac{e^x + e^{-x}}{2}\right)^2 + \left(\frac{e^x - e^{-x}}{2}\right)^2$$

$$= \frac{1}{4}\left\{(e^{2x} + 2 + e^{-2x}) + (e^{2x} - 2 + e^{-2x})\right\}$$

$$= \frac{1}{2}(e^{2x} + e^{-2x})$$

$$= \cosh 2x.$$

From item 1 it can be readily seen that $x = a\cosh u, y = b\sinh u$ are parametric equations for the hyperbola $\dfrac{x^2}{a^2} - \dfrac{y^2}{b^2} = 1$, hence the name "hyperbolic functions". (Compare this with the parametric representation $x = \cos u, y = \sin u$ of the circle $x^2 + y^2 = 1$.) You could try to sketch the graph of this hyperbola using the parametric equations. ■

The similarity between the familiar trigonometric identities and corresponding hyperbolic identities can be seen from Table 4.1. Do you notice any pattern?

You *should* notice that corresponding to each trigonometric identity there is a hyperbolic identity which has the same form except for a possible change of sign. The change of sign occurs whenever a product of two sinh functions occurs (for example, in $\sinh x \sinh y$ or $\sinh^2 x$) or an implied product of two sinh functions (for example, in $\tanh x \tanh y$ which is just the ratio $\frac{\sinh x \sinh y}{\cosh x \cosh y}$).

Given the value of one of the hyperbolic functions it is a relatively simple matter to obtain the values of the other hyperbolic functions using the identity $\cosh^2 x - \sinh^2 x = 1$; see the following worked example.

Worked Example 4.2.2

1. *If* $\cosh x = 4$ *find* exact *values for* $\sinh x$ *and* $\tanh x$.
2. *If* $\sinh x = 2$ *find* exact *values for* $\cosh x$ *and* $\tanh x$.
3. *If* $\tanh x = 0.2$ *find the* exact *value of* $\cosh x$.

1. Since $\cosh x = 4$ it follows that $16 - \sinh^2 x = 1$ so that $\sinh x = \pm\sqrt{15}$.

   Consequently, $\tanh x = \dfrac{\sinh x}{\cosh x} = \pm\dfrac{\sqrt{15}}{4}$.
2. Since $\sinh x = 2$ it follows that $\cosh^2 x - 4 = 1$ so that $\cosh x = \sqrt{5}$. (Note that $\cosh x$ cannot be negative.)

   Hence $\tanh x = \dfrac{\sinh x}{\cosh x} = \dfrac{2}{\sqrt{5}} = \dfrac{2\sqrt{5}}{5}$.
3. From the identity $\cosh^2 x - \sinh^2 x = 1$ we can obtain the identity

   $$1 - \tanh^2 x = \operatorname{sech}^2 x$$

   on dividing through by $\cosh^2 x$.

   Now $\tanh x = 0.2$ so that $\operatorname{sech}^2 x = 0.96$, which implies

   $$\cosh x = \frac{10}{4\sqrt{6}} = \frac{5}{2\sqrt{6}}.$$

■

The similarity between the trigonometric and hyperbolic functions even carries over to the calculus of such functions as indicated in the next section.

Trigonometric	Hyperbolic
$\cos^2 x + \sin^2 x = 1$	$\cosh^2 x - \sinh^2 x = 1$
$1 + \tan^2 x = \sec^2 x$	$1 - \tanh^2 x = \operatorname{sech}^2 x$
$\sin 2x = 2 \sin x \cos x$	$\sinh 2x = 2 \sinh x \cosh x$
$\cos 2x = \cos^2 x - \sin^2 x$	$\cosh 2x = \cosh^2 x + \sinh^2 x$
$\cos 2x = 2 \cos^2 x - 1$	$\cosh 2x = 2 \cosh^2 x - 1$
$\cos 2x = 1 - 2 \sin^2 x$	$\cosh 2x = 1 + 2 \sinh^2 x$
$\cos(x + y) = \cos x \cos y - \sin x \sin y$	$\cosh(x + y) = \cosh x \cosh y + \sinh x \sinh y$
$\sin(x + y) = \sin x \cos y + \cos x \sin y$	$\sinh(x + y) = \sinh x \cosh y + \cosh x \sinh y$
$\tan(x + y) = \dfrac{\tan x + \tan y}{1 - \tan x \tan y}$	$\tanh(x + y) = \dfrac{\tanh x + \tanh y}{1 + \tanh x \tanh y}$

Table 4.1: Similarity between trigonometric and hyperbolic functions.

4.2.2 The Derivatives of Hyperbolic Functions

The derivatives of hyperbolic functions are easily obtained as follows:

$$\frac{d}{dx}\cosh x = \frac{d}{dx}\left(\frac{e^x + e^{-x}}{2}\right) = \frac{e^x - e^{-x}}{2} = \sinh x$$

$$\frac{d}{dx}\sinh x = \frac{d}{dx}\left(\frac{e^x - e^{-x}}{2}\right) = \frac{e^x + e^{-x}}{2} = \cosh x$$

and

$$\begin{aligned}\frac{d}{dx}\tanh x &= \frac{d}{dx}\frac{\sinh x}{\cosh x} \\ &= \frac{\cosh x \cosh x - \sinh x \sinh x}{\cosh^2 x} \\ &= \frac{1}{\cosh^2 x} = \operatorname{sech}^2 x\end{aligned}$$

using the quotient rule. The derivatives of

$$\operatorname{sech} x, \operatorname{cosech} x \text{ and } \coth x$$

are respectively

$$-\operatorname{sech} x \tanh x, -\operatorname{cosech} x \coth x \text{ and } -\operatorname{cosech}^2 x.$$

(See Exercises 14 and 15.)

From Table 4.2 it is clear the derivatives of the trigonometric and hyperbolic functions have similar forms except for the signs; for the trigonometric functions those with a minus sign are the co-functions (cos, cosec and cot) whereas for the hyperbolic functions those with the minus sign are the reciprocal functions (sech , cosech and coth).

Using the chain rule, we obtain Table 4.3.

Worked Example 4.2.3 *Determine the derivative of:*

1. $x^2 \tanh 4x$ *2.* $e^{-2x}\sinh 3x$ *3.* $\cosh^3 2x$

1. Using the product rule

$$\begin{aligned}\frac{d}{dx}(x^2 \tanh 4x) &= 2x \tanh 4x + x^2 \operatorname{sech}^2 4x \cdot 4 \\ &= 2x(\tanh 4x + 2x \operatorname{sech}^2 4x).\end{aligned}$$

Function	Derivative
$\sin x$	$\cos x$
$\sinh x$	$\cosh x$
$\cos x$	$-\sin x$
$\cosh x$	$\sinh x$
$\tan x$	$\sec^2 x$
$\tanh x$	$\text{sech}^2 x$
$\sec x$	$\sec x \tan x$
$\text{sech}\, x$	$-\text{sech}\, x \tanh x$
$\text{cosec}\, x$	$-\text{cosec}\, x \cot x$
$\text{cosech}\, x$	$-\text{cosech}\, x \coth x$
$\cot x$	$-\text{cosec}^2 x$
$\coth x$	$-\text{cosech}^2 x$

Table 4.2: Derivatives of trigonometric and hyperbolic functions.

Function	Derivative
$\sinh f(x)$	$\cosh f(x) \cdot f'(x)$
$\cosh f(x)$	$\sinh f(x) \cdot f'(x)$
$\tanh f(x)$	$\text{sech}^2 f(x) \cdot f'(x)$

Table 4.3: Derivatives of hyperbolic functions of general argument.

2. Once again we use the product rule to give

$$\begin{aligned}\frac{d}{dx}&(e^{-2x}\sinh 3x)\\ &= e^{-2x}(-2)\sinh 3x + e^{-2x}\cosh 3x\cdot 3\\ &= e^{-2x}(3\cosh 3x - 2\sinh 3x).\end{aligned}$$

3. Writing $\cosh^3 2x$ as $(\cosh 2x)^3$ it is a simple matter to write down the derivative (using the chain rule) as

$$\frac{d}{dx}\cosh^3 2x = 3(\cosh 2x)^2\cdot\sinh 2x\cdot 2 = 6\cosh^2 2x\sinh 2x.$$

■

Self-help exercises

1. Write down the derivatives of:

(a) $\sinh 3x$	$[3\cosh 3x]$	(d) $\cosh 4x^2$	$[8x\sinh 4x^2]$
(b) $\sinh 3x^2$	$[6x\cosh 3x^2]$	(e) $\tanh 2x$	$[2\text{sech}^2 2x]$
(c) $\cosh 4x$	$[4\sinh 4x]$	(f) $\tanh 2x^2$	$[4x\ \text{sech}^2 2x^2]$

2. (a) If $\cosh x = 2$, write down the values of $\sinh x$. $[\pm\sqrt{3}]$

   (b) If $\sinh x = 4$, write down the value of $\cosh x$. $[\sqrt{17}]$

4.3 Inverse Functions

4.3.1 Introduction

Suppose that $f : X \to \mathbf{R}$ is a function with domain X and range Y. If f is a 1–1 function[1] then there is a function g which maps Y back into X according to the rule $g(f(x)) = x$; also note that

- $f(g(x)) = x$

---

[1] A function $f : X \to \mathbf{R}$ is 1–1 if different elements of the domain map onto different images; that is, if $x_{\cdot} \neq x_{\cdot}$ implies $f(x_{\cdot}) \neq f(x_{\cdot})$. This is equivalent to

$$f(x_{\cdot}) = f(x_{\cdot}) \Rightarrow x_{\cdot} = x_{\cdot}\,.$$

A function f can also be shown to be 1–1 by sketching its graph.

- If f is an increasing function on its domain (or $f'(x) > 0$) then f is 1–1.
- If f is a decreasing function on its domain (or $f'(x) < 0$) then f is 1–1.

- the domain of g is simply the range of f
- the range of g is the domain of f.

Such a function is called the inverse of f and is denoted by f^{-1}. Can you explain why f needs to be 1–1?

Beware! f^{-1} denotes the inverse function of f and not the reciprocal, which is usually denoted by $\frac{1}{f}$.

In other words, the inverse f^{-1} of a 1–1 function f is such that

- $f^{-1}(y) = x$ if $y = f(x)$ and conversely (thus the two equations are *equivalent*)
- the domain of f^{-1} is the range of f
- the domain of f is the range of f^{-1}
- $f^{-1}(f(x)) = x$ for all x in the domain of f
- $f(f^{-1}(x)) = x$ for all x in the domain of f^{-1}.

Worked Example 4.3.1 *If the function f is defined by*

$$f(x) = 3x^3 - 6$$

for all $x \in \mathbf{R}$ determine $f^{-1}(x)$.
It is obvious that f^{-1} does exist since f is 1–1 (evident either by sketching its graph or by noting that since $f'(x) = 9x^2 \geq 0$ for all x the function is increasing).
Using the defining property $f(f^{-1}(x)) = x$ in the rule for this function, we have

$$\begin{aligned} 3\{f^{-1}(x)\}^3 - 6 &= x \\ \Rightarrow \qquad \{f^{-1}(x)\}^3 &= \frac{x+6}{3} \\ \Rightarrow \qquad f^{-1}(x) &= \sqrt[3]{\frac{x+6}{3}} \end{aligned}$$

Alternatively, since $y = f(x)$ is equivalent to $x = f^{-1}(y)$ we can determine f^{-1} by simply solving $y = f(x)$ for x as follows:
If $y = 3x^3 - 6$ then $x^3 = \frac{y+6}{3}$.
Thus $x = \sqrt[3]{\frac{y+6}{3}}$ so that $f^{-1}(y) = \sqrt[3]{\frac{y+6}{3}}$.
That is, $f^{-1}(x) = \sqrt[3]{\frac{x+6}{3}}$. ■

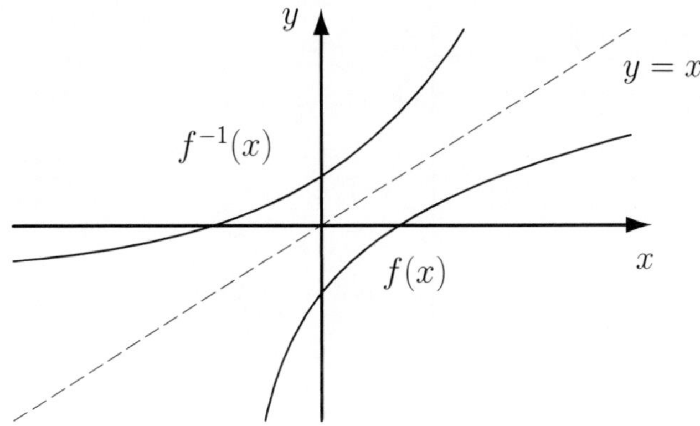

Figure 4.10: Graph of a function $y = f(x)$ and its inverse $y = f^{-1}(x)$.

Note that if the point (a, b) lies on the graph of $y = f(x)$ then $b = f(a)$ so that $a = f^{-1}(b)$; that is, the point (b, a) lies on the graph of $y = f^{-1}(x)$. It follows that the graph of $y = f^{-1}(x)$ is obtained by reflecting the graph of $y = f(x)$ in the line $y = x$. (See Figure 4.10.)

Worked Example 4.3.2 *Consider the function defined by*

$$f(x) = x^2 - 2x - 3.$$

Explain why, although this function has no inverse on **R**, *by suitably restricting the domain of the function we can ensure that it does have an inverse on the restricted domain. Find this inverse.*

It can be shown that f is not 1–1 on **R** by either sketching its graph (see Figure 4.11) or by determining $f'(x) = 2x - 2$ and noting that this is not always of the same sign (that is, the function decreases for some values of x and increases for other values of x).

Thus f has no inverse.

Note however, either by considering the sign of $f'(x)$ or by looking at the graph, that the function is 1–1 on the interval $[1, \infty)$ (and also on $(-\infty, 1]$).

First consider the function $f_1 : [1, \infty) \to \mathbf{R}$ where $f_1(x) = x^2 - 2x - 3$.

Since the range of f_1 is $[-4, \infty)$ the domain of f_1^{-1} (which we have already shown to exist since f is 1–1 on $[1, \infty)$) is $[-4, \infty)$.

To determine $f_1^{-1}(x)$ we shall solve $y = x^2 - 2x - 3$ for x in $x \geq 1$.

Writing $x^2 - 2x - 3 = y$ and using the formula for a quadratic we obtain

$$x = 1 \pm \sqrt{y + 4}.$$

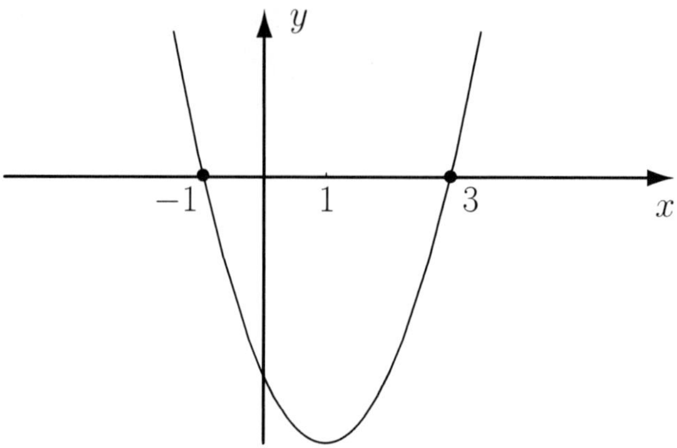

Figure 4.11: Graph of the function $y = x^2 - 2x - 3$.

It remains to decide which of these two functions we require.
Since

- f_1^{-1} has range $[1, \infty)$ (being the domain of f_1), and
- $1 + \sqrt{y+4} \geq 1$ (but $1 - \sqrt{y+4} \leq 1$)

we must have

$$f_1^{-1}(y) = 1 + \sqrt{y+4}$$

or equivalently

$$f_1^{-1}(x) = 1 + \sqrt{x+4} \text{ for } x \geq -4.$$

How can we explain the other solution $x = 1 - \sqrt{y+4}$?
If we define $f_2 : (-\infty, 1] \to \mathbf{R}$ where $f_2(x) = x^2 - 2x - 3$ we can show that this other solution leads to the inverse of f_2. (You should verify this.) ■

Self-help exercises

1. Determine the inverse of each of the functions:

   (a) $f : \mathbf{R} \to \mathbf{R}$ where $f(x) = 3x + 7$ $[\frac{1}{3}(x-7)]$

   (b) $f : \mathbf{R} \to \mathbf{R}$ where $f(x) = 3x^3 + 7$ $[\sqrt[3]{\frac{x-7}{3}}]$

   (c) $f : [0, \infty) \to \mathbf{R}$ where $f(x) = 3\sqrt{x}$ $[\frac{x^2}{9}$ for $x \geq 0]$

   (d) $f : [1, \infty) \to \mathbf{R}$ where $f(x) = \frac{x+1}{x-1}$ $[\frac{x+1}{x-1}]$

2. Explain why the function $f : \mathbf{R} \to \mathbf{R}$ where $f(x) = \sin x$ has no inverse. [The function is not 1−1]

4.4 The Logarithm Function

4.4.1 Definition

The logarithm function $\log x$ or $\ln x$ (the natural log or log to the base e) is defined to be the inverse of the exponential function e^x.

You should note that since the exponential function has domain $\mathbf{R}$ and range $(0, \infty)$ the logarithm function is defined for $x > 0$ by

$$\boxed{y = \log x \text{ if and only if } x = e^y}$$

and satisfies

$$\boxed{\log e^x = x \text{ for all } x \quad \text{and} \quad e^{\log x} = x \text{ for all } x > 0}$$

The graph of $y = \log x$ is simply obtained by reflecting the graph of $y = e^x$ about the line $y = x$ as in Figure 4.12.

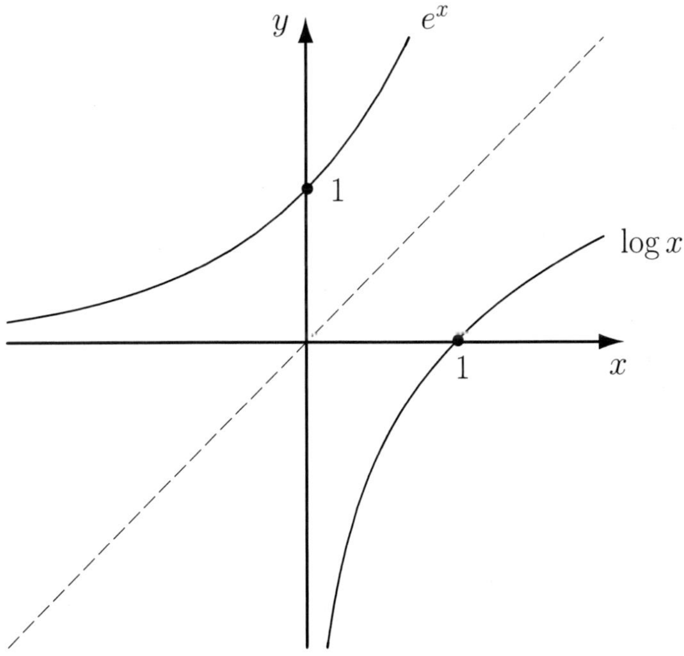

Figure 4.12: Graph of the function $y = e^x$ and its inverse $y = \log x$.

Self-help exercises

1. Write down the *exact* values of

   (a) $e^{\log 7}$ [7] (c) $e^{-\log 7}$ [1/7]

   (b) $e^{2\log 7}$ [49] (d) $e^{\log 4} + e^{-\log 4}$ [$4\frac{1}{4}$]

2. For what values of x are the following defined

   (a) $\log(-2x)$ [$(-\infty, 0)$] (d) $\log(1 - x^2)$ [$(-1, 1)$]

   (b) $\log x^2$ [$(-\infty, 0) \cup (0, \infty)$]

   (c) $\log(x - 1)$ [$(1, \infty)$] (e) $\log(x^2 + 2x + 2)$ [**R**]

4.4.2 Differentiation of Logarithmic Functions

To obtain the derivative of $\log x$ we proceed as follows. Let $y = \log x$ so that $e^y = x$. On differentiating both sides with respect to x we obtain $e^y \frac{dy}{dx} = 1$ so that $\frac{dy}{dx} = \frac{1}{e^{\cdot}}$. Therefore

$$\boxed{\frac{d}{dx} \log x = \frac{1}{x}}$$

It follows by an application of the chain rule that

$$\frac{d}{dx} \log f(x) = \frac{1}{f(x)} \cdot f'(x) = \frac{f'(x)}{f(x)} \text{ for } f(x) > 0.$$

(You should verify this!)

In fact (see Exercise 20) it is possible to show

$$\boxed{\frac{d}{dx} \log |f(x)| = \frac{1}{f(x)} \cdot f'(x) = \frac{f'(x)}{f(x)}}$$

For example

- $\dfrac{d}{dx} \log(x^2 + 2x + 10) = \dfrac{2x + 2}{x^2 + 2x + 10}$

- $\dfrac{d}{dx} \log^2(x^2 + 2x + 10)$
  $= \dfrac{d}{dx} \{\log(x^2 + 2x + 10)\}^2$
  $= 2\log(x^2 + 2x + 10) \dfrac{2x + 2}{x^2 + 2x + 10}$

- $\dfrac{d}{dx}(x^2 \log 3x) = 2x \log 3x + x^2 \dfrac{3}{3x} = x(2\log 3x + 1)$ using the product rule.

(The function $\log^2 f(x)$ used above means $\log f(x) \cdot \log f(x)$.)

It is often useful to use the following logarithm laws to simplify expressions involving logarithms prior to differentiation:

$$\log(ab) = \log a + \log b \text{ for all real } a > 0, b > 0$$
$$\log(\frac{a}{b}) = \log a - \log b \text{ for all real } a > 0, b > 0$$
$$\log a^b = b \log a \text{ for all real } b \text{ and real } a > 0$$

These rules are employed in Worked Example 4.4.1.

You should note however that the above logarithm laws do *not* apply to expressions such as $\log(a+b)$ and $\frac{\log a}{\log b}$. In particular

$$\log(a+b) \neq \log a + \log b.$$

Worked Example 4.4.1 *Determine the derivatives of:*

1. $\log\sqrt{\dfrac{1+x}{1-x}}$

2. $\log(x + \sqrt{1+x^2})$

1. First use the logarithm laws to write

$$\log\sqrt{\frac{1+x}{1-x}} = \frac{1}{2}(\log(1+x) - \log(1-x))$$

so that its derivative can easily be written down as

$$\frac{1}{2}\left(\frac{1}{1+x} + \frac{1}{1-x}\right) = \frac{1}{1-x^2}.$$

2. As the logarithm laws do not apply to $\log(x + \sqrt{1+x^2})$ (Why?) we just apply the chain rule as follows:

$$\begin{aligned}
&\frac{d}{dx}\log(x+\sqrt{1+x^2}) \\
&= \frac{1}{x+\sqrt{1+x^2}}\left\{1 + \frac{1}{2}(1+x^2)^{-1/2}2x\right\} \\
&= \frac{1}{x+\sqrt{1+x^2}}\left\{1 + \frac{x}{\sqrt{1+x^2}}\right\} \\
&= \frac{1}{x+\sqrt{1+x^2}}\left\{\frac{x+\sqrt{1+x^2}}{\sqrt{1+x^2}}\right\} \\
&= \frac{1}{\sqrt{1+x^2}}
\end{aligned}$$

■

Self-help exercises

1. Write down the derivatives of

   (a) $\log 3x$ $[\frac{1}{x}]$

   (b) $\log 3x^2$ $[\frac{2}{x}]$

   (c) $\log(1+3x)$ $[\frac{3}{1+3x}]$

   (d) $\log(1+3x^2)$ $[\frac{6x}{1+3x^{\cdot}}]$

   (e) $x\log x$ $[1+\log x]$

2. If y is a function of x then which of the following expressions represents $\frac{d}{dx}(\log y)$?

   (a) $\dfrac{1}{y}$ (b) $\dfrac{1}{\frac{dy}{dx}}$ (c) $\dfrac{1}{y}\dfrac{dy}{dx}$ [(c)]

3. Use the log laws to simplify each of the following:

   (a) $\log\dfrac{\sqrt{1+x^2}}{(2+\sin x)^3}$ $[\frac{1}{2}\log(1+x^2) - 3\log(2+\sin x)]$

   (b) $\log\sqrt{\dfrac{1+x^2}{1-x^2}}$ for $|x|<1$

   $[\frac{1}{2}\log(1+x^2) - \frac{1}{2}\log(1-x) - \frac{1}{2}\log(1+x)]$

   (c) $\log\dfrac{e^{-2x^{\cdot}}x^4}{(1+x^2)^2(2+x^4)^3}$

   $[-2x^2 + 4\log x - 2\log(1+x^2) - 3\log(2+x^4)]$

4.4.3 Logarithmic Differentiation

Whenever it is necessary to work with exponentials or logarithms to a base other than e it is usual to either rely on the definitions of such functions

- $a^x = e^{x\log a}$
- $y = \log_a x$ if and only if $a^y = x$

or to use the technique of logarithmic differentiation which is described below. Both these methods are illustrated in the following worked examples.

Worked Example 4.4.2 *Determine the derivative of* $5^{3x^{\cdot}-1}$.

Method 1

Using the definition of a^x we put

$$5^{3x^{\cdot}-1} = e^{(3x^{\cdot}-1)\log 5}.$$

Then, with an application of the chain rule, we get

$$\begin{aligned}
\frac{d}{dx}(5^{3x^2-1}) &= \frac{d}{dx}e^{(3x^2-1)\log 5} \\
&= e^{(3x^2-1)\log 5}\frac{d}{dx}[(3x^2-1)\log 5] \\
&= e^{(3x^2-1)\log 5}6x\log 5 \\
&= 5^{3x^2-1}6x\log 5
\end{aligned}$$

Method 2

An alternative approach is to use logarithmic differentiation; that is, starting with $y = 5^{3x^2-1}$ we take logarithms of both sides and then differentiate both sides with respect to x.
Taking logarithms gives

$$\log y = (3x^2-1)\log 5.$$

Differentiating both sides with respect to x yields

$$\frac{1}{y}y' = 6x\log 5 \text{ so that } y' = 6xy\log 5 = 5^{3x^2-1}6x\log 5$$

as before. ■

Worked Example 4.4.3 *Determine the derivatives of*

1. $\log_{10}(1+x^2)$ *2.* $(1+\frac{1}{x})^x$ *for* $x > 0$.

1. Let $y = \log_{10}(1+x^2)$ so that $10^y = 1+x^2$; now take logarithms (to base e) of both sides so that $y\log 10 = \log(1+x^2)$. It is now straightforward to obtain $y'\log 10 = \dfrac{2x}{1+x^2}$ or $y' = \dfrac{2x}{(1+x^2)\log 10}$.
2. Let $y = \left(1+\dfrac{1}{x}\right)^x$.

   Taking logarithms we get

$$\log y = \log\left(1+\frac{1}{x}\right)^x = x\log\left(1+\frac{1}{x}\right).$$

   Using logarithmic differentiation we then get

$$\begin{aligned}
\frac{1}{y}y' &= \log\left(1+\frac{1}{x}\right) + x\frac{1}{1+\frac{1}{x}}\cdot\left(-\frac{1}{x^2}\right) \\
&= \log\left(1+\frac{1}{x}\right) - \frac{1}{x+1}
\end{aligned}$$

Therefore

$$\begin{aligned} y' &= y\left\{\log\left(1+\frac{1}{x}\right)-\frac{1}{x+1}\right\} \\ &= \left(1+\frac{1}{x}\right)^x\left\{\log\left(1+\frac{1}{x}\right)-\frac{1}{x+1}\right\}. \end{aligned}$$

■

Logarithmic differentiation is often useful in differentiating complicated products and quotients such as

$$\frac{(1-2x)^3\sqrt{1+x^2}}{1+4e^{x^2}}.$$

Although such functions can be differentiated using the quotient rule together with the product rule it is much simpler to proceed as follows. First, let

$$y=\frac{(1-2x)^3\sqrt{1+x^2}}{1+4e^{x^2}}.$$

Taking logs (to base e) of both sides and using the log laws gives

$$\log y = 3\log(1-2x)+\frac{1}{2}\log(1+x^2)-\log(1+4e^{x^2}).$$

Differentiating with respect to x gives

$$\frac{1}{y}\frac{dy}{dx}=\frac{-6}{1-2x}+\frac{x}{1+x^2}-\frac{8xe^{x^2}}{1+4e^{x^2}}.$$

so that

$$\frac{dy}{dx}=y\left(\frac{-6}{1-2x}+\frac{x}{1+x^2}-\frac{8xe^{x^2}}{1+4e^{x^2}}\right)$$

where

$$y=\frac{(1-2x)^3\sqrt{1+x^2}}{1+4e^{x^2}}.$$

4.5 The Inverse Trigonometric Functions

4.5.1 The Inverse Sine and Cosine Functions

Since the sine and cosine functions are not 1−1 they have no inverses; however it is possible to restrict their domains so that they become 1−1 on that domain and hence have an inverse on the corresponding domain. (See Figure 4.13.)

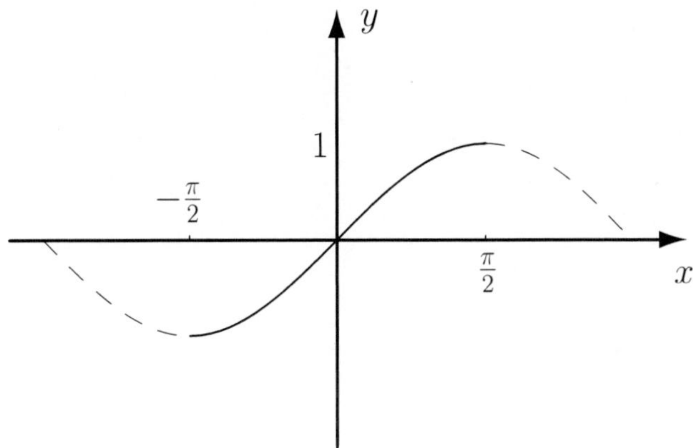

Figure 4.13: Graph of the function $y = \sin x$ with domain restricted to yield a 1−1 function.

Definition 4.2 *The inverse sine function, denoted by* $\sin^{-1} x$ *or* $\arcsin x$ *is defined to be the inverse of* $f : [-\frac{\pi}{2}, \frac{\pi}{2}] \to [-1, 1]$ *where* $f(x) = \sin x$. *(See Figure 4.14.)*

Beware! In the above definition the superscript -1 does not refer to an exponent of -1; that is, $\sin^{-1} x$ does not mean $\frac{1}{\sin x}$; similarly $\cos^{-1} x$ does not mean $\frac{1}{\cos x}$. Consequently do not confuse $\sin^{-1} x$ with $(\sin x)^{-1}$ and $\cos^{-1} x$ with $(\cos x)^{-1}$.

From Definition 4.2 it follows that

- Since the domain of f is $[-\frac{\pi}{2}, \frac{\pi}{2}]$ the range of $\arcsin x$ is $[-\frac{\pi}{2}, \frac{\pi}{2}]$.
- Since the range of f is $[-1, 1]$ the domain of $\arcsin x$ is $[-1, 1]$.
- $\sin(\arcsin x) = x$ for all $x \in [-1, 1]$.
- $\arcsin(\sin x) = x$ for all $x \in [-\frac{\pi}{2}, \frac{\pi}{2}]$.

Definition 4.3 *The inverse cosine function, denoted by* $\cos^{-1} x$ *or* $\arccos x$, *is defined to be the inverse of* $g : [0, \pi] \to [-1, 1]$ *where* $g(x) = \cos x$. *(See Figures 4.15 and 4.16. Also refer to the warning that follows Definition 4.2.)*

From Definition 4.3 it follows that

- Since the domain of g is $[0, \pi]$ the range of $\arccos x$ is $[0, \pi]$.
- Since the range of g is $[-1, 1]$ the domain of $\arccos x$ is $[-1, 1]$.
- $\cos(\arccos x) = x$ for all $x \in [-1, 1]$.
- $\arccos(\cos x) = x$ for all $x \in [0, \pi]$.

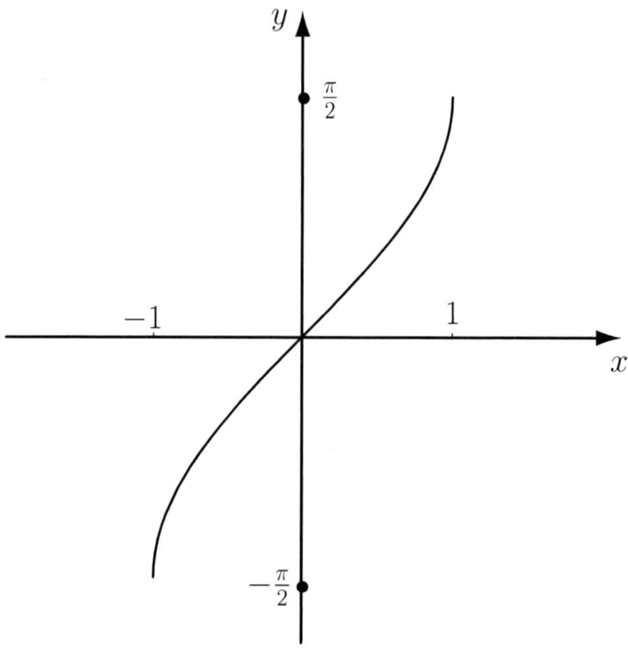

Figure 4.14: Graph of the function $y = \arcsin x$.

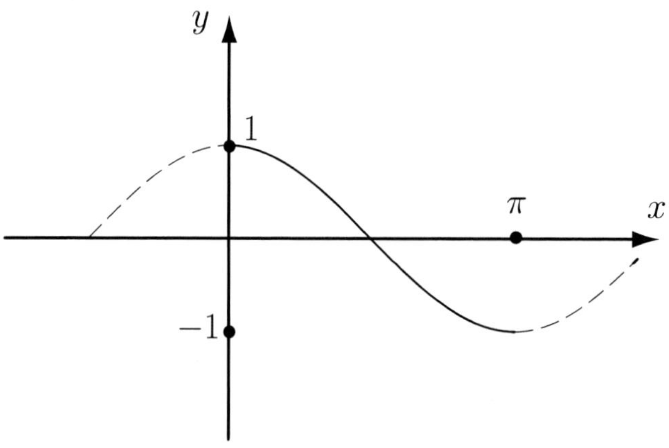

Figure 4.15: Graph of the function $y = \cos x$ with the domain restricted to yield a 1−1 function.

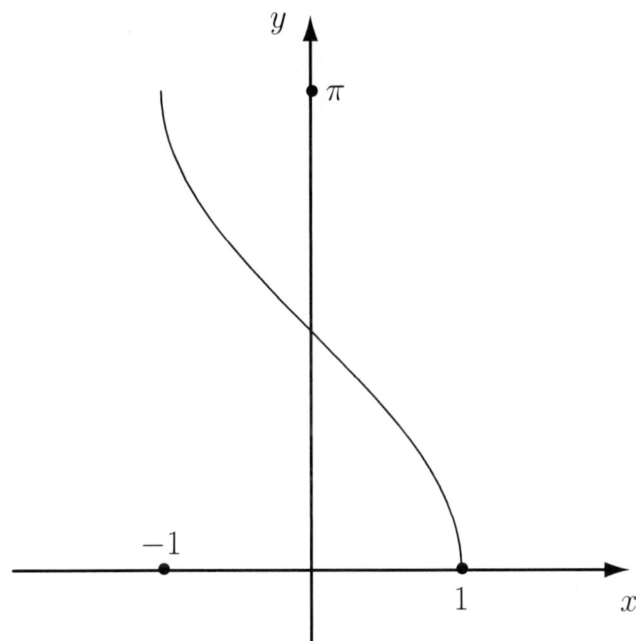

Figure 4.16: Graph of the function $y = \arccos x$.

Just as there is a simple but important relation between the sine and cosine functions, namely

$$\sin\left(\frac{\pi}{2} - x\right) = \cos x \text{ or } \cos\left(\frac{\pi}{2} - x\right) = \sin x$$

there is a corresponding relation between the inverse functions:

$$\boxed{\arcsin x + \arccos x = \frac{\pi}{2} \text{ for all } x \in [-1, 1]}$$

To prove this, let $y = \arcsin x$ where $x \in [-1, 1]$ and $y \in [-\frac{\pi}{2}, \frac{\pi}{2}]$. Then $\sin y = x$ which can be rewritten as $\cos(\frac{\pi}{2} - y) = x$. Since $y \in [-\frac{\pi}{2}, \frac{\pi}{2}]$ it follows that $(\frac{\pi}{2} - y) \in [0, \pi]$; thus $\frac{\pi}{2} - y = \arccos x$ and the result follows.

4.5.2 The Inverse Tangent Function

Definition 4.4 *The inverse tangent function, denoted by* $\tan^{-1} x$ *or* $\arctan x$, *is defined to be the inverse of* $f : (-\frac{\pi}{2}, \frac{\pi}{2}) \to \mathbf{R}$ *where* $f(x) = \tan x$. *(Also refer to the warning that follows Definition 4.2.)*

From Definition 4.4 it follows that

- Since the domain of f is $(-\frac{\pi}{2}, \frac{\pi}{2})$ the range of $\arctan x$ is $(-\frac{\pi}{2}, \frac{\pi}{2})$.

- Since the range of f is $\mathbf{R}$ the domain of $\arctan x$ is $\mathbf{R}$.
- $\tan(\arctan x) = x$ for all real x.
- $\arctan(\tan x) = x$ for all $x \in (-\frac{\pi}{2}, \frac{\pi}{2})$.

(See Figures 4.17 and 4.18.)

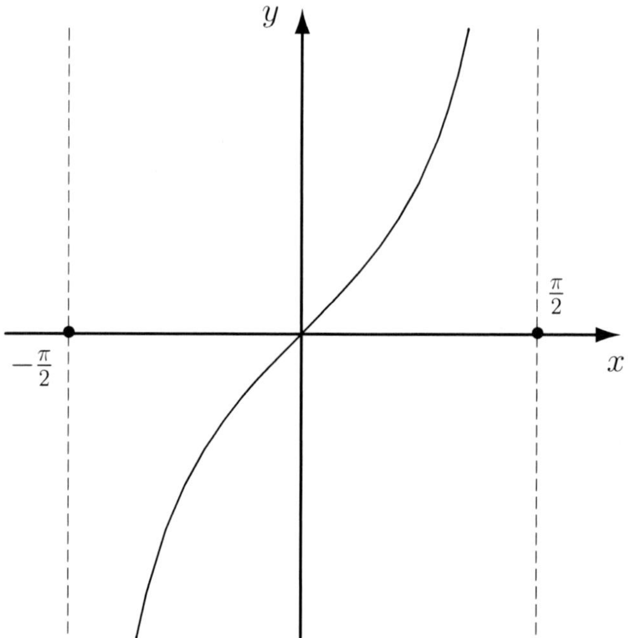

Figure 4.17: Graph of the function $y = \tan x$, $-\frac{\pi}{2} < x < \frac{\pi}{2}$.

Self-help exercises

1. Determine the *exact* value of each of the following:

   (a) i. $\sin(\arcsin(0.2))$; iii. $\sin(\arccos(-0.4))$;
   ii. $\cos(\arccos(-0.4))$; iv. $\cos(\arcsin(-0.4))$.

   $[0.2, -0.4, \frac{1}{5}\sqrt{21}, \frac{1}{5}\sqrt{21}]$

   (b) i. $\arcsin(\sin(\frac{\pi}{4}))$; iii. $\arccos(\cos(5\pi/3)$;
   ii. $\arccos(\cos(\pi/3))$; iv. $\arctan(\tan(3\pi/4))$.

   $[\frac{1}{4}\pi, \frac{1}{3}\pi, \frac{1}{3}\pi, -\frac{1}{4}\pi]$

2. Which of the following expressions are not defined: $\arcsin(1.2)$, $\cos 2$, $\arcsin(\pi)$ and $\cos(\arcsin(4))$? [All but the second]

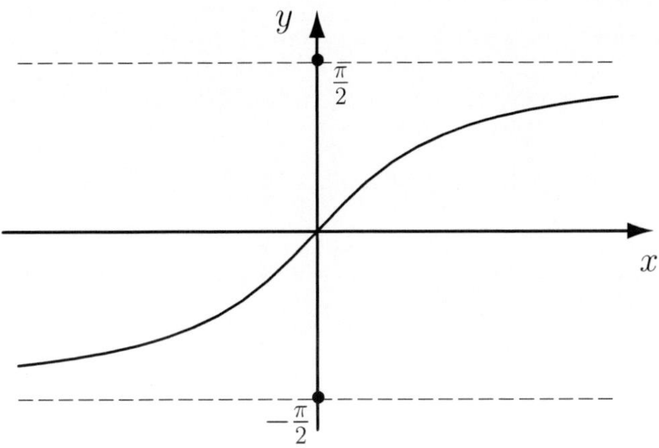

Figure 4.18: Graph of the function $y = \arctan x$.

4.5.3 Derivatives of Inverse Circular Functions

We shall derive the derivative of $\arcsin x$ and then merely state the corresponding results for $\arccos x$ and $\arctan x$, leaving their derivations as exercises. (See Exercise 27.)

If $y = \arcsin x, -1 < x < 1$ then $x = \sin y$ where $-\frac{\pi}{2} < y < \frac{\pi}{2}$. Differentiating with respect to x yields

$$1 = \cos y \cdot \frac{dy}{dx} \text{ so that } \frac{dy}{dx} = \frac{1}{\cos y} = \frac{1}{\sqrt{1 - \sin^2 y}}.$$

Note that the positive square root is taken here since $\cos y > 0$ because $-\frac{\pi}{2} < y < \frac{\pi}{2}$. We have thus shown that

$$\boxed{\frac{d}{dx} \arcsin x = \frac{1}{\sqrt{1 - x^2}} \quad \text{for } -1 < x < 1}$$

By using a similar approach you should show that

$$\frac{d}{dx} \arccos x = -\frac{1}{\sqrt{1 - x^2}} \quad \text{for } -1 < x < 1$$

and

$$\boxed{\frac{d}{dx} \arctan x = \frac{1}{1 + x^2}}$$

Alternatively, the derivative of $\arccos x$ can be obtained from the derivative of $\arcsin x$ by differentiating the identity $\arcsin x + \arccos x = \frac{\pi}{2}$. In summary, we have:

$$\begin{aligned}
\frac{d}{dx}\arcsin x &= \frac{1}{\sqrt{1-x^2}} \text{ for } -1 < x < 1 \\
\frac{d}{dx}\arccos x &= -\frac{1}{\sqrt{1-x^2}} \text{ for } -1 < x < 1 \\
\frac{d}{dx}\arctan x &= \frac{1}{1+x^2}
\end{aligned}$$

These results are needed so frequently in practice that you should endeavour to remember them; failing this you can always extract them from the formulae in Appendix C (put $a = 1$ into Formulae (28) and (29) on page 494).

Using the chain rule we can obtain the following more general results:

$$\begin{aligned}
\frac{d}{dx}\arcsin f(x) &= \frac{1}{\sqrt{1-f(x)^2}}f'(x) \text{ for } |f(x)| < 1 \\
\frac{d}{dx}\arccos f(x) &= -\frac{1}{\sqrt{1-f(x)^2}}f'(x) \text{ for } |f(x)| < 1 \\
\frac{d}{dx}\arctan f(x) &= \frac{1}{1+f(x)^2}f'(x)
\end{aligned}$$

Worked Example 4.5.1 *Determine the derivatives of*

1. $\arctan \dfrac{2}{x}$ *2.* $\arctan \dfrac{3x-1}{x+3}$.

3. $x \arcsin x + \sqrt{1-x^2}$.

1. By the chain rule

$$\begin{aligned}
\frac{d}{dx}\arctan\frac{2}{x} &= \frac{1}{1+(\frac{2}{x})^2}.(-2x^{-2}) \\
&= -\frac{x^2}{4+x^2}\frac{2}{x^2} \\
&= -\frac{2}{4+x^2}
\end{aligned}$$

2. Using the chain rule

$$\begin{aligned}
&\frac{d}{dx}\arctan\frac{3x-1}{x+3}\\
&= \frac{1}{1+(\frac{3x-1}{x+3})^2}\cdot\frac{(x+3)\cdot 3-(3x-1)1}{(x+3)^2}\\
&= \frac{10}{(x+3)^2+(3x-1)^2}\\
&= \frac{1}{1+x^2}
\end{aligned}$$

3. Using both the product rule and the chain rule

$$\begin{aligned}
&\frac{d}{dx}(x\arcsin x+\sqrt{1-x^2})\\
&= \arcsin x+\frac{x}{\sqrt{1-x^2}}+\frac{1}{2}(1-x^2)^{-1/2}(-2x)\\
&= \arcsin x
\end{aligned}$$

■

4.6 The Inverse Hyperbolic Functions

The hyperbolic functions $\sinh x$ and $\tanh x$ are both 1–1 functions and hence have inverse functions denoted by $\operatorname{arsinh} x$ (or $\sinh^{-1} x$) and $\operatorname{artanh} x$ (or $\tanh^{-1} x$) respectively.

Properties of arsinh and artanh

1. $y = \operatorname{arsinh} x$ if and only if $\sinh y = x$.
2. Since the sinh function has $\mathbf{R}$ as its domain and range, $\mathbf{R}$ is also the domain and range of arsinh.
3. The graph of $y = \operatorname{arsinh} x$ is given in Figure 4.19.
4. $y = \operatorname{artanh} x, -1 < x < 1$ if and only if $\tanh y = x$.
5. Since the domain of $\tanh x$ is $\mathbf{R}$, the range of $\operatorname{artanh} x$ is $\mathbf{R}$; the range of $\tanh x$ is $(-1, 1)$ so that $(-1, 1)$ is also the domain of $\operatorname{artanh} x$.
6. The graph of $y = \operatorname{artanh} x$ is given in Figure 4.20.

In order to define an inverse hyperbolic cosine function we need to restrict the domain of the cosh function. (Why?) We define $\operatorname{arcosh} x$ (or $\cosh^{-1} x$) to be the inverse of the function defined by $\cosh x, x \geq 0$.

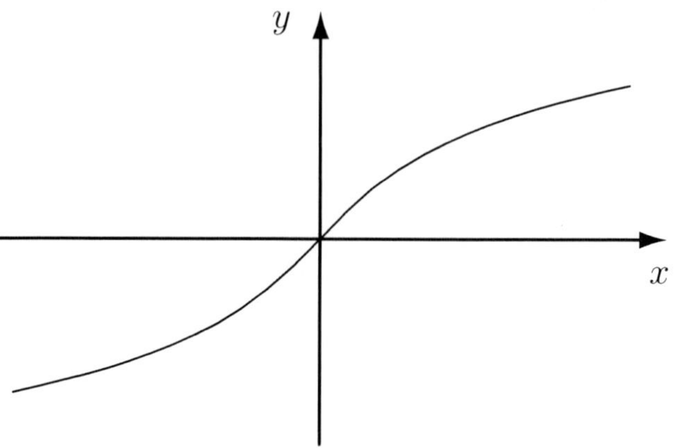

Figure 4.19: Graph of the function $y = \operatorname{arsinh} x$.

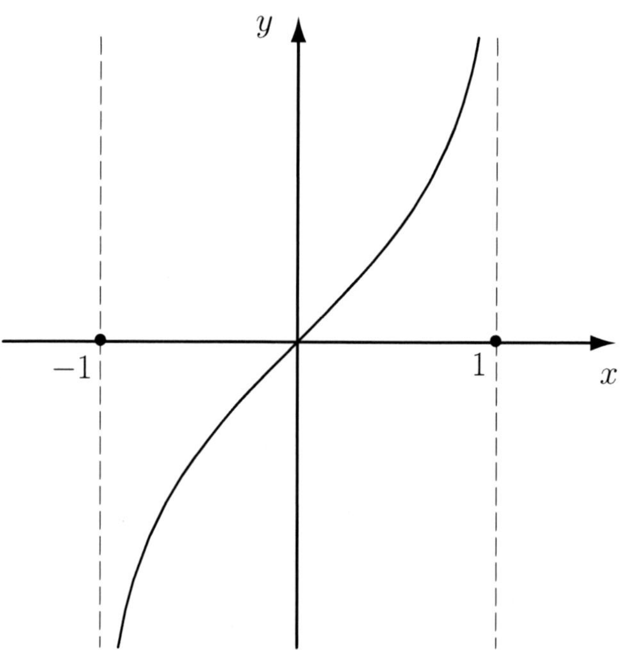

Figure 4.20: Graph of the function $y = \operatorname{artanh} x$.

Properties of arcosh

1. $\cosh(\operatorname{arcosh} x) = x$ for $x \geq 1$ so that if $y = \operatorname{arcosh} x$ then $\cosh y = x$.
2. The inverse function arcosh has domain $[1, \infty)$ and range $[0, \infty)$.
3. The graph of $y = \operatorname{arcosh} x$ is given in Figure 4.21.

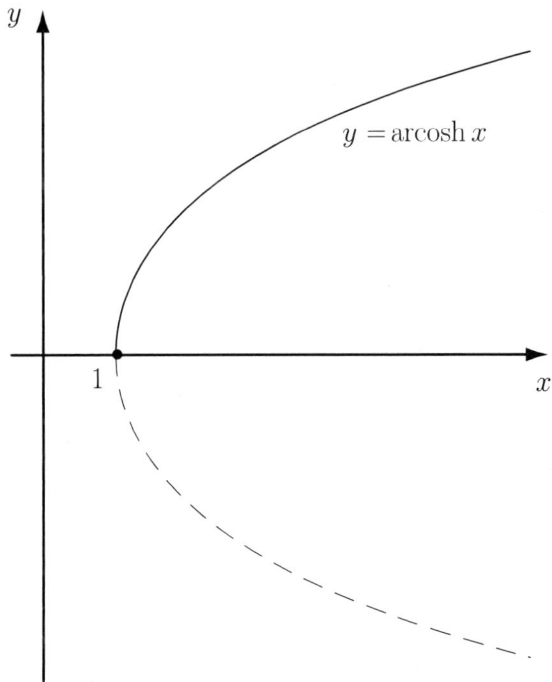

Figure 4.21: Graph of the function $y = \operatorname{arcosh} x$.

Each of the inverse functions $\operatorname{arsinh} x$, $\operatorname{arcosh} x$ and $\operatorname{artanh} x$ can be expressed in terms of the natural logarithm function. This is not surprising when it is considered that the hyperbolic functions are defined in terms of exponential functions and the logarithm function is the inverse of the exponential function.

For example, if $y = \operatorname{arsinh} x$, then $\sinh y = x$ or $e^y - e^{-y} = 2x$. On multiplying both sides by e^y we obtain $e^{2y} - 2xe^y - 1 = 0$ which is a quadratic in e^y. The solutions are

$$e^y = \frac{2x \pm \sqrt{4x^2 + 4}}{2} = x \pm \sqrt{x^2 + 1}.$$

But $x - \sqrt{x^2 + 1}$ is negative, so the only possible solution (why?) is $e^y = x + \sqrt{x^2 + 1}$, so that $y = \log(x + \sqrt{x^2 + 1})$.

Thus

$$\boxed{\operatorname{arsinh} x = \log(x + \sqrt{x^2+1})}$$

Similarly, logarithmic expressions can be obtained for $\operatorname{arcosh} x$ and $\operatorname{artanh} x$ but these are left as exercises. (See Exercises 28 and 30.)

$$\begin{array}{|rcll|}\hline \operatorname{arsinh} x &=& \log(x + \sqrt{x^2+1}), & \text{all } x \\ \operatorname{arcosh} x &=& \log(x + \sqrt{x^2-1}), & x \geq 1 \\ \operatorname{artanh} x &=& \dfrac{1}{2}\log\left(\dfrac{1+x}{1-x}\right), & -1 < x < 1 \\ \hline \end{array}$$

(These results are included in Appendix B on Page 487.)

These identities are particularly useful for obtaining *exact* values of inverse hyperbolic functions as follows:

$$\operatorname{arsinh} 3 = \log(3+\sqrt{10}), \operatorname{arcosh} 3 = \log(3+\sqrt{8}) \text{ and } \operatorname{artanh} 0.3 = \frac{1}{2}\log\frac{13}{7}.$$

Worked Example 4.6.1 *Determine all solutions of*

1. $\sinh x = 4$ *2.* $\tanh x = 4$ *3.* $4\tanh x = 1$.

1. $\sinh x = 4 \Rightarrow e^x - e^{-x} = 8$ so that $e^{2x} - 8e^x - 1 = 0$, a quadratic in e^x.
   Solving, using the formula for a quadratic, leads to $e^x = 4 \pm \sqrt{17}$. However, since e^x is never negative, the only valid solution is $e^x = 4 + \sqrt{17}$. Therefore
   $$x = \log(4+\sqrt{17}).$$
   Alternatively we could have noted that
   $$x = \operatorname{arsinh} 4 = \log(4+\sqrt{17})$$
   using the logarithm equivalent for $\operatorname{arsinh} x$ given in Appendix B on Page 487.
2. The equation $\tanh x = 4$ has no solution since 4 lies outside the range of the tanh function.

3. To solve $\tanh x = \frac{1}{4}$ write $\dfrac{e^x - e^{-x}}{e^x + e^{-x}} = \dfrac{1}{4}$.
   Thus, cross multiplying $4(e^x - e^{-x}) = e^x + e^{-x}$.
   Therefore $3e^x = 5e^{-x} \Rightarrow e^{2x} = \frac{5}{3}$ or $x = \frac{1}{2}\log\frac{5}{3}$.
   Alternatively, we could use the fact that

$$x = \operatorname{artanh}\frac{1}{4} = \frac{1}{2}\log\frac{1+\frac{1}{4}}{1-\frac{1}{4}} = \frac{1}{2}\log\frac{5}{3}.$$

■

It is a relatively simple matter to obtain derivatives of these inverse hyperbolics as indicated in the following worked example.

Worked Example 4.6.2 *Show that* $\dfrac{d}{dx}\operatorname{arcosh} x = \dfrac{1}{\sqrt{x^2-1}}$.

Method 1

Let $y = \operatorname{arcosh} x$. Then $\cosh y = x$.
Differentiating both sides with respect to x gives

$$\begin{aligned} \sinh y\,\frac{dy}{dx} &= 1 \\ \Rightarrow \qquad \frac{dy}{dx} &= \frac{1}{\sinh y} \\ \Rightarrow \qquad \frac{dy}{dx} &= \frac{1}{\sqrt{x^2-1}} \qquad \text{since} \qquad \sinh y = \sqrt{\cosh^2 y - 1}. \end{aligned}$$

Method 2

Using the logarithmic equivalent from Appendix B on Page 487 we have

$$\operatorname{arcosh} x = \log(x + \sqrt{x^2-1})$$

so that

$$\begin{aligned} \frac{d}{dx}\operatorname{arcosh} x &= \frac{d}{dx}\log(x + \sqrt{x^2-1}) \\ &= \frac{1}{x+\sqrt{x^2-1}}\left(1 + \frac{x}{\sqrt{x^2-1}}\right) \\ &= \frac{1}{\sqrt{x^2-1}}. \end{aligned}$$

■

The derivation of the other two are left as exercises. Thus

$$\begin{aligned}\frac{d}{dx}\operatorname{arsinh} x &= \frac{1}{\sqrt{x^2+1}}\\ \frac{d}{dx}\operatorname{arcosh} x &= \frac{1}{\sqrt{x^2-1}} \text{ for } x>1\\ \frac{d}{dx}\operatorname{artanh} x &= \frac{1}{1-x^2} \text{ for } |x|<1\end{aligned}$$

As with the inverse trigonometric (circular) functions, these results are needed so frequently in practice that you should endeavour to remember them; failing this, you can always extract them from the Tables (put $a = 1$ into Formulae (31), (30) and (27) in Appendix C on Page 494).

Using the chain rule, we can obtain the following more general results:

$$\begin{aligned}\frac{d}{dx}\operatorname{arsinh} f(x) &= \frac{1}{\sqrt{f(x)^2+1}}f'(x)\\ \frac{d}{dx}\operatorname{arcosh} f(x) &= \frac{1}{\sqrt{f(x)^2-1}}f'(x) \text{ for } f(x)>1\\ \frac{d}{dx}\operatorname{artanh} f(x) &= \frac{1}{1-f(x)^2}f'(x) \text{ for } |f(x)|<1\end{aligned}$$

Worked Example 4.6.3 *Determine the derivatives of*

1. $x \operatorname{arsinh} 3x$ *2.* $\operatorname{arcosh}\left(\frac{1}{x}\right)$ *3.* $x \operatorname{artanh} x$.

1. Using the product rule

$$\begin{aligned}\frac{d}{dx}x\operatorname{arsinh} 3x &= \operatorname{arsinh} 3x + x\frac{3}{\sqrt{9x^2+1}}\\ &= \operatorname{arsinh} 3x + \frac{3x}{\sqrt{9x^2+1}}\end{aligned}$$

2. Using the chain rule

$$\begin{aligned}\frac{d}{dx}\operatorname{arcosh}\frac{1}{x} &= \frac{1}{\sqrt{(\frac{1}{x})^2-1}}\cdot\left(-\frac{1}{x^2}\right)\\ &= -\frac{1}{x\sqrt{1-x^2}} \quad \text{for } 0<x<1\end{aligned}$$

3. Using the product rule

$$\begin{aligned}\frac{d}{dx}x\operatorname{artanh} 2x &= \operatorname{artanh} 2x + x\frac{1}{1-4x^2}2\\ &= \operatorname{artanh} 2x + \frac{2x}{1-4x^2}\end{aligned}$$

■

Self-help exercises

1. Use the tables in Appendix C to write down the derivative of

   (a) $\arctan x$ $[\frac{1}{x^{\cdot}+1}]$

   (b) $\arcsin x$ $[\frac{1}{\sqrt{1-x^{\cdot}}}]$

   (c) $\operatorname{arsinh} x$ $[\frac{1}{\sqrt{1+x^{\cdot}}}]$

   (d) $\operatorname{arcosh} x$ $[\frac{1}{\sqrt{x^{\cdot}-1}}]$

   (e) $\operatorname{artanh} x$ $[\frac{1}{1-x^{\cdot}}]$

2. Use the chain rule to write down the derivative of

   (a) $\arctan 2x$ $[\frac{2}{4x^{\cdot}+1}]$

   (b) $\arcsin x^2$ $[\frac{2x}{\sqrt{1-x^{\cdot}}}]$

   (c) $\operatorname{arsinh} 3x$ $[\frac{3}{\sqrt{1+9x^{\cdot}}}]$

   (d) $\operatorname{arcosh} \sqrt{x}$ $[\frac{1}{2\sqrt{x^{\cdot}-x}}]$

   (e) $\operatorname{artanh} \frac{1}{x}$ $[-\frac{1}{x^{\cdot}-1}]$

In this topic we have

- Introduced the exponential and hyperbolic functions
- Reviewed inverse functions.
- Examined the logarithm, inverse trigonometric and hyperbolic functions; and analysed their properties, graphs and derivatives.
- Discussed logarithmic differentiation.

4.7 Quick Test Number 4

Functions and their Properties — Part 1

Question **Selection**

---

1. Which of the following functions are *not* one-to-one?
   (a) $y = x^2$ on $[0, \infty)$ (b) $y = \sin x$ on $[0, \pi]$
   (c) $y = x^3$ on $(-\infty, \infty)$ (d) $y = \cos x$ on $[0, \pi]$

2. Which of the following functions have an inverse?
   (a) $y = x^2$ on $[0, \infty)$ (b) $y = \sin x$ on $[0, \pi]$
   (c) $y = x^3$ on $(-\infty, \infty)$ (d) $y = \cos x$ on $[0, \pi]$

3. $\log(ab) = \log a \log b$? (a) True (b) False

4. $\log(a + b) = \log a + \log b$? (a) True (b) False

5. Which of the following is correct?
   (a) $\log(3x^2) = \log 3 + \log x^2$ (b) $\log(3 + x^2) = \log 3 \log x^2$
   (c) $\log(3x^2) = \log 3 \log x^2$

6. $e^{ab} = e^a e^b$? (a) True (b) False

7. $e^{a+b} = e^a + e^b$? (a) True (b) False

8. Which of the following is correct?
   (a) $e^{3x^2} = e^3 + e^{x^2}$ (b) $e^{3+x^2} = e^3 e^{x^2}$ (c) $e^{3x^2} = e^3 e^{x^2}$

9. $\log(e^x) = x$ for all x?
   (a) True, because e^x and $\log x$ are inverse functions (b) False

10. $\log(e^{x^2-4x+2})$ equals
    (a) $x^2 - 4x + 2$ for all x (b) x (c) $x^2 - 4x + 2$ only for $x > 0$

11. $e^{\log x} = x$ for all $x > 0$?
    (a) True, because e^x and $\log x$ are inverse functions (b) False

12. When is "$e^{\log(4-x^2)} = 4 - x^2$" true?
    (a) for all x (b) only if $x > 0$ (c) only if $|x| < 2$

13. $e^{-\log x}$ equals (a) $-x$ (b) $\dfrac{1}{x}$ (c) neither

---

Functions and their Properties — Part 2

Question **Selection**

1. $e^{2\log x}$ equals (a) $2x$ (b) x^2 (c) neither

2. $\frac{\log x^{\cdot}}{\log x}$ equals (a) x (b) $\log x$ (c) 2

3. If $y = x + c$ then (a) $e^y = e^x + e^c$ (b) $e^y = ke^x$ (c) neither

4. If $\log y = \log 2x + \log 3$ then (a) $y = 2x + 3$ (b) $y = 6x$ (c) neither

5. Simplify $\log\left\{\dfrac{2^x\, x^3}{(x+5)\, e^x}\right\}$ as much as possible:

   (a) $\log 2^x + \log x^3 - \log(x+5) - \log e^x$

   (b) $x\log 2 + 3\log x - \log(x+5) - \log e^x$

   (c) $2\log x + x\log 3 - \log x + \log 5 - x\log e$

   (d) $x\log 2 + 3\log x - \log(x+5) - x$

6. $\log(x + \sqrt{1+x^2})$ equals

   (a) $\log x + \frac{1}{2}\log(1+x^2)$ (b) $\log x \cdot \log\sqrt{1+x^2}$ (c) neither

7. $\arcsin(\sin\dfrac{5\pi}{2})$ equals (a) 1 (b) $\dfrac{5\pi}{2}$ (c) $\dfrac{\pi}{2}$

8. Match the items in column 1 with the definitions in column 2.

   (a) $\sinh x$ (A) $(e^x - e^{-x})/2$

   (b) $\cosh x$ (B) 1

   (c) $\cosh^2 x + \sinh^2 x$ (C) $1 + 2\sinh^2 x$

   (d) $\operatorname{arccosh} x$ (D) $(e^x + e^{-x})/2$

   (e) $\cosh^2 x - \sinh^2 x$ (E) $\log(x + \sqrt{x^2 - 1})$

9. $e^{-\log 5}$ equals (a) -5 (b) $-\dfrac{1}{5}$ (c) $\dfrac{1}{5}$

10. $y = \arctan a + \arctan b \Rightarrow \tan y = a + b$? (a) True (b) False

Functions and their Derivatives

Question **Selection**

---

1. $\dfrac{d}{dx}(\log 3x)$ equals (a) $\dfrac{1}{x}$ (b) $\dfrac{3}{x}$ (c) $\dfrac{1}{3x}$ (d) $\dfrac{1}{3}+\dfrac{1}{x}$

2. If y is a function of x, $\dfrac{d}{dx}(\log y)$ equals

   (a) $\dfrac{1}{y}$ (b) $1/\dfrac{dy}{dx}$ (c) $\dfrac{1}{y}\dfrac{dy}{dx}$

3. $\dfrac{d}{dx}\{\log(x^2+1)\}$ equals (a) $\dfrac{1}{x^2+1}$ (b) $\dfrac{2x}{x^2+1}$ (c) neither

4. $\dfrac{d}{dx}(\arcsin x) = \arccos x$? (a) True (b) False

5. $\arcsin x = \dfrac{1}{\sqrt{1-x^2}}$? (a) True (b) False

6. $\dfrac{d}{dx}(\arctan x)$ equals (a) $\operatorname{arcsec}^2 x$ (b) $\dfrac{1}{1-x^2}$ (c) $-\dfrac{1}{\sin^2 x}$

   (d) $\dfrac{1}{1+x^2}$ (e) none of these

7. $\dfrac{d}{dx}(\arctan 3x)$ equals

   (a) $\dfrac{1}{1+3x^2}$ (b) $\dfrac{1}{1+9x^2}$ (c) $\dfrac{3}{1+9x^2}$ (d) $\dfrac{3}{1+x^2}$

8. $\dfrac{d}{dx}(\sinh x)$ equals (a) $\arcsin x$ (b) $\cosh x$ (c) $-\cosh x$

9. $\dfrac{d}{dx}(\sinh^3 4x)$ equals

   (a) $3\cosh^2 4x$ (b) $3\sinh^2 4x + 4\cosh 4x$
   (c) $\cosh^3 4x$ (d) $12\sinh^2 4x\cosh 4x$

10. $\dfrac{d}{dx}(\cosh x)$ equals (a) $\arctan x$ (b) $\sinh x$ (c) $-\sinh x$

11. $\dfrac{d}{dx}\{\cosh(x^2+1)\}$ equals

    (a) $\sinh(x^2+1)$ (b) $2x\sinh(x^2+1)$ (c) $\sinh(2x^3+2x)$

12. $\dfrac{d}{dx}\{\sinh(x^2+1)\}$ equals

    (a) $\cosh(2x^3+2x)$ (b) $\cosh(x^2+1)$ (c) $2x\cosh(x^2+1)$

13. $\dfrac{d}{dx}\{\arctan(x^2+1)\}$ equals

(a) $\dfrac{1}{1+(x^2+1)^2}$ (b) $\dfrac{2x}{1+(x^2+1)^2}$ (c) $\dfrac{2x}{1+x^2}$ (d) $\dfrac{1}{1+(2x)^2}$

14. $\dfrac{d}{dx}\{\arcsin(x^2-1)\}$ equals

(a) $\dfrac{1}{\sqrt{1-(x^2-1)^2}}$ (b) $\dfrac{2x}{\sqrt{1-(x^2-1)^2}}$

(c) $\dfrac{2x}{\sqrt{1-x^2}}$ (d) $\dfrac{1}{\sqrt{1-(2x)^2}}$

4.8 Exercises

Before attempting any of these miscellaneous exercises, make sure that you have successfully answered all the self-help exercises appearing throughout the chapter.

1. Use a calculator to determine the value of $\frac{e^{\cdot}-1}{h}$ when

   (a) $h = 0.1$ (b) $h = 0.001$ (c) $h = 0.000001$.

2. Sketch the graphs of:

   (a) $y = 1 - 2^{-x}$
   (b) $y = 3^{-x} + 1$
   (c) $y = 2(1 - 3^{-x})$
   (d) $y = 1 + (\frac{1}{2})^x$
   (e) $y = 3 - 2^x$
   (f) $y = 2(1 + e^{-x})$
   (g) $y = 2(1 - e^{-x})$
   (h) $y = 3e^x + 1$
   (i) $y = 3 - 2e^x$
   (j) $y = \dfrac{1}{3 - 2e^{-x}}$
   (k) $y = \dfrac{1}{1 + 3e^x}$.

3. Determine the derivatives of each of the following functions:

   (a) $e^{x^{\cdot}+2}$
   (b) $e^{\sin x}$
   (c) xe^{-2x}
   (d) x^3e^{2x}
   (e) $e^{\sqrt{x^{\cdot}+1}}$
   (f) $e^{3x}\sin 2x$
   (g) $\sqrt{1 + e^{2x}}$
   (h) $\dfrac{1}{\sqrt{1 + e^{-3x}}}$
   (i) $(e^{2x} - 2e^x)^2$
   (j) $e^{-2x}\tan 3x$
   (k) $e^{2x}\cos^2 3x$
   (l) $\dfrac{e^x + e^{-x}}{e^x - e^{-x}}$.

4. Verify that $y = Ae^{-2x} + Be^{3x}$ satisfies the differential equation

$$y'' - y' - 6y = 0.$$

5. Verify that $y = (Ax + B + x^4)e^{-2x}$ satisfies the differential equation

$$y'' + 4y' + 4y = 12x^2e^{-2x}.$$

6. Consider the function defined by $f(x) = e^{-x} - 1 + x$. Show that

   (a) $f(0) = 0$

   (b) $f(x)$ is an increasing function on $[0, \infty)$.

   Deduce that $e^{-x} > 1 - x$ for $x > 0$.

7. The population $P(t)$ at time t hours of a certain colony of bacteria is given by

$$P(t) = P_0 e^{kt} \text{ where } P_0 \text{ and } k \text{ are constants.}$$

Initially (that is, when $t = 0$) the population is 120 and after three hours it is 200.

   (a) Determine P_0 and k.

   (b) Find the size of the population after a further one hour.

   (c) At what rate does the population increase for each of the following times?

      i. initially

      ii. at time $t = 3$

8. Consider the function defined by $f(x) = xe^{-2x}$.

   (a) Determine the intervals on which

      i. $f(x)$ is monotonic increasing

      ii. $f(x)$ is monotonic decreasing

      iii. the graph of $f(x)$ is concave up

      iv. the graph of $f(x)$ is concave down.

   (b) Sketch the graph of $y = f(x)$.

9. Using the definitions of $\cosh x$ and $\sinh x$, prove each of the following:

   (a) $\cosh 2x = 2\cosh^2 x - 1$

   (b) $\cosh 2x = 1 + 2\sinh^2 x$

   (c) $\sinh 2x = 2\sinh x \cosh x$

   (d) $\sinh(x + y) = \sinh x \cosh y + \cosh x \sinh y$.

10. Sketch the graphs of

   (a) $y = \operatorname{sech} x$ (Hint: First sketch the graph of $y = \cosh x$.)

   (b) $y = \coth x$ (Hint: First sketch the graph of $y = \tanh x$.)

11. Prove that $\lim_{x \to -\infty} \tanh x = -1$.

(Hint: Write $\tanh x = \frac{e^{\cdot\cdot}-1}{e^{\cdot\cdot}+1}$.)

12. Starting with the identity $\cosh^2 x - \sinh^2 x = 1$, prove that

$$\tanh^2 x + \operatorname{sech}^2 x = 1.$$

13. Determine the derivatives of each of the following functions:

    (a) $e^{2x}\sinh 3x$
    (b) $x^3\cosh 2x$
    (c) $\cosh^2 3x$
    (d) $(\sinh x - \cosh x)^2$
    (e) $\tanh^3 2x$
    (f) $\dfrac{1+\sinh x}{1-\sinh x}$
    (g) $e^{-3x}\cosh 2x$
    (h) $\sinh^3 4x$.

14. By writing $\operatorname{sech} x$ as $(\cosh x)^{-1}$ show that

$$\frac{d}{dx}(\operatorname{sech} x) = -\operatorname{sech} x \tanh x.$$

15. Prove each of the following:

    (a) $\dfrac{d}{dx}(\operatorname{cosech} x) = -\operatorname{cosech} x \coth x$
    (b) $\dfrac{d}{dx}(\coth x) = -\operatorname{cosech}^2 x$.

16. Use the identities for $\sinh(x+y)$ and $\cosh(x+y)$ to prove that

$$\tanh(x+y) = \frac{\tanh x + \tanh y}{1 + \tanh x \tanh y}.$$

17. (a) Determine exact values of $\cosh(\log 3)$, $\sinh(\log 5)$ and $\tanh(\log 2)$.
    (b) If $\cosh x = 3$, determine exact values of $\sinh x$ and $\tanh x$.
    (c) If $\sinh x = 3$, determine exact values of $\cosh x$ and $\tanh x$.
    (d) If $3\tanh x = 1$, determine exact values of $\cosh x$ and $\sinh x$.

18. Determine which of the following functions have inverses.

    For those which do have inverses find the inverse function together with its domain, range and graph.

    (a) $f : [-\frac{1}{2}, \infty) \to \mathbf{R}$, where $f(x) = \sqrt{2x+1}$
    (b) $f : \mathbf{R} \to \mathbf{R}$, where $f(x) = \frac{1}{3}(x^3 - 2)$
    (c) $f : \mathbf{R} \to \mathbf{R}$, where $f(x) = x^2 + 2x - 8$
    (d) $f : [-1, \infty) \to \mathbf{R}$, where $f(x) = x^2 + 2x - 8$
    (e) $f : (-\infty, -1] \to \mathbf{R}$, where $f(x) = x^2 + 2x - 8$.

19. Sketch the graphs of

   (a) $y = \log(x-1)$
   (b) $y = \log x^2$
   (c) $y = 3 + \log x$
   (d) $y = \frac{1}{\log x}$.

20. Assuming the result

$$\frac{d}{dx}(\log f(x)) = \frac{f'(x)}{f(x)} \text{ for } f(x) > 0$$

   prove that

$$\frac{d}{dx}(\log |f(x)|) = \frac{f'(x)}{f(x)} \text{ whenever } f(x) \neq 0.$$

   (Hint: Note that for $x < 0$, $\log |x| = \log(-x)$.)

21. Determine the inverse function of each of the following functions:

   (a) $f : \mathbf{R} \to \mathbf{R}$, where $f(x) = 2e^{-3x} + 1$
   (b) $f : (2, \infty) \to \mathbf{R}$, where $f(x) = 2\log(x-2)$
   (c) $f : \mathbf{R} \to \mathbf{R}$, where $f(x) = \log(x + \sqrt{x^2+1})$.

22. Determine the derivatives of each of the following functions:

   (a) $x^2 \log x$
   (b) $\log(1+x^3)$
   (c) $\log(1+2x)^7$
   (d) $\log \dfrac{2}{x}$
   (e) $\log(x^3 + \cosh^2 x)$
   (f) $\log x^6$
   (g) $\log^6 x$
   (h) $(\log x)^6$
   (i) $\log \frac{1+6x}{1-2x}$
   (j) $\log \sqrt{\frac{1+6x^{\cdot}}{1+2x^{\cdot}}}$
   (k) $\log \frac{e^{-\cdot\cdot} \sin x}{\sqrt{1+2x^{\cdot}}}$
   (l) $\frac{\log x}{1+\log x}$.

23. If $y = a\cos(\log x) + b\sin(\log x)$ where a and b are constants, show that $x^2y'' + xy' + y = 0$.

24. Use logarithmic differentiation to find $\dfrac{dy}{dx}$ for each of the following:

   (a) $y = 2^{-x}$
   (b) $y = 4^{x^{\cdot}}$
   (c) $y = x^{\sin x}$
   (d) $y = \dfrac{(1+x^3)^2(1+\sin x)^3}{1+2x^4}$
   (e) $y = \sqrt{\dfrac{1+x^2}{1-x^2}}$
   (f) $y = \dfrac{e^{-x^{\cdot}}\sqrt{1+x^2}}{(1+\sin^2 x)(1+e^{2x})}$.

25. Determine the derivatives of

(a) $\arcsin \dfrac{1}{x}$

(b) $\arcsin \sqrt{x}$

(c) $\arccos \dfrac{1}{\sqrt{x}}$

(d) $\arccos 2x$

(e) $\arctan \dfrac{2}{x}$

(f) $e^{-2x} \arcsin 4x$

(g) $\arctan \dfrac{x}{3} + \arctan \dfrac{3}{x}$

(h) $\arctan \dfrac{x+2}{2x-1}$

(i) $x \arccos x - \sqrt{1-x^2}$

(j) $x \arctan x - \log \sqrt{1+x^2}$.

26. If $\log(x^2 + y^2) = 2 \arctan \dfrac{y}{x}$ show that $(x - y)y' = x + y$.

27. (a) Prove that $\dfrac{d}{dx}(\arctan x) = \dfrac{1}{1+x^2}$.
(First write $y = \arctan x$ as $\tan y = x$ and then differentiate with respect to x.)

(b) Prove that $\dfrac{d}{dx}(\arccos x) = -\dfrac{1}{\sqrt{1-x^2}}$.
(First write $y = \arccos x$ as $\cos y = x$ and then differentiate with respect to x.)

28. (a) Show that $\operatorname{arcosh} x = \log(x + \sqrt{x^2 - 1})$.
(Hint: Write $\cosh y = x$ and, using the definition of cosh, solve for y.)

(b) Use the result from (a) to show that

$$\frac{d}{dx} \operatorname{arcosh} x = \frac{1}{\sqrt{x^2 - 1}}.$$

29. Use the tables in Appendix B to determine **exact** values of

(a) $\operatorname{arsinh} 2$ (b) $\operatorname{arcosh} 2$ (c) $\operatorname{artanh} 0.2$.

30. (a) Show that $\operatorname{artanh} x = \dfrac{1}{2} \log \dfrac{1+x}{1-x}$.
(Hint: Write $\tanh y = x$, express $\tanh y$ in terms of exponentials, and solve for y.)

(b) Use the result from (a) to show that

$$\frac{d}{dx} \operatorname{artanh} x = \frac{1}{1 - x^2}.$$

31. Determine the derivatives of

(a) $x^2 \operatorname{arsinh} 2x$

(b) $\operatorname{arcosh} \dfrac{1}{x}$

(c) $\operatorname{arcosh} \sqrt{x}$

(d) $e^{-2x} \operatorname{arsinh} x^2$

(e) $\operatorname{artanh} \dfrac{2x}{1+x^2}$

(f) $x \operatorname{arsinh} x - \sqrt{x^2+1}$

(g) $2x \operatorname{arcosh} 2x - \sqrt{4x^2-1}$.

Topic 5

Integration and its Applications

5.1 Antiderivatives

Up until now we have been mainly concerned with the problem of differentiation; that is, given a function f, find its derivative f'. We now focus our attention on the converse problem; that is, given a function f, find a function F such that $F' = f$, or in other words, given the derivative find the function. The function F is then called an **antiderivative** of f.

For example:

- If $f(x) = 2x$ then the functions $F(x)$ defined by x^2 or $x^2 + 2$ or $x^2 - 3$ or $x^2 + \pi$ are all antiderivatives; in fact, if $F(x) = x^2 + c$ where c is any constant, then F is an antiderivative of f.
- $F(x) = \sin x$ is an antiderivative of $f(x) = \cos x$.
- $F(x) = \frac{1}{3}(1+x)^3$ is an antiderivative of $f(x) = (1+x)^2$.
- $F(x) = \frac{1}{2}\tan 2x$ is an antiderivative of $f(x) = \sec^2 2x$.

Definition 5.1 *If $F(x)$ is any antiderivative of the function defined by $f(x)$ then the most general antiderivative of $f(x)$ is $F(x) + c$ where c is an arbitrary constant and we often write*

$$\boxed{\int f(x)\,dx = F(x) + c}$$

Here we are only interested in determining antiderivatives of reasonably simple functions; methods of determining antiderivatives of more involved functions are deferred until Topic 6.

For ease of reference, Table 5.1 gives the antiderivatives of some of the commonly occurring functions. In this table, a and b are constants and c is the arbitrary constant of integration.

Function	Antiderivative
$(ax+b)^n$ where $n \neq -1$	$\dfrac{(ax+b)^{n+1}}{a(n+1)} + c$
$\dfrac{1}{ax+b}$	$\dfrac{1}{a}\log\lvert ax+b\rvert + c$
e^{ax+b}	$\dfrac{1}{a}e^{ax+b} + c$
$\sin(ax+b)$	$-\dfrac{1}{a}\cos(ax+b) + c$
$\cos(ax+b)$	$\dfrac{1}{a}\sin(ax+b) + c$
$\sec^2(ax+b)$	$\dfrac{1}{a}\tan(ax+b) + c$
$\sinh(ax+b)$	$\dfrac{1}{a}\cosh(ax+b) + c$
$\cosh(ax+b)$	$\dfrac{1}{a}\sinh(ax+b) + c$
$\operatorname{sech}^2(ax+b)$	$\dfrac{1}{a}\tanh(ax+b) + c$

Table 5.1: Antiderivatives for some simple functions.

Self-help exercises

1. Write down antiderivatives for the following functions:

   (a) $3x^2+2x-\frac{1}{x^{\cdot}}$ $[x^3+x^2+\frac{1}{x}+c]$ (c) $3+\frac{1}{x}+\frac{3}{x^{\cdot}}$ $[3x+\log|x|-\frac{1}{x^{\cdot}}+c]$

   (b) $\sqrt{x}-\frac{1}{\sqrt{x}}$ $[\frac{2}{3}x^{3/2}-2\sqrt{x}+c]$

2. Write down antiderivatives for the following functions:

   (a) $(3x+5)^2$ $[\frac{1}{9}(3x+5)^3+c]$ (d) $\frac{4}{2x-3}$ $[2\log|2x-3|+c]$

   (b) $\sqrt{2x+7}$ $[\frac{1}{3}(2x+7)^{3/2}+c]$ (e) $\frac{1}{2-3x}$ $[-\frac{1}{3}\log|2-3x|+c]$

   (c) $\frac{1}{\sqrt{2x+1}}$ $[\sqrt{2x+1}+c]$ (f) $\frac{1}{(x-2)^{\cdot}}$ $[-\frac{1}{2(x-2)^{\cdot}}+c]$

3. Write down antiderivatives for the following functions:

   (a) $\sin 3x+\cos 5x$ $[-\frac{1}{3}\cos 3x+\frac{1}{5}\sin 5x+c]$

   (b) $\sec^2 3x-e^{2x}-e^{-3x}+\frac{1}{4e^{\cdot\cdot}}$ $[\frac{1}{3}\tan 3x-\frac{1}{2}e^{2x}+\frac{1}{3}e^{-3x}-\frac{1}{8}e^{-2x}+c]$

5.2 The Definite Integral

5.2.1 Definition of a Definite Integral

Consider a function f which is defined on the closed interval $[a,b]$. We do not require the function to be continuous but insist that it is bounded and does not have too many discontinuities (in fact we shall assume that f has at most a finite number of discontinuities). Note that the function defined by $f(x)=\frac{1}{x}$ on $[0,1]$ does not satisfy our conditions (why?), but the function depicted in Figure 5.1 does satisfy these conditions on $[a,b]$.

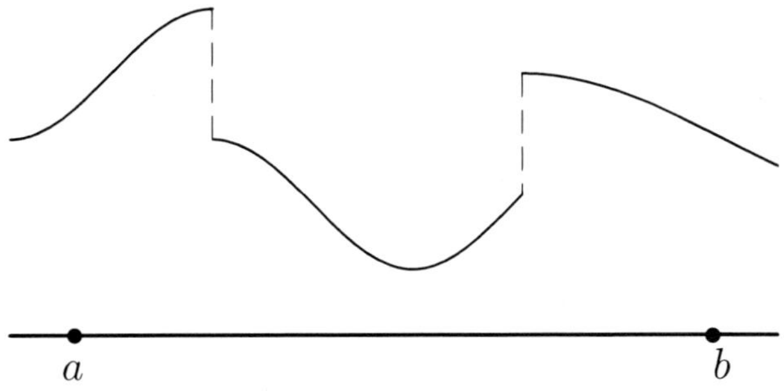

Figure 5.1: Piecewise continuous function.

Consider a partition $\mathcal{P}$ of $[a,b]$ into n subintervals defined by

$$a=x_0<x_1<x_2<\cdots<x_{i-1}<x_i<\cdots<x_{n-1}<x_n=b.$$

The lengths $\Delta x_1, \Delta x_2, \Delta x_3, \ldots, \Delta x_n$ of the subintervals $[x_0, x_1]$, $[x_1, x_2]$, $[x_2, x_3]$, ..., $[x_{n-1}, x_n]$ associated with the partition $\mathcal{P}$ need not be the same. In each subinterval $[x_{i-1}, x_i]$ choose **any** point ξ_i; that is, $\xi_i \in [x_{i-1}, x_i]$. The sum

$$\sum_{i=1}^{n} f(\xi_i)\Delta x_i = \sum_{i=1}^{n} f(\xi_i)(x_i - x_{i-1})$$

is called a **Riemann sum** for the function f on $[a, b]$ corresponding to the partition $\mathcal{P}$. Figure 5.2 gives a geometrical interpretation of this Riemann sum.

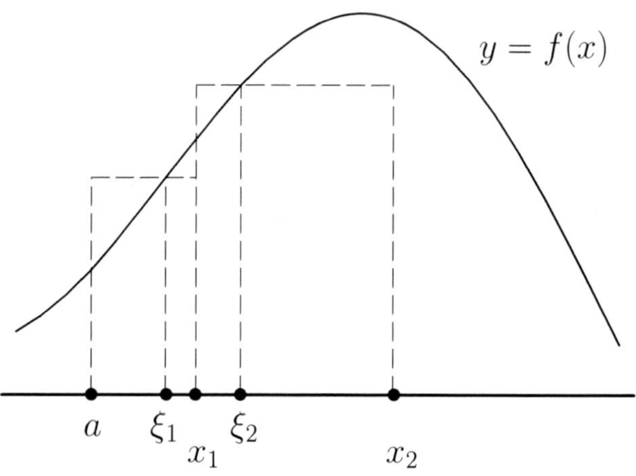

Figure 5.2: Geometrical interpretation of Riemann sum.

Note that

- If the function is always positive on $[a, b]$ this Riemann sum gives an approximation to the area between the graph of $y = f(x)$, the x-axis and the ordinates $x = a$ and $x = b$.
- If the function is always negative on $[a, b]$ the **negative** value of this Riemann sum gives an approximation to the area between the graph of $y = f(x)$, the x-axis and the ordinates $x = a$ and $x = b$.

Worked Example 5.2.1 *Consider the function defined by*

$$f(x) = x$$

on $[1, 2]$.
Determine Riemann sums corresponding to the partition of $[1, 2]$ *into* n **equal** *subintervals and where*

1. ξ_i *is the left-hand endpoint of each subinterval*

2. *ξ_i is the right-hand endpoint of each subinterval.*

The partition $\mathcal{P}$ of $[1, 2]$ into n equal subintervals is defined by

$$1 = x_0 < x_1 < \cdots < x_{i-1} < x_i < \cdots < x_{n-1} < x_n = 2$$

where

$$x_i = 1 + \frac{i}{n}.$$

1. For the subinterval $[x_{i-1}, x_i]$ the left-hand endpoint is

$$x_{i-1} = 1 + \frac{i-1}{n}$$

so that the Riemann sum with $\xi_i = x_{i-1} = 1 + \frac{i-1}{n}$ is

$$\begin{aligned}
&\sum_{i=1}^{n} f(\xi_i)(x_i - x_{i-1}) \\
&\quad = \sum_{i=1}^{n} \xi_i \frac{1}{n} \\
&\quad = \frac{1}{n} \sum_{i=1}^{n} \left(1 + \frac{i-1}{n}\right) \\
&\quad = \frac{1}{n} \left(\sum_{i=1}^{n} 1 + \sum_{i=1}^{n} \frac{i-1}{n}\right) \\
&\quad = \frac{1}{n} \left[n + \frac{1}{n} \{1 + 2 + 3 + \cdots + (n-1)\}\right] \\
&\quad = \frac{1}{n} \left\{n + \frac{1}{n} \frac{n}{2} (n-1)\right\} \\
&\quad = 1 + \frac{1}{2n}(n-1) \\
&\quad = \frac{3}{2} - \frac{1}{2n}
\end{aligned}$$

(The Riemann sum we have just calculated is often referred to as the **lower Riemann sum** as it uses the value of ξ_i in each subinterval corresponding to the minimum value of $f(x)$ in that subinterval; see Figure 5.3.)

2. For the subinterval $[x_{i-1}, x_i]$ the right-hand endpoint is

$$x_i = 1 + \frac{i}{n}$$

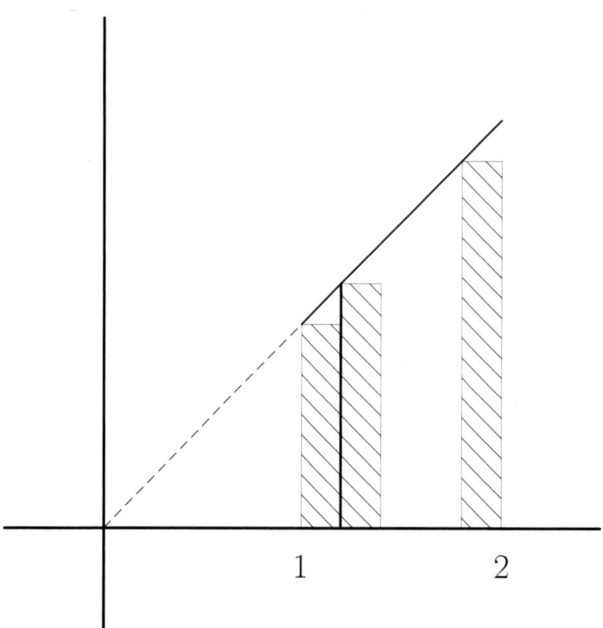

Figure 5.3: The lower Riemann sum.

so that the Riemann sum with $\xi_i = x_i = 1 + \frac{i}{n}$ is

$$\begin{aligned}
\sum_{i=1}^{n} f(\xi_i)(x_i - x_{i-1}) &= \sum_{i=1}^{n} \xi_i \frac{1}{n} \\
&= \frac{1}{n} \sum_{i=1}^{n} \left(1 + \frac{i}{n}\right) \\
&= \frac{1}{n} \left(\sum_{i=1}^{n} 1 + \sum_{i=1}^{n} \frac{i}{n} \right) \\
&= \frac{1}{n} \left\{ n + \frac{1}{n}(1 + 2 + 3 + \cdots + n) \right\} \\
&= \frac{1}{n} \left\{ n + \frac{1}{n} \frac{n}{2}(n+1) \right\} \\
&= 1 + \frac{1}{2n}(n+1) \\
&= \frac{3}{2} + \frac{1}{2n}
\end{aligned}$$

(This Riemann sum is often referred to as the **upper Riemann sum** as it uses the value of ξ_i in each subinterval corresponding to the maximum value of $f(x)$ in that subinterval; see Figure 5.4.)

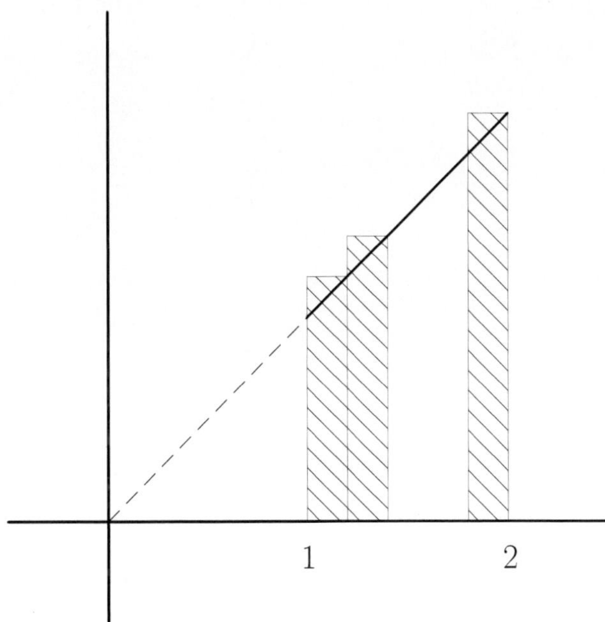

Figure 5.4: The upper Riemann sum.

■

Note that in the preceding example both Riemann sums tend to the same value $\frac{3}{2}$ as $n \to \infty$; that is, as we increase the number of subintervals so that the width of each tends to 0, the Riemann sums both tend to the same value.

We return now to the general case of a function defined by $f(x)$ on $[a, b]$. For each partition $\mathcal{P}$ of $[a, b]$ (that is, for each choice of

$$a = x_0, x_1, x_2, \ldots, x_{n-1}, x_n = b$$

and for each choice of $\xi_k \in [x_{k-1}, x_k]$) there is a corresponding Riemann sum

$$\sum_{k=1}^{n} f(\xi_k)(x_k - x_{k-1}) = \sum_{k=1}^{n} f(\xi_k)\Delta x_k.$$

If all these Riemann sums tend to the same value L as $n \to \infty$ (that is, as the width of the widest strip — corresponding to the largest Δx_k — tends to 0) we say that the function f is **integrable** on $[a, b]$ and we then write

$$L = \int_a^b f(x)\, dx.$$

The expression $\int_a^b f(x)\, dx$ is then called the **definite integral** of $f(x)$ from a to b. The quantity b is usually referred to as the **upper limit** or the **upper**

terminal of integration; a as the **lower limit** or the **lower terminal** of integration.

If f is continuous on a closed interval $[a, b]$ then it is certainly integrable; however, continuity is not a necessary condition for integrability. In fact if f is continuous on $[a, b]$ except possibly at a finite number of points at which it has finite discontinuities then f is integrable.

The process of integration can be thought of in fairly simple terms as follows: add up all elements and take the limit as the number of elements tends to infinity (or as the widths of all elements tend to 0).

Worked Example 5.2.2 *Explain why the function defined by $f(x) = x^2$ is integrable on $[a, b]$.*

Evaluate $\int_a^b x^2\,dx$ by considering appropriate Riemann sums.

Since the function is continuous on the closed interval $[a, b]$ it is integrable.

Consequently, to evaluate $\int_a^b x^2\,dx$ we can evaluate the limit of a Riemann sum corresponding to any partition of $[a, b]$.

To simplify matters we shall partition $[a, b]$

- into n equal subintervals defined by

  $$a = x_0 < x_1 < \cdots < x_i < \cdots < x_{n-1} < x_n = b$$

  where

  $$x_i = a + i\frac{b-a}{n} = a + i\Delta$$

  and where $\Delta = \frac{b-a}{n}$ is the width of each subinterval,

- and select ξ_i to be the right-hand endpoint of each subinterval; that is, for the subinterval $[x_{i-1}, x_i]$ we shall take $\xi_i = x_i = a + i\Delta$.

The corresponding Riemann sum (actually the Riemann upper

sum; see Figure 5.5) is then

$$
\begin{aligned}
\sum_{i=1}^{n} & f(\xi_i)(x_i - x_{i-1}) \\
&= \sum_{i=1}^{n} \xi_i^2 \Delta \text{ since } f(x) = x^2 \\
&= \Delta \sum_{i=1}^{n} (a + i\Delta)^2 \\
&= \Delta \sum_{i=1}^{n} (a^2 + 2ai\Delta + i^2\Delta^2) \\
&= \Delta \left(\sum_{i=1}^{n} a^2 + \sum_{i=1}^{n} 2ai\Delta + \sum_{i=1}^{n} i^2\Delta^2 \right) \\
&= \Delta \left(na^2 + 2a\Delta \sum_{i=1}^{n} i + \Delta^2 \sum_{i=1}^{n} i^2 \right) \\
&= \Delta \left\{ na^2 + 2a\frac{n}{2}(n+1)\Delta + \frac{n}{6}(n+1)(2n+1)\Delta^2 \right\} \\
&= \frac{b-a}{n} \left\{ na^2 + na(n+1)\frac{b-a}{n} + \right. \\
&\qquad \left. \frac{n}{6}(n+1)(2n+1)\frac{(b-a)^2}{n^2} \right\} \\
&= (b-a) \left\{ a^2 + a(1 + \frac{1}{n})(b-a) + \right. \\
&\qquad \left. \frac{1}{6}(1 + \frac{1}{n})(2 + \frac{1}{n})(b-a)^2 \right\}
\end{aligned}
$$

As $n \to \infty$ the above Riemann sum tends to

$$
\begin{aligned}
(b-a) & \left\{ a^2 + a(b-a) + \frac{1}{3}(b-a)^2 \right\} \\
&= \frac{1}{3}(b-a)(b^2 + ab + a^2) = \frac{b^3}{3} - \frac{a^3}{3}.
\end{aligned}
$$

Thus we have shown that

$$\int_a^b x^2 \, dx = \frac{b^3}{3} - \frac{a^3}{3}.$$

■

The observant reader may have noticed that the right-hand side is expressed in terms of an antiderivative of the integrand x^2 evaluated at the lower and upper limits a and b.

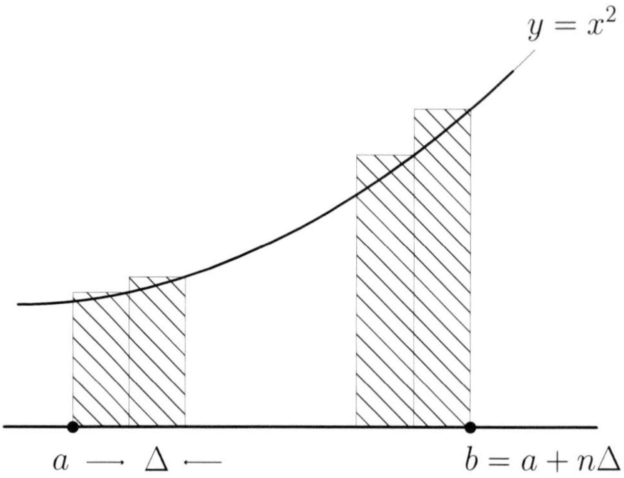

Figure 5.5: The upper Riemann sum.

This idea is followed up in the next section.

Self-help exercises

1. For **any** partition of $[a, b]$ and any choice of ξ_i in each subinterval, determine a Riemann sum for the function defined by $f(x) = k$ where k is a constant.

   Deduce that $\int_a^b k\,dx = k(b - a)$.

2. For the function defined by $f(x) = 4 + 2x$ on $[0, 6]$, calculate the Riemann sum $\sum f(\xi_i)\Delta x_i$ corresponding to the partition $0 < 2 < 2.5 < 4 < 5 < 6$ and the choice of points

$$\xi_1 = 1, \xi_2 = 2.25, \xi_3 = 3, \xi_4 = 4.5, \xi_5 = 6.$$

[60.25]

5.2.2 The Fundamental Theorem of Calculus

Calculating the value of $\int_a^b f(x)\,dx$ by evaluating Riemann sums (as in Worked Example 5.2.2) is extremely difficult for all but very simple functions. However, all is not lost! It turns out that the definite integral is related to an antiderivative of $f(x)$.

Theorem 5.1 *If f is a continuous function on $[a, b]$ and F is an antiderivative of f, then*

$$\boxed{\int_a^b f(x)\,dx = F(b) - F(a)}$$

(Readers who are interested in the proof of this important result can see an introductory calculus text such as [8].)

It is usual to write $F(b) - F(a)$ as $[F(x)]_a^b$; this notation means that the value of $F(x)$ at $x = a$ is subtracted from its value at $x = b$. That is

$$\int_a^b f(x)\,dx = [F(x)]_a^b = F(b) - F(a).$$

For example:

- $\displaystyle\int_1^2 x^2\,dx = \left[\frac{x^3}{3}\right]_1^2 = \frac{8}{3} - \frac{1}{3} = \frac{7}{3}$
- $\displaystyle\int_0^2 (3x^2 - 4x + 1)\,dx = \left[x^3 - 2x^2 + x\right]_0^2 = 2 - 0 = 2$
- $\displaystyle\int_1^2 \frac{1}{x}\,dx = [\log x]_1^2 = \log 2$
- $\displaystyle\int_0^{\pi/4} \cos 2x\,dx = \left[\frac{1}{2}\sin 2x\right]_0^{\pi/4} = \frac{1}{2}\sin\frac{\pi}{2} = \frac{1}{2}.$

5.2.3 Some Properties of Definite Integrals

The following properties of the definite integral are stated without proof; they are extremely useful in evaluating integrals.

1. If f is any integrable function then

$$\boxed{\int_a^a f(x)\,dx = 0}$$

2. (Abutting intervals)

If $f(x)$ is integrable on any interval containing the three points a, b and c then

$$\boxed{\int_a^b f(x)\,dx = \int_a^c f(x)\,dx + \int_c^b f(x)\,dx}$$

3. If $f(x) \le g(x)$ for all $x \in [a, b]$ then

$$\int_a^b f(x)\,dx \le \int_a^b g(x)\,dx.$$

(See Figure 5.6.)

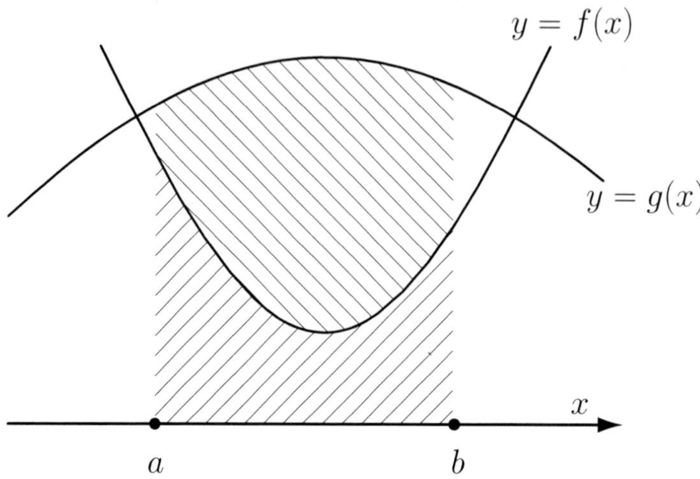

Figure 5.6: Comparison of areas.

4. If $f(x)$ is an even function — that is, if $f(-x) = f(x)$ for all x (see Figure 5.7) — then

$$\int_{-a}^{a} f(x)\,dx = 2\int_{0}^{a} f(x)\,dx.$$

Note that f is an even function if its graph is symmetrical about the y-axis.

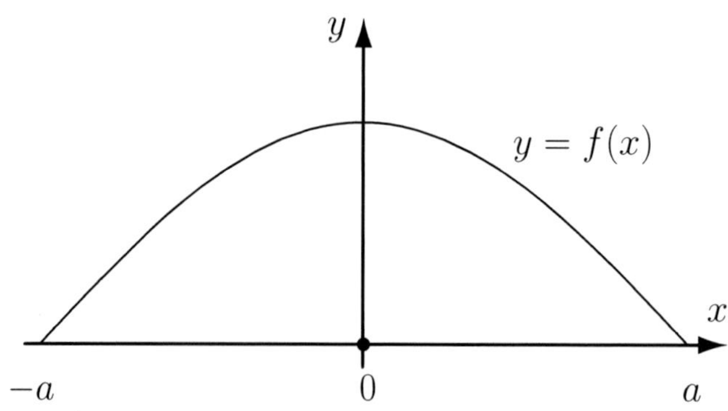

Figure 5.7: Example of an even function.

If $f(x)$ is an odd function; that is, if $f(-x) = -f(x)$ for all x (see Figure 5.8) then

$$\int_{-a}^{a} f(x)\,dx = 0.$$

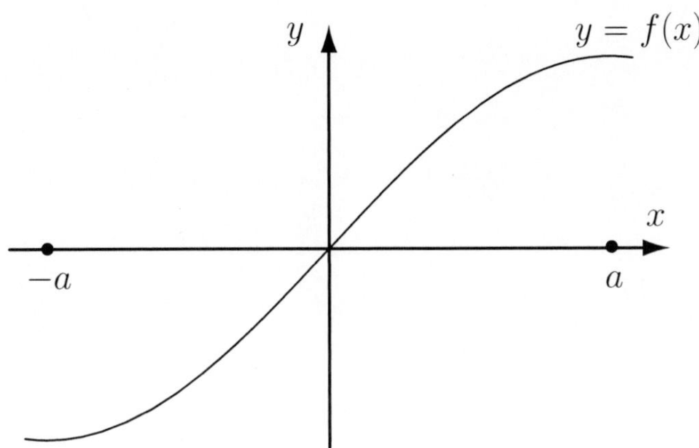

Figure 5.8: Example of an odd function.

For example:

(a) $\displaystyle\int_{-\pi/2}^{\pi/2} \cos x\, dx = 2\int_0^{\pi/2} \cos x\, dx$ since $\cos x$ is an even function.

(b) $\displaystyle\int_{-\pi/2}^{\pi/2} \sin x\, dx = 0$ since $\sin x$ is an odd function.

(c) $\displaystyle\int_{-2}^{2} (3x^2 - 6)\, dx = 2\int_0^{2} (3x^2 - 6)\, dx$ since $3x^2 - 6$ is an even function.

Self-help exercises

1. Indicate which of the following functions are even or odd functions.

(a) $2x^3 - 3x$ (b) $2x^3 + 3$ (c) $\sin 2x$ (d) $\sin x^2$

[odd, neither, odd, even]

2. Evaluate

(a) $\int_1^2 (3x^2 - 2)\,dx$ [5]

(b) $\int_0^2 \frac{6}{3x+4}\,dx$ [2 log 2.5]

(c) $\int_1^4 \frac{1}{2\sqrt{x}}\,dx$ [1]

(d) $\int_0^{\pi/4} \sin 2x\,dx$ [0.5]

(e) $\int_{-\pi/2}^{\pi/2} x \sin x^2\,dx$ [0]

(f) $\int_{-\pi/2}^{\pi/2} \cos x\,dx$ [2]

5.3 Applications of the Definite Integral

5.3.1 Areas of Regions Between Two Curves

In Section 5.2.1 we saw how definite integrals could be used to determine the area of the region between the graph of a function and the x-axis.

Recall that:

- if $f(x) \geq 0$ for $a \leq x \leq b$ then the area of the region bounded by the curve $y = f(x)$, the x-axis and the lines $x = a$ and $x = b$ (that is, the shaded region in Figure 5.9) is $\int_a^b f(x)\,dx$

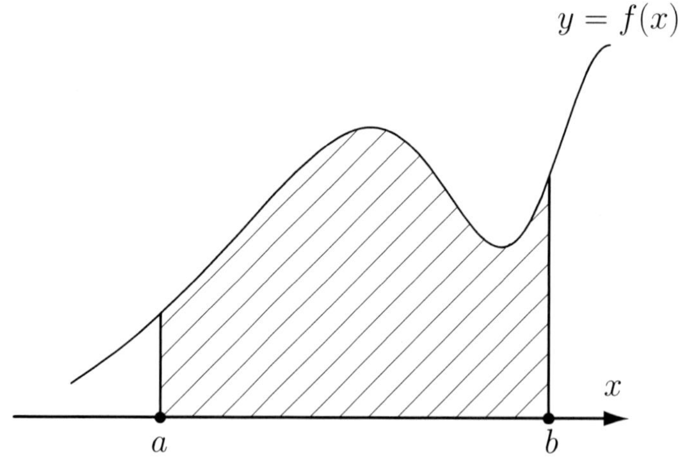

Figure 5.9: Area of region above the x-axis.

- if $f(x) \leq 0$ for $a \leq x \leq b$ then the area of the region bounded by the curve $y = f(x)$, the x-axis and the lines $x = a$ and $x = b$ (that is, the shaded region in Figure 5.10) is $-\int_a^b f(x)\,dx$.

In either case the area is $\int_a^b |f(x)|\,dx$.

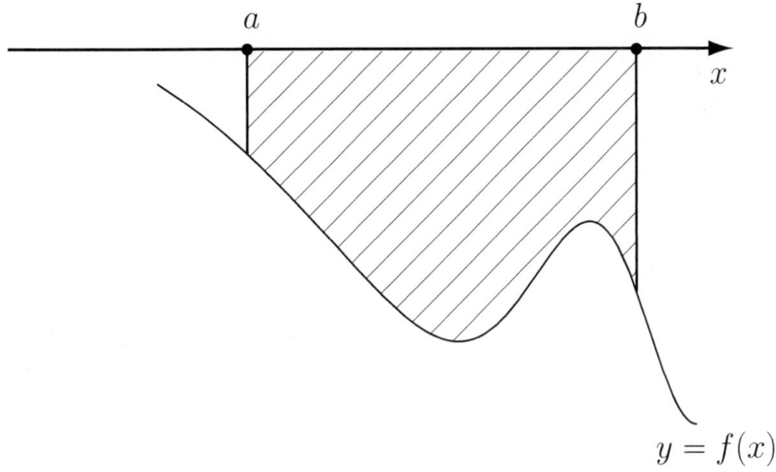

Figure 5.10: Area of region below the x-axis.

We shall now turn our attention to the more general problem of finding the area of regions bounded by two curves as illustrated in Figures 5.11 and 5.12. In this case the functions f and g are such that $f(x) \leq g(x)$ for all $x \in [a, b]$.

In Case 1 where both functions are positive it is clear that the area A of the shaded region is given by

$$\begin{aligned} A &= \text{area of region ABCD} - \text{area of region ABEF} \\ &= \int_a^b g(x)\,dx - \int_a^b f(x)\,dx \\ &= \int_a^b \{g(x) - f(x)\}\,dx \end{aligned}$$

In Case 2 it is less obvious that the area of the shaded region is given by the very same integral

$$\int_a^b \{g(x) - f(x)\}\,dx.$$

To see this we shall subdivide the region into thin vertical slices of width Δx as shown and calculate a Riemann sum which approximates the required area. On letting $\Delta x \to 0$ we then obtain an expression for the area as a definite integral.

Each element is approximately a rectangle having length $y_1 - y_2$ and width Δx so its area is approximately $(y_1 - y_2)\Delta x$. (Note that even in Case 2, the length of the rectangle is $y_1 - y_2$ — why? — and not $y_1 + y_2$ as some might expect.)

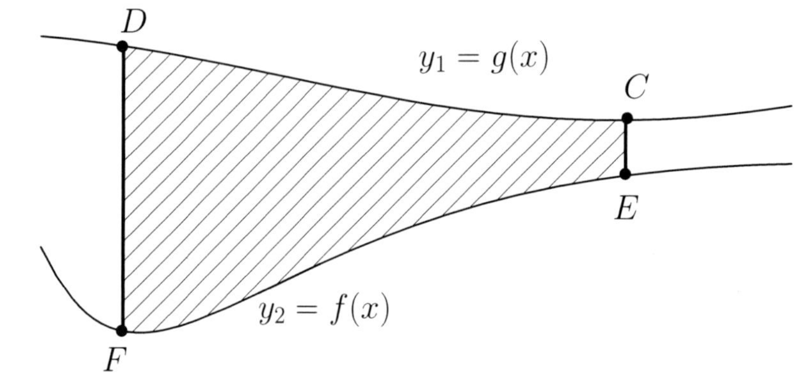

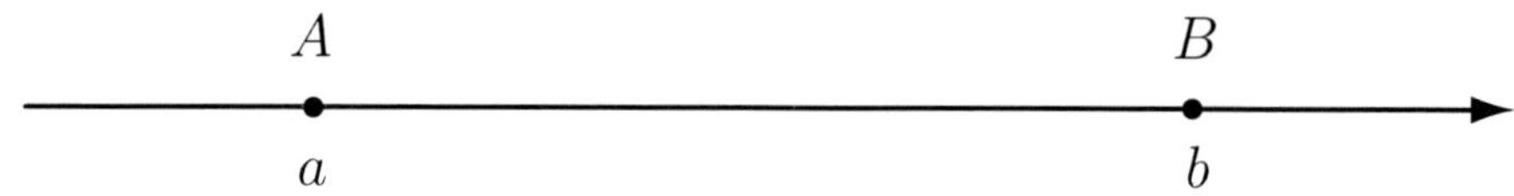

Figure 5.11: Area of general region; $0 < f(x) \leq g(x)$.

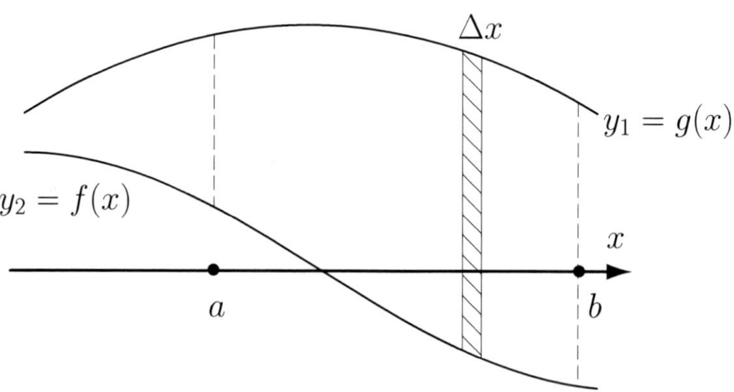

Figure 5.12: Area of general region.

Thus in each case the area of the required region is approximately the sum over all elements of all terms of the type $(y_1 - y_2)\Delta x$; that is, the area A of the region is approximated by

$$A \approx \sum_{\text{all elements}} (y_1 - y_2)\Delta x.$$

Letting $\Delta x \to 0$ gives

$$A = \int_a^b (y_1 - y_2)\, dx = \int_a^b \{g(x) - f(x)\}\, dx.$$

To determine the area A of the region between two curves as illustrated in Figure 5.13 it is necessary to consider separately the intervals $[a, c]$ and $[c, b]$ to obtain the area of the shaded region as

$$A = \int_a^c \{g(x) - f(x)\}\, dx + \int_c^b \{f(x) - g(x)\}\, dx.$$

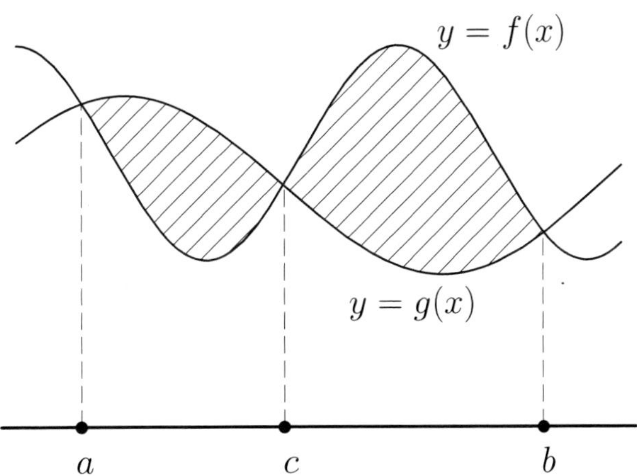

Figure 5.13: Area between general curves.

When setting up integrals which give the area of regions between curves it is sometimes not always the most convenient to subdivide the region into vertical elements (that is, elements parallel to the y-axis) but may be simpler to slice into horizontal elements (that is, elements parallel to the x-axis).

The following worked examples illustrate these ideas.

Worked Example 5.3.1 *Determine the area of the region between the curves $y = x^2$ and $y = 3x$ using*

1. vertical elements *2. horizontal elements.*

1. We first slice the region into vertical elements of width Δx as shown in Figure 5.14; the figure only shows a typical element.

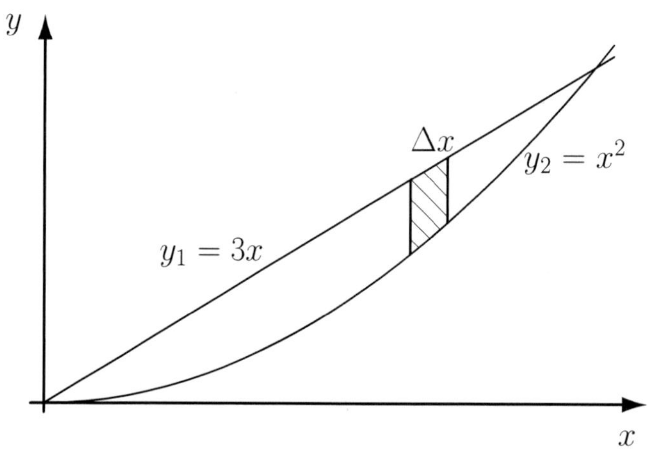

Figure 5.14: Vertical elements.

The area of this typical element is given by

$$\Delta A \approx (y_1 - y_2)\Delta x$$

where y_1 is the distance from the x-axis to the straight line $y = 3x$ and y_2 is the distance from the x-axis to the parabola $y = x^2$.
That is, $\Delta A \approx (3x - x^2)\Delta x$ so that the area A of the required region is approximated by

$$A \approx \sum_{\text{all elements}} (3x - x^2)\Delta x.$$

Letting $\Delta x \to 0$ gives

$$A = \int_0^3 (3x - x^2)\, dx = \left[\frac{3x^2}{2} - \frac{x^3}{3}\right]_0^3 = \frac{9}{2}.$$

2. Now we slice the region into horizontal elements or strips of width Δy as in Figure 5.15; once again the figure only shows a typical element.
The area of this typical element is given by

$$\Delta A \approx (x_1 - x_2)\Delta y$$

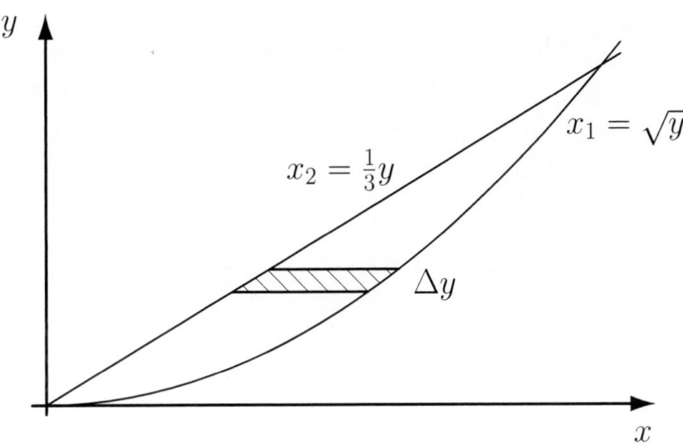

Figure 5.15: Horizontal elements.

where x_1 is the distance from the y-axis to the parabola $y = x^2$ and x_2 is the distance from the y-axis to the straight line $y = 3x$.
That is, $\Delta A \approx (\sqrt{y} - \frac{y}{3})\Delta y$ so that the area A of the required region is approximated by

$$A \approx \sum_{\text{all elements}} \left(\sqrt{y} - \frac{y}{3}\right) \Delta y.$$

Letting $\Delta y \to 0$ gives

$$A = \int_0^9 \left(\sqrt{y} - \frac{y}{3}\right) dy = \left[\frac{2}{3}y^{3/2} - \frac{y^2}{6}\right]_0^9 = \frac{9}{2}.$$

■

Worked Example 5.3.2 *Determine the area of the region between the curves $y = x^2 - 4x$ and $y = 2x$.*
From Figure 5.16 it is evident that it would be easier to slice the required region into vertical elements of width Δx. (What happens if we slice into horizontal elements?)
The area of a typical element is given by

$$\Delta A \approx (y_1 - y_2)\Delta x$$

where $y_1 = 2x$ and $y_2 = x^2 - 4x$.
That is, $\Delta A \approx (2x - x^2 + 4x)\Delta x = (6x - x^2)\Delta x$ so that the area A of the required region is approximated by

$$A \approx \sum_{\text{all elements}} (6x - x^2)\Delta x$$

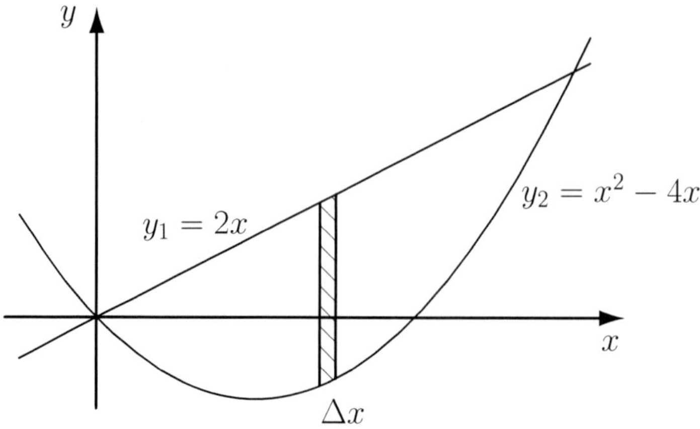

Figure 5.16: Graph to support Worked Example 5.3.2.

Letting $\Delta x \to 0$ gives

$$A = \int_0^6 (6x - x^2)\,dx = \left[3x^2 - \frac{x^3}{3}\right]_0^6 = 36.$$

■

In the preceding worked example, if we had sliced the region into horizontal elements why would it have been necessary to consider separately the regions corresponding to $y \geq 0$ and $y < 0$?

Self-help exercise

Write down (but **do not evaluate**) the integrals which give the areas of the following regions:

1. the region between the curve $y = x^2 - 4$, the x-axis and the lines $x = 0$ and $x = 4$

   $$[\int_0^2 (4 - x^2)\,dx + \int_2^4 (x^2 - 4)\,dx]$$

2. the region between the curve $y = x^2 - 4$ and the line $x + y = 2$

   $$[\int_{-3}^2 (6 - x - x^2)\,dx]$$

3. the region between the curves $x = y^2 - 5$ and $x = 1 - y^2$.

   $$[\int_{-\sqrt{3}}^{\sqrt{3}} (6 - 2y^2)\,dy]$$

5.3.2 Volumes of Revolution

Consider the region bounded by the parabola $y = x^2$ and the straight line $y = 3x$; see Figure 5.15.

Suppose we rotate this region about the y-axis. In order to determine the volume of the solid formed, we shall slice the volume perpendicular to the y-axis into thin slices. Each of these slices has the shape of a thin cylindrical washer (with beveled edges) which has thickness Δy, inner radius r and outer radius R; see Figure 5.17.

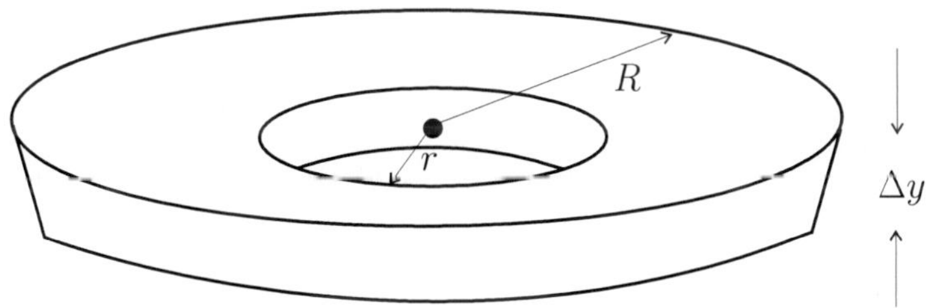

Figure 5.17: Thin cylindrical washer.

Its volume is approximately

$$\Delta V = \pi R^2 \Delta y - \pi r^2 \Delta y = \pi(R^2 - r^2)\Delta y$$

so that the volume V of the solid is approximated by

$$\boxed{V \approx \sum_{\text{all elements}} (\pi R^2 - \pi r^2)\Delta y}$$

But R is the distance from the y-axis to the parabola $y = x^2$ and r is the distance from the y-axis to the straight line $y = 3x$.

Note that we now need to express R and r in terms of y. (Why?) Thus $R = \sqrt{y}$ and $r = \frac{y}{3}$ so that

$$V \approx \sum_{\text{all elements}} \pi\left(y - \frac{y^2}{9}\right)\Delta y.$$

Letting $\Delta y \to 0$ gives

$$V = \pi\int_0^9 \left(y - \frac{y^2}{9}\right) dy = \pi\left[\frac{y^2}{2} - \frac{y^3}{27}\right]_0^9 = \frac{27\pi}{2}.$$

How would we need to modify our working in the previous discussion if the region had been rotated about the line $x = -2$ or even the line $x = 3$? In each case the volume of a typical element would still be $\pi R^2 \Delta y - \pi r^2 \Delta y$ although here R and r would now refer to the outer and inner radii of the cylindrical washer which are not simply distances from the y-axis.

Worked Example 5.3.3 *Determine the volume of the solid obtained by rotating the bounded region in the first quadrant defined by*

$$\{(x, y) : x \geq 0, 2x^2 \leq y \leq 2\}$$

about

1. *the x-axis*
2. *the line $y = 2$*
3. *the line $y = -3$.*

1. Taking vertical slices, a typical element is approximately the shape of a thin cylindrical washer of width Δx, inner radius r and outer radius R. Its volume is approximately $\pi R^2 \Delta x - \pi r^2 \Delta x$.

   We now have to obtain expressions for R and r in terms of y. Clearly, from Figure 5.18, $R = 2$ and $r = y = 2x^2$.

   Thus the volume V of the solid of revolution is given by

$$V = \pi \int_0^1 (4 - 4x^4)\, dx = \pi \left[4x - \frac{4}{5}x^5\right]_0^1 = \frac{16\pi}{5}.$$

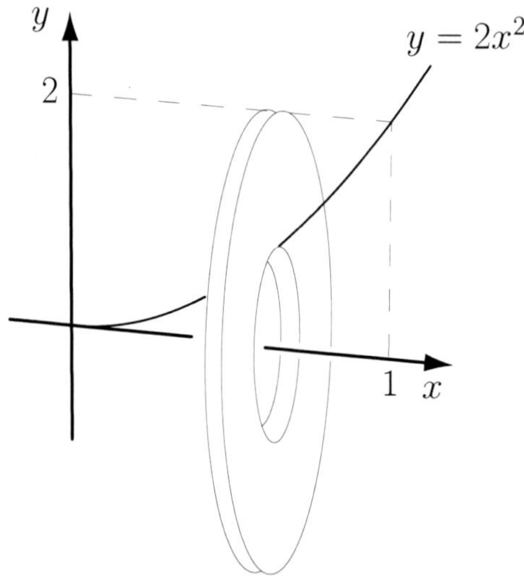

Figure 5.18: Graph in support of Worked Example 5.3.3; rotation about the x-axis.

2. When the region is rotated about the line $y = 2$ and the resulting solid is sliced using vertical elements, each element is approximately a thin cylindrical disc of width Δx and radius R (that is, there is no "hole" in the washer; see Figure 5.19).

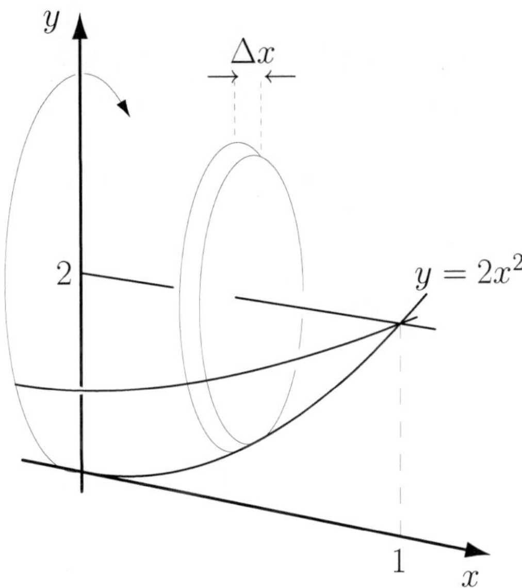

Figure 5.19: Graph in support of Worked Example 5.3.3; rotation about the line $y = 2$.

Using arguments similar to those used before, we obtain the volume of a typical element to be $\pi R^2 \Delta x$ where R is the distance from the line $y = 2$ to the parabola; that is, $R = 2 - y = 2 - 2x^2$.
This leads to the approximation

$$V \approx \sum_{\text{all elements}} \pi (2 - 2x^2)^2 \Delta x.$$

Letting $\Delta x \to 0$ then gives

$$V = \pi \int_0^1 (2 - 2x^2)^2 \, dx = \pi \int_0^1 (4 - 8x^2 + 4x^4) \, dx = \frac{32\pi}{15}.$$

3. When the region is rotated about the line $y = -3$ and the resulting solid is sliced using vertical elements, each element is approximately a thin cylindrical washer of width Δx, inner radius r and outer radius R.
The volume of a typical element is $\pi(R^2 - r^2)\Delta x$ where R is the distance from the line $y = -3$ to the line $y = 2$ and r is the distance from the line $y = -3$ to the parabola; see Figure 5.20. That is, $R = 5$ and $r = 3 + y = 3 + 2x^2$.
This leads to the approximation

$$V \approx \sum_{\text{all elements}} \pi \{5^2 - (3 + 2x^2)^2\} \Delta x$$

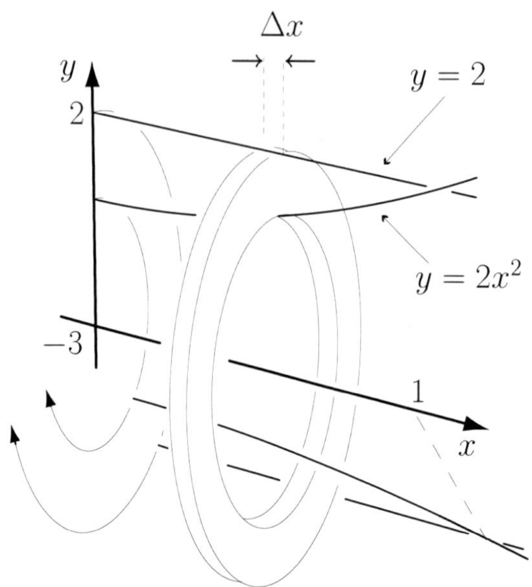

Figure 5.20: Graph in support of Worked Example 5.3.3; rotation about the line $y = -3$.

Letting $\Delta x \to 0$ gives

$$\begin{aligned} V &= \pi \int_0^1 \{25 - (3 + 2x^2)^2\}\, dx \\ &= \pi \int_0^1 (16 - 12x^2 - 4x^4)\, dx \\ &= \pi \left[16x - 4x^3 - \frac{4}{5}x^5\right]_0^1 \\ &= \frac{56\pi}{5}. \end{aligned}$$

■

The methods described in the preceding examples for determining volumes of revolutions rely on taking slices perpendicular to the axis of rotation. This does not always lead to an integral which is manageable, as is shown in the following worked example.

Worked Example 5.3.4 *The region bounded by the curve $y = 3x^2 - x^3$ and the x-axis is rotated about the y-axis. Determine the volume of the solid of revolution which is obtained.* Taking horizontal slices (that is, slices perpendicular to the axis of rotation as was done in all the previous examples; see Figure 5.21), a typical element is approximately the shape of a thin cylindrical washer of width Δy, inner radius r and outer radius R. Its volume is approximately $\pi R^2 \Delta y - \pi r^2 \Delta y$.

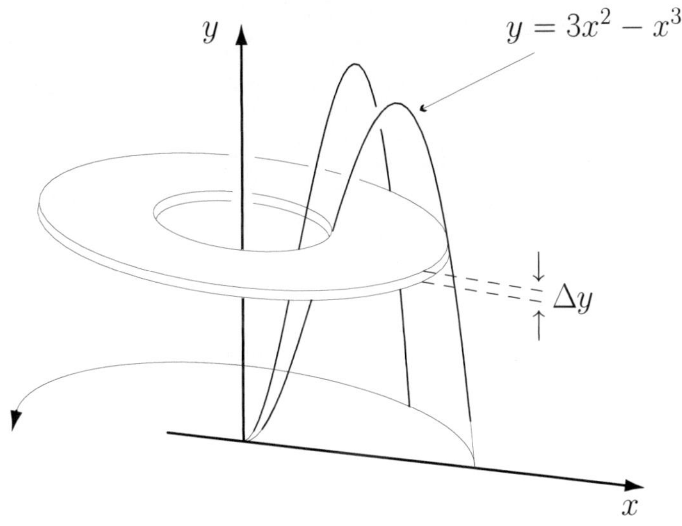

Figure 5.21: Graph in support of Worked Example 5.3.4; rotation about the y-axis — cylindrical washer.

This leads to the result

$$\boxed{V = \int_0^4 \pi(R^2 - r^2)\, dy}$$

In order to evaluate the integral we need to express R and r in terms of y; this is equivalent to solving $y = 3x^2 - x^3$ for x in terms of y and this is extremely difficult.

Fortunately the same problem does not arise if we slice the region into *vertical elements*. The element of volume obtained when a typical slice is rotated about the y-axis has the shape of a thin cylindrical shell (see Figure 5.22) having height y, inner radius r and outer radius R and consequently has volume given approximately by $\pi(R^2 - r^2)y$.

Now r is the distance from the y-axis to the curve; that is,

$$r = x$$

and

$$R = r + \Delta x = x + \Delta x.$$

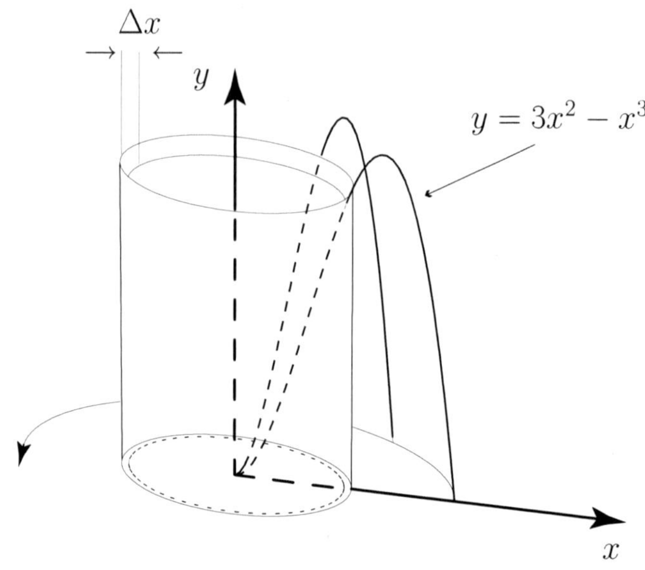

Figure 5.22: Graph in support of Worked Example 5.3.4; rotation about the y-axis — cylindrical shell.

Thus the volume of the element is approximately

$$\begin{aligned}\Delta V &= \pi\{(x + \Delta x)^2 - x^2\}y \\ &= \pi\{2x\Delta x + (\Delta x)^2\}y \\ &\approx 2\pi xy\Delta x \\ &= 2\pi x(3x^2 - x^3)\Delta x\end{aligned}$$

Letting $\Delta x \to 0$ gives

$$\begin{aligned}V &= \int_0^3 2\pi x(3x^2 - x^3)\,dx \\ &= 2\pi \int_0^3 (3x^3 - x^4)\,dx \\ &= 2\pi \left[\frac{3}{4}x^4 - \frac{1}{5}x^5\right]_0^3 = \frac{243\pi}{10}\end{aligned}$$

■

Self-help exercises

1. Determine integrals (do not evaluate these integrals) which give the volume of the solid obtained by rotating the following regions about

the given line.

(a) The region between the curve $y = 2x - x^2$ and the x-axis; rotated about the x-axis $[\pi \int_0^2 (2x - x^2)^2 \, dx]$

(b) The region between the curve $y = 2x - x^2$ and the line $y = x$; rotated about the x-axis

$[\pi \int_0^1 \{(2x - x^2)^2 - x^2\} \, dx]$

(c) The region in the first quadrant bounded by the curve $y = x^2$, the line $y = 2 - x$ and the x-axis; rotated about the x-axis.

$[\pi \int_0^1 x^4 \, dx + \pi \int_1^2 (2 - x)^2 \, dx]$

2. The region bounded by the curve $y = 2x - x^2$ and the line $y = x$ is rotated about the y-axis. Determine an integral which gives the volume of the solid of revolution which is obtained. (Hint: Use the shell method.) $[2\pi \int_0^1 x(x - x^2) \, dx]$

5.3.3 Other Applications

The previous sections have indicated how definite integrals can be used to determine areas of regions and volumes of solids of revolutions. Definite integrals are also used in numerous other situations arising in applied mathematics and engineering. For example integration can be used to determine

- the position of the centroid of a solid
- the moment of inertia of a solid about a given axis
- the length of a curve
- the surface area of a volume of revolution
- the work done by an expanding gas

or even to find the force exerted on the retaining wall of a dam due to the enclosed water.

Some examples of these applications are included in the exercises.

5.4 Approximate Integration

Generally it is not possible to express an integral in terms of elementary functions so that approximate methods are needed to evaluate certain definite integrals. Such methods are also necessary when the function being integrated is only known at a finite number of points.

Throughout this section we discuss two simple methods of finding numerical approximations to integrals of the form $\int_a^b f(x) \, dx$. There is no loss of generality in assuming that $f(x) \geq 0$ on $[a, b]$.

Usually the interval $[a, b]$ is subdivided into n subintervals by partitioning the interval as follows (see Section 5.2.1)

$$a = x_0 < x_1 < x_2 < \cdots < x_n = b.$$

On each of these subintervals we *approximate* the given function $f(x)$ by a low order polynomial.

Although it is not necessary, we shall assume that each subinterval $[x_i, x_{i+1}]$ has the same length $h = x_{i+1} - x_i$; the value of h is then called the **stepsize**.

For ease of notation, throughout this section we shall denote the value of the function at x_i by y_i so that $y_i = f(x_i)$. (See Figure 5.23.)

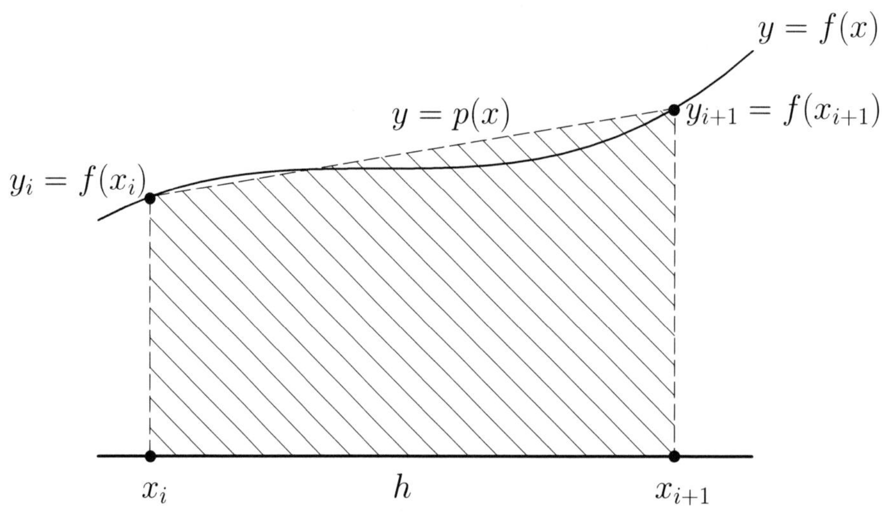

Figure 5.23: Notation for the trapezoidal rule approximation.

5.4.1 The Trapezoidal Rule

On each of the subintervals $[x_i, x_{i+1}]$ of $[a, b]$ we shall approximate the function $f(x)$ by a linear function $p(x)$ which agrees with $f(x)$ at both endpoints x_i and x_{i+1}. (See Figure 5.23.)

In Figure 5.23, $\int_{x_i}^{x_{i+1}} p(x)\,dx$ represents the area of the shaded region bounded by the trapezium. But the value of this integral is given by

$$\begin{aligned}\int_{x_i}^{x_{i+1}} p(x)\,dx &= \tfrac{h}{2}[f(x_i) + f(x_{i+1})] \\ &= \tfrac{h}{2}(y_i + y_{i+1})\end{aligned}$$

Thus we have the trapezoidal rule approximation

$$\boxed{\int_{x_i}^{x_{i+1}} f(x)\,dx \approx \frac{h}{2}(y_i + y_{i+1})}$$

Applying this approximation to each subinterval of $[a, b]$ (see Figure 5.24) we obtain

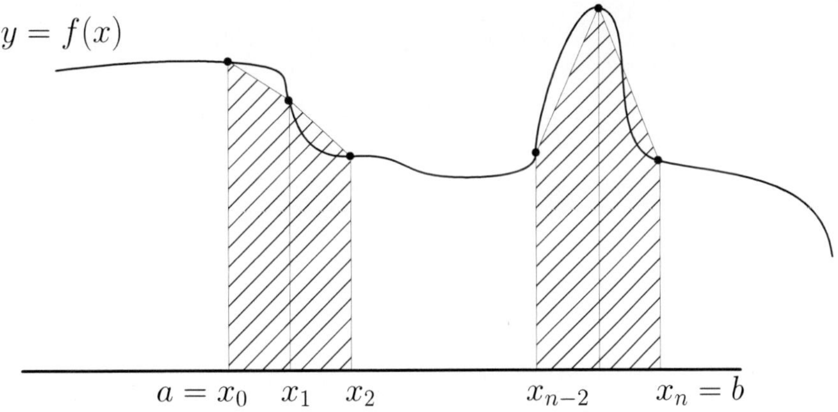

Figure 5.24: Trapezoidal rule.

$$\begin{aligned}
&\int_a^b f(x)\,dx \\
&= \int_{x_0}^{x_1} f(x)\,dx + \int_{x_1}^{x_2} f(x)\,dx + \cdots + \int_{x_{n-1}}^{x_n} f(x)\,dx \\
&\approx \frac{h}{2}(y_0 + y_1) + \frac{h}{2}(y_1 + y_2) + \cdots + \frac{h}{2}(y_{n-1} + y_n)
\end{aligned}$$

so that we have the **trapezoidal rule** approximation

$$\boxed{\int_a^b f(x)\,dx \quad \approx \quad \tfrac{1}{2}h(y_0 + 2y_1 + 2y_2 + 2y_3 + \cdots + 2y_{n-1} + y_n)}$$

Worked Example 5.4.1 *Use the trapezoidal rule with*

1. *five strips* 2. *ten strips*

to obtain an approximate value to the integral

$$\int_0^1 \frac{1}{1+x^2}\,dx.$$

Compare both these results with the exact *value of the integral.*

1. Using five strips implies that the step size h is given by

$$h = 0.2.$$

The values of

$$f(x) = \frac{1}{1+x^2}$$

at the points $x = 0$, 0.2, 0.4, 0.6, 0.8 and 1.0 (that is, the values of y_i for $i = 0, 1, 2, 3, 4, 5$) are given in Table 5.2. Therefore the five-strip trapezoidal rule approximation is given by

$$\begin{aligned}
&\int_0^1 \frac{dx}{1+x^2} \\
&\approx \frac{h}{2}(y_0 + 2y_1 + 2y_2 + 2y_3 + 2y_4 + y_5) \\
&= \frac{1}{10}(1.0000 + 2 \times 3.1687 + 0.5000) \\
&= 0.1 \times 7.8374 \\
&= 0.7837
\end{aligned}$$

where the result has been rounded to four decimal places.

x	$1+x^2$	$f(x) = \frac{1}{1+x^2}$	
0	1.0000	1.0000	
0.2	1.0400		0.9615
0.4	1.1600		0.8621
0.6	1.3600		0.7353
0.8	1.6400		0.6098
1.0	2.0000	0.5000	
		1.5000	3.1687

Table 5.2: Tabulated results for five-strip trapezoidal rule approximation.

2. Using ten strips implies the step size h is given by

$$h = 0.1.$$

The values of

$$f(x) = \frac{1}{1+x^2}$$

at the points $x = 0,\ 0.1,\ 0.2, \ldots\ ,\ 1.0$ (that is the values of y_i for $i = 0, 1, 2, \ldots, 10$) are given in Table 5.3.

Therefore the ten-strip trapezoidal rule approximation is given by

$$\begin{aligned}
&\int_0^1 \frac{dx}{1+x^2} \\
&\quad \approx \frac{h}{2}(y_0 + 2y_1 + 2y_2 + 2y_3 + \cdots + 2y_9 + y_{10}) \\
&\quad = \frac{1}{20}(1.0000 + 2 \times 7.0998 + 0.5000) \\
&\quad = 0.05 \times 15.6996 \\
&\quad = 0.7850 \text{ (rounded to four decimal places)}
\end{aligned}$$

The exact value of the integral is $\frac{1}{4}\pi$ which has the decimal approximation 0.7854 to the same four decimal places.

■

5.4.2 Simpson's Rule

Rather than approximate the function $f(x)$ by linear functions on each subinterval of $[a, b]$, as in the trapezoidal rule, a better approximation to the value of the integral $\int_a^b f(x)\,dx$ may be obtained by approximating the function by a higher degree polynomial such as a quadratic.

Here we shall approximate $f(x)$ over **two** adjacent subintervals $[x_{i-1}, x_i]$ and $[x_i, x_{i+1}]$ by a quadratic polynomial $p(x)$ which agrees with $f(x)$ at the three points where $x = x_{i-1}$, x_i and x_{i+1} (see Figure 5.25) so that

$$f(x_{i-1}) = p(x_{i-1}), f(x_i) = p(x_i) \text{ and } f(x_{i+1}) = p(x_{i+1}).$$

Note that there is only one second degree polynomial which has these properties.

There is no loss of generality in assuming that $x_{i-1} = -h, x_i = 0$ and $x_{i+1} = h$. Let $p(x) = Ax^2 + Bx + C$. Then

$$\begin{aligned}
\int_{-h}^{h} p(x)\,dx &= \left[\frac{1}{3}Ax^3 + \frac{1}{2}Bx^2 + Cx\right]_{-h}^{h} \\
&= \frac{2h}{3}(Ah^2 + 3C).
\end{aligned} \qquad (5.1)$$

x	$1+x^2$	$f(x)=\dfrac{1}{1+x^2}$	
0	1.0000	1.0000	
0.1	1.0100		0.9901
0.2	1.0400		0.9615
0.3	1.0900		0.9174
0.4	1.1600		0.8621
0.5	1.2500		0.8000
0.6	1.3600		0.7353
0.7	1.4900		0.6711
0.8	1.6400		0.6098
0.9	1.8100		0.5525
1.0	2.0000	0.5000	
		1.5000	7.0998

Table 5.3: Tabulated results for ten-strip trapezoidal rule approximation.

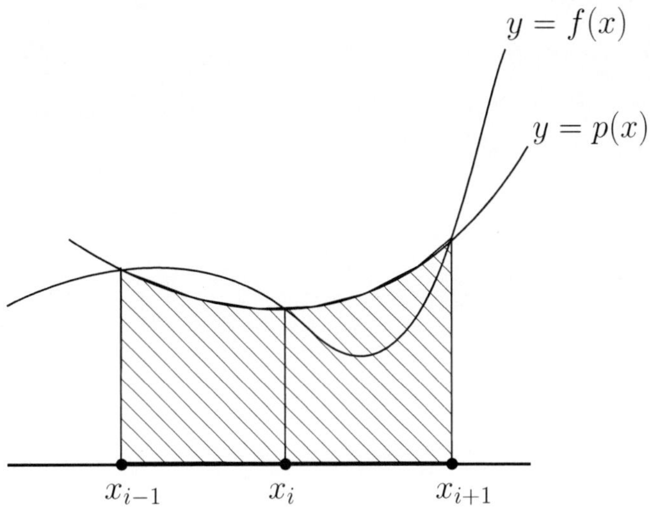

Figure 5.25: Notation for Simpson's rule approximation.

But, by definition, $p(-h) = f(-h)$ so that $Ah^2 - Bh + C = f(-h)$; $p(0) = f(0)$ so that $C = f(0)$ and $p(h) = f(h)$ so that $Ah^2 + Bh + C = f(+h)$. These may be thought of as three equations for the three unknowns A, B and C. Solving, we get $C = f(0)$ and $Ah^2 = \frac{1}{2}\{f(-h) - 2f(0) + f(h)\}$.

Thus, from Equation 5.1

$$\begin{aligned} \int_{-h}^{h} p(x)\,dx &= \frac{h}{3}\{f(-h) - 2f(0) + f(h) + 6f(0)\} \\ &= \frac{h}{3}\{f(-h) + 4f(0) + f(h)\}. \end{aligned}$$

Hence

$$\begin{aligned} \int_{x_{i-1}}^{x_{i+1}} f(x)\,dx &\approx \int_{x_{i-1}}^{x_{i+1}} p(x)\,dx \\ &= \frac{h}{3}\{f(x_{i-1}) + 4f(x_i) + f(x_{i+1})\}. \end{aligned}$$

That is, we have the Simpson's rule approximation

$$\boxed{\int_{x_{i-1}}^{x_{i+1}} f(x)\,dx \approx \tfrac{1}{3}h(y_{i-1} + 4y_i + y_{i+1})}$$

(Clearly this will give the exact value of $\int_{x_{i-1}}^{x_{i+1}} f(x)\,dx$ if $f(x)$ is a second degree polynomial; for then $p(x) = f(x)$.)

To obtain the Simpson rule approximation to $\int_a^b f(x)\,dx$, we partition $[a, b]$ into an **even** number of equal subintervals and apply the above estimate to each pair of consecutive subintervals. That is, we partition $[a, b]$ as follows:

$$a = x_0 < x_1 < x_2 < \cdots < x_{2n-2} < x_{2n-1} < x_{2n} = b.$$

The step size h then takes the value $\frac{1}{2n}(b - a)$. Consequently, the approximation on each pair of intervals is given by

$$\int_{x.}^{x.} f(x)\,dx \approx \frac{h}{3}(y_0 + 4y_1 + y_2)$$

$$\int_{x.}^{x.} f(x)\,dx \approx \frac{h}{3}(y_2 + 4y_3 + y_4)$$

$$\int_{x.}^{x.} f(x)\,dx \approx \frac{h}{3}(y_4 + 4y_5 + y_6)$$

$$\vdots$$

$$\int_{x.._{-.}}^{x..} f(x)\,dx \approx \frac{h}{3}(y_{2n-2} + 4y_{2n-1} + y_{2n})$$

Adding these results gives the **Simpson's rule** for approximating $\int_a^b f(x)dx$ using $2n$ strips.

$$\int_a^b f(x)dx$$
$$= \frac{h}{3}(y_0 + 4y_1 + 2y_2 + 4y_3 + 2y_4 + \cdots + 2y_{2n-2} + 4y_{2n-1} + y_{2n})$$
$$= \frac{h}{3}\left(y_0 + y_{2n} + 4\sum_{k=0}^{n-1} y_{2k+1} + 2\sum_{k=1}^{n-1} y_{2k}\right)$$

▌**Worked Example 5.4.2** *Approximate the value of*

$$\int_0^1 \frac{dx}{1+x^2}$$

using Simpson's rule with ten strips.

Here $h = 0.1$ and the required values of $f(x) = \frac{1}{1+x^{\cdot}}$ (that is, the values of y_i for $i = 0, 1, 2, \ldots, 10$) are given in Table 5.4. Therefore the ten-strip Simpson's rule approximation is given by

$$\begin{aligned}
\int_0^1 \frac{dx}{1+x^2} &\approx \frac{h}{3}(y_0 + 4y_1 + 2y_2 + 4y_3 + 2y_4 + 4y_5 + 2y_6 + \\
&\quad 4y_7 + 2y_8 + 4y_9 + y_{10}) \\
&= \frac{0.1}{3}(1.5000 + 4 \times 3.9311 + 2 \times 3.1687) \\
&= 0.7854 \text{ (rounded to four decimal places)}
\end{aligned}$$

x	$1+x^2$		$f(x)=\dfrac{1}{1+x^2}$	
0	1.0000	1.0000		
0.1	1.0100		0.9901	
0.2	1.0400			0.9615
0.3	1.0900		0.9174	
0.4	1.1600			0.8621
0.5	1.2500		0.8000	
0.6	1.3600			0.7353
0.7	1.4900		0.6711	
0.8	1.6400			0.6098
0.9	1.8100		0.5525	
1.0	2.0000	0.5000		
		1.5000	3.9311	3.1687

Table 5.4: Tabulated results for ten-strip Simpson's rule approximation.

(Note that the above approximation agrees with the exact result $\frac{1}{4}\pi$ to four decimal places.) ■

5.4.3 Errors in Approximate Integration

Whenever an approximation is made it is important to be able to give an estimate of the error involved. We would certainly like to be able to give estimates for the error involved when using either the trapezoidal rule or Simpson's rule to evaluate $\int_a^b f(x)dx$. Moreover, we would like to know how the accuracy of our estimate can be improved. In this course we shall only make some general remarks concerning errors.

In each of the two methods considered above we approximated $f(x)$ over a subinterval $[x_i, x_{i+1}]$ by a polynomial $p(x)$. The error due to approximating the function by the polynomial is given by

$$\boxed{\left|\int_{x_i}^{x_{i+1}} f(x)\,dx - \int_{x_i}^{x_{i+1}} p(x)\,dx\right| = \left|\int_{x_i}^{x_{i+1}} [f(x) - p(x)]\,dx\right|}$$

and is called the **truncation error**.

When the trapezoidal rule is applied to $\int_a^b f(x)\,dx$ using n strips of equal width h, the truncation error over each strip is at most $\frac{1}{12}M_S h^3$ where M_S is an upper bound for $|f''(x)|$ over that strip. If it can be shown that $|f''(x)| \leq M$ for x in the complete interval $[a, b]$, then the total truncation error is at most

$$n\frac{1}{12}Mh^3 = \frac{1}{12}(b-a)Mh^2.$$

▌**Worked Example 5.4.3** *Give an upper estimate of the truncation error involved in evaluating each of the following integrals using the trapezoidal rule with the indicated number of strips:*

1. $\int_1^{1.1} x \log x\,dx$ *using a single strip*

2. $\int_1^{5} x \log x\,dx$ *using 40 strips*

3. $\int_2^{6} x \log x\,dx$ *using 40 strips.*

1. Here $f''(x) = \frac{1}{x}$ which implies $|f''(x)| \leq 1$ on $[1, 1.1]$ so that $M = 1$ and $h = 0.1$.

   Thus the truncation error is at most $\frac{1}{12}(1)(0.1)^3 < 0.0001$.

2. Once again $|f''(x)| \leq 1$ on $[1, 5]$ so that $M = 1$. Also, as there are 40 strips, $h = \frac{4}{40} = 0.1$ as before.

   Thus the truncation error is at most $\frac{1}{12}(4)(1)(0.1)^2 \approx 0.003$.

3. Here $|f''(x)| \le \frac{1}{2} = 0.5$ on $[2,6]$ so that $M = 0.5$ and $h = \frac{4}{40} = 0.1$.
   Thus the truncation error is at most
   $$\frac{1}{12}(4)(0.5)(0.1)^2 < 0.002.$$

■

It should be noted that in most cases the *actual* truncation error is likely to be smaller than the above estimate. This is because

- the errors over the strips may tend to cancel rather than accumulate (see Figures 5.26 and 5.27)
- the upper bound M of $|f''(x)|$ over the complete interval $[a,b]$ may be considerably larger than the upper bounds of $|f''(x)|$ in many of the individual strips $[x_i, x_{i+1}]$.

A similar analysis to that for the trapezoidal rule gives the following upper estimate for the truncation error when using Simpson's rule.

> When using Simpson's rule with two strips to estimate $\int_a^b f(x)\,dx$, the truncation error is at most
> $$\frac{M_S h^5}{90}$$
> where $|f^{(4)}(x)| \le M_S$ over the two strips.
> If $2n$ strips are used the truncation error is at most
> $$\frac{1}{180}(b-a)Mh^4$$
> where M is an upper bound for $|f^{(4)}(x)|$ over the complete interval $[a,b]$.

In obtaining an estimate of $\int_a^b f(x)\,dx$ we might expect to improve the accuracy by approximating $f(x)$ by a higher degree polynomial through a larger number of points. Generally Simpson's rule, which uses a second degree polynomial through three points, gives a better approximation than the trapezoidal rule which uses a first degree polynomial through two points.

Alternatively, greater accuracy may be obtained by increasing the number of strips (or equivalently by decreasing the step size). Certainly this decreases the truncation error; however this involves a larger amount of calculation which, in turn, may introduce another source of possible error — an accumulation of rounding errors.

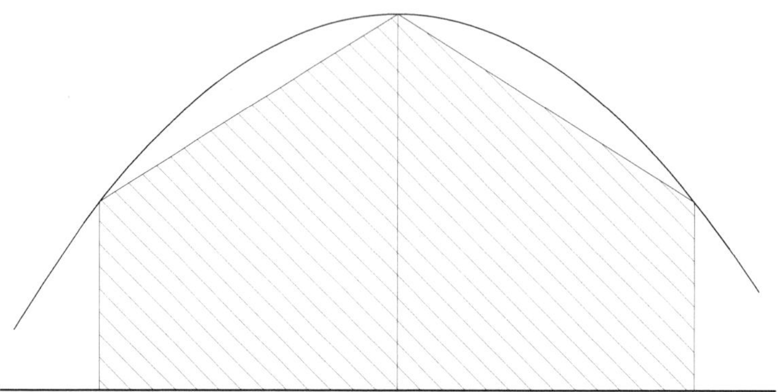

Figure 5.26: Accumulation of errors.

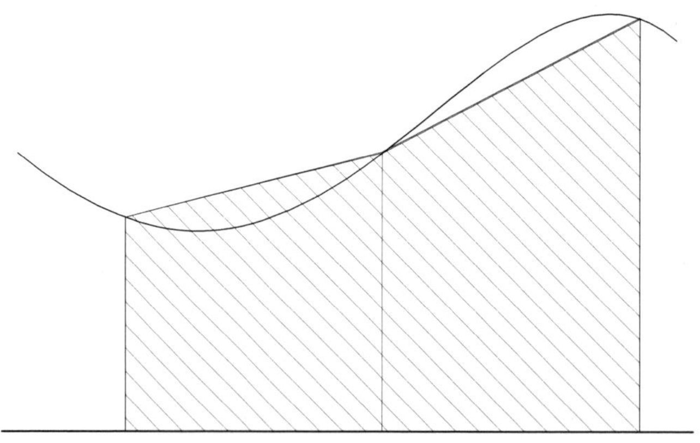

Figure 5.27: Cancellation of errors.

In this topic we have

- Reviewed antidifferentiation.
- Examined definite integration and the fundamental theorem.
- Used integration techniques to evaluate areas between curves and volumes of solids of revolution.
- Applied the trapezoidal and Simpson's rules to find numerical estimates of definite integrals.

5.5 Quick Test Number 5

Basic Mathematical Skills

Question	Selection
1. $y = \sqrt{a} + \sqrt{b} \Rightarrow y^2 = a + b$	(a) True (b) False
2. $y = \log a + \log b \Rightarrow e^y = a + b$	(a) True (b) False
3. $e^{2x+c} = e^{2x} + e^c$	(a) True (b) False

Integration

Question **Selection**

---

1. $\frac{d}{dx}\left(x^4\right) = 4x^3$ means that

   (a) $\int x^3\,dx = 4x^4 + c$ (b) $\int \frac{1}{4}x^4\,dx = x^3 + c$ (c) $\int x^3\,dx = \frac{1}{4}x^4 + c$

2. $\int x^{1/2}\,dx$ is

   (a) $\frac{1}{2}x^{3/2} + c$ (b) $\frac{1}{2}x^{-1/2} + c$ (c) $\frac{2}{3}x^{3/2} + c$ (d) $\frac{3}{2}x^{3/2} + c$

3. $\int \frac{1}{x}\,dx$ is (a) $-\frac{1}{x^2} + c$ (b) $\log|x| + c$ (c) $-\frac{x^{-2}}{2} + c$

4. $\displaystyle\int \sin x\, dx$ is (a) $\cos x + c$ (b) $-\cos x + c$ (c) $-\sin x + c$

5. $\displaystyle\int \cos x\, dx$ is (a) $\sin x + c$ (b) $-\sin x + c$ (c) $-\cos x + c$

6. $\displaystyle\int \frac{1}{5}x^2\sqrt{x+2}\, dx$ equals

   (a) $\displaystyle\frac{1}{5}\int x^2\sqrt{x+2}\, dx + c$ (b) $\displaystyle\frac{1}{5}x^2\int \sqrt{x+2}\, dx + c$

   (c) $\displaystyle\frac{1}{5}x^2\sqrt{x+2}\int dx + c$

7. $\displaystyle\frac{d}{dx}\left(\int \frac{x}{\sqrt{25-x^2}}\, dx\right)$ equals

   (a) $\displaystyle\frac{x}{\sqrt{25-x^2}}$ (b) something complicated! (c) 0

8. $\displaystyle\int_0^3 \frac{x}{\sqrt{25-x^2}}\, dx$ is (a) a function of x (b) a number (c) neither

9. $\displaystyle\frac{d}{dx}\left(\int_0^3 \frac{x}{\sqrt{25-x^2}}\, dx\right)$ equals

   (a) $\displaystyle\frac{x}{\sqrt{25-x^2}}$ (b) something complicated! (c) 0

5.6 Exercises

Before attempting any of these miscellaneous exercises, make sure that you have successfully answered all the self-help exercises appearing throughout the chapter.

1. Write down antiderivatives for each of the following functions:

   (a) $(2x-3)^3$

   (b) $(2x^2-3)^3$

   (c) $\frac{1}{2x-3}$

   (d) e^{2x-3}

   (e) $\frac{2x}{1+x^2}$

   (f) $\sin 3x$

   (g) $\cos 2x$

   (h) $\sec^2 5x$.

2. Determine upper and lower Riemann sums using n equal subintervals for the function defined by $f(x) = 4x + 2$ on $[0, 2]$.

   Hence write down the value of

   $$\int_0^2 (4x+2)\,dx.$$

3. (a) Explain why $|x^2 - 2x|$ equals $2x - x^2$ if $0 \leq x \leq 2$ but equals $x^2 - 2x$ if $x \geq 2$. Hence evaluate

   $$\int_0^4 |x^2 - 2x|\,dx.$$

   (b) Evaluate $\displaystyle\int_0^{3\pi/2} |\sin x|\,dx$. (Note that you need to consider the intervals $[0, \pi]$ and $[\pi, \frac{3}{2}\pi]$ separately. Why?)

4. Determine the area of the region between the following pairs of curves:

   (a) $y = x^2$ and $y = 3x$

   (b) $y = 2 - x^2$ and $y = x$

   (c) $y = 2 - x^2$ and $y = x^2$.

5. Determine the area of the region between the curves $y = 3x - x^2$ and $y = -x$.

6. Find the area of the region enclosed by the two curves $y = x^3 - 3x^2 + 3x$ and $y = x$.

7. Determine the area of the region enclosed by the curves $y = x^3 + 2x^2 + x - 3$ and $y = 2x - 1$.

8. (a) Derive the formula for the volume of a cone of height h and base radius r by rotating the region bounded by the straight line $y = \frac{r}{h}x$ and the x-axis between $x = 0$ and $x = h$ about the x-axis.

   (b) Derive the formula for the volume of a sphere of radius r by rotating the region bounded by the semicircle $y = \sqrt{r^2 - x^2}$ and the x-axis about the x-axis.

9. Determine the volume of the solid obtained by rotating the

region bounded by the curves $y = 5x$ and $y = x^2$ about:

(a) the x-axis

(b) the y-axis

(c) the line $x = -2$

(d) the line $y = -1$.

10. Determine the volume of the solid obtained by rotating the region bounded by the curve $y = 3x^2$ and the lines $x = 2$, $y = 0$ about

(a) the x-axis

(b) the y-axis

(c) the line $x = 2$

(d) the line $x = 4$

(e) the line $y = 12$.

11. The region bounded by the curve $y = x^4 - x^5$ and the x-axis is rotated about the y-axis. Determine the volume of the solid of revolution which is obtained.

12. The region bounded by the curve $y = x - x^2$ and the x-axis is rotated about the y-axis. Determine the volume of the solid of revolution which is obtained using

(a) the shell method

(b) the washer method.

13. **Arc-length**

The length of the curve having equation $y = f(x), a \leq x \leq b$ is

$$L = \int_a^b \sqrt{1 + y'^2}\, dx.$$

(a) Determine the length of the curve $y = x^{3/2}$ between the points $(0, 0)$ and $(4, 8)$.

(b) Determine the length of the curve $y = \frac{x^{\cdot}}{4} + \frac{1}{8x^{\cdot}}$ between the points where $x = 1$ and $x = 2$.

14. **Surface Area**

The surface area of the volume of revolution formed by rotating the curve having equation $y = f(x), a \leq x \leq b$ about the x-axis is given by

$$S = \int_a^b 2\pi y \sqrt{1 + y'^2}\, dx.$$

(a) Determine the surface area of a sphere of radius r.

(Hint: Rotate the semicircle $y = \sqrt{r^2 - x^2}$ about the x-axis.)

(b) A surface is formed by rotating the curve $y = \cosh x, 0 \leq x \leq 2$ about the x-axis. Determine an integral (**do not evaluate the integral**) which represents its surface area.

15. **Moment of Inertia**

The moment of inertia of a body about an axis is the second moment of mass of the body about this axis.

The diagram in Figure 5.28 indicates a long thin uniform rod having density ρ and length $2L$ lying along the x-axis between $x = -L$ and $x = L$. For the element of the rod which

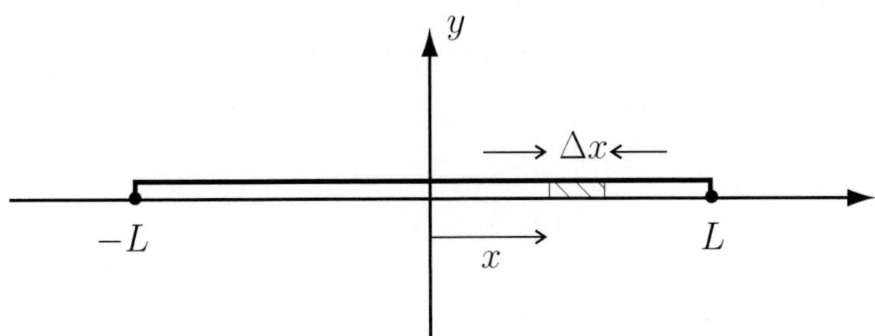

Figure 5.28: Long thin rod.

is shown (that is, the element having length Δx and distant x from the origin)

(a) What is the mass of the element?

(b) What is the second moment of mass of the element about the y-axis?

Deduce that the moment of inertia of a uniform rod of length $2L$ about an axis through its centre perpendicular to the rod is $I = \frac{1}{3}ML^2$ where M is the mass of the rod.

16. **The Force on a Dam Wall**

The diagram in Figure 5.29 shows the front view of the retaining wall of a reservoir which is full of water having density ρ.

For the element of the dam wall which is shown (that is, the element having length y, width Δx and at a depth x)

(a) What is the approximate area of the element?

(b) What is the water pressure on the element?

(c) Deduce an expression for the approximate force on the element due to the water pressure.

Hence find the total force exerted on the dam wall by the water.

(Hint: The pressure at depth h in a fluid of density ρ is proportional to ρh and pressure is force per unit area.)

17. Approximate $\int_0^1 \frac{dx}{\sqrt{1+x}}$ using Simpson's rule with ten strips.

18. Obtain approximations to

$$\int_0^2 \frac{dx}{\sqrt{1+x^2}}$$

using

(a) the trapezoidal rule

(b) Simpson's rule

with ten strips and a stepsize $h = 0.2$.

Compare the accuracy of your results with the exact result $\log(2+\sqrt{5})$.

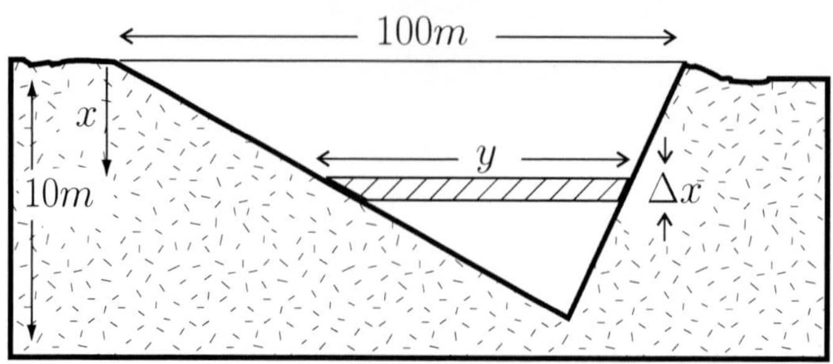

Figure 5.29: Dam wall.

19. The work done in compressing a piston is given by

$$W = \frac{1}{12}\int_0^4 F(s)\,ds.$$

Use Simpson's rule with four strips to calculate W using the following tabulated results

s	0	1	2	3	4
$F(s)$	2.8	8.9	17.4	30.1	50.9

20. Use

(a) the trapezoidal rule (b) Simpson's rule

with ten strips to obtain approximate values for

$$\int_0^1 \sin x^2\,dx.$$

21. Estimate the maximum truncation error when using

(a) the trapezoidal rule (b) Simpson's rule

to approximate $\int_1^4 \log x\,dx$ using eight strips.

22. Use the following data to estimate $\int_0^2 f(x)\,dx$.

x	0	0.2	0.4	0.6	0.8	1.0	1.2
$f(x)$	1.0000	0.8187	0.6703	0.5488	0.4493	0.3679	0.3012

x	1.4	1.6	1.8	2.0
$f(x)$	0.2466	0.2019	0.1653	0.1353

23. The period T of a pendulum of length l is given by

$$T = 2\pi c\sqrt{\frac{l}{g}}$$

where g is the acceleration due to gravity and c is a constant related to the maximum angle of swing from the mean (downward) position.

If the maximum angle of swing is 60°, the constant c is given by

$$c = \int_0^1 \frac{1}{\sqrt{1 - \frac{1}{4}\sin^2(\frac{\pi t}{2})}}\,dt.$$

Use Simpson's rule with four strips to estimate c given

t	0	0.25	0.5	0.75	1
$f(t)$	1.000	1.019	1.069	1.128	1.155

24. A homogeneous flexible cable suspended between two fixed points at the same height forms a curve called a *catenary*, having equation

$$y = a\cosh\frac{x}{a}.$$

(The constant a appearing in this equation depends on the weight per unit length of the cable as well as the tension in the cable.)

Determine the length of the cable between $x = -a$ and $x = a$.

Topic 6

Methods of Integration

6.1 Basic Integration Formulae

Many of the functions that need to be integrated have one of the standard forms given in Table 6.1 or can be reduced to one of these forms by means of a suitable substitution (see Section 6.2). In each of these formulae, a and b are constants with $a \neq 0$ and c is the arbitrary constant of integration.

With reference to Table 6.1, note that:

- the first tabulated result gives the integral of a **power of a linear function**; it cannot be applied to powers of other types of functions such as $(2x^2+1)^3$
- the second result in Table 6.1 gives the integral of the **reciprocal of a linear function** and does not apply to integrals such as

  $$\int \frac{1}{1+x^2}\,dx.$$

  The value of this integral is *not* $\log(1+x^2)$. (Verify this by differentiation!)

The following worked examples illustrate the use of these basic integration formulae.

▌**Worked Example 6.1.1** *Determine each of the following integrals:*

1. $\displaystyle\int (2x+1)^5\,dx$

2. $\displaystyle\int_0^1 \frac{1}{3x+1}\,dx$

3. $\displaystyle\int_0^3 \frac{1}{\sqrt{5x+1}}\,dx$

4. $\displaystyle\int \frac{1}{(2x+1)^2}\,dx.$

$\int (ax+b)^n\,dx$	$=$	$\dfrac{(ax+b)^{n+1}}{a(n+1)}+c$ for $n\neq -1$
$\int \dfrac{1}{ax+b}dx$	$=$	$\dfrac{1}{a}\log\lvert ax+b\rvert+c$
$\int e^{ax+b}\,dx$	$=$	$\dfrac{1}{a}e^{ax+b}+c$
$\int \sin(ax+b)\,dx$	$=$	$-\dfrac{1}{a}\cos(ax+b)+c$
$\int \cos(ax+b)\,dx$	$=$	$\dfrac{1}{a}\sin(ax+b)+c$
$\int \sec^2(ax+b)\,dx$	$=$	$\dfrac{1}{a}\tan(ax+b)+c$
$\int \cosh(ax+b)\,dx$	$=$	$\dfrac{1}{a}\sinh(ax+b)+c$
$\int \sinh(ax+b)\,dx$	$=$	$\dfrac{1}{a}\cosh(ax+b)+c$
$\int \mathrm{sech}^2(ax+b)\,dx$	$=$	$\dfrac{1}{a}\tanh(ax+b)+c$

Table 6.1: Standard integral forms.

1. Using the first result in Table 6.1
$$\int (2x+5)^5\,dx = \frac{1}{12}(2x+1)^6 + c.$$
2. Using the second result in Table 6.1
$$\int_0^1 \frac{1}{3x+1}\,dx = \left[\frac{1}{3}\log|3x+1|\right]_0^1 = \frac{1}{3}\log 4.$$
3. Using the first result in Table 6.1 we obtain
$$\int_0^3 \frac{1}{\sqrt{5x+1}}dx = \int_0^3 (5x+1)^{-1/2}\,dx = \left[\frac{2}{5}\sqrt{5x+1}\right]_0^3 = \frac{6}{5}.$$
4. Writing $1/(2x+1)^2$ in the form $(2x+1)^{-2}$, it is clear that this is just a power of a linear function so the first result in Table 6.1 applies again.
   Thus
$$\int \frac{1}{(2x+1)^2}\,dx$$
$$= \int (2x+1)^{-2}\,dx = \frac{(2x+1)^{-1}}{-2} + c = -\frac{1}{2(2x+1)} + c.$$

■

The exercises in the next worked example, while not any of the basic integration formulae in Table 6.1, are easily transformed into one of these forms.

Worked Example 6.1.2 *Determine the value of each of the following integrals:*

1. $\int e^{-2x}\cosh 4x\,dx$

2. $\int (2x^2+1)^3\,dx$

3. $\int (2\sqrt{x} - \frac{1}{\sqrt{x}})^2\,dx$

4. $\int \cos 3x \cos x\,dx.$

1. Using the identity
$$\cosh x = \frac{1}{2}(e^x + e^{-x})$$
   we can rewrite the integral as
$$\begin{aligned}\frac{1}{2}\int e^{-2x}(e^{4x}+e^{-4x})\,dx &= \frac{1}{2}\int (e^{2x}+e^{-6x})\,dx \\ &= \frac{1}{2}(\frac{1}{2}e^{2x} - \frac{1}{6}e^{-6x}) + c \\ &= \frac{1}{12}(3e^{2x} - e^{-6x}) + c.\end{aligned}$$

2. Since $(2x^2+1)^3$ is not a *power of a linear function* the integrand needs to be expanded using the binomial theorem as follows:

$$\begin{aligned}\int (2x^2+1)^3\,dx &= \int (8x^6+12x^4+6x^2+1)\,dx\\ &= \frac{8}{7}x^7+\frac{12}{5}x^5+2x^3+x+c.\end{aligned}$$

3. As in (2) we need to expand the integrand:

$$\begin{aligned}\int \left(2\sqrt{x}-\frac{1}{\sqrt{x}}\right)^2 dx &= \int \left(4x-4+\frac{1}{x}\right) dx\\ &= 2x^2-4x+\log|x|+c.\end{aligned}$$

4. Using formula (20) on Page 485 in Appendix B

$$\cos x\cos y = \frac{1}{2}\{\cos(x+y)+\cos(x-y)\}$$

we obtain

$$\begin{aligned}\int \cos 3x\cos x\,dx &= \frac{1}{2}\int (\cos 4x+\cos 2x)\,dx\\ &= \frac{1}{8}(\sin 4x+2\sin 2x)+c.\end{aligned}$$

■

6.2 Substitution Techniques

Frequently, integrals can be evaluated using a simple substitution which reduces the integral to one of the standard forms given in Section 6.1. While there are certain standard techniques available, deciding which substitution to employ comes largely from experience and this can only be gained by attempting a large number of exercises.

A commonly occurring integral has the form $\int f(g(x))g'(x)\,dx$. For example the integrals

- $\int (1+x^3)^9 3x^2\,dx$,
- $\int \frac{(\log x)^3}{x}\,dx = \int \frac{1}{x}(\log x)^3\,dx$
- $\int x\sin x^2\,dx = \frac{1}{2}\int 2x\sin x^2\,dx$,
- $\int \frac{2x}{\sqrt{1+x^2}}\,dx$

all have this form.

In such cases the substitution $v = g(x)$ will reduce the integral to the form $\int f(v)\, dv$ since

$$\int f(g(x))g'(x)\, dx \overset{v=g(x)}{=} \int f(v)\frac{dv}{dx}\, dx = \int f(v)\, dv.$$

If this latter integral is of standard form, or can be evaluated by some other method, then the original integral is easily determined. In particular, the integral $\int (f'(x)/f(x))\, dx$ can be seen to equal $\log|f(x)|$ by inspection; alternatively the substitution $v = f(x)$ will yield the same result. Namely

$$\boxed{\int \frac{f'(x)}{f(x)}\, dx \quad = \quad \log|f(x)| + c}$$

Examples:

1. $\displaystyle\int \frac{x+2}{x^2+4x-8}\, dx$

   $\displaystyle = \frac{1}{2}\int \frac{2x+4}{x^2+4x-8}\, dx = \frac{1}{2}\log|x^2+4x-8| + c.$

2. $\displaystyle\int \frac{2x}{1+x^2}\, dx = \log(1+x^2) + c$

3. $\displaystyle\int \tanh x\, dx = \int \frac{\sinh x}{\cosh x}\, dx = \log\cosh x + c$

Note that in the last two examples above, the modulus sign is superfluous as $1+x^2$ and $\cosh x$ are always positive.

Worked Example 6.2.1 *Evaluate the integral*

$$\int \frac{c^x}{\sqrt{2+3e^x}}\, dx.$$

Try the substitution $v = 2 + 3e^x$. (Why?)
Then $\dfrac{dv}{dx} = 3e^x$ so that

$$\int \frac{e^x}{\sqrt{2+3e^x}}\, dx = \frac{1}{3}\int \frac{dv}{\sqrt{v}} = \frac{2}{3}\sqrt{v} + c = \frac{2}{3}\sqrt{2+3e^x} + c.$$

■

Worked Example 6.2.2 *Determine the value of the integral*

$$\int_0^1 \frac{(\arctan x)^2}{1+x^2}\, dx.$$

Try the substitution $v = \arctan x$. (Why?)

Then $\dfrac{dv}{dx} = \dfrac{1}{1+x^2}$ so that

$$\int_0^1 \frac{(\arctan x)^2}{1+x^2}\,dx = \int_0^{\pi/4} v^2\,dv = \Big[v^3/3\Big]_0^{\pi/4} = \pi^3/192.$$

■

Worked Example 6.2.3 *Find the integral*

$$\int \frac{\sin\sqrt{x}}{\sqrt{x}}\,dx.$$

Whenever $\sqrt{x}$ occurs in the integrand the substitution $v = \sqrt{x}$, or equivalently $x = v^2$ will often lead to a simpler integral; it will certainly remove the square root. More generally, if $\sqrt{ax+b}$ occurs in the integrand the substitution $v = \sqrt{ax+b}$ is often worth trying.
Now, if $v = \sqrt{x}$ then $\dfrac{dv}{dx} = \dfrac{1}{2\sqrt{x}}$ so that

$$\begin{aligned}\int \frac{\sin\sqrt{x}}{\sqrt{x}}\,dx &= 2\int \sin\sqrt{x}\left(\frac{1}{2\sqrt{x}}\right)dx\\ &= 2\int \sin v\,dv\\ &= -2\cos v + c\\ &= -2\cos\sqrt{x} + c.\end{aligned}$$

Equivalently the substitution $x = v^2$ leads to $\dfrac{dx}{dv} = 2v$ so that

$$\int \frac{\sin\sqrt{x}}{\sqrt{x}}\,dx = \int \frac{\sin v}{v} 2v\,dv = -2\cos v + c = -2\cos\sqrt{x} + c.$$

■

Self-help exercises

1. Determine which of the following statements are true:

   (a) $\displaystyle\int \frac{1}{1+x}\,dx = \log|1+x| + c$

   (b) $\displaystyle\int \frac{1}{1+2x}\,dx = \log|1+2x| + c$

   (c) $\displaystyle\int \frac{1}{1+x^2}\,dx = \log(1+x^2) + c$

(d) $\displaystyle\int \frac{2x}{1+x^2}\,dx = \log(1+x^2) + c$

(e) $\displaystyle\int \frac{x^3}{1+x^4}\,dx = \log(1+x^4) + c$

(f) $\displaystyle\int \frac{x^3}{1+x^4}\,dx = \frac{1}{4}\log(1+x^4) + c$

(g) $\displaystyle\int \frac{\sin x}{\cos x}\,dx = \log|\cos x| + c$

[(a), (d), (f)]

2. In each of the following integrals write down a substitution which you might use in order to determine the integral.

(a) $\displaystyle\int x^3(1+x^4)^{10}\,dx$ $[u = 1 + x^4]$

(b) $\displaystyle\int x^2\sqrt{1+x^3}\,dx$ $[u = 1 + x^3]$

(c) $\displaystyle\int \frac{\sin x}{(1+\cos x)^4}\,dx$ $[u = 1 + \cos x]$

(d) $\displaystyle\int \frac{(\log x)^3}{x}\,dx$ $[u = \log x]$

(e) $\displaystyle\int \frac{1+x^2}{\sqrt{1+x}}\,dx$ $[u = 1 + x \text{ or } 1 + x = u^2\ ]$

(f) $\displaystyle\int \sin^3 x\cos x\,dx$ $[u = \sin x]$

(g) $\displaystyle\int \frac{e^x}{\sqrt{1+2e^x}}\,dx$ $[u = e^x \text{ or } u = 1 + 2e^x]$

(h) $\displaystyle\int x^2\sin x^3\,dx$ $[u = x^3]$

(i) $\displaystyle\int x^2(1-x)^{10}\,dx$ $[u = 1 - x]$

6.3 Integration by Parts

We have seen that products such as $x^2e^{-x^{\cdot}}$ and $x\sin x^2$ can be integrated by simple substitution techniques. However for more general products such as xe^{2x}, $x^2\sin x$, $x\arctan x$, $e^{3x}\sin 2x$ and $x\log x$ we need a different approach based on the familiar product rule for differentiation

$$\frac{d}{dx}(uv) = u'v + uv'.$$

This can be rewritten as

$$uv' = \frac{d}{dx}(uv) - u'v$$

which yields, on integrating both sides with respect to x, the result

$$\boxed{\int uv'\,dx = uv - \int u'v\,dx}$$

This provides us with a useful method (called *integration by parts*) of integrating certain products.

Of the two terms in the product uv' to be integrated, u needs to be differentiated and v' needs to be integrated to obtain v. Thus we choose v' to be the factor which is more easily integrated. For example, to find

1. $\int x^2 \log x\,dx$ we choose $v' = x^2$ and hence $u = \log x$

2. $\int \arctan x\,dx = \int 1.\arctan x\,dx$ we choose $v' = 1$ and hence $u = \arctan x$.

Note that when integrating functions such as $\arctan x$, $\log x$, $\arcsin x$ we write

$$\int f(x)\,dx \text{ as } \int 1.f(x)\,dx$$

and choose $v' = 1$ and $u = f(x)$.

The test of whether integration by parts is an appropriate method of evaluating a given integral or not is based on the new integral to be determined. If this new integral is no more difficult than the original one then integration by parts will usually be successful.

Worked Example 6.3.1 *Determine the value of the following integrals:*

1. $\int x^2 \log x\,dx$ *2.* $\int_0^1 \arctan x\,dx.$

1. Since x^2 is the more easily integrated of the two terms in the product, let $v' = x^2$ and $u = \log x$. (Thus $v = \frac{1}{3}x^3$ and $u' = \frac{1}{x}$.)

   On applying the integration by parts formula we get

$$\begin{aligned}\int x^2 \log x\,dx &= \frac{x^3}{3}\log x - \int \frac{x^3}{3}\frac{1}{x}\,dx \\ &= \frac{x^3}{3}\log x - \frac{1}{3}\int x^2\,dx \\ &= \frac{x^3}{3}\log x - \frac{1}{9}x^3 + c.\end{aligned}$$

2. Writing

$$\int_0^1 \arctan x\,dx = \int_0^1 1.\arctan x\,dx$$

we choose $v' = 1$ so that $v = x$ and $u = \arctan x$ so that $u' = \dfrac{1}{1+x^2}$.

Thus

$$\begin{aligned}\int_0^1 \arctan x\,dx &= [x\arctan x]_0^1 - \int_0^1 \frac{x}{1+x^2}\,dx \\ &= \arctan 1 - \left[\tfrac{1}{2}\log(1+x^2)\right]_0^1 \\ &\qquad \text{using the substitution } u = 1+x^2 \\ &= \tfrac{\pi}{4} - \tfrac{1}{2}\log 2.\end{aligned}$$

Comment: This device of writing the given integrand as a product of one times the function to be integrated is also useful for most of the inverse trigonometric and inverse hyperbolic functions.

■

For integrals such as $\int xe^{2x}\,dx$, $\int x^2\sin x\,dx$ and $\int e^{-x}\sin x\,dx$, the choice of v' is less obvious since each term in the integrand is easily integrated. For example, to find $\int x\sin x\,dx$ there are two possibilities:

1. $u = x, v' = \sin x \Rightarrow u' = 1$ and $v = -\cos x$
2. $u = \sin x, v' = x \Rightarrow u' = \cos x$ and $v = \frac{1}{2}x^2$.

However the latter case leads to

$$\int x\sin x\,dx = \frac{1}{2}x^2\sin x - \frac{1}{2}\int x^2\cos x\,dx$$

where the integral on the right is *more complicated* than the original one. On the other hand, the first of these leads to

$$\int x\sin x\,dx = -x\cos x + \int \cos x\,dx = -x\cos x + \sin x + c.$$

In general, if both factors are easily integrated and one of the factors is a polynomial, we choose u to be the polynomial factor — then u' will be a simpler polynomial. Sometimes integration by parts may need to be carried out more than once as in the following example.

Worked Example 6.3.2 *Determine the value of the integral*

$$\int (x^2 - 2x + 3) \cos x \, dx.$$

We choose $u = x^2 - 2x + 3$ and $v' = \cos x$ so that $u' = 2x - 2$ and $v = \sin x$. Thus

$$\int (x^2 - 2x + 3) \cos x \, dx = (x^2 - 2x + 3) \sin x - \int (2x - 2) \sin x \, dx.$$

In the integral on the right, take

$$u = 2x - 2, v' = \sin x \Rightarrow u' = 2, v = -\cos x$$

so that

$$\begin{aligned}
&\int (x^2 - 2x + 3) \cos x \, dx \\
&\quad = (x^2 - 2x + 3) \sin x - \left\{ -(2x - 2) \cos x - \int 2(-\cos x) \, dx \right\} \\
&\quad = (x^2 - 2x + 3) \sin x + (2x - 2) \cos x - 2 \sin x + c.
\end{aligned}$$

■

For the integral $\int e^{2x} \cos x \, dx$ the choice of u and v' is not important. If we take $u = e^{2x}, v' = \cos x$ so that $u' = 2e^{2x}$, $v = \sin x$ and define

$$I = \int e^{2x} \cos x \, dx$$

we obtain

$$\begin{aligned}
I &= \int e^{2x} \cos x \, dx \\
&= e^{2x} \sin x - \int 2e^{2x} \sin x \, dx \\
&= e^{2x} \sin x - \left(-2e^{2x} \cos x + \int 4e^{2x} \cos x \, dx \right) \\
&= e^{2x} \sin x + 2e^{2x} \cos x - 4I.
\end{aligned}$$

(You should note carefully what happens if instead we had taken $v' = 2e^{2x}$, and $u = \sin x$ in the second step of the above calculation.) This last equation is simply an algebraic equation for the unknown I. Solving for I, we get

$$I = \int e^{2x} \cos x \, dx = \frac{1}{5} e^{2x} (\sin x + 2 \cos x) + c.$$

(This integral can also be simply evaluated using complex numbers by noting that $e^{2x}\cos x = \Re e^{(2+i)x}$.)

This last example is a particular case of the following result

$$\int e^{ax}\cos bx\,dx = \frac{e^{ax}}{a^2+b^2}(a\cos bx + b\sin bx)$$

which is included, along with a similar result for $\int e^{ax}\sin bx\,dx$, on Page 495 in Appendix C. These results can be used when needed as follows:

$$\begin{aligned}\int e^{3x}\cos x\,dx &= \tfrac{1}{10}e^{3x}(3\cos x + \sin x) + c\\ \int e^{-x}\sin 2x\,dx &= -\tfrac{1}{5}e^{-x}(2\cos 2x + \sin 2x) + c\end{aligned}$$

Self-help exercises

1. Each of the following integrals can be evaluated using integration by parts.

   In each case write down your choice for u and for v'.

   (a) $\int x^3 \log x\,dx$ $[u = \log x, v' = x^3]$

   (b) $\int x \arctan x\,dx$ $[u = \arctan x, v' = x]$

   (c) $\int x^2 e^{-x}\,dx$ $[u = x^2, v' = e^{-x}]$

   (d) $\int x \sin 3x\,dx$ $[u = x, v' = \sin 3x]$

   (e) $\int \arcsin x\,dx$ $[u = \arcsin x, v' = 1]$

   (f) $\int (2x-1)\cos x\,dx$ $[u = 2x-1, v' = \cos x]$

   (g) $\int (\log x)^2\,dx$ $[u = (\log x)^2, v' = 1]$

   (h) $\int \operatorname{arsinh} x\,dx.$ $[u = \operatorname{arsinh} x, v' = 1]$

2. Use the results

$$\int e^{ax}\cos bx\,dx = \frac{e^{ax}}{a^2+b^2}(a\cos bx + b\sin bx)$$

   and

$$\int e^{ax}\sin bx\,dx = \frac{e^{ax}}{a^2+b^2}(a\sin bx - b\cos bx)$$

   to evaluate each of the integrals

(a) $\int e^{2x} \cos 3x \, dx$ $[\frac{e^{\cdot x}}{13}(2\cos 3x + 3\sin 3x)]$

(b) $\int e^{-x} \cos 2x \, dx$ $[\frac{e^{-x}}{5}(-\cos 2x + 2\sin 2x)]$

(c) $\int 2\cosh 2x \cos 3x \, dx.$ $[\frac{2}{13}(2\sinh 2x \cos 3x + 3\cosh 2x \sin 3x)]$

6.4 Reduction Formulae

Consider the integral $\int x^n e^x \, dx$ where n is a positive integer.

If we integrate by parts with $v' = e^x$ and $u = x^n$ we obtain

$$\int x^n e^x \, dx = x^n e^x - n \int x^{n-1} e^x \, dx. \tag{6.1}$$

Note that the integral on the right-hand side has the same form as the original integral but the power of x has been reduced by 1. If we denote $\int x^n e^x \, dx$ by I_n, where the subscript n denotes the power of x in the integrand, we can represent Equation 6.1 by

$$I_n = x^n e^x - nI_{n-1}.$$

This is an example of a *reduction formula*. If we apply this result repeatedly, we will eventually arrive at I_0. That is, we will reduce the given integral to one that can be easily evaluated, namely $\int e^x \, dx$.

Worked Example 6.4.1 *Use the reduction formula*

$$\int x^n e^x \, dx = x^n e^x - n \int x^{n-1} e^x \, dx$$

to evaluate the integrals

1. $\int x^2 e^x \, dx$ *2.* $\int_0^1 x^3 e^x \, dx$ *3.* $\int x^2 e^{-2x} \, dx.$

1. The required integral is the value of I_2. Using the recurrence formula developed above we have

$$\begin{aligned} \int x^2 e^x \, dx &= I_2 \\ &= x^2 e^x - 2I_1 \\ &= x^2 e^x - 2(xe^x - I_0) \\ &= x^2 e^x - 2xe^x + 2\int e^x \, dx \\ &= x^2 e^x - 2xe^x + 2e^x + c. \end{aligned}$$

2. For the definite integral given, the reduction formula has the form

$$\begin{aligned}\int_0^1 x^n e^x\,dx &= [x^n e^x]_0^1 - n\int_0^1 x^{n-1}e^x\,dx \\ &= e - n\int_0^1 x^{n-1}e^x\,dx \quad (n \geq 1).\end{aligned}$$

Thus, denoting $\int_0^1 x^n e^x\,dx$ by J_n we have

$$\begin{aligned}J_3 &= e - 3J_2 \\ &= e - 3(e - 2J_1) \\ &= -2e + 6J_1 \\ &= -2e + 6(e - J_0) \\ &= 4e - 6J_0.\end{aligned}$$

But $J_0 = \int_0^1 e^x\,dx = e - 1$ so that $J_3 = 6 - 2e$.

3. In order to apply the reduction formula to $\int x^2 e^{-2x}$ we must first transform this integral into the required form by using the substitution $v = -2x$ so that

$$\int x^2 e^{-2x}\,dx = \int (-v/2)^2 e^v(-\frac{1}{2}\,dv) = -\frac{1}{8}\int v^2 e^v\,dv.$$

$$\begin{aligned}\text{Now } I_2 &= \int v^2 e^v\,dv \\ &= v^2e^v - 2I_1 \\ &= v^2e^v - 2(ve^v - I_0) \\ &= v^2e^v - 2ve^v + 2e^v + c\end{aligned}$$

Therefore $\displaystyle\int x^2 e^{-2x}\,dx = -\frac{1}{8}(4x^2 + 4x + 2)e^{-2x} + c.$

■

Other important reduction formulae which are included in Appendix C on Page 495 are

$$\int \sin^m x\cos^n x\,dx = \frac{\sin^{m+1} x\cos^{n-1} x}{m+n} + \frac{n-1}{m+n}\int \sin^m x\cos^{n-2} x\,dx \quad (6.2)$$

$$\int \sin^m x \cos^n x\,dx = -\frac{\sin^{m-1} x \cos^{n+1} x}{m+n} + \frac{m-1}{m+n}\int \sin^{m-2} x \cos^n x\,dx \tag{6.3}$$

$$\int \tan^n x\,dx = \frac{1}{n-1}\tan^{n-1} x - \int \tan^{n-2} x\,dx \tag{6.4}$$

and

$$\int \sec^n x\,dx = \frac{1}{n-1}\sec^{n-2} x \tan x + \frac{n-2}{n-1}\int \sec^{n-2} x\,dx. \tag{6.5}$$

Note that

- In Equation 6.2, the power of $\cos x$ is reduced by 2 on each application of the reduction formula; consequently, successive applications of this will ultimately either remove the cosine term from the integral (if n is even) or reduce it to $\cos x$ (if n is odd). Similar comments apply to Equation 6.3 in relation to the power of $\sin x$.

- The power of the term $\tan x$ is reduced by 2 on each application of Equation 6.4 so that $\int \tan^n x\,dx$ can always be reduced to either

  1. $\displaystyle\int 1\,dx = x$ if n is even
  2. $\displaystyle\int \tan x\,dx = \int \frac{\sin x}{\cos x}\,dx = -\log|\cos x|$ if n is odd.

- The power of $\sec x$ is reduced by 2 on each application of Equation 6.5, so that $\int \sec^n x\,dx$ can ultimately be reduced to either

  1. $\displaystyle\int 1\,dx = x$ if n is even
  2. $\displaystyle\int \sec x\,dx = \log|\sec x + \tan x|$ if n is odd.

- These reduction formulae are often significantly simplified for certain definite integrals (see Worked Example 6.4.2).

▮ **Worked Example 6.4.2** *Evaluate each of the following integrals:*

1. $\displaystyle\int_0^{\pi/2} \sin^6 x \cos^3 x\,dx$

2. $\displaystyle\int_0^{\pi} \sin^6 x\,dx$

3. $\displaystyle\int_0^{\pi/4} \tan^4 x\,dx.$

1. First note that the reduction formula 6.2 reduces to

$$\int_0^{\pi/2} \sin^m x \cos^n x\,dx = \frac{n-1}{m+n}\int_0^{\pi/2} \sin^m x \cos^{n-2} x\,dx$$

as the first term on the right-hand side is zero when either $x = 0$ or $x = \pi/2$.

Thus

$$\int_0^{\pi/2} \sin^6 x \cos^3 x \, dx = \frac{2}{9} \int_0^{\pi/2} \sin^6 x \cos x \, dx.$$

Although we could now apply the reduction formula 6.3 to successively reduce the power of the sine term in the integrand, it is much simpler to now evaluate this latter integral using the substitution $u = \sin x$ to obtain

$$\int_0^{\pi/2} \sin^6 x \cos^3 x \, dx$$

$$= \frac{2}{9} \int_0^{\pi/2} \sin^6 x \cos x \, dx = \frac{2}{9} \int_0^1 u^6 \, du = \frac{2}{63}.$$

2. Once again the reduction formula 6.3 simplifies, on putting $n = 0$, to

$$\int_0^{\pi} \sin^m x \, dx = \frac{m-1}{m} \int_0^{\pi} \sin^{m-2} x \, dx.$$

Thus

$$\begin{aligned} \int_0^{\pi} \sin^6 x \, dx &= \frac{5}{6} \int_0^{\pi} \sin^4 x \, dx \\ &= \frac{5}{6} \frac{3}{4} \int_0^{\pi} \sin^2 x \, dx \\ &= \frac{5}{6} \frac{3}{4} \frac{1}{2} \int_0^{\pi} 1 \, dx \\ &= \frac{5}{6} \frac{3}{4} \frac{1}{2} \pi \\ &= \frac{5}{16} \pi. \end{aligned}$$

3. The reduction formula 6.4 applied to a definite integral over the interval $[0, \pi/4]$ becomes

$$\int_0^{\pi/4} \tan^n x \, dx = \frac{1}{n-1} - \int_0^{\pi/4} \tan^{n-2} x \, dx.$$

Therefore

$$\begin{aligned} \int_0^{\pi/4} \tan^4 x \, dx &= \frac{1}{3} - \int_0^{\pi/4} \tan^2 x \, dx \\ &= \frac{1}{3} - (1 - \int_0^{\pi/4} 1 \, dx) \\ &= -\frac{2}{3} + \frac{1}{4} \pi. \end{aligned}$$

■

Self-help exercises

Use the reduction formulae in Appendix C on Page 495 to write down the values of

1. $\displaystyle\int_0^{\pi/2} \sin^2 x\, dx$ $[\pi/4]$
2. $\displaystyle\int_0^{\pi} \cos^2 x\, dx$ $[\pi/2]$
3. $\displaystyle\int_0^{2\pi} \cos^4 x\, dx$ $[3\pi/4]$
4. $\displaystyle\int_0^{\pi} \sin^2 x \cos^4 x\, dx$ $[\pi/16]$
5. $\displaystyle\int_0^{\pi/2} \sin^6 x \cos^3 x\, dx$ $[2/63]$
6. $\displaystyle\int_0^{\pi/4} \tan^2 x\, dx$ $[1-\pi/4]$
7. $\displaystyle\int_0^{\pi/4} \sec^4 x\, dx.$ $[4/3]$

6.5 Some Further Standard Forms

The six integrals appearing in Table 6.2 occur so frequently in practice that they are considered standard and are tabulated in Appendix C on Pages 494 and 495.

The first three of these arise out of the differentiation of the functions $\operatorname{arcsin} x$, $\operatorname{arcosh} x$ and $\operatorname{arsinh} x$ respectively. The last three can easily be obtained by using a substitution which "removes" the square root. For example, the substitution $x = a\sinh\theta$ will remove the square root in $\sqrt{x^2+a^2}$ as in the following example.

Example:

$$\begin{aligned}
\int \sqrt{x^2+a^2}\, dx &= \int \sqrt{a^2\sinh^2\theta + a^2}\; a\cosh\theta\, d\theta \\
&= a^2 \int \cosh^2\theta\, d\theta \quad \text{taking } a > 0 \\
&= \frac{1}{2}a^2 \int (1+\cosh 2\theta)\, d\theta \\
&= \frac{1}{2}a^2(\theta + \frac{1}{2}\sinh 2\theta) + c \\
&= \frac{1}{2}a^2(\theta + \sinh\theta\cosh\theta) + c \\
&= \frac{1}{2}a^2 \operatorname{arsinh}\frac{x}{a} + \frac{1}{2}a^2\frac{x}{a}\sqrt{\frac{x^2}{a^2}+1} + c \\
&= \frac{1}{2}x\sqrt{x^2+a^2} + \frac{1}{2}a^2 \operatorname{arsinh}\frac{x}{a} + c.
\end{aligned}$$

Function	Integral
$\dfrac{1}{\sqrt{a^2 - x^2}}$	$\arcsin \dfrac{x}{a}$
$\dfrac{1}{\sqrt{x^2 - a^2}}$	$\operatorname{arcosh} \dfrac{x}{a}$
$\dfrac{1}{\sqrt{x^2 + a^2}}$	$\operatorname{arsinh} \dfrac{x}{a}$
$\sqrt{a^2 - x^2}$	$\dfrac{1}{2}x\sqrt{a^2 - x^2} + \dfrac{1}{2}a^2 \arcsin \dfrac{x}{a}$
$\sqrt{x^2 - a^2}$	$\dfrac{1}{2}x\sqrt{x^2 - a^2} - \dfrac{1}{2}a^2 \operatorname{arcosh} \dfrac{x}{a}$
$\sqrt{x^2 + a^2}$	$\dfrac{1}{2}x\sqrt{x^2 + a^2} + \dfrac{1}{2}a^2 \operatorname{arsinh} \dfrac{x}{a}$

Table 6.2: Frequently occurring integrals.

Similarly the substitutions $x = a\cosh\theta$ and $x = a\sin\theta$ can be used to obtain the tabulated results in Table 6.2 for $\int \sqrt{x^2 - a^2}\,dx$ and $\int \sqrt{a^2 - x^2}\,dx$ respectively.

Worked Example 6.5.1 *Use the tabulated results (Table 6.2) to find*

1. $\int \sqrt{4x^2 + 9}\,dx$ *2.* $\int_1^2 \sqrt{x^2 - 1}\,dx.$

1.
$$\begin{aligned}\int \sqrt{4x^2+9}\,dx &= 2\int \sqrt{x^2 + \frac{9}{4}}\,dx \\ &= 2\left\{\frac{x}{2}\sqrt{x^2+\frac{9}{4}} + \frac{1}{2}\left(\frac{9}{4}\text{arsinh}\,\frac{2x}{3}\right)\right\} \\ &= \frac{x}{2}\sqrt{4x^2+9} + \frac{9}{4}\text{arsinh}\,\frac{2x}{3}.\end{aligned}$$

You should note, in the application of the tabulated result, the need to first make the coefficient of x^2 equal to 1. This integral could have been evaluated in the following alternative manner by using the substitution $v = 2x$ so that

$$\begin{aligned}\int \sqrt{4x^2+9}\,dx &= \frac{1}{2}\int \sqrt{v^2+9}\,dv \\ &= \frac{1}{2}\left(\frac{v}{2}\sqrt{v^2+9} + \frac{9}{2}\text{arsinh}\,\frac{v}{3}\right) \\ &= \frac{x}{2}\sqrt{4x^2+9} + \frac{9}{4}\text{arsinh}\,\frac{2x}{3}.\end{aligned}$$

2.
$$\begin{aligned}\int_1^2 \sqrt{x^2-1}\,dx &= \left[\tfrac{1}{2}x\sqrt{x^2-1} - \tfrac{1}{2}\text{arcosh}\,x\right]_1^2 \\ &= \sqrt{3} - \frac{1}{2}\text{arcosh}\,2 + \frac{1}{2}\text{arcosh}\,1 \\ &= \sqrt{3} - \frac{1}{2}\log(2+\sqrt{3}).\end{aligned}$$

■

When integrating functions involving $\sqrt{ax^2+bx+c}$ it is frequently useful to modify the integrand (for example, to complete the square) or to use a substitution suggested by the standard forms discussed above.

6.5.1 Integrals of Various Special Forms

Integrals of the First Type

These are integrals of the form

- $\displaystyle\int \sqrt{ax^2+bx+c}\,dx$
- $\displaystyle\int \frac{dx}{\sqrt{ax^2+bx+c}}.$

In order to reduce either of these integrals to standard form it may be necessary to complete the square.

Worked Example 6.5.2 *Determine the value of the two integrals*

1. $\displaystyle\int_0^1 \sqrt{2x-x^2}\,dx$
2. $\displaystyle\int \frac{1}{\sqrt{x^2+2x-4}}\,dx.$

1.
$$\begin{aligned}
\int_0^1 \sqrt{2x-x^2}\,dx &= \int_0^1 \sqrt{1-(x^2-2x+1)}\,dx \\
&= \int_0^1 \sqrt{1-(x-1)^2}\,dx \\
&= \int_{-1}^0 \sqrt{1-v^2}\,dv \quad \text{where } v = x-1 \\
&= \left[\frac{1}{2}v\sqrt{1-v^2}+\frac{1}{2}\arcsin v\right]_{-1}^0 \\
&= 0-(0+\frac{1}{2}\arcsin(-1)) \\
&= \frac{1}{4}\pi.
\end{aligned}$$

2.
$$\begin{aligned}
\int \frac{dx}{\sqrt{x^2+2x-4}} &= \int \frac{dx}{\sqrt{x^2+2x+1-5}} \\
&= \int \frac{dx}{\sqrt{(x+1)^2-5}} \\
&= \int \frac{dv}{\sqrt{v^2-5}} \quad \text{where } v = x+1 \\
&= \operatorname{arcosh} \frac{1}{\sqrt{5}}v + c \\
&= \operatorname{arcosh} \frac{1}{\sqrt{5}}(x+1) + c.
\end{aligned}$$

■

Integrals of the Second Type

These are integrals of the form

- $\displaystyle\int \frac{Ax+B}{\sqrt{ax^2+bx+c}}\,dx$

- $\displaystyle\int (Ax+B)\sqrt{ax^2+bx+c}\,dx.$

If $Ax+B$ is the derivative of ax^2+bx+c these integrals are easily found using the substitution $v = ax^2+bx+c$; if $Ax+B$ is not the derivative of ax^2+bx+c we rewrite the integrals as the sum of two integrals, one of which has $Ax+B$ replaced by a multiple of the derivative of ax^2+bx+c whilst the other is of type 1. See the following examples.

Worked Example 6.5.3 *Evaluate each of the integrals*

1. $\displaystyle\int \frac{3x+2}{\sqrt{x^2+2x-4}}\,dx$ *2.* $\displaystyle\int 2x\sqrt{2x-x^2}\,dx.$

1. Since $2x+2$ is the derivative of x^2+2x-4 we rewrite $3x+2$ as $\frac{3}{2}(2x+2)-1$ so that

$$\begin{aligned}
\int \frac{3x+2}{\sqrt{x^2+2x-4}}\,dx &= \frac{3}{2}\int \frac{2x+2}{\sqrt{x^2+2x-4}}\,dx \\
&\quad -\int \frac{dx}{\sqrt{x^2+2x-4}} \\
&= \frac{3}{2}\int \frac{du}{\sqrt{u}} - \int \frac{dx}{\sqrt{(x+1)^2-5}} \\
&\quad \text{where } u = x^2+2x-4 \\
&= 3\sqrt{u} - \operatorname{arcosh}\frac{1}{\sqrt{5}}(x+1)+c \\
&= 3\sqrt{x^2+2x-4} \\
&\quad -\operatorname{arcosh}\frac{1}{\sqrt{5}}(x+1)+c.
\end{aligned}$$

An alternative approach is to complete the square first as

follows:

$$\begin{aligned}
\int \frac{3x+2}{\sqrt{x^2+2x-4}}\,dx &= \int \frac{3x+2}{\sqrt{x^2+2x+1-5}}\,dx \\
&= \int \frac{3x+2}{\sqrt{(x+1)^2-5}}\,dx \\
&= \int \frac{3u-1}{\sqrt{u^2-5}}\,du \\
&\quad \text{where } u = x+1 \\
&= \int \frac{3u}{\sqrt{u^2-5}}\,du - \int \frac{du}{\sqrt{u^2-5}} \\
&= 3\sqrt{u^2-5} - \operatorname{arcosh} \frac{1}{\sqrt{5}}u + c \\
&= 3\sqrt{x^2+2x-4} \\
&\quad -\operatorname{arcosh} \frac{1}{\sqrt{5}}(x+1) + c.
\end{aligned}$$

2. Here we rewrite $2x\sqrt{2x-x^2}$ as

$$-(2-2x)\sqrt{2x-x^2} + 2\sqrt{2x-x^2}$$

(why?) so that

$$\begin{aligned}
&\int 2x\sqrt{2x-x^2}\,dx \\
&= 2\int \sqrt{2x-x^2}\,dx - \int (2-2x)\sqrt{2x-x^2}\,dx \\
&= 2\int \sqrt{1-(x-1)^2}\,dx - \int \sqrt{u}\,du \\
&\quad \text{where } u = 2x-x^2 \\
&= (x-1)\sqrt{2x-x^2} + \arcsin(x-1) - \frac{2}{3}u^{3/2} + c \\
&= (x-1)\sqrt{2x-x^2} + \arcsin(x-1) \\
&\quad -\frac{2}{3}(2x-x^2)^{3/2} + c
\end{aligned}$$

■

Integrals of the Third Type

These are integrals which are reduced to standard forms by making a suitable trigonometric or hyperbolic substitution. Note that

- the substitution $x = a\sin\theta$ (or $x = a\cos\theta$) removes the square root from expressions such as $\sqrt{a^2 - x^2}$

- the substitution $x = a\sinh\theta$ (or $x = a\tan\theta$) removes the square root from expressions such as $\sqrt{a^2 + x^2}$

- the substitution $x = a\cosh\theta$ (or $x = a\sec\theta$) removes the square root from expressions such as $\sqrt{x^2 - a^2}$.

Consequently integrals that involve one or the other of these square roots (or more generally *any* fractional power such as $\frac{3}{2}$, $\frac{5}{2}$ etc.) will generally lead to simpler integrals involving trigonometric or hyperbolic functions.

▌**Worked Example 6.5.4** *Use a suitable trigonometric or hyperbolic substitution to evaluate the following integrals:*

1. $\displaystyle\int \frac{\sqrt{x^2-4}}{x^2}\,dx$ *2.* $\displaystyle\int \frac{dx}{x^2\sqrt{4+x^2}}.$

1.
$$\begin{aligned}
\int \frac{\sqrt{x^2-4}}{x^2}\,dx &= \int \frac{2\sinh\theta . 2\sinh\theta}{4\cosh^2\theta}\,d\theta \\
&\qquad \text{where } x = 2\cosh\theta \\
&= \int \tanh^2\theta\,d\theta \\
&= \int (1-\operatorname{sech}^2\theta)\,d\theta \\
&= \theta - \tanh\theta + c \\
&= \operatorname{arcosh}\frac{1}{2}x - \frac{\sqrt{\frac{x^2}{4}-1}}{\frac{x}{2}} + c \\
&= \operatorname{arcosh}\frac{1}{2}x - \frac{1}{x}\sqrt{x^2-4} + c.
\end{aligned}$$

(Note that an alternative approach would have been to use the substitution

$$x = 2\sec\theta.$$

Try it!)

2. $$\int \frac{dx}{x^2\sqrt{4+x^2}} = \int \frac{2\cosh\theta}{4\sinh^2\theta(4\cosh^2\theta)^{1/2}}\,d\theta$$
where $x = 2\sinh\theta$
$$= \frac{1}{4}\int \operatorname{cosech}^2\theta\,d\theta$$
$$= -\frac{1}{4}\coth\theta + c$$
$$= -\frac{1}{4}\frac{\sqrt{1+\frac{x^2}{4}}}{\frac{x}{2}} + c$$
$$= -\frac{1}{4x}\sqrt{x^2+4} + c.$$
(Try using the alternative substitution $x = 2\tan\theta$.)

■

Self-help exercises

1. Use the appropriate formulae in Appendix C (Pages 494–495) to write down the values of

(a) $\int_0^1 \sqrt{1-x^2}\,dx$ $[\pi/4]$

(b) $\int \frac{1}{\sqrt{4-x^2}}\,dx$ $[\operatorname{arcsin}\frac{x}{2}]$

(c) $\int \sqrt{x^2+9}\,dx$ $[\frac{9}{2}\operatorname{arsinh}\frac{x}{3} + \frac{x}{2}\sqrt{x^2+9}]$

(d) $\int \frac{1}{\sqrt{x^2-2}}\,dx$ $[\operatorname{arcosh}\frac{x}{\sqrt{2}}]$

(e) $\int \sqrt{1-x^2}\,dx$ $[\frac{1}{2}\operatorname{arcsin}x + \frac{1}{2}x\sqrt{1-x^2}]$

2. In each of the following cases write down a substitution you might use to evaluate the integral

(a) $\int \frac{\sqrt{1-x^2}}{x}\,dx$ $[x = \sin\theta]$

(b) $\int \frac{\sqrt{4+x^2}}{x}\,dx$ $[x = 2\sinh\theta$ or $x = 2\tan\theta]$

(c) $\int \frac{\sqrt{x^2-9}}{x^2}\,dx$ $[x = 3\cosh\theta$ or $x = 3\sec\theta]$

(d) $\int \frac{1}{x\sqrt{4-9x^2}}\,dx$ $[x = \frac{2}{3}\sin\theta]$

6.5.2 Integrals of Rational Functions

Partial Fractions

You should all be familiar with the methods used to combine two or more algebraic fractions into a single fraction having a common denominator. For example

- the three fractions
$$\frac{2}{x}+\frac{3}{x-1}-\frac{1}{x+2}$$
can be combined to obtain the single expression
$$\frac{4x^2+9x-4}{x(x-1)(x+2)}$$

- the two fractions
$$\frac{x}{x^2+1}-\frac{1}{x+1}$$
can be combined to obtain the single expression
$$\frac{x-1}{(x^2+1)(x+1)}$$

- the three fractions
$$\frac{2}{x+2}+\frac{3}{x-1}-\frac{1}{(x-1)^2}$$
can be combined to obtain the single expression
$$\frac{5x^2-2x-6}{(x-1)^2(x+2)}.$$

Often it is necessary to be able to carry out the reverse process; that is, to convert a rational function (a quotient of two polynomials) into two or more simpler fractions called partial fractions. This technique is important in many areas of mathematics such as integration, Laplace transforms and series expansions of rational functions.

Rational functions $p(x)/q(x)$ can either be improper fractions (where the degree of the polynomial $p(x)$ is greater than or equal to the degree of the polynomial $q(x)$) such as

$$\frac{x^2}{x+2} \text{ and } \frac{x^2-4}{x^2-x-2}$$

or proper fractions such as

$$(2x-1)/(x^2-x-2).$$

Before decomposing a rational function into partial fractions it may be necessary to divide $p(x)$ by $q(x)$ if the rational function is improper. Having done this $q(x)$ needs to be factorized. Assuming that we are only interested in factorization over the real numbers, $q(x)$ will only contain linear and/or quadratic factors with possibly repeated factors occurring. When expressing a rational function into partial fractions the form of the expansion depends on the type and multiplicity of the factors of $q(x)$. For example, the partial fraction decomposition for

- $\dfrac{2x+1}{x(x-1)(x+2)}$ is $\dfrac{A}{x}+\dfrac{B}{x-1}+\dfrac{C}{x+2}$ (linear factors)
- $\dfrac{2x+1}{x(x-1)^2}$ is $\dfrac{A}{x}+\dfrac{B}{x-1}+\dfrac{C}{(x-1)^2}$ (repeated linear factor)
- $\dfrac{2x+1}{(x^2+1)(x+2)}$ is $\dfrac{A}{x+2}+\dfrac{Bx+C}{x^2+1}$ (quadratic factor)
- $\dfrac{3x+1}{(x^2-2x+2)(x^2+4)^2}$ is $\dfrac{Ax+B}{x^2-2x+2}+\dfrac{Cx+D}{x^2+4}+\dfrac{Ex+F}{(x^2+4)^2}$ (repeated quadratic factor).

To integrate such rational functions we proceed as follows:

- divide $p(x)$ by $q(x)$ if the degree of $q(x)$ is less than or equal to the degree of $p(x)$
- factorize $q(x)$ and express the quotient (after division if necessary) in partial fraction form
- integrate each term using the techniques previously developed.

Worked Example 6.5.5 *Determine the value of the integral*

$$\int \frac{x^3+2}{x^3-x}\,dx.$$

Since the degree of the numerator is not less than that of the denominator we first divide out to obtain

$$\frac{x^3+2}{x^3-x} = 1+\frac{x+2}{x^3-x} = 1+\frac{x+2}{x(x-1)(x+1)}.$$

Now

$$\frac{x+2}{x(x-1)(x+1)} = \frac{A}{x}+\frac{B}{x-1}+\frac{C}{x+1}$$

so that

$$x+2 = A(x-1)(x+1)+Bx(x+1)+Cx(x-1).$$

Putting $x = 0$, 1 and -1 in turn gives respectively $A = -2$, $B = \frac{3}{2}$ and $C = \frac{1}{2}$. Hence

$$\frac{x^3+2}{x^3-x} = 1 - \frac{2}{x} + \frac{1}{2}\left(\frac{3}{x-1} + \frac{1}{x+1}\right)$$

and thus

$$\int \frac{x^3+2}{x^3-x}\,dx = x - 2\log|x| + \frac{3}{2}\log|x-1| + \frac{1}{2}\log|x+1| + c.$$

■

Worked Example 6.5.6 *Evaluate the integral*

$$\int \frac{3x^2+9}{(2x-1)(x^2+2x+2)}\,dx.$$

Here division is not necessary. (Why?) Now

$$\frac{3x^2+9}{(2x-1)(x^2+2x+2)} = \frac{A}{2x-1} + \frac{Bx+C}{x^2+2x+2}$$

so that

$$3x^2+9 = A(x^2+2x+2) + (Bx+C)(2x-1).$$

Putting $x = 1/2$ gives $A = 3$.
Putting $x = 0$ gives $9 = 2A - C \Rightarrow C = -3$.
Putting $x = -1$ gives $12 = A - 3(C - B) \Rightarrow B = 0$.
Thus

$$\begin{aligned}
&\int \frac{3x^2+9}{(2x-1)(x^2+2x+2)}\,dx \\
&= \int \frac{3}{2x-1}\,dx - \int \frac{3}{x^2+2x+2}\,dx \\
&= \frac{3}{2}\log|2x-1| - 3\int \frac{dx}{(x+1)^2+1} \\
&\qquad \text{on completing the square} \\
&= \frac{3}{2}\log|2x-1| - 3\arctan(x+1) + c.
\end{aligned}$$

■

Worked Example 6.5.7 *Determine the value of each of the following integrals:*

1. $\displaystyle\int_0^1 \frac{2x+2}{x^2+2x+10}\,dx$ *2.* $\displaystyle\int \frac{3x+4}{x^2+2x+10}\,dx.$

1. Since $2x+2$ is the derivative of $x^2+2x+10$ the substitution $v = x^2+2x+10$ gives

$$\begin{aligned}\int_0^1 \frac{2x+2}{x^2+2x+10}\,dx &= \int_{10}^{13} \frac{dv}{v} \\ &= \log 13 - \log 10 \\ &= \log 1.3\end{aligned}$$

2. If we write $3x+4$ as $\frac{3}{2}(2x+2)+1$ (why?) we have

$$\begin{aligned}&\int \frac{3x+4}{x^2+2x+10}\,dx \\ &= \frac{3}{2}\int \frac{2x+2}{x^2+2x+10}\,dx + \int \frac{dx}{x^2+2x+10} \\ &= \frac{3}{2}\log(x^2+2x+10) + \int \frac{dx}{(x+1)^2+9}\end{aligned}$$

on completing the square in the second integral

$$= \frac{3}{2}\log(x^2+2x+10) + \frac{1}{3}\arctan\frac{x+1}{3} + c$$

where use has been made of the standard result

$$\int \frac{dx}{x^2+a^2} = \frac{1}{a}\arctan\frac{x}{a}.$$

■

Self-help exercises

1. In each of the following cases indicate whether the correct partial fraction decomposition has been used:

(a) $\displaystyle\frac{2}{(x-1)(x+2)} = \frac{A}{x-1} + \frac{B}{x+2}$ [True]

(b) $\displaystyle\frac{2x^2}{(x-1)(x+2)} = \frac{A}{x-1} + \frac{B}{x+2}$ [False (Divide out!)]

(c) $\displaystyle\frac{2x+7}{(x-1)^2(x+2)} = \frac{A}{x-1} + \frac{B}{(x-1)^2} + \frac{C}{x+2}$ [True]

(d) $\displaystyle\frac{2x-1}{x^3-x} = \frac{A}{x-1} + \frac{B}{x+1}$ [False]

(e) $\dfrac{x^3}{x^3 - x} = 1 + \dfrac{A}{x-1} + \dfrac{B}{x+1} + \dfrac{C}{x}$ [True]

(f) $\dfrac{2x+11}{(x-1)(x^2+2)} = \dfrac{A}{x-1} + \dfrac{B}{x^2+2}$ [False]

(g) $\dfrac{2x+11}{(x-1)(x^2+2)} = \dfrac{A}{x-1} + \dfrac{Bx+C}{x^2+2}$ [True]

(h) $\dfrac{2}{(x^2-1)(x+2)} = \dfrac{A}{x^2-1} + \dfrac{B}{x+2}$ [False]

2. Write down the following integrals:

(a) $\displaystyle\int \frac{3}{x-1}\,dx$ [$3\log|x-1|$]

(b) $\displaystyle\int \frac{4}{2x-1}\,dx$ [$2\log|2x-1|$]

(c) $\displaystyle\int \frac{3}{(x-1)^2}\,dx$ [$-\frac{3}{x-1}$]

(d) $\displaystyle\int \frac{3x}{x^2+1}\,dx$ [$\frac{3}{2}\log(x^2+1)$]

(e) $\displaystyle\int \frac{2x+4}{x^2+4x+12}\,dx$ [$\log(x^2+4x+12)$]

(f) $\displaystyle\int \frac{3}{x^2+9}\,dx$ [$\arctan\frac{x}{3}$]

In this topic we have

- Reviewed basic integration methods including substitution techniques.
- Emphasized the use of tables of integrals.
- Discussed integration by parts, emphasized standard forms and introduced partial fractions.

6.6 Quick Test Number 6

Methods of Integration

Question **Selection**

Simple substitution:

1. $\displaystyle\int \frac{\log x}{x}\,dx$ can be evaluated using

   (a) $u = x$ (b) $u = \log x$ (c) $u = \dfrac{1}{x}$

2. $\displaystyle\int \frac{x}{\sqrt{25 - x^2}}\,dx$ can be evaluated using

   (a) $u = x$ (b) $u = \sqrt{25 - x^2}$ (c) $u = 25 - x^2$

3. $\displaystyle\int 3x^2\,(x^3 + 2)^{100}\,dx$ can be evaluated using

   (a) $u = x$ (b) $u = 3x^2$ (c) $u = x^3 + 2$

4. $\displaystyle\int \cos x^2\,dx$ is (a) $\sin x^2$ (b) $\dfrac{\sin x^2}{2x}$

   (c) impossible using u-substitution (d) $-2x \sin x^2$

5. $\displaystyle\int \sin^4 x\,dx$ equals (a) $\dfrac{1}{5}\sin^5 x$ (b) $-\cos^4 x$ (c) $-\dfrac{1}{5}\cos^5 x$

   (d) $-\dfrac{1}{5}\dfrac{\sin^5 x}{\cos x}$ (e) none of these

---

Substitution:

For the following integrals a substitution $u = g(x)$ is given.
Choose the correct du and substitute u and du into the integral.
Change the bounds of integration accordingly.

6. $\displaystyle\int_0^1 3x\sqrt{4 - x^2}\,dx,\ u = 4 - x^2.$

   Choose: (a) $du = -2x$ (b) $du = -2xdx$ (c) $du = (4 - 2x)dx$

   To obtain (a) $-\dfrac{3}{2}\displaystyle\int_0^1 \sqrt{u}\,du$ (b) $\displaystyle\int_0^1 3x\sqrt{u}$

   (c) $-\dfrac{3}{2}\displaystyle\int_4^3 \sqrt{u}\,du$ (d) $-\dfrac{3}{2}\displaystyle\int_4^3 \sqrt{4 - x^2}\,du$

7. $\int_0^{\pi} \sin\sqrt{x}\,dx,\ u = \sqrt{x}$ i.e. $x = u^2$.

Choose: (a) $2udu = dx$ (b) $du = \sqrt{x}dx$ (c) $du = \frac{1}{2}x^{-1/2}$

To obtain (a) $\int_0^{\pi} \sin u$ (b) $\int_0^{\pi} 2u \sin u\,du$

(c) $\int_0^{\sqrt{\pi}} 2u \sin u\,du$ (d) $\int_0^{\pi} 2u \sin\sqrt{x}\,du$

Integration by parts:

The integration by parts formula is

$$\int u\,dv = uv - \int v\,du.$$

In the following "integration by parts" problems, determine the correct choices for u and dv.

8. $\int x^3 \log x\,dx$ (a) $u = x^3,\ dv = \log x\,dx$ (b) $u = \log x,\ dv = x^3\,dx$

9. $\int x^3 \sin x\,dx$ (a) $u = x^3,\ dv = \sin x\,dx$ (b) $u = \sin x,\ dv = x^3\,dx$

10. $\int x^3 e^{5x}\,dx$ (a) $u = x^3,\ dv = e^{5x}\,dx$ (b) $u = e^{5x},\ dv = x^3\,dx$

11. $\int x \arcsin x\,dx$

(a) $u = x,\ dv = \arcsin x\,dx$ (b) $u = \arcsin x,\ dv = x\,dx$

Partial fraction expansions:

12. $\dfrac{2x+3}{(x+3)(x+2)(x+1)} = \dfrac{A}{x+1} + \dfrac{B}{x+3} + \dfrac{C}{x+2}$ (a) True (b) False

13. $\dfrac{x^3+3x^2+5}{(x+1)(x+4)} = \dfrac{A}{x+1} + \dfrac{B}{x+4}$ (a) True (b) False

14. $\dfrac{x+1}{(x+2)^2(x+5)} = \dfrac{A}{x+2} + \dfrac{B}{(x+2)^2} + \dfrac{C}{x+5}$ (a) True (b) False

15. $\dfrac{3}{x^2-7x+6} = \dfrac{A}{x-6} + \dfrac{B}{x-1}$ (a) True (b) False

16. $\dfrac{2x+5}{(x-3)(x^2+3x+5)} = \dfrac{A}{x-3} + \dfrac{B}{x^2+3x+5}$ (a) True (b) False

17. Complete the square for $x^2 - 10x + 36$:

(a) $(x-10)^2 - 64$ (b) $(x-5)^2 + 61$
(c) $(x-5)^2 + 11$ (d) $(x-6)^2 + 2x$

18. $x^2 + 2x + 5 = A(x-2) + B(x+1) + C(x+1)^2$ for some constants A, B and C. The best way to find A is to:

    (a) expand out the RHS and equate coefficients of x^2, x etc.
    (b) set $x = -1$ in both sides and solve for A
    (c) set $x = 2$ in both sides and solve for A

19. Writing $3x + 7$ in the form $A(x-2) + B$ where A and B are constants gives:

    (a) $3(x-2)+1$ (b) $3(x-2)-1$ (c) $3(x-2)+7$ (d) $3(x-2)+13$

20. $\int \frac{3(x+1)}{(x+1)^2+4}\,dx$ equals (a) $\frac{3}{2}\log|(x+1)^2+4|$ (b) $\frac{3}{2}\log|\frac{x+1}{2}|$
    (c) $\frac{3}{2}\arctan\frac{x+1}{2}$ (d) $\frac{3}{2}\arctan|(x+1)^2+4|$

21. $\int \frac{5}{(x+1)^2+4}\,dx$ equals (a) $\frac{5}{2}\log|(x+1)^2+4|$ (b) $\frac{5}{2}\log|\frac{x+1}{2}|$
    (c) $\frac{5}{2}\arctan\frac{x+1}{2}$ (d) $\frac{5}{2}\arctan|(x+1)^2+4|$

Use of tables:

22. Using the formula $\displaystyle\int \frac{1}{u^2+a^2}\,du = \frac{1}{a}\arctan\frac{u}{a}$ to evaluate

$$\int \frac{1}{1+4x^2}\,dx$$

gives

(a) $\dfrac{1}{2x}\arctan\dfrac{1}{2x}$ (b) $\dfrac{1}{2}\arctan 2x$ (c) $\arctan 2x$

23. Using the formula $\displaystyle\int \frac{1}{\sqrt{a^2-u^2}}\,du = \arcsin\frac{u}{a}$ to evaluate

$$\int \frac{1}{\sqrt{1-9x^2}}\,dx$$

gives

(a) $\dfrac{1}{3}\arcsin 3x$ (b) $\arcsin 3x$ (c) $\operatorname{arccosh}\dfrac{1}{3x}$

6.7 Exercises

Before attempting any of these miscellaneous exercises, make sure that you have successfully answered all the self-help exercises appearing throughout the chapter.

1. Use the basic integration formulae to determine
   (a) $\int (3x-5)^3\,dx$
   (b) $\int_2^3 \sqrt{3x-5}\,dx$
   (c) $\int \frac{1}{(3x-5)^3}\,dx$
   (d) $\int_0^1 \frac{1}{3x-5}\,dx$
   (e) $\int \frac{1}{\sqrt{3-5x}}\,dx$
   (f) $\int \frac{1}{3-5x}\,dx$
2. Determine
   (a) $\int \sin(3x-2)\,dx$
   (b) $\int \sec^2(3x-2)\,dx$
   (c) $\int \sinh(3x-2)\,dx$
   (d) $\int \operatorname{sech}^2(3x-2)\,dx$
   (e) $\int \cos(3x-2)\,dx$
   (f) $\int \cosh(3x-2)\,dx$
3. Determine each of the following integrals.
   (a) $\int e^{-3x}\sinh 2x\,dx$
   (b) $\int e^{-3x}\cosh 2x\,dx$
   (c) $\int e^{-3x}\sinh^2 x\,dx$
   (d) $\int \cosh 3x \sinh 2x\,dx$
   (e) $\int \sin 3x \sin 2x\,dx$
   (f) $\int \sin 3x \cos 2x\,dx$
   (g) $\int \tan^2 x\,dx$
4. Find
   (a) $\int \left(2x - \frac{3}{x^2}\right)^2 dx$
   (b) $\int (e^{2x} - 3e^{-x})^2\,dx$
   (c) $\int (3x^2-2)^3\,dx$
5. Using an appropriate substitution, determine each of the following integrals.
   (a) $\int_0^1 \frac{2x+7}{x^2+7x+1}\,dx$
   (b) $\int \frac{x^2+3}{\sqrt{x^3+9x+1}}\,dx$
   (c) $\int (1-x)\sqrt{2x-x^2}\,dx$
   (d) $\int \frac{\arcsin 2x}{\sqrt{1-4x^2}}\,dx$
   (e) $\int_0^3 \frac{2x+7}{\sqrt{x+1}}\,dx$
   (f) $\int \frac{e^x}{(3e^x+7)^2}\,dx$
   (g) $\int_0^{\pi/8} \tan 2x\,dx$
   (h) $\int \tan^4 x \sec^2 x\,dx$

(i) $\int_0^{\pi/2} \sin^3 x\,dx$

(j) $\int \frac{\cos x}{\sqrt{\sin x}}\,dx$

(k) $\int x \sec^2 x^2\,dx$

(l) $\int (x+\sin x)^2\,dx$

6. Evaluate each of the following integrals.

   (a) $\int \arcsin x\,dx$

   (b) $\int x^2 \log x\,dx$

   (c) $\int \log x\,dx$

   (d) $\int_0^{\pi} x \sin x\,dx$

   (e) $\int_0^1 x e^{-2x}\,dx$

   (f) $\int x^2 \sin x\,dx$

   (g) $\int (2x+1)e^{-x}\,dx$

   (h) $\int \frac{\log(\log x)}{x}\,dx$

   (i) $\int_1^4 \log\sqrt{x}\,dx$

   (j) $\int \sin\sqrt{x}\,dx$

   (k) $\int x \sinh x\,dx$

7. Using an appropriate reduction formula determine each of the following integrals.

   (a) $\int_0^{\pi} \sin^4 x \cos^2 x\,dx$

   (b) $\int \sin^4 x \cos^3 x\,dx$

   (c) $\int_0^{\pi} \sin^8 x\,dx$

   (d) $\int_0^{\pi/2} \cos^4 x\,dx$

   (e) $\int_0^{\pi/4} \cos^4 x\,dx$

   (Be careful!)

   (f) $\int_0^{\pi/4} \tan^6 x\,dx$

   (g) $\int \sec^4 x\,dx$

   (h) $\int \tan^5 x\,dx$

   (i) $\int_0^{\pi/4} \sec^4 x \tan^2 x\,dx$

   (j) $\int_0^1 x^4 e^{-x}\,dx$

   (k) $\int_0^1 x^4 e^{-2x}\,dx$

8. (a) By writing $\sin^n x = \sin^{n-1} x \sin x$ obtain the reduction formula

$$\int_0^{\pi/2} \sin^n x\,dx = \frac{n-1}{n}\int_0^{\pi/2} \sin^{n-2} x\,dx$$

   where $n \geq 2$ is an integer.

   (b) Hence determine

   i. $\int_0^{\pi/2} \sin^7 x\,dx$

   ii. $\int_0^{\pi/2} \sin^6 x\,dx$

   iii. $\int_0^{\pi/2} \sin^6 x \cos^2 x\,dx.$

(c) Using a method similar to that used in (a) determine a reduction formula for
$$\int_0^{\pi/2} \cos^n x\,dx.$$

9. Use the tables in Appendix C to determine

(a) $\int \sqrt{9x^2+1}\,dx$

(b) $\int \frac{1}{\sqrt{4x^2-1}}\,dx$

(c) $\int \sqrt{1-2x^2}\,dx$

(d) $\int \sqrt{2x+x^2}\,dx$

(e) $\int \frac{1}{\sqrt{3x^2-1}}\,dx$

(f) $\int \sqrt{x^2+2x+10}\,dx$

(g) $\int \sqrt{x^2+2x-3}\,dx$

10. Evaluate each of the following integrals.

(a) $\int_0^1 \frac{x+3}{\sqrt{x^2+2x+10}}\,dx$

(b) $\int_0^1 (x+3)\sqrt{x^2+2x+10}\,dx$

(c) $\int_0^1 \frac{3x-4}{\sqrt{x^2-4x+5}}\,dx$

(d) $\int_0^1 x\sqrt{x^2+2x+2}\,dx$

11. Use an appropriate trigonometric or hyperbolic substitution to determine

(a) $\int \frac{\sqrt{x^2+4}}{x^2}\,dx$

(b) $\int \frac{1}{x\sqrt{x^2+4}}\,dx$

(c) $\int \frac{\sqrt{4-x^2}}{x}\,dx$

(d) $\int \frac{\sqrt{x^2-4}}{x^2}\,dx$

12. Determine each of the following integrals.

(a) $\int_0^1 \frac{x+3}{x+1}\,dx$

(b) $\int \frac{x^2}{x+1}\,dx$

(c) $\int \frac{x^2}{x^2-1}\,dx$

(d) $\int_0^2 \frac{x^2}{x^2+2x+2}\,dx$

(e) $\int \frac{2x+3}{x^2-x-2}\,dx$

(f) $\int \frac{3}{x^2+x}\,dx$

(g) $\int \frac{2x+3}{x^3+4x^2+3x}\,dx$

(h) $\int \frac{x^3+1}{x^3-4x}\,dx$

(i) $\int \frac{7x-2}{(x-2)^2(x+2)}\,dx$

(j) $\int \frac{5-3x}{(x-1)^2(x+1)}\,dx$

(k) $\int \frac{2x^2+x+7}{x^3+x^2-2}\,dx$

(l) $\int_1^{\sqrt{3}} \frac{x^2+x+1}{x^3+x}\,dx$

(m) $\int \frac{1}{x^{1/2}+x^{1/3}}\,dx$

(n) $\int_0^1 \frac{e^x}{e^{2x}+3e^x+2}\,dx$

(o) $\int_0^2 \frac{x+3}{(x+1)(x^2+1)}\,dx$

(p) $\displaystyle\int \frac{x+5}{x(x^2+2x+5)}\,dx$

(q) $\displaystyle\int \frac{2x+5}{x^2+2x+2}\,dx$

(r) $\displaystyle\int \frac{x+5}{x^2+4x+13}\,dx$

(s) $\displaystyle\int \frac{dx}{1+e^x}$

13. Verify that

$$\frac{x^4(1-x)^4}{1+x^2} = x^6 - 4x^5 + 5x^4 - 4x^2 + 4 - \frac{4}{1+x^2}.$$

Hence show that

$$\int_0^1 \frac{x^4(1-x)^4}{1+x^2}\,dx = \frac{22}{7} - \pi.$$

Explain why this result implies that $\pi < \frac{22}{7}$.

14. Use the substitution

$$x = 4\sin^2\theta - 2\cos^2\theta$$

to determine

$$\int \sqrt{\frac{4-x}{2+x}}\,dx.$$

15. By writing $\sec^n x$ as $\sec^{n-2} x \sec^2 x$ obtain the reduction formula

$$\int \sec^n x\,dx = \frac{\sec^{n-2} x \tan x}{n-1} + \frac{n-2}{n-1}\int \sec^{n-2} x\,dx$$

for integral $n \geq 2$. Hence determine

(a) $\displaystyle\int \sec^6 x\,dx$

(b) $\displaystyle\int_0^{\pi/4} \sec^7 x\,dx.$

16. (a) If $t = \tan\frac{x}{2}$ verify that

i. $\sin x = \dfrac{2t}{1+t^2}$

ii. $\cos x = \dfrac{1-t^2}{1+t^2}$

iii. $\dfrac{dx}{dt} = \dfrac{2}{1+t^2}.$

(b) Using the substitution $t = \tan\frac{x}{2}$ together with the results from (a), determine each of the following:

i. $\displaystyle\int \frac{1}{2+\sin x}\,dx$

ii. $\displaystyle\int \frac{1}{1+\sin x+\cos x}\,dx.$

Miscellaneous Exercises

Determine each of the following integrals.

17. $\displaystyle\int e^{-2x} \sin 3x \, dx$

18. $\displaystyle\int e^{-\cos 2x} \sin 2x \, dx$

19. $\displaystyle\int e^{-2x} \sinh 3x \, dx$

20. $\displaystyle\int (e^{-2x} - x)^2 \, dx$

21. $\displaystyle\int \frac{x^2}{2x+1} \, dx$

22. $\displaystyle\int (2x+3)e^{-x} \, dx$

23. $\displaystyle\int x \cosh x \, dx$

24. $\displaystyle\int_0^{\log 2} \frac{\sinh x}{e^x} \, dx$

25. $\displaystyle\int \cos \sqrt{x} \, dx$

26. $\displaystyle\int \frac{3}{(1+x^2)(4+x^2)} \, dx$

27. $\displaystyle\int_0^1 \frac{x^3}{x+1} \, dx$

28. $\displaystyle\int \frac{dx}{(1+x)\sqrt{x}}$

29. $\displaystyle\int \cosh 2x \cosh x \, dx$

30. $\displaystyle\int e^x \sqrt{1+e^{2x}} \, dx$

31. $\displaystyle\int \sinh^2 3x \, dx$

32. $\displaystyle\int_0^1 \frac{x^2+3}{x+2} \, dx$

33. $\displaystyle\int \frac{x^2-2x+5}{x(x-1)^2} \, dx$

34. $\displaystyle\int x^3 \log x \, dx$

35. $\displaystyle\int \cos^4 2x \, dx$

36. $\displaystyle\int_0^1 \frac{x^2}{4-x^2} \, dx$

37. $\displaystyle\int \operatorname{arsinh} x \, dx$

38. $\displaystyle\int \tan^2 4x \, dx$

39. $\displaystyle\int x \sec^2 x \, dx$

40. $\displaystyle\int \frac{\cos^3 x}{\sqrt{\sin x}} \, dx$

41. $\displaystyle\int \frac{x^2}{\sqrt{2x+1}} \, dx$

42. $\displaystyle\int \frac{dx}{(1+x^2)^2}$

43. $\displaystyle\int \frac{dx}{\sqrt{x^2+4x+8}}$

44. $\displaystyle\int x^2 e^{-x^{\cdot}} \, dx$

45. $\displaystyle\int x^3 e^{-x^{\cdot}} \, dx$

46. $\displaystyle\int \log(1+x^2) \, dx$

47. $\displaystyle\int_0^1 x(1-x)^8 \, dx$

48. $\displaystyle\int \frac{2x+7}{x+3} \, dx$

49. $\displaystyle\int \frac{x^3+4}{x^3-x^2} \, dx$

50. $\int \cos 3x \cos 5x \, dx$

51. $\int x \arcsin x \, dx$

52. $\int \frac{dx}{x + \sqrt{x}}$

53. $\int \frac{\sin x}{\sqrt{2 + \cos x}} \, dx$

54. $\int_0^1 \frac{x}{\sqrt{4 - x^2}} \, dx$

55. $\int \frac{\sqrt{x} + 1}{\sqrt{x} - 1} \, dx$

56. $\int \sqrt{4x - x^2} \, dx$

57. $\int \frac{dx}{\sqrt{x^2 + 4x - 1}}$

58. $\int_0^2 \frac{x^3}{\sqrt{4 + x^2}} \, dx$

59. $\int (x^2 + 1) \cos x \, dx$

60. $\int \frac{x + 1}{x^2 + 2x + 10} \, dx$

61. $\int \frac{x + 2}{x^2 + 2x + 10} \, dx$

62. $\int \frac{\arctan \sqrt{x}}{\sqrt{x}} \, dx$

63. $\int \frac{e^{2x}}{e^x + 1} \, dx$

64. $\int \frac{2}{x^3 + x^2 + x + 1} \, dx$

65. $\int \frac{2}{x^3 - x} \, dx$

66. $\int e^{3x} \cos 5x \, dx$

67. $\int \frac{x}{x^2 + 4x + 5} \, dx$

68. $\int \frac{\sqrt{9 - x^2}}{x} \, dx$

Part II

Second Semester

Topic 7

Differential Equations

7.1 Introduction

An ordinary differential equation is a relation between two variables x, y and the derivatives $y', y'', y''' \ldots$. Differential equations occur in numerous situations where physical phenomena are represented by a mathematical model. For example,

1. the current $i(t)$ flowing in an R-L circuit with applied e.m.f. $E(t)$ is given by the solution of the differential equation

$$L\frac{di}{dt} + Ri = E(t)$$

2. the speed of a certain chemical reaction $A + B \to C$ can be modelled by

$$\frac{dx}{dt} - k(a - x)(b - x)$$

where a and b are the initial amounts of the reactants A and B respectively, k is a known constant and x is the amount of product C formed at time t

3. the angular displacement $\theta(t)$ at any time t of a rigid body pendulum can be modelled by the differential equation

$$I\frac{d^2\theta}{dt^2} + mgh\sin\theta = 0$$

where I, m, g and h are constants.

The *order* of a differential equation is simply the order of the highest derivative which occurs in the equation. For example,

1. $2x\dfrac{dy}{dx} + x^2y = 4$ is a first order differential equation and

2. $3\dfrac{d^2y}{dx^2} + 4\dfrac{dy}{dx} + 7y = \sin x$ is a second order differential equation.

Before discussing methods of solving a differential equation we need to make the distinction between the *general solution* and a *particular solution* of a differential equation. The general solution of a differential equation is the most general function $y = y(x)$ which satisfies the equation. For example, if the differential equation is $y' = 2$ then $y = 2x + c$ is the general solution. It is evident that the general solution of a first order differential equation should contain ONE arbitrary constant (effectively one integration needs to be carried out to obtain the solution) and the general solution of a second order differential equation should contain TWO arbitrary constants. Returning to the simple example $y' = 2$, the solutions $y = 2x$, $y = 2x - 3$ and $y = 2x + 7$ are all examples of particular solutions.

Worked Example 7.1.1 *Show that $y = ae^{2x} + be^{-x}$ $(a, b$ constants) is a solution of the differential equation*

$$y'' - y' - 2y = 0.$$

Find the particular solution which satisfies the initial conditions $y(0) = 0$ and $y'(0) = 6$.
Since $y = ae^{2x} + be^{-x}$ then $y' = 2ae^{2x} - be^{-x}$ and $y'' = 4ae^{2x} + be^{-x}$.
Hence $y'' - y' - 2y = 0$.
Thus $y = ae^{2x} + be^{-x}$ is a solution (in fact the general solution, as above).
Since $y(0) = 0$ we have $0 = a + b$ and since $y'(0) = 6$ we have $6 = 2a - b$. This implies $a = 2$ and $b = -2$.
Thus the particular solution which satisfies $y(0) = 0$ and $y'(0) = 6$ is

$$y = 2e^{2x} - 2e^{-x}.$$

■

7.2 First Order Differential Equations

The most general first order differential equation has the form

$$\boxed{f(x, y, y') = 0}$$

We shall consider two classes of such equations which frequently arise in the modelling of physical situations:

- separable type $y' = F(x)G(y)$ and

- first order linear type $y' + p(x)y = q(x)$,

both of which allow an analytical representation of their solution. In fact in most other cases, numerical (approximate) techniques will often need to be used (see Section 7.4 and the subject MA011 Numerical Methods A).

7.2.1 Separable Differential Equations

The differential equation $y' = f(x,y)$ is said to be separable if the right-hand side $f(x,y)$ can be written as the product of two factors, one containing x only and one containing y only (that is, $f(x,y) = F(x)G(y)$); consequently the x and y variables can be "separated out".

In such cases we can write

$$\boxed{\frac{dy}{dx} = F(x)G(y) \Rightarrow \frac{1}{G(y)}\frac{dy}{dx} = F(x) \quad \text{if} \quad G(y) \neq 0}$$

Integrating both sides with respect to x yields the solution in implicit form as

$$\boxed{\int \frac{dy}{G(y)} = \int F(x)\,dx}$$

Worked Example 7.2.1 *Find the general solution of the differential equation*

$$xyy' = 1 + y^2.$$

We write the differential equation as $\dfrac{y}{1+y^2}\dfrac{dy}{dx} = \dfrac{1}{x}, \quad x \neq 0.$
Integrating both sides with respect to x gives

$$\int \frac{y\,dy}{1+y^2} = \int \frac{dx}{x} \Rightarrow \tfrac{1}{2}\log(1+y^2) = \log|x| + \log c.$$

(It is convenient to write the constant of integration as $\log c$ when the other expressions are log functions.)
Thus $\log\sqrt{1+y^2} = \log c|x|$ or $\sqrt{1+y^2} = c|x|$, so that $1 + y^2 = Ax^2$ where A is a constant.

■

Worked Example 7.2.2 *Newton's law of cooling states that the rate at which a body cools is proportional to the excess of its temperature above that of its surroundings. A body, initially at 80° C, is cooling in surroundings maintained at 20° C. After 10 minutes the temperature of the body is 60° C.*
Find the body's temperature after a further 10 minutes.

If $\theta = \theta(t)$ is the body's temperature after t minutes, Newton's law of cooling gives

$$\frac{d\theta}{dt} = -k(\theta - 20) \text{ where } k \text{ is a constant.}$$

Separating the variables and integrating gives

$$\int \frac{d\theta}{\theta - 20} = -\int kdt \quad \Rightarrow \quad \log(\theta - 20) = -kt + c \text{ since } \theta > 20.$$

When $t = 0, \theta = 80$ so that $c = \log 60$. Thus $\log(\theta - 20) = \log 60 - kt$ or

$$\log \frac{\theta - 20}{60} = -kt. \tag{7.1}$$

When $t = 10, \theta = 60 \quad \Rightarrow \quad -10k = \log 2/3$.
From Equation 7.1, $\theta - 20 = 60e^{-kt}$ or $\theta = 20 + 60e^{-kt}$.
Thus when $t = 20$, $\theta = 20 + 60e^{-20k} = 20 + 60e^{2\log 2/3} = 20 + 60(\frac{2}{3})^2 \approx 46.7$.
Thus, after a further 10 minutes, the temperature of the body is approximately 46.7°C. ■

7.2.2 First Order Linear Differential Equations

A first order differential equation $f(x, y, y') = 0$ is said to be of the linear type if it is linear in both y and y'. For example, $2xy' - x^2y = \sin x$ is of the linear type but $2xy' - xy^2 = \sin x$ and $2xy'^{\,2} + xy = \sin x$ are not of the linear type.

The general first order linear differential equation has the form

$$\boxed{y' + p(x)y = q(x)} \tag{7.2}$$

In order to motivate the general discussion of such equations, consider the particular example $x^2y' + 2xy = 4$. Since the left-hand side is an exact derivative, namely $d(x^2y)/dx$, the differential equation is easily solved by integrating both sides with respect to x as follows

$$\frac{d}{dx}(x^2y) = 4 \Rightarrow x^2y = 4x + c \Rightarrow y = \frac{4}{x} + \frac{c}{x^2} \text{ if } x \neq 0.$$

If we can express the left-hand side of Equation 7.2 as an exact derivative, a similar technique can be adopted to solve the general first order linear differential equation. In order to do this we multiply both sides of Equation 7.2 by some function $I(x)$ which is to be chosen *in order to make the resulting left-hand side an exact derivative*.

Consider the equation obtained by multiplying Equation 7.2 by the factor $I(x)$; that is

$$I(x)\frac{dy}{dx} + I(x)p(x)y = I(x)q(x).$$

We now *choose* $I = I(x)$ so that the left-hand side is exactly

$$\frac{d}{dx}(I(x)y) = I(x)\frac{dy}{dx} + \frac{dI}{dx}y.$$

Clearly this requires

$$\frac{dI}{dx} \equiv I(x)p(x)$$

so that

$$\int \frac{dI}{I} = \int p(x)dx \quad \text{and} \quad \boxed{I = e^{\int p(x)dx}}$$

The function $I = I(x)$ is called an *integrating factor* of the differential equation 7.2 because on multiplying Equation 7.2 by I it becomes

$$\frac{d}{dx}(Iy) = I(x)q(x)$$

which is easily solved for y.

Note that in the above derivation, the factor $p(x)$ is the coefficient of y when *the coefficient of* y' *is* 1. If the coefficient of y' is not 1 it must be made so. For example, if the differential equation to be solved is

$$xy' - 2y = 4$$

we need to first rewrite the differential equation as

$$y' - \frac{2}{x}y = \frac{4}{x}$$

so that $p(x) = -2/x$ and $q(x) = 4/x$. An integrating factor is then

$$I(x) = e^{-2\log x} = e^{\log x^{-2}} = \frac{1}{x^2}.$$

Worked Example 7.2.3 *Find the general solution of*

$$xy' - y = x^3e^x \tag{7.3}$$

and determine the particular solution which satisfies $y'(0) = 0$.

We first rewrite the differential equations as

$$y' - \frac{1}{x}y = x^2e^x \tag{7.4}$$

so that $p(x) = -1/x$ and $q(x) = x^2e^x$.

An integrating factor is $I = e^{-\int dx/x} = e^{-\log x} = 1/x$.
Multiplying Equation 7.4 by I yields $\dfrac{d}{dx}\left(\dfrac{y}{x}\right) = xe^x$.
Integrating both sides gives

$$\frac{y}{x} = \int xe^x\,dx = xe^x - e^x + c$$

so that the general solution is $y = x^2e^x - xe^x + cx$.
Since $y' = 2xe^x + x^2e^x - xe^x - e^x + c$, $y'(0) = 0$ implies $0 = -1 + c$ or $c = 1$.
Thus the required particular solution is $y = x(xe^x - e^x + 1)$. ■

Self-help exercises

1. Determine which of the following differential equations are of the separable type and which are of the linear type.

   (a) $y' - xy = xy^2$ [separable]

   (b) $y' - xy = x^2$ [linear]

   (c) $(1 + x^2)(1 + y^2)y' = 4x$ [separable]

   (d) $xy' = 2 - xy$. [linear]

2. Determine an integrating factor for each of the following linear differential equations.

   (a) $y' - 2y = e^{-2x}$ $[e^{-2x}]$

   (b) $xy' + 3y = x^2e^{-2x}$ $[x^3]$

   (c) $xy' - 2y = x^3e^{2x}$ $[1/x^2]$

   (d) $\cos x\, y' + \sin x\, y = \sin 2x$ $[1/\cos x]$

   (e) $\tan x\, y' + 2y = \sin^3 x$. $[\sin^2 x]$

3. Verify that, for any constants A and B

   (a) $y = Ae^{4x} + Be^{-2x}$ is the general solution of $y'' - 2y' - 8y = 0$

   (b) $y = A\sinh 2x + B\cosh 2x$ is the general solution of $y'' - 4y = 0$ and

   (c) $y = e^{-3x}(A\cos x + B\sin x)$ is the general solution of $y'' + 6y' + 10y = 0$.

7.3 Second Order Differential Equations

We consider the linear differential equation

$$ay'' + by' + cy = f(x), \tag{7.5}$$

where the coefficients a, b and c are constants. The general solution of this equation is closely related to the solution of the so-called *homogeneous equation*

$$ay'' + by' + cy = 0,$$

which we investigate first.

7.3.1 The Homogeneous Equation $ay'' + by' + cy = 0$

In order to motivate the following discussion it is instructive to consider a particular example such as $y'' + 5y' + 6y = 0$. If we write this as

$$(y'' + 3y') + 2(y' + 3y) = 0$$

and make the change of dependent variable $z = y' + 3y$, we obtain the linear first order differential equation $z' + 2z = 0$ which, by using the methods of the previous section, has the general solution $z = Ae^{-2x}$. Thus

$$y' + 3y = Ae^{-2x}$$

which is another first order linear differential equation with integrating factor $I = e^{3x}$. Hence

$$\frac{d}{dx}(e^{3x}y) = Ae^{x}$$

which on integration yields

$$e^{3x}y = Ae^{x} + B.$$

Thus the general solution of $y'' + 5y' + 6y = 0$ is

$$y = Ae^{-2x} + Be^{-3x}.$$

Note that the exponents -2 and -3 which appear in the terms e^{-2x} and e^{-3x} are precisely the roots of the quadratic equation

$$m^2 + 5m + 6 = 0,$$

which has the same coefficients as the differential equation. It turns out that this is not pure coincidence.

Returning to the general case

$$ay'' + by' + cy = 0, \tag{7.6}$$

the foregoing example suggests that we look for solutions of the form $y = e^{mx}$. If $y = e^{mx}$ then $y' = me^{mx}$ and $y'' = m^2e^{mx}$ so that e^{mx} is a solution of Equation 7.6 provided

$$am^2e^{mx} + bme^{mx} + ce^{mx} = 0 \text{ or}$$

$$am^2 + bm + c = 0 \text{ since } e^{mx} \neq 0.$$

This quadratic, called *the auxiliary equation* (or *characteristic equation*) of Equation 7.6, has the same coefficients as Equation 7.6. For example, the auxiliary equation of

- $y'' - 3y' - 4y = 0$ is $m^2 - 3m - 4 = 0$,
- $2y'' - y' + 7y = 0$ is $2m^2 - m + 7 = 0$ and
- $y'' + 3y' = 0$ is $m^2 + 3m = 0$.

To complete the solution of such second order differential equations we need to consider the three cases corresponding to

1. $b^2 - 4ac > 0$ 2. $b^2 - 4ac < 0$ 3. $b^2 - 4ac = 0$.

1. If $b^2 - 4ac > 0$ the auxiliary equation has two distinct real roots m_1 and m_2 giving the solutions $e^{m_\cdot x}$ and $e^{m_\cdot x}$. The general solution of Equation 7.6 is then

$$y = Ae^{m_\cdot x} + Be^{m_\cdot x}.$$

2. If $b^2 - 4ac < 0$ the auxiliary equation has complex conjugate roots $\alpha \pm i\beta$ giving the solutions $e^{(\alpha+i\beta)x}$ and $e^{(\alpha-i\beta)x}$. The general solution of Equation 7.6 is then

$$y = A_1e^{(\alpha+i\beta)x} + B_1e^{(\alpha-i\beta)x}.$$

Noting that $e^{i\theta} = \cos\theta + i\sin\theta$ and $e^{-i\theta} = \cos\theta - i\sin\theta$ this solution can be expressed in a more convenient form as follows

$$\begin{aligned} y &= e^{\alpha x}(A_1e^{i\beta x} + B_1e^{-i\beta x}) \\ &= e^{\alpha x}(A_1(\cos\beta x + i\sin\beta x) + B_1(\cos\beta x - i\sin\beta x)) \\ &= e^{\alpha x}(A\cos\beta x + B\sin\beta x), \end{aligned}$$

where A and B are constants.

3. If $b^2 - 4ac = 0$ the auxiliary equation has two equal real roots $m_1 = m_2 = -b/2a$. The solution

   $$y = Ae^{m.\,x} + Be^{m.\,x} = Ae^{m.\,x} + Be^{m.\,x}$$

   cannot possibly be the general solution of Equation 7.6. (Why?)

   To find a second independent solution of Equation 7.6 we try $y = xe^{m.\,x}$.

   Then $y' = e^{m.\,x} + m_1 xe^{m.\,x}$ and $y'' = 2m_1 e^{m.\,x} + m_1^2 xe^{m.\,x}$ so that it is a solution of Equation 7.6 if

   $$2am_1 + am_1^2 x + b + bm_1 x + cx = 0,$$

   that is, if

   $$(2am_1 + b) + (am_1^2 + bm_1 + c)x = 0,$$

   and this is certainly so since m_1 is a double root of $am^2 + bm + c = 0$ and $m_1 = -\frac{b}{2a}$.

   Thus the general solution of Equation 7.6 is then

   $$\begin{aligned} y &= Axe^{m.\,x} + Be^{m.\,x} \\ &= (Ax + B)e^{m.\,x}. \end{aligned}$$

Consequently, we can write down the general solution of Equation 7.6 once we have determined the roots of the auxiliary equation. A summary of the solution procedure is given in Figure 7.1.

▌Worked Example 7.3.1 *Find the general solution of each of the following differential equations.*

1. $2y'' + 5y' - 3y = 0$

2. $y'' - 4y' + 4y = 0$

3. $\ddot{x} + 4\dot{x} + 13x = 0.$

1. The auxiliary equation is

   $$2m^2 + 5m - 3 = 0 \Rightarrow (2m - 1)(m + 3) = 0 \Rightarrow m = 1/2, -3.$$

   Thus the general solution is $y = Ae^{x/2} + Be^{-3x}$.
2. The auxiliary equation is

   $$m^2 - 4m + 4 = 0 \Rightarrow (m - 2)^2 = 0 \Rightarrow m = 2, 2.$$

   Thus the general solution is $y = (Ax + B)e^{2x}$.
3. The auxiliary equation is

   $$m^2 + 4m + 13 = 0 \Rightarrow m = \frac{-4 \pm \sqrt{16 - 52}}{2} = -2 \pm 3i,$$

   so that the general solution is $x = e^{-2t}(A \cos 3t + B \sin 3t)$.

■

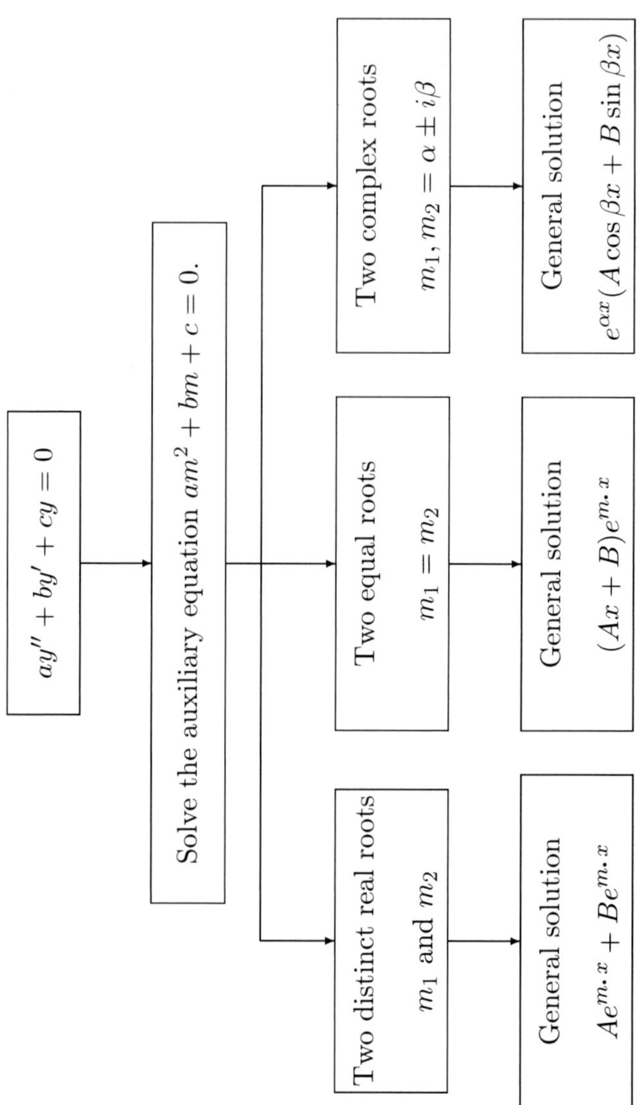

Figure 7.1: Solution procedure for a homogeneous differential equation.

Self-help exercises

1. Write down the general solution of the differential equation $ay''+by'+cy=0$ if the roots of the auxiliary equation are

   (a) -2 and 4 $[y = Ae^{-2x} + Be^{4x}]$

   (b) 0 and $1/2$ $[y = A + Be^{x/2}]$

   (c) -3 and -3 $[y = (Ax + B)e^{-3x}]$

   (d) $3i$ and $-3i$ $[y = A\sin 3x + B\cos 3x]$

   (e) $-2+5i$ and $-2-5i$. $[y = e^{-2x}(A\sin 5x + B\cos 5x)]$

2. Find the general solution of each of the following differential equations.

   (a) $4y'' - y = 0$ $[y = Ae^{x/2} + Be^{-x/2}]$

   (b) $2y'' + 3y' - 2y = 0$ $[y = Ae^{-2x} + Be^{x/2}]$

   (c) $y'' - 6y' + 25y = 0$ $[y = e^{3x}(A\sin 4x + B\cos 4x)]$

   (d) $9y'' - 12y' + 4y = 0$. $[y = (Ax + B)e^{2x/3}]$

7.3.2 The General Equation $ay'' + by' + cy = f(x)$

If y_C is the general solution of the equation

$$ay'' + by' + cy = 0$$

and y_P is *any* solution of

$$ay'' + by' + cy = f(x)$$

then we can show that

$$\boxed{y_G = y_C + y_P}$$

is the general solution of

$$\boxed{ay'' + by' + cy = f(x)}$$

This result can be established as follows. By the definitions of y_P and y_G

$$ay_P'' + by_P' + cy_P = f(x) \tag{7.7}$$

$$ay_G'' + by_G' + cy_G = f(x). \tag{7.8}$$

Subtracting Equation 7.7 from Equation 7.8 gives

$$a(y_G - y_P)'' + b(y_G - y_P)' + c(y_G - y_P) = 0.$$

But y_C is the general solution of this homogeneous differential equation. Thus

$$y_G - y_P = y_C \text{ or}$$

$$y_G = y_C + y_P.$$

Note that

y_C is called the complementary function (CF) of Equation 7.5

y_P is called a particular integral (PI) of Equation 7.5.

In the previous section we discussed how to find the complementary function y_C. For the range of right-hand sides discussed in these notes, a particular integral y_P is usually found by intelligent guesswork (called the method of undetermined coefficients) and depends on the form of the function $f(x)$ appearing on the right-hand side. (Other techniques — such as "Variation of Parameters" — are available for more general right-hand sides; the student is referred to the standard references for a discussion of such methods.)

The techniques for finding y_P are illustrated in this text by considering a comprehensive range of examples.

Case 1 *$f(x)$ a polynomial.*

Example: Consider the particular example $y'' + 5y' + 6y = 12x + 4$.

Solution: The complementary function is $y_C = Ae^{-2x} + Be^{-3x}$. To find y_P we ask the question:

> What function when differentiated twice and substituted into the left-hand side $y'' + 5y' + 6y$ will yield the right-hand side $12x + 4$?

An obvious candidate is $y = ax + b$ for suitable values of a and b.

If $y = ax + b$ then $y' = a$ and $y'' = 0$ so that it is a solution of $y'' + 5y' + 6y = 12x + 4$ if

$$0 + 5a + 6(ax + b) = 12x + 4 \Rightarrow 6ax + (5a + 6b) = 12x + 4.$$

Equating coefficients gives $6a = 12$ and $5a + 6b = 4$; that is, $a = 2$ and $b = -1$.

Thus $y_P = 2x - 1$ is a particular integral so that the general solution of the differential equation $y'' + 5y' + 6y = 12x + 4$ is

$$y = Ae^{-2x} + Be^{-3x} + 2x - 1.$$

Note that

- To obtain a particular integral for:

$$y'' + 2y' + 3y = x^2 \text{ we would try } y = ax^2 + bx + c$$

$$y'' + 2y' + 5y = 3x \text{ we would try } y = ax + b.$$

- If $f(x)$ is a constant, a particular integral can be found immediately by inspection.

  For example, $y = 7$ is an obvious solution of $2y'' - 3y' + 8y = 56$.

- For the differential equation $y'' + 5y' + 6y = 12x + 4$ we tried $y = ax + b$ to obtain a particular solution.

  If the differential equation had instead been $y'' + 5y' = 12x + 4$ you may be inclined to once again try $y = ax + b$ to obtain a particular integral.

  However this will not succeed. (Try it!)

  In order for $y'' + 5y'$ to equal $12x + 4$ we need to try a polynomial of degree 2; that is, to try $y = ax^2 + bx + c$. (See the following Worked Example 7.3.2.)

Worked Example 7.3.2 *For the differential equation*

$$y'' + 3y' = 6x,$$

find

1. *the general solution*
2. *the particular solution for which* $y(0) = 0$ *and* $y'(0) = 0$.

1. We first find the complementary function.
   The auxiliary equation $m^2 + 3m = 0$ has roots $m = 0, -3$ so that the complementary function is

$$y_C = A + Be^{-3x}.$$

   To find a particular integral we ask the question:

   > What function when differentiated twice and substituted into $y'' + 3y'$ will yield $6x$?

   A quadratic! (NOT a linear function of x.)
   Try $y = ax^2 + bx + c$.
   Then $y' = 2ax + b$ and $y'' = 2a$ so that it is a solution if $2a + 3(2ax + b) = 6x$.

Equating coefficients gives

$$6a = 6, 2a + 3b = 0 \Rightarrow a = 1 \text{ and } b = -\frac{2}{3}.$$

Note that c cannot be determined; we take $c = 0$ so that $y_P = x^2 - \frac{2}{3}x$ is a particular integral. (Why does any value of c lead to the same general solution?)
The general solution is thus $y = y_C + y_P$; that is,

$$y = A + Be^{-3x} + x^2 - \frac{2}{3}x.$$

2. $y(0) = 0 \Rightarrow 0 = A + B$ and $y'(0) = 0 \Rightarrow 0 = -3B - \frac{2}{3}$.
Thus $B = -\frac{2}{9}$ and $A = \frac{2}{9}$ so that the solution which satisfies the given conditions is

$$y = \frac{2}{9} - \frac{2}{9}e^{-3x} + x^2 - \frac{2}{3}x.$$

■

Case 2 $f(x)$ *an exponential.*

Example: Consider the particular example $y'' + 5y' + 6y = 12e^{2x}$.
Solution: As before the complementary function is given by $y_C = Ae^{-2x} + Be^{-3x}$.
To find a particular integral we ask the question:

> What function when differentiated twice and substituted into the left-hand side $y'' + 5y' + 6y$ will yield the right-hand side $12e^{2x}$?

(Why is $y = ae^{2x}$ a likely function?)
If $y = ae^{2x}$ then $y' = 2ae^{2x}$ and $y'' = 4ae^{2x}$ so that it is a solution if

$$4ae^{2x} + 10ae^{2x} + 6ae^{2x} = 12e^{2x},$$

that is, if $20a = 12$ or $a = 3/5$ since $e^{2x} \neq 0$.
Thus $y_P = \frac{3}{5}e^{2x}$ is a particular integral and the general solution is

$$y = Ae^{-2x} + Be^{-3x} + \frac{3}{5}e^{2x}.$$

Example: Now consider the equation $y'' + 5y' + 6y = 12e^{-2x}$.
Solution: This time the function on the RHS is part of the complementary function $Ae^{-2x} + Be^{-3x}$ and trying the expression $y = ae^{-2x}$ for a particular integral will not work. (Try it!)

Instead,[1] we need to try $y = axe^{-2x}$.
Then $y' = a(1-2x)e^{-2x}$ and $y'' = a(-4+4x)e^{-2x}$ so that axe^{-2x} is a solution if

$$a(-4+4x)e^{-2x} + 5a(1-2x)e^{-2x} + 6axe^{-2x} = 12e^{-2x}.$$

Therefore $a(-4+4x) + 5a(1-2x) + 6ax = 12$, so that $a = 12$.
Hence $y_P = 12xe^{-2x}$ is a particular integral and the general solution is

$$y = Ae^{-2x} + Be^{-3x} + 12xe^{-2x}.$$

In summary then, if the RHS ae^{mx} is part of the complementary function, try $y = axe^{mx}$ (instead of $y = ae^{mx}$) for a particular integral.

But what if xe^{mx} is also part of the complementary function as in $y'' - 4y' + 4y = 3e^{2x}$? (Here the complementary function is $y = (Ax+B)e^{2x}$.) Trying $y = ae^{2x}$ for a particular integral will not work and neither will $y = axe^{2x}$. (Try it!) In such cases we try the expression $y = ax^2e^{2x}$.

You should note that when finding the general solution it is important that the complementary function is found first as this can affect the choice of a trial function for a particular integral.

Case 3 $f(x) = a\cos cx + b\sin cx$.

Example: Consider the particular example $y'' + 5y' + 6y = 26\cos 2x$.

Solution: Once again the complementary function is $y_C = Ae^{-2x} + Be^{-3x}$.
To find a particular integral, an initial guess might be to try $y = a\cos 2x$ but this will not work since differentiation of a cosine function yields a sine function.
We need to try $y = a\cos 2x + b\sin 2x$.
Then $y' = -2a\sin 2x + 2b\cos 2x$ and $y'' = -4a\cos 2x - 4b\sin 2x$ so that the above choice is a solution provided

$$(-4a\cos 2x - 4b\sin 2x) + 5(-2a\sin 2x + 2b\cos 2x) +$$

$$6(a\cos 2x + b\sin 2x) = 26\cos 2x.$$

Equating coefficients of $\cos 2x$ and $\sin 2x$ gives

$$2a + 10b = 26 \text{ and } -10a + 2b = 0 \Rightarrow a = \frac{1}{2}, b = \frac{5}{2}.$$

[1] This is a general procedure. If the usual "guess" does not work — because part of the assumed form lies in the complementary function — then multiply the original form by the independent variable x. This process is repeated until no part of the guess for the particular integral lies in the complementary function.

Note the similarity of this procedure to the form the independent solutions of the complementary function take when the roots of the auxiliary equation are repeated.

Thus $y_P = \frac{1}{2}(\cos 2x + 5\sin 2x)$ and the general solution is

$$y = Ae^{-2x} + Be^{-3x} + \frac{1}{2}(\cos 2x + 5\sin 2x).$$

▌**Worked Example 7.3.3** *Find the general solution of $y'' + 4y = 26\cos 2x$.*
We first find the complementary function.
The auxiliary equation $m^2 + 4 = 0$ has roots $\pm 2i$.
Thus $y_C = e^{0x}(A\cos 2x + B\sin 2x) = A\cos 2x + B\sin 2x$.
To find a particular integral we might be tempted to try $y = a\cos 2x + b\sin 2x$ but this will not work because part of it (in fact all in this case) is in the complementary function. (Verify that the above choice does indeed fail!)
Instead we try $y = x(a\cos 2x + b\sin 2x)$.
Then $y' = a\cos 2x + b\sin 2x + x(-2a\sin 2x + 2b\cos 2x)$ and $y'' = -4a\sin 2x + 4b\cos 2x + x(-4a\cos 2x - 4b\sin 2x)$ so that this new choice will be a solution of $y'' + 4y = 26\cos 2x$ if

$$-4a\sin 2x + 4b\cos 2x = 26\cos 2x$$

that is, if $-4a = 0$ and $4b = 26$. Thus $y_P = \frac{13}{2}x\sin 2x$ and the general solution is

$$y = A\cos 2x + B\sin 2x + \tfrac{13}{2}x\sin 2x.$$

■

Case 4 $f(x) = g(x)e^{mx}$.

Differential equations such as $y'' + 5y' + 6y = xe^{mx}$ and $y'' + 5y' + 6y = e^{mx}\sin 3x$ can be reduced to the cases previously considered by putting $y = v(x)e^{mx}$ (i.e. reformulate the differential equation in terms of a new dependent variable v) which effectively removes the exponential.

▌**Worked Example 7.3.4** *Find the general solution of $y'' - y' - 2y = 4e^{2x}\sin x$.*
The auxiliary equation $m^2 - m - 2 = 0$ has the two roots $m = -1, 2$ so that the CF is $y_C = Ae^{-x} + Be^{2x}$.
To obtain a particular integral we put $y = ve^{2x}$ so that $y' = (v' + 2v)e^{2x}$ and $y'' = (v'' + 4v' + 4v)e^{2x}$.
The differential equation $y'' - y' - 2y = 4e^{2x}\sin x$ becomes

$$v'' + 4v' + 4v - v' - 2v - 2v = 4\sin x \text{ or}$$

$$v'' + 3v' = 4\sin x.$$

To determine the particular integral for this equation, try $v = a\sin x + b\cos x$.
Then $v' = a\cos x - b\sin x$ and $v'' = -a\sin x - b\cos x$ so that the trial function is a solution if

$$-a\sin x - b\cos x + 3a\cos x - 3b\sin x = 4\sin x.$$

This equation is satisfied if $-a - 3b = 4$ and $-b + 3a = 0$; that is, if $a = -2/5$ and $b = -6/5$.
Thus $y_P = -\frac{1}{5}(2\sin x + 6\cos x)e^{2x}$ so that the general solution is

$$y = Ae^{-x} + Be^{2x} - \frac{1}{5}(2\sin x + 6\cos x)e^{2x}.$$

■

Case 5 *$f(x)$ is a sum of functions considered in the previous cases.*

It is easily shown that if $y = y_1$ is a particular integral of the differential equation $ay'' + by' + cy = f_1(x)$ and $y = y_2$ is a particular integral of the differential equation $ay'' + by' + cy = f_2(x)$ then $y = y_1 + y_2$ is a particular integral of the differential equation $ay'' + by' + cy = f_1(x) + f_2(x)$.

For example, to find a particular integral for

$$y'' - y' - 2y = 4 + 6e^x + \sin 2x, \tag{7.9}$$

we first find particular integrals for each of the equations

$$y'' - y' - 2y = 4$$
$$y'' - y' - 2y = 6e^x$$
$$y'' - y' - 2y = \sin 2x$$

separately, and then the particular integral for Equation 7.9 is just the sum of the three individual particular integrals.

Worked Example 7.3.5 *Find the general solution of*

$$y'' - y' - 6y = 4\cosh^2 x. \tag{7.10}$$

The auxiliary equation $m^2 - m - 6 = 0$ has the roots -2 and 3 so that the complementary function is $y_C = Ae^{-2x} + Be^{3x}$.
To determine the particular integral, note that using the definition of the hyperbolic function cosh, the RHS can be written as $e^{2x} + 2 + e^{-2x}$.

We thus consider separately the equations

$$y'' - y' - 6y = e^{2x}$$
$$y'' - y' - 6y = 2$$
$$y'' - y' - 6y = e^{-2x}.$$

The first has the particular integral $-\frac{1}{4}e^{2x}$. (Verify this!)
The second has the particular integral $-\frac{1}{3}$. (Verify this!)
The third has the particular integral $-\frac{1}{5}xe^{-2x}$. (Verify this!)
Thus a particular integral of Equation 7.10 is

$$y_P = -\frac{1}{4}e^{2x} - \frac{1}{3} - \frac{1}{5}xe^{-2x},$$

so that the general solution is

$$y = Ae^{-2x} + Be^{3x} - \frac{1}{4}e^{2x} - \frac{1}{3} - \frac{1}{5}xe^{-2x}.$$

■

Worked Example 7.3.6 *A particle of mass m, attached to the free end of a vertical spring of stiffness k, is oscillating vertically subject to a viscous resistance λ times its speed. If the upper end of the spring is constrained to oscillate with angular frequency ω, the displacement $x(t)$ of the mass at time t, is given by the solution of*

$$m\frac{d^2x}{dt^2} + \lambda\frac{dx}{dt} + kx = mg + a\cos\omega t.$$

Determine $x(t)$ if $\lambda^2 < 4mk$.
The auxiliary equation $mc^2 + \lambda c + k = 0$ has roots given by

$$\begin{aligned} c &= \frac{-\lambda \pm \sqrt{\lambda^2 - 4mk}}{2m} \\ &= -\frac{\lambda}{2m} \pm \frac{\Omega i}{2m} \end{aligned}$$

where $\Omega = \sqrt{4mk - \lambda^2}$.
Thus the complementary function is given by

$$x = e^{-\frac{\lambda t}{\cdot m}}(A\cos\tfrac{\Omega t}{2m} + B\sin\tfrac{\Omega t}{2m}).$$

The particular integral corresponding to the term mg on the RHS is mg/k and the particular integral corresponding to the term $a\cos\omega t$ is

$$x = \frac{a}{(k - m\omega^2)^2 + \lambda^2\omega^2}((k - m\omega^2)\cos\omega t + \lambda\omega\sin\omega t)$$

(obtained by trying a solution of the form $x = c\cos\omega t + d\sin\omega t$). Thus the general solution is

$$x = \frac{mg}{k} + e^{-\frac{\lambda t}{\cdot m}}(A\cos\frac{\Omega t}{2m} + B\sin\frac{\Omega t}{2m}) + \frac{a}{(k-m\omega^2)^2+\lambda^2\omega^2}((k-m\omega^2)\cos\omega t + \lambda\omega\sin\omega t).$$

■

Self-help exercises

1. In each of the following differential equations, indicate the form of the function you would try for a particular integral (you may have to first find the complementary function).

   (a) $y'' - 2y' - 3y = 3x + 6$ $[y = ax + b]$
   (b) $y'' - 2y' = 3x + 6$ $[y = ax^2 + bx + c]$
   (c) $y'' - 4y = 5e^x$ $[y = ae^x]$
   (d) $y'' - 4y = 5e^{2x}$ $[y = axe^{2x}]$
   (e) $y'' - 6y' + 9y = 2e^{3x}$. $[y = ax^2e^{3x}]$

2. The general solution of the differential equation $y'' + 4y = 0$ is $y = A\sin 2x + B\cos 2x$.

   Hence write down the form of the function you would try for a particular integral of

   (a) $y'' + 4y = 4\sin 3x$ $[y = a\sin 3x + b\cos 3x]$
   (b) $y'' + 4y = 4\sin 2x$. $[y = x(a\sin 2x + b\cos 2x)]$

7.4 Numerical Solution of Differential Equations

Many differential equations that arise in practice cannot be solved exactly in terms of elementary functions. However, a large variety of methods have been devised to obtain approximate numerical estimates of the solutions of such equations. Here we shall restrict our attention to a very simple method to "solve" first order differential equations expressible in the form

$$y' = f(x, y) \text{ where } y(x_0) = y_0. \tag{7.11}$$

More general and better methods are available but are not discussed in this text.

We shall try to determine solutions of Equation 7.11 corresponding to values of x beginning at $x = x_0$ (which is the position in which the initial condition is imposed) and proceed in steps of h; that is, for $x = x_0, x_0 + h, x_0 + 2h, x_0 + 3h, \ldots$ where the value of h is predetermined.

Notation: Throughout this section we shall denote

- the value of $x_0 + nh$ by x_n
- the corresponding value of y, that is $y(x_0 + nh)$, by y_n
- the corresponding numerical approximations to y_n by Y_n.

Using the notation introduced in Figure 7.2 it is clear that

$$y_1 = SN \approx QN = TN + QT = y_0 + PT \times y'(x_0) = y_0 + hf(x_0, y_0).$$

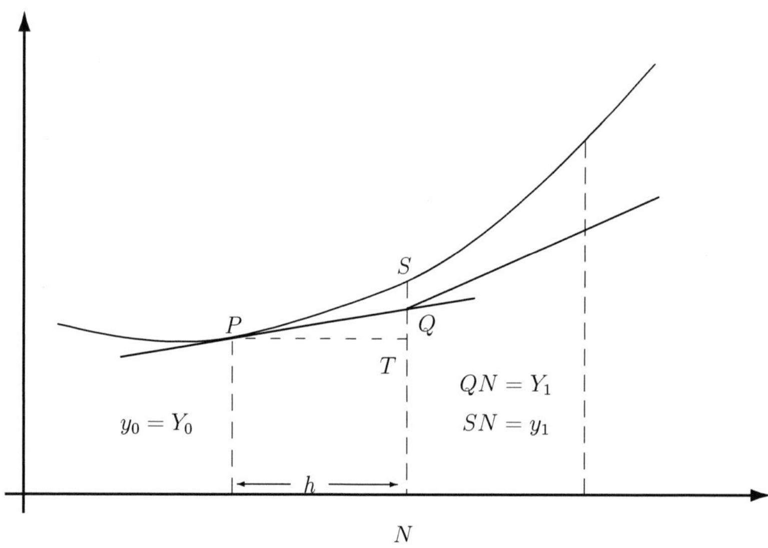

Figure 7.2: Euler's method of obtaining an approximate solution to Equation 7.11.

In general,

$$y_{n+1} \approx y_n + hf(x_n, y_n).$$

Once an estimate for y_n is known, y_{n+1} can be approximated.

Using Figure 7.2, we see that

$$Y_1 = Y_0 + hY_0' = Y_0 + hf(x_0, Y_0) \qquad (7.12)$$

gives an approximation for the solution y_1 in terms of $Y_0(= y_0)$ and in general

$$Y_{n+1} = Y_n + hY_n' = Y_n + hf(x_n, Y_n) \qquad (7.13)$$

gives an approximation for the solution y_{n+1} of Equation 7.11 in terms of the approximation for y_n. The algorithm 7.13 is called *Euler's method.*

At each step in the procedure an error is involved due to approximating the length of ST by QT. Consequently there is an *accumulation* of these errors the further this procedure is carried out. Clearly, at each step the size of the error can be reduced by choosing a smaller value of the step size h.

Worked Example 7.4.1 *Use Euler's method to find approximate numerical estimates of the solution of the initial value problem*

$$y' = x + y \text{ where } y(0) = 1$$

on the interval $[0, 1]$ *and compare the solutions with the exact (that is analytical) solution.*

You should verify that this differential equation can be solved analytically with solution $y = 2e^x - x - 1$.

Euler's formula is

$$\begin{aligned} Y_{n+1} &= Y_n + hf(x_n, Y_n) \\ &= Y_n + h(x_n + Y_n) \text{ with } Y_0 = 1. \end{aligned}$$

If we take $h = 0.1$, then

$$Y_{n+1} = 0.1x_n + 1.1Y_n.$$

Putting $n = 0$ gives an approximation to $y(0.1)$ as

$$Y_1 = 0.1x_0 + 1.1Y_0 = 1.1000 \text{ since } x_0 = 0 \text{ and } Y_0 = 1.$$

To four decimal places the analytic solution at $x = 0.1$ is 1.1103 so that the error in this numerical estimate at the point $x = 0.1$ is 0.0103.

Putting $n = 1$ gives the following approximation to $y(0.2)$:

$$Y_2 = 0.1x_1 + 1.1Y_1 = (0.1)^2 + (1.1)^2 = 1.2200.$$

Proceeding in this way, we successively obtain $Y_3, Y_4, \ldots, Y_{10}$. These results have been tabulated in Table 7.1 together with the corresponding analytic results. Note that the error increases as n increases. The accuracy could have been improved by taking a smaller value of h, but this involves more calculation and a possible accumulation of rounding errors. The values obtained by Euler's method using $h = 0.05$ are given in the last column of Table 7.1. You should compare the solutions using $h = 0.1$ and $h = 0.05$, noting that the smaller value of h gives better accuracy. ■

x	Y_n	$2e^x - x - 1$	Error	$Y_{2n}\,(h = 0.05)$
0.0	1.0000	1.0000	0.0000	1.0000
0.1	1.1000	1.1103	0.0103	1.1050
0.2	1.2200	1.2428	0.0228	1.2311
0.3	1.3620	1.3997	0.0377	1.3803
0.4	1.5282	1.5836	0.0554	1.5550
0.5	1.7210	1.7974	0.0764	1.7579
0.6	1.9431	2.0442	0.1011	1.9918
0.7	2.1974	2.3275	0.1301	2.2600
0.8	2.4871	2.6511	0.1640	2.5659
0.9	2.8158	3.0192	0.2034	2.9134
1.0	3.1874	3.4366	0.2492	3.3068

Table 7.1: Comparison of analytical and numerical estimates of Worked Example 7.4.1.

In this topic we have

- Introduced first order differential equations of the separable and linear types.
- Discussed applications to Newton's law of cooling.
- Examined homogeneous and non-homogeneous second order linear constant coefficient differential equations.
- Generated particular integrals corresponding to right-hand sides which are (combinations of) polynomials, exponentials, cosine and sine.
- Applied this technique to analyse mechanical oscillations.
- Discussed Euler's method for finding numerical approximations.

7.5 Quick Test Number 7

Basic Mathematical Skills

Question **Selection**

Logarithms and exponentials:

1. $\int -\frac{2}{x}\,dx$ equals (a) $-2\log|x|$ (b) $\frac{2}{x^2}$ (c) $-\frac{2}{x^2}$
2. $-2\log x$ equals
   (a) $\log(-2x)$ (b) $\log(-2)+\log x$ (c) $\log(-x^2)$ (d) $\log(1/x^2)$
3. $\mathrm{e}^{-2\log x}$ equals (a) e^{-2x} (b) $-2x$ (c) $\frac{1}{x^2}$ (d) $-\frac{1}{x^2}$ (e) $-x^2$

Quadratic equations:

4. $m^2+4m+5=0$; m equals
   (a) -1 or -4 (b) 1 or 4 (c) $-2+i$ or $-2-i$
5. $m^2+4m=0$; m equals (a) 0 or 4 (b) 0 or -4 (c) $2i$ or $-2i$
6. $m^2+9=0$; m equals (a) $3i$ or $-3i$ (b) 3 or -3 (c) 3 repeated

---

First Order Differential Equations

Question **Selection**

---

1. Which of the following are differential equations?

(a) $y + 3xy = 5x$ (b) $y' + 3xy = 5x$
(c) $y^2 + 3y + 4 = \cos 2x$ (d) $y'' + 3y' + 4y = \cos 2x$

Classify the following first order differential equations as separable or linear or both.

2. $y' + xy^2 = 0$ (a) separable (b) linear (c) both
3. $y' + xy = 0$ (a) separable (b) linear (c) both
4. $y' + xy = \cos x$ (a) separable (b) linear (c) both
5. $x^2y' = 4x - 2y$ (a) separable (b) linear (c) both
6. $x^2y' = x^4e^y$ (a) separable (b) linear (c) both
7. $y' - xy = xy^2$ (a) separable (b) linear (c) both

Solution of linear differential equations:

8. An integrating factor for the linear DE $y' + 10y = \cos x$ is

(a) e^{10x} (b) $10x$ (c) 10 (d) e^{10}

9. An integrating factor for the linear DE $xy' - 2y = x^5$ is

(a) e^{-2x} (b) $-2x$ (c) $\frac{1}{x^2}$ (d) $-\frac{1}{x^2}$ (e) $-x^2$

Homogeneous Second Order Differential Equations

Question **Selection**

---

1. For the DE $y''-4y'-5y = 0$, the characteristic equation $m^2-4m-5 = 0$ is obtained by trying

(a) $y = m$ (b) $y = Ae^{mx}$ (c) $y = me^{Ax}$ (d) $y = Ame^{mx}$

Choose the correct characteristic equation:

2. $y'' - 2y' - 3y = 0$ (a) $m^2 - 2m - 3 = 0$ (b) $m^2 - 2m - 3m = 0$

3. $y'' - 5y' = 0$ (a) $m^2 - 5 = 0$ (b) $m^2 - 5m = 0$

4. $y'' + 4y = 0$ (a) $m^2 + 4m = 0$ (b) $m^2 + 4 = 0$

Given the roots of the characteristic equation, choose the correct solution:

5. $m = -1$ and $m = -4$

   (a) $y = Ae^{-1} + Be^{-4}$ (b) $y = Ae^{-x} + Be^{-4x}$ (c) $y = -e^{x} - 4e^{x}$

6. $m = 3$ and $m = 3$

   (a) $y = Ae^{3} + Bxe^{3}$ (b) $y = Ae^{3x} + Be^{3x}$ (c) $y = Ae^{3x} + Bxe^{3x}$

7. $m = -3i$ and $m = 3i$

   (a) $y = Ae^{-3i} + Be^{3i}$ (b) $y = A\cos 3x + B\sin 3x$

   (c) $y = A\cos 3x + Bx\cos 3x$

8. $m = 0$ and $m = 2$

   (a) $y = A + Be^{2}$ (b) $y = Ae^{x} + Be^{2x}$ (c) $y = A + Be^{2x}$

9. $m = 2 - 5i$ and $m = 2 + 5i$

   (a) $y = Ae^{5x}\cos 2x + Be^{5x}\sin 2x$ (b) $y = Ae^{2x}\cos 5x + Be^{2x}\sin 5x$

General Second Order Differential Equations

Question **Selection**

If part of the usual guess for y_P is in the complementary function y_C, then this guess for y_P must be multiplied by x.
Choose y_P in the following:

1. For $y'' - 4y' - 5y = 7e^{2x}$ the complementary function is $y_C = Ae^{-x} + Be^{5x}$.

   (a) $y_P = ae^{2x}$ (b) $y_P = axe^{2x}$ (c) $y_P = ae^{5x}$ (d) $y_P = 7e^{2x}$

2. For $y'' - 2y' - 3y = 7e^{3x}$ the complementary function is $y_C = Ae^{-x} + Be^{3x}$.

   (a) $y_P = ae^{3x}$ (b) $y_P = axe^{3x}$ (c) $y_P = ax^2e^{3x}$ (d) $y_P = 7e^{3x}$

3. For $y'' + 9y = 3\cos 5x$ the complementary function is $y_C = A\cos 3x + B\sin 3x$.

(a) $y_P = 3\cos 5x$ (b) $y_P = ax\cos 5x$

(c) $y_P = a\cos 5x + b\sin 5x$ (d) $y_P = ax\cos 3x + bx\sin 3x$

4. For $y'' - 4y' + 4y = 5e^{2x}$ the complementary function is $y_C = Ae^{2x} + Bxe^{2x}$.

(a) $y_P = ae^{2x}$ (b) $y_P = axe^{2x}$ (c) $y_P = ax^2e^{2x}$ (d) $y_P = 5e^{2x}$

5. For $y'' + 16y = 6\cos 4x$ the complementary function is $y_C = A\cos 4x + B\sin 4x$.

(a) $y_P = a\cos 4x$ (b) $y_P = ax\cos 4x$

(c) $y_P = a\cos 4x + b\sin 4x$ (d) $y_P = ax\cos 4x + bx\sin 4x$

6. For $y'' - 4y = 3x + 1$ the complementary function is $y_C = Ae^{-2x} + Be^{2x}$.

(a) $y_P = ax^2 + bx$ (b) $y_P = ax + b$

(c) $y_P = 3x + 1$ (d) $y_P = 3ax + b$

7. For $y'' + 2y' = x + 5$ the complementary function is $y_C = A + Be^{-2x}$.

(a) $y_P = ax^2 + bx$ (b) $y_P = ax + b$

(c) $y_P = x + 5$ (d) $y_P = ae^{-2x}$

7.6 Exercises

> Before attempting any of these miscellaneous exercises, make sure that you have successfully answered all the self-help exercises appearing throughout the chapter.

1. Find the general solutions of the following differential equations:

   (a) $yy' = x$

   (b) $y' - 2xy = 0$

   (c) $y(xy' + y) = 2$

   (d) $y' = 10y$

   (e) $xyy' = 1 + x$.

2. Determine the solution of

   (a) $y' + 2y = 4$, with $y(0) = 0$

   (b) $xy' - y(y - 1) = 0$, with $y(1) = 2$

   (c) $\sqrt{1 - x^2}y' = \sqrt{1 - y^2}$, with $y(0) = 1$

   (d) $\sqrt{1 + x^2}y' = xe^{-y}$, with $y(1) = 0$.

3. Find the solution of

   $$y' = \frac{x^2 + y^2}{2xy},$$

   by putting $y = vx$.

4. Use the substitution $v = x + y$ to determine the solution of

   $$y' = \frac{x + y - 2}{x + y + 2}.$$

5. Obtain the general solution of

   (a) $y' - 2y = x$

   (b) $xy' + y = e^x$

   (c) $xy' - 4y = x^5e^{2x}$

   (d) $\cos x\, y' - y \sin x = \sin x$

   (e) $y' + 3y = e^{-x}$

   (f) $y' + 2y \tan x = \sin x$.

6. Determine the solutions of

   (a) $(1 - x^2)y' + xy = 0$, with $y(0) = 3$

   (b) $xy' = (y - 1)(y - 2)$, with $y(4) = 0$

   (c) $y' = \cos^2 y$, with $y(1) = 0$

   (d) $xy' + y = x\sqrt{x}$, with $y(1) = 0$.

7. Solve each of the following second order differential equations:

   (a) $y'' \quad y' = x$

   (b) $y'' - 2y' = e^x$

   (c) $y'' + yy' = 0$.

   (Note that each can be reduced to a first order equation by putting $v = y'$.)

8. The tangent to a curve at an arbitrary point $P(x, y)$ cuts the y-axis at a point Q. If the product of the ordinates at P and Q is 1 and the curve passes through the point $(1, 2)$, find the equation of the curve.

9. A tank contains 100 litres of solution in which is dissolved 200 kg of salt. A salt solution of concentration 1 kg/L flows into the tank at the rate of 3 L/min. and the mixture, kept uniform by stirring, flows out at the rate of 2 L/min. Find the amount of salt in the tank after t minutes.

10. A student carrying a flu virus returns to an isolated school campus consisting of 1000 students. It is observed that a total of 50 students are infected after four days. If the rate at which the virus spreads is assumed to be proportional to both the number of infected students and the number of students not infected, determine the number of infected students after six days.

11. A body is projected vertically downward with velocity V in a medium which is assumed to offer a resistance that is proportional to the speed.

    Determine the velocity as a function of time and find the terminal velocity.

12. A body is cooling in surroundings maintained at 10°C. Its temperature θ°C after t minutes is given by

$$\frac{d\theta}{dt} = -k(\theta - 10)$$

    where k is a constant. If the temperature of the body is initially 70°C and 10 minutes later is 40°C

    (a) show that $k = \frac{1}{10}\log 2$
    (b) find the body's temperature after a further 15 minutes.

13. An e.m.f. $E(t)$ is applied to an electrical circuit containing a resistance R in series with an inductance L. The current $i(t)$ at time t in the circuit is given by the solution of

$$L\frac{di}{dt} + Ri = E(t)$$

    where $i(0) = 0$. Determine $i(t)$ and sketch the i-t graph when

    (a) $E(t) = E_0$
    (b) $E(t) = E_0 \sin \omega t$
    (c) $E(t) = E_0 e^{-at}, \quad a \neq \frac{R}{L}$

    where E_0 is a constant.

14. Determine the general solution of each of the following differential equations:

    (a) $y'' - y' - 6y = 0$
    (b) $y'' + 2y' - 8y = 0$
    (c) $y'' - 6y' + 9y = 0$
    (d) $y'' + 2y' + 10y = 0$
    (e) $y'' + 4y' + 9y = 0$
    (f) $4y'' + 4y' + y = 0$
    (g) $y'' - 9y = 0$
    (h) $y'' - 9y' = 0$
    (i) $\ddot{y} - 2\dot{y} + 2y = 0$
    (j) $\ddot{y} + 4y = 0.$

15. Find the general solution of each of the following differential equations:

    (a) $y'' - y' - 6y = 12$

(b) $y'' - y' - 6y = 12x$

(c) $\ddot{y} - \dot{y} - 2y = 4e^t$

(d) $y'' - y' - 2y = 4e^{2x}$

(e) $y'' + y' - 2y = 4x^2$

(f) $\ddot{y} - 2\dot{y} = 2t$

(g) $y'' - 8y' + 16y = 24e^{4x}$

(h) $y'' - 2y' + 5y = 20 \sin 2x$

(i) $y'' + 4y = 20 \sin 2x$.

16. Solve $y'' - 5y' + 6y = 6e^{ax}$ for the cases

    (a) $a = 4$

    (b) $a = 2$.

17. Determine the solution of

$$y'' - 2y' - 3y = 8e^{3x},$$

if $y(0) = y'(0) = 3$.

18. Determine the general solution of each of the following differential equations

    (a) $y'' - 4y = 8x^2 + 3e^x$

    (b) $y'' - y' - 2y = 4 \cosh 2x$

    (c) $y'' + 2y' + 2y = 4 \sin^2 x$

    (d) $y'' + 2y' + y = e^x \cos x$

    (e) $y'' - 3y' + 2y = 6xe^{-x}$

    (f) $3y'' - y' - 2y = xe^{-x}$

    (g) $y'' + 4y = e^x \sin 2x$.

19. If $y = ve^{x^{\cdot}}$ show that

$$y' = (v' + 2xv)e^{x^{\cdot}}$$

and

$$y'' = (v'' + 4xv' + 4x^2v + 2v)e^{x^{\cdot}}.$$

Use this substitution to transform the differential equation

$$y'' - 4xy' + (4x^2 - 1)y = xe^{x^{\cdot}} \qquad (7.14)$$

to one having constant coefficients.

Hence obtain the general solution of Equation 7.14.

20. Use the substitution $x = e^t$ to simplify the differential equation

$$x^2y'' + 2xy' = (\log x)^2 \qquad (7.15)$$

to one having constant coefficients.

Hence determine the solution of Equation 7.15.

21. Bessel's equation of order 1/2 is

$$x^2y'' + xy' + (x^2 - \frac{1}{4})y = 0. \qquad (7.16)$$

Show that the substitution $y = v/\sqrt{x}$ simplifies this equation to

$$\frac{d^2v}{dx^2} + v - 0.$$

Hence find the general solution of Equation 7.16.

22. A constant e.m.f. E_0 volts is applied to a circuit containing, in series, a resistance R ohms, an inductance L henries and a condenser having capacitance C farads. The charge $q(t)$ on the condenser (which is initially uncharged) is given by

$$L\frac{d^2q}{dt^2} + R\frac{dq}{dt} + \frac{q}{C} = E_0.$$

If the values of resistance, conductance and capacitance take the values

$$R = 100, \quad L = 0.005$$

$$C = 10^{-6}, \ E_0 = 1000$$

show that

$$q = 0.001 + e^{-\alpha t}(A\cos\alpha t + B\sin\alpha t),$$

where $\alpha = 10\,000$.

23. A body of unit mass moves along the x-axis under the action of

- a force of magnitude $\omega^2 x$ directed towards the origin
- a driving force $e^{-\lambda t}\cos bt$
- a frictional resistance $2\lambda\dot{x}$ where $\dot{x}$ is the speed of the body.

The displacement $x(t)$ at time t is given by

$$\ddot{x} + 2\lambda\dot{x} + \omega^2 x = e^{-\lambda t}\cos bt.$$

If the body starts from rest at the origin and $\omega > \lambda$ obtain $x(t)$ for the two cases

(a) $b = \sqrt{\omega^2 - \lambda^2}$

(b) $b \neq \sqrt{\omega^2 - \lambda^2}$.

24. The angular displacement $\theta(t)$ of a rigid body pendulum oscillating about a fixed axis is given by

$$I\frac{d^2\theta}{dt^2} = -mgh\sin\theta,$$

where m, g, h and I are constants.

For small oscillations (that is, when θ is small) we can replace $\sin\theta$ by θ.

Show that for small oscillations the motion is simple harmonic in nature with period $2\pi\sqrt{\frac{I}{mgh}}$.

25. An e.m.f. $E_0 \sin\omega t$ is applied to an electrical circuit comprising a resistance R, an inductance L and a capacitance C in series.

The current $i(t)$ in the circuit at time t satisfies the differential equation

$$L\frac{d^2 i}{dt^2} + R\frac{di}{dt} + \frac{i}{C} = E_0\omega\cos\omega t.$$

If $R^2C < 4L$, determine $i(t)$ for the case when $i(0) = 0$ and $\frac{di}{dt}(0) = 0$.

26. Obtain numerical estimates of the solution of the differential equation

$$y' = -2xy, \ y(0) = 1,$$

on the interval $[0, 1]$, using the method of Euler with step sizes of $h = 0.1$ and $h = 0.05$.

Compare your results with the exact solution.

27. Obtain approximate solutions of the differential equation

$$y' = x + y^2, \ y(0) = 1,$$

on the interval $[0, 1]$ using the method of Euler with a step size of $h = 0.1$.

28. A body moves so that its position $x(t)$ at time t is given by

$$\ddot{x} + 2\mu\dot{x} + m^2 x = a \sin \omega t,$$

where μ, m, a and ω are constants.

Determine the general solution if

(a) $\mu > m$

(b) $\mu < m$

(c) $\mu = m$.

29. Eliminate x from the equations

$$\begin{aligned} \ddot{x} + \ddot{y} - 3x - 3y - 7 &= 0 \\ 2y - 3x - 7 &= 0 \end{aligned}$$

and hence find $x(t), y(t)$ given that

$$x(0) = \dot{y}(0) = 0.$$

30. The phenomenon of *beating* is known to occur whenever a periodic forcing term has a frequency close to that of a natural frequency of a given system.

An example of this is given in a system that is modelled by the initial-value problem

$$\ddot{x} + \Omega^2 x = F_0 \cos\{(\Omega + \epsilon)t\}$$

with $x(0) = \dot{x}(0) = 0$, where F_0, Ω and ϵ are given positive constants (with ϵ small compared with Ω), t is the time, x is the displacement and dots refer to differentiation with respect to time.

(a) Deduce that a general solution is expressible in the form

$$A \cdot \sin(\alpha t) \cdot \sin(\beta t),$$

for suitable values of A, α and β which are to be determined.

(b) i. State an approximation valid for ϵ/Ω small and show it has the form

$$B(t; \epsilon) \sin \Omega t.$$

ii. Using the formula for $B(t; \epsilon)$ derived above, deduce that its period is

$$4\pi/\epsilon,$$

the so-called *beat frequency*.

Topic 8

Infinite Series

8.1 Introduction

In order to motivate the ideas to be introduced in this chapter consider the following series:

$$\sum_{n=1}^{\infty} \frac{1}{n(n+1)} = \frac{1}{1.2} + \frac{1}{2.3} + \frac{1}{3.4} + \cdots + \frac{1}{n(n+1)} + \cdots$$

If we try to "add up" the terms of this series we obtain:

$S_1 =$ the sum of the first term $= \frac{1}{1.2} = \frac{1}{2}$

$S_2 =$ the sum of the first two terms $= \frac{1}{2} + \frac{1}{2.3} = \frac{2}{3}$

$S_3 =$ the sum of the first three terms $= \frac{2}{3} + \frac{1}{3.4} = \frac{3}{4}$

$S_4 =$ the sum of the first four terms $= \frac{3}{4} + \frac{1}{4.5} = \frac{4}{5}$.

Continuing in this manner we would obtain $S_{10} = \frac{10}{11}, S_{100} = \frac{100}{101}, S_{1000} = \frac{1000}{1001}$ and so on. The terms $S_1, S_2, S_3, \ldots S_n, \ldots$ are referred to as the partial sums of the series. It appears that there is a pattern to these "partial sums".

You should note that:

- The sum of the first n terms of the series, that is the nth partial sum, appears to have the form $S_n = \frac{n}{n+1}$.
- As more and more terms of the series are added the partial sums appear to be "approaching" the value 1.

In fact, as we shall prove later, both of these observations are valid.

Since $S_n = \frac{n}{n+1}$ tends to 1 as n tends to infinity, which we write as $\lim_{n\to\infty} \frac{n}{n+1} = 1$ or $\frac{n}{n+1} \to 1$ as $n \to \infty$ and which are read as "the limit of $\frac{n}{n+1}$ as n approaches infinity is 1", we say that the series $\sum_{n=1}^{\infty} \frac{1}{n(n+1)}$ converges and has sum 1.

Now we shall consider two examples of a special series which may already be familiar to you, namely geometric series.[1]

▌Example: Consider $\sum_{n=0}^{\infty}(-1)^n(\frac{1}{3})^n = 1 - \frac{1}{3} + \frac{1}{9} - \frac{1}{27} + \cdots$
Solution: The sum of the first n terms is

$$S_n = \frac{1-(-\frac{1}{3})^n}{1-(-\frac{1}{3})} = \frac{3}{4}\left\{1 - \left(-\frac{1}{3}\right)^n\right\}.$$

As n approaches infinity the value of S_n approaches $3/4$ since the term $(-\frac{1}{3})^n \to 0$ as $n \to \infty$.
We say that the series converges and has sum $3/4$.

▌Example: Consider $\sum_{n=1}^{\infty} 2^{n-1} = 1 + 2 + 4 + 8 + 16 + \cdots$
Solution: The sum of the first n terms is $S_n = 2^n - 1$. (Verify!)
As n approaches infinity the value of S_n increases without bound; that is, S_n does not tend to a limit as $n \to \infty$ and we say that the series $\sum_{n=1}^{\infty} 2^{n-1}$ diverges (or does not converge).

We now make some of the ideas introduced in the preceding discussion more precise.

8.2 Convergence of a Series

An infinite series has the form $\sum_{n=1}^{\infty} a_n = a_1 + a_2 + a_3 + \cdots + a_n + \cdots$ where a_n is called the nth term of the series.

Frequently $\sum_{n=1}^{\infty} a_n$ is abbreviated to simply $\sum a_n$.

The nth partial sum S_n of the series is the sum of the first n terms. That is,

$$S_1 = a_1$$

$$S_2 = a_1 + a_2$$

$$\vdots$$

and generally

$$\boxed{S_n = a_1 + a_2 + \cdots + a_n = \sum_{m=1}^{n} a_m \text{ for } n = 1, 2, 3, \ldots}$$

The series $\sum_{n=1}^{\infty} a_n$ is said to **converge** (or be convergent) if the sequence of partial sums S_n tends to a finite limit L as n tends to infinity; that is, if $S_n \to L$ as $n \to \infty$. We then say that L is the **sum** of the series. If S_n does not tend to a limit we say that the series **diverges** (or is divergent).

---

[1] A geometric series is a series having the form $\sum_{n=0}^{\infty} ar^n = a + ar + ar^2 + ar^3 + \cdots$. The term r is often referred to as the common ratio. It is well known that $a + ar + ar^2 + \cdots + ar^{n-1} = \frac{a(1-r^n)}{1-r}$ if $r \neq 1$.

Worked Example 8.2.1 *Show that the series* $\sum_{n=1}^{\infty} \frac{1}{n(n+1)}$ *converges and has sum 1.*

Here $S_n = \dfrac{1}{1.2} + \dfrac{1}{2.3} + \dfrac{1}{3.4} + \cdots + \dfrac{1}{n(n+1)}$.

Using partial fractions we have $\frac{1}{m(m+1)} = \frac{1}{m} - \frac{1}{m+1}$ for $m = 1, 2, 3, \ldots$ so that we can rewrite S_n as

$$S_n = \left(\frac{1}{1} - \frac{1}{2}\right) + \left(\frac{1}{2} - \frac{1}{3}\right) + \left(\frac{1}{3} - \frac{1}{4}\right) + \cdots + \left(\frac{1}{n} - \frac{1}{n+1}\right) = 1 - \frac{1}{n+1}$$

since all but the first and last terms cancel.

Clearly $S_n \to 1$ as $n \to \infty$ since $\frac{1}{n+1} \to 0$ so that the series $\sum_{n=1}^{\infty} \frac{1}{n(n+1)}$ converges and has sum 1. ■

Worked Example 8.2.2 *Obtain a simple expression for the nth partial sum of the series* $\sum_{n=1}^{\infty} \frac{1}{\sqrt{n+1}+\sqrt{n}}$.
Deduce that this series is divergent.

Here

$$\begin{aligned} S_n &= \frac{1}{\sqrt{2}+\sqrt{1}} + \frac{1}{\sqrt{3}+\sqrt{2}} + \cdots + \frac{1}{\sqrt{n+1}+\sqrt{n}} \\ &= (\sqrt{2}-\sqrt{1}) + (\sqrt{3}-\sqrt{2}) + \cdots + (\sqrt{n+1}-\sqrt{n}) \\ &\qquad \text{(on rationalizing each term)} \\ &= \sqrt{n+1} - 1 \\ &\qquad \text{(since all other terms cancel).} \end{aligned}$$

But since $\sqrt{n+1} \to \infty$ as $n \to \infty$ it follows that S_n does not tend to a limit as $n \to \infty$.

Thus the series is divergent. ■

Worked Example 8.2.3 *Obtain the nth partial sum of the geometric series* $\sum_{n=1}^{\infty} \frac{3}{2^{n-1}}$.
Deduce that the series converges.
What is its sum?

$$\begin{aligned} \text{Here } S_n &= 3 + \frac{3}{2} + \frac{3}{4} + \frac{3}{8} + \cdots + \frac{3}{2^{n-1}} \\ &= \frac{3\{1-(1/2)^n\}}{1-1/2} \quad = \quad 6\{1-(0.5)^n\} \end{aligned}$$

As $n \to \infty$, we have $(0.5)^n \to 0$ so that $S_n \to 6$.

Thus the series converges and has sum 6. ■

Self-help exercises

1. Determine the nth partial sum S_n of the geometric series

$$2 - \frac{1}{2} + \frac{1}{8} - \frac{1}{32} + \frac{1}{128} - \frac{1}{512} + \cdots$$

Does this series converge?

What is its sum if it does converge?

$[\frac{8}{5}\left\{1-(-\frac{1}{4})^n\right\}$; yes; 8/5]

2. Using the identity

$$\frac{n}{(n+1)!} = \frac{1}{n!} - \frac{1}{(n+1)!}$$

determine the nth partial sum of the series

$$\sum_{n=1}^{\infty} \frac{n}{(n+1)!}.$$

Does this series converge?

If so, what is its sum?

$[1 - \frac{1}{(n+1)!}$; yes; 1]

3. Writing $\log \frac{n+1}{n} = \log(n+1) - \log n$ determine the nth partial sum S_n of the series $\sum_{n=1}^{\infty} \log(1+\frac{1}{n})$.

Does this series converge?

If so, what is its sum?

$[S_n = \log(n+1)$; no]

8.2.1 A Test for Divergence

Suppose that the series $\sum a_n$ converges and has sum L. Therefore

$$S_n = a_1 + a_2 + a_3 + \cdots + a_n \to L \text{ as } n \to \infty$$

and

$$S_{n-1} = a_1 + a_2 + a_3 + \cdots + a_{n-1} \to L \text{ as } n \to \infty.$$

Now the nth term of the series, a_n, satisfies $a_n = S_n - S_{n-1}$. Therefore, since S_n and S_{n-1} both tend to the same limit L as n tends to infinity, a_n must tend to zero. Thus a **necessary condition** for the series $\sum a_n$ to converge is that $\lim_{n\to\infty} a_n = 0$; that is, its nth term must tend to zero as n tends to infinity. More importantly, this implies that if the terms of a series do not tend to zero then the series must diverge.

If $a_n \not\to 0$ then $\sum a_n$ diverges.

Beware!

Contrary to what you might first expect, if the terms of a series tend to zero, we *cannot* deduce that the series $\sum a_n$ converges. The example in Worked Example 8.2.2 indicates a series which *diverges* even though its terms tend to zero.

For example:

1. $\sum_{n=1}^{\infty} \frac{n}{n+1}$ diverges since its nth term $a_n = \frac{n}{n+1} \not\to 0$ as $n \to \infty$.
2. $\sum_{n=1}^{\infty} (-1)^n (\frac{5}{4})^n$ diverges since its nth term does not tend to zero.
3. The series $\sum_{n=1}^{\infty} \frac{1}{n}$ is such that its nth *term* tends to zero; however, we cannot make any deduction from this concerning whether the *series* converges or diverges.

Self-help exercises

1. Determine the nth term of each of the following series:

   (a) $\frac{2}{1.3} + \frac{4}{3.5} + \frac{6}{5.7} + \frac{8}{7.9} + \cdots$

   (b) $\frac{\sqrt{2}}{1} + \frac{\sqrt{4}}{5} + \frac{\sqrt{6}}{9} + \frac{\sqrt{8}}{13} + \cdots$

   (c) $\frac{2}{3} - \frac{4}{5} + \frac{6}{7} - \frac{8}{9} + \cdots$

   $\left[\frac{2n}{(2n-1)(2n+1)}; \frac{\sqrt{2n}}{4n-3}; (-1)^{n-1}\frac{2n}{2n+1}\right]$

2. For each of the series in the above exercise, determine whether the nth term tends to zero or not.

   What, if anything, can immediately be said about the convergence behaviour of each series?

   [nth term tends to zero, nothing; nth term tends to zero, nothing; $a_n \not\to 0$, so series must diverge]

8.3 Positive Term Series

Unlike the series in Worked Examples 8.2.1–8.2.3, for most series that arise in practice, it is either not possible or very difficult to obtain simple closed expressions for S_n, so that we cannot determine whether the series converges or diverges by simply investigating $\lim_{n\to\infty} S_n$. Fortunately there are other ways of determining the convergence behaviour of a given series. Initially we shall restrict our discussion to so-called **positive term series**; that is, series of the form $\sum a_n$ where every $a_n > 0$.

If $\sum a_n$ is a positive term series the sequence of partial sums S_n is monotonically increasing (that is, $S_n < S_{n+1}$ for all n).

There are only two possibilities to consider:

- S_n increases without bound (see Figure 8.1), in which case $S_n \to \infty$ as $n \to \infty$, so that the series must diverge. (See Worked Example 8.2.2.)

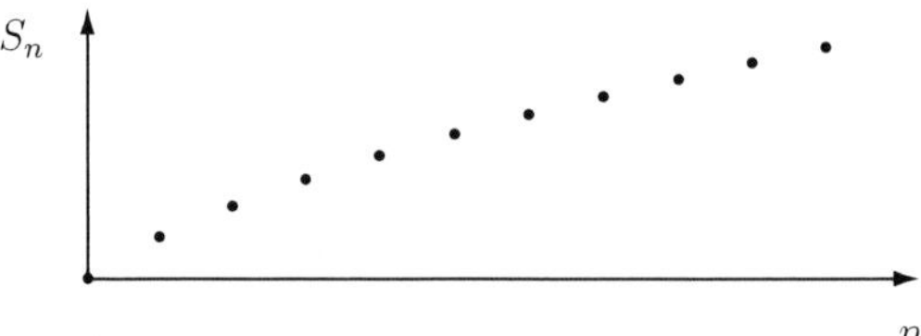

Figure 8.1: The terms of the sequence $S_n = \sqrt{n+1} - 1$.

- The partial sums are bounded above (see Figure 8.2); that is, there exists some constant K such that $S_n < K$ for all n. For such cases S_n must tend to some limit ($\leq K$) and hence the series must converge.[2] (See Worked Examples 8.2.1 and 8.2.3.)

Figure 8.2: The terms of the sequence $S_n = 1 - \frac{1}{n+1} = \frac{n}{n+1}$.

Using these ideas we shall derive two simple tests which can be used to determine whether a given positive term series converges or diverges.

8.3.1 The Comparison Test

A powerful method for determining whether a given positive series converges or diverges is to compare it with a series whose convergence behaviour is already known.

In order to help understand and motivate the way in which the comparison test works we consider some examples.

1. Consider the two series

$$\sum_{n=1}^{\infty} \frac{1}{\sqrt{n+1}+\sqrt{n}} = \frac{1}{\sqrt{2}+\sqrt{1}} + \frac{1}{\sqrt{3}+\sqrt{2}} + \frac{1}{\sqrt{4}+\sqrt{3}} + \cdots \quad (8.1)$$

[2] A sequence which is both monotonic increasing and bounded above converges. (This result is referred to as the monotonic convergence theorem.)

and

$$\sum_{n=1}^{\infty} \frac{1}{\sqrt{n}} = \frac{1}{\sqrt{1}} + \frac{1}{\sqrt{2}} + \frac{1}{\sqrt{3}} + \cdots \tag{8.2}$$

We have already shown (Worked Example 8.2.2) that the series 8.1 diverges with

$$S_n = \sqrt{n+1} - 1 \to \infty \text{ as } n \to \infty.$$

Each term of the series 8.2 is greater than the corresponding term of the series 8.1 since $\sqrt{n} < \sqrt{n+1} + \sqrt{n}$.

Consequently the nth partial sum T_n of the series 8.2 must exceed S_n.

But $S_n \to \infty$ as $n \to \infty$ so that T_n must also increase without bound as $n \to \infty$.

That is, $\sum \frac{1}{\sqrt{n}}$ must diverge.

2. In a similar manner we shall compare the following two series:

$$\sum_{n=1}^{\infty} \frac{1}{n(n+1)} = \frac{1}{1.2} + \frac{1}{2.3} + \frac{1}{3.4} + \cdots \tag{8.3}$$

and

$$\sum_{n=1}^{\infty} \frac{1}{(n+1)^2} = \frac{1}{2^2} + \frac{1}{3^2} + \frac{1}{4^2} + \cdots \tag{8.4}$$

We have already shown (Worked Example 8.2.1) that the series 8.3 converges with nth partial sum $S_n = 1 - \frac{1}{n+1}$. Since each term of the series 8.4 is less than the corresponding term of the series 8.3 (verify this!), the nth partial sum T_n of the series 8.4 must be less than S_n.

Thus $T_n < S_n < 1$ for each n.

Hence T_n must tend to a limit as $n \to \infty$; that is, the series 8.4 must converge.

Note that the series

$$\sum_{n=1}^{\infty} \frac{1}{n^2} = 1 + \frac{1}{2^2} + \frac{1}{3^2} + \frac{1}{4^2} + \cdots$$

is essentially the same as the series $\sum_{n=1}^{\infty} \frac{1}{(n+1)^{\cdot}}$. How precisely do they differ?

We shall now formalize the ideas introduced in the previous two examples.

Suppose $\sum a_n$ and $\sum b_n$ are two positive term series such that

$$0 \le a_n \le b_n \text{ for all } n.$$

(It is actually sufficient that $0 \le a_n \le b_n$ for all n sufficiently large.)

We shall loosely describe this situation by saying that $\sum a_n$ is a *smaller series* than $\sum b_n$, meaning that each term of the series $\sum a_n$ is less than (or equal to) the corresponding term of the series $\sum b_n$.

Equivalently we may loosely say that $\sum b_n$ is a *larger series* than $\sum a_n$.

We shall show that:

- if $\sum b_n$ is convergent then $\sum a_n$ must converge; that is, if the *larger series* converges then the *smaller series* must also converge
- if $\sum a_n$ is divergent then $\sum b_n$ must diverge; that is, if the *smaller series* diverges then the *larger series* must also diverge.

Theorem 8.1 (The Comparison Test) *If $\sum a_n$ and $\sum b_n$ are positive term series which satisfy*

$$0 \leq a_n \leq b_n \text{ for all } n$$

then

1. *if $\sum b_n$ is convergent then $\sum a_n$ must converge*
2. *if $\sum a_n$ is divergent then $\sum b_n$ must diverge.*

Proof: Suppose that S_n and T_n denote the nth partial sums of $\sum a_n$ and $\sum b_n$ respectively.
Thus

$$S_n = a_1 + a_2 + a_3 + \cdots + a_n \leq b_1 + b_2 + b_3 + \cdots + b_n = T_n \qquad (8.5)$$

so that

$$0 \leq S_n \leq T_n \text{ for all } n. \qquad (8.6)$$

1. Suppose $\sum b_n$ converges so that T_n tends to some finite limit L as $n \to \infty$.

   This means that there is some constant K such that $T_n \leq K$ for all n. From Equation 8.6 $S_n \leq T_n$ so that $S_n \leq K$ for all n.

   Since the partial sums S_n are bounded above (and increasing) the series $\sum a_n$ must converge.

2. Suppose that $\sum a_n$ diverges so that $S_n \to \infty$ as $n \to \infty$.

   From Equation 8.6 $T_n \geq S_n$ so it follows that $T_n \to \infty$ as $n \to \infty$; that is, the series $\sum b_n$ must diverge.

In order for the comparison test to be useful we need to have at our disposal some standard series which are known to be convergent or divergent and which can be used for comparison. Two types of series which are often used are:

- geometric series — which converge when the common ratio r is less than 1 and diverge when r is greater than or equal to 1. (We only need consider $r > 0$ here. Why?)
- the so-called p-series

$$\sum_{n=1}^{\infty} \frac{1}{n^p} = \frac{1}{1^p} + \frac{1}{2^p} + \frac{1}{3^p} + \cdots \tag{8.7}$$

which is convergent if $p > 1$ and divergent if $p \le 1$. (See Exercises 7 and 8.)

That is

> The "p-series" $\displaystyle\sum_{n=1}^{\infty} \frac{1}{n^p}$ is
>
> - convergent if $p > 1$
> - divergent if $p \le 1$.

The particular case of $p = 1$, that is the *divergent* series $\sum_{n=1}^{\infty} \frac{1}{n}$, is referred to as the **harmonic series**.

Worked Example 8.3.1 *Determine whether the following series converge or diverge:*

1. $\displaystyle\sum \frac{\sin^2 n}{n^2}$ *2.* $\displaystyle\sum \frac{3 + \cos 2n}{n^2}$ *3.* $\displaystyle\sum \frac{3 + \cos 2n}{n}$.

1. Since $\sin^2 n \le 1$ it follows that $\frac{\sin^2 n}{n^2} \le \frac{1}{n^2}$. But $\sum \frac{1}{n^2}$ converges (a "p-series" with $p = 2$) so that the "smaller" series $\sum \frac{\sin^2 n}{n^2}$ must converge by the comparison test.
2. Since $-1 \le \cos 2n \le 1$ we can use the inequality $\frac{3+\cos 2n}{n^2} \le \frac{4}{n^2}$ to deduce that $\sum \frac{3+\cos 2n}{n^2}$ converges since the "larger" series $\sum \frac{4}{n^2}$ converges.
3. $0 \le \frac{3+\cos 2n}{n} \le \frac{4}{n}$.
   We know $\sum \frac{4}{n}$ diverges, but we cannot use the above inequality to deduce that $\sum \frac{3+\cos 2n}{n}$ diverges. (Why?)
   However, we can use the inequality
   $$\frac{3 + \cos 2n}{n} \ge \frac{2}{n}$$
   and the comparison test to deduce that $\sum \frac{3+\cos 2n}{n}$ diverges.

■

In the following worked example we use an intuitive argument to establish the convergence and divergence of two series. This approach is formalised as the limit comparison theorem immediately following the example.

Worked Example 8.3.2 *Determine whether the following series converge or diverge:*

1. $\sum \dfrac{n^2+3n-2\sqrt{n}+3}{n^4+6}$ *2.* $\sum \dfrac{3n^2+4}{n^2\sqrt{n}+1}$.

1. Although it is possible to obtain the inequality

$$\frac{n^2+3n-2\sqrt{n}+3}{n^4+6} \leq \frac{5}{n^2}$$

and deduce the convergence of $\sum \frac{n^{\cdot}+3n-2\sqrt{n}+3}{n^{\cdot}+6}$ by comparing it with the convergent series $\sum \frac{5}{n^{\cdot}}$ a simpler approach can be adopted as follows.

For large values of n the terms of the series behave like $\frac{1}{n^{\cdot}}$, as the dominant term in the numerator is n^2 while, in the denominator, it is n^4.

Thus we would expect the given series to have a convergence behaviour similar to that of the series $\sum \frac{1}{n^{\cdot}}$. But this latter series converges, so the given series also converges.

2. For large values of n the terms of the series behave like $\frac{3n^{\cdot}}{n^{\cdot}\sqrt{n}} = \frac{3}{\sqrt{n}}$.

The series $\sum \frac{3}{\sqrt{n}}$ diverges (a p-series with $p = 1/2$), so the series $\sum \frac{3n^{\cdot}+4}{n^{\cdot}\sqrt{n}+1}$ also diverges.

■

Theorem 8.2 (Limit Comparison Theorem) *Let* $\sum a_n$ *and* $\sum b_n$ *be two series of positive terms. If*

$$\lim_{n\to\infty} \frac{a_n}{b_n} = c > 0$$

then the two series either both converge or both diverge.

Self-help exercises

1. Explain why the comparison test cannot be applied to the series $\sum \frac{\sin n}{n^{\cdot}}$

[Series is not positive-term]

2. Which of the inequalities $\frac{2}{n^{\cdot}} \leq \frac{3+\sin n}{n^{\cdot}} \leq \frac{4}{n^{\cdot}}$ can be used to prove the convergence of the series $\sum \dfrac{3+\sin n}{n^2}$? [$\frac{3+\sin n}{n^{\cdot}} \leq \frac{4}{n^{\cdot}}$]

3. Which of the inequalities $\frac{2}{n} \leq \frac{3+\sin n}{n} \leq \frac{4}{n}$ can be used to prove the divergence of the series $\sum \frac{3+\sin n}{n}$? [$\frac{2}{n} \leq \frac{3+\sin n}{n}$]

4. For large values of n the nth term of the series $\sum \frac{1}{n\sqrt{n^{\cdot}+1}}$ behaves like which of the following

   (a) $\dfrac{1}{n^4}$ (b) $\dfrac{1}{n^3}$ (c) $\dfrac{1}{n^{5/2}}$.

   Based on your answer indicate whether you would expect the series to converge or diverge. [(c); series converges]

5. For large values of n the nth term of the series $\sum \frac{n+2}{\sqrt{n}(n+1)^{\cdot}}$ behaves like which of the following

   (a) $\dfrac{1}{n^2}$ (b) $\dfrac{1}{\sqrt{n}}$ (c) $\dfrac{1}{n^{3/2}}$.

   Based on your answer indicate whether you would expect the series to converge or diverge. [(c); series converges]

6. For large values of n the nth term of the series $\sum \frac{n}{100n^{\cdot}+1}$ behaves like which of the following

   (a) $\dfrac{1}{n^2}$ (b) $\dfrac{1}{100n}$ (c) $\dfrac{1}{100n^2}$.

   Based on your answer indicate whether you would expect the series to converge or diverge. [(b); series diverges]

8.3.2 The Ratio Test

Theorem 8.3 *If* $\sum a_n$ *is a positive term series such that* $\lim_{n\to\infty} \frac{a_{n\cdot\,\cdot}}{a_n}$ *exists and equals* L *then* $\sum a_n$ *converges if* $L < 1$ *and diverges if* $L > 1$.

If $L = 1$ *the test gives no indication of whether* $\sum a_n$ *converges or diverges.*

The ratio test is extremely powerful in determining the convergence behaviour of series whose terms involve

- factorials; for example $\displaystyle\sum \frac{n!}{(2n)!}$
- factorial-like expressions; for example $\sum \frac{1.3.5...(2n-1)}{3.6.9...3n}$
- exponentials; that is, nth powers such as 3^n and 2^{3n}, for example $\sum \frac{n^{\cdot}}{3^n}$.

Worked Example 8.3.3 *Determine which of the following series converge:*

1. $\sum \dfrac{n!n!}{(2n)!}$
2. $\sum \dfrac{3.6.9\ldots 3n}{1.3.5\ldots(2n-1)}$
3. $\sum \dfrac{n^2}{3^{2n-1}}$.

1. $$\lim_{n\to\infty} \frac{a_{n+1}}{a_n} = \lim_{n\to\infty} \frac{(n+1)!(n+1)!}{(2n+2)!}\frac{(2n)!}{n!n!} \quad \text{since } a_n = \tfrac{n!n!}{(2n)!}$$
$$= \lim_{n\to\infty} \frac{(n+1)(n+1)}{(2n+2)(2n+1)} = 1/4$$
Since this limit is less than 1 the series must converge by the ratio test.

2. $$\lim_{n\to\infty} \frac{a_{n+1}}{a_n} = \lim_{n\to\infty} \frac{3.6.9\ldots(3n+3)}{1.3.5\ldots(2n+1)}\frac{1.3.5\ldots(2n-1)}{3.6.9\ldots 3n} \quad \text{since } a_n = \tfrac{3.6.9\ldots 3n}{1.3.5\ldots(2n-1)}$$
$$= \lim_{n\to\infty} \frac{3n+3}{2n+1} = \frac{3}{2}$$
Since this limit exceeds 1 the series must diverge by the ratio test.

3. $$\lim_{n\to\infty} \frac{a_{n+1}}{a_n} = \lim_{n\to\infty} \frac{(n+1)^2}{3^{2n+1}}\frac{3^{2n-1}}{n^2} \quad \text{since } a_n = \frac{n^2}{3^{2n-1}}$$
$$= \lim_{n\to\infty} \frac{(n+1)^2}{9n^2} = \frac{1}{9}$$
Since this limit is less than 1 the series must converge by the ratio test.

■

Self-help exercises

1. Indicate to which of the following series you would apply the ratio test in order to determine its convergence behaviour.

(a) $\sum \dfrac{n10^n}{(n+1)!}$ (b) $\sum \dfrac{4+\cos^2 n}{\sqrt{n}}$ (c) $\sum (-1)^n \dfrac{3^n}{n!}$ (d) $\sum \dfrac{n^2}{3^n}$ (e) $\sum \dfrac{e^n}{(n+1)!}$.

[(a), (d), (e)]

2. For each of the following series, write down the nth term a_n, the $(n+1)$th term a_{n+1}, and calculate $\dfrac{a_{n+1}}{a_n}$.

   In each case indicate what the ratio test tells you about the series.

   (a) $\sum \dfrac{n3^n}{2^{2n-1}}$ $\quad [\frac{a_n\cdot\cdot}{a_n} = \frac{3(n+1)}{4n}$; series converges (L $= 3/4$)]

   (b) $\sum \dfrac{3^n}{n!}$ $\quad [\frac{a_n\cdot\cdot}{a_n} = \frac{3}{n+1}$; series converges (L $= 0$)]

   (c) $\sum \dfrac{1.3.5\cdots(2n-1)}{1.5.9\cdots(4n-3)}$. $\quad [\frac{a_n\cdot\cdot}{a_n} = \frac{2n+1}{4n+1}$; series converges (L $= 1/2$)]

8.4 Alternating Series

A series of the form

$$\sum_{n=1}^{\infty}(-1)^{n+1}a_n = a_1 - a_2 + a_3 - a_4 + \cdots + (-1)^{n+1}a_n + \cdots$$

where each $a_n > 0$ (that is, where the terms are alternately positive and negative) is called an alternating series.

For example,

1. $$\sum_{n=2}^{\infty}(-1)^n\frac{1}{n\log n} = \frac{1}{2\log 2} - \frac{1}{3\log 3} + \frac{1}{4\log 4} - \frac{1}{5\log 5} + \cdots$$

2. $$\sum_{n=1}^{\infty}(-1)^{n+1}\frac{n}{2n+1} = \frac{1}{3} - \frac{2}{5} + \frac{3}{7} - \frac{4}{9} + \cdots$$

are examples of alternating series.

However the series $\sum_{n=2}^{\infty}(-1)^n\frac{\sin n}{n}$ is not alternating. (Why?)

In Section 8.2.1 we saw that if the terms of a series tend to zero it was not possible to deduce the convergence of that series. For alternating series, however, if the terms tend to zero **and** decrease in magnitude we can deduce that the series converges.

Theorem 8.4 *Consider the series $\sum(-1)^{n+1}a_n$ where each $a_n > 0$.*
If $\lim_{n\to\infty} a_n = 0$ and $a_{n+1} < a_n$ for all n, then $\sum(-1)^{n+1}a_n$ converges.

Proof: The $(2n)$th partial sum is

$$\begin{aligned} S_{2n} &= a_1 - a_2 + a_3 - a_4 + \cdots + a_{2n-1} - a_{2n} \\ &= (a_1 - a_2) + (a_3 - a_4) + \cdots + (a_{2n-1} - a_{2n}). \end{aligned}$$

Since $a_{n+1} < a_n$ each expression in brackets is positive. Thus, the sequence of partial sums $S_2, S_4, S_6, \ldots, S_{2n}$ is increasing.

Also $S_{2n} = a_1 - (a_2 - a_3) - (a_4 - a_5) - \cdots - a_{2n}$ where, again, each term in brackets is positive so that $S_{2n} < a_1$; that is, the sequence of even partial sums S_n is bounded above.
Thus S_{2n} converges to some limit L.
Also $S_{2n+1} = a_1 - a_2 + a_3 - \cdots - a_{2n} + a_{2n+1} = S_{2n} + a_{2n+1}$ so that $S_{2n+1} \to L$ as $n \to \infty$ since $a_{2n+1} \to 0$.
Consequently the sequence of partial sums, S_n (n even and odd) converges, showing that the alternating series converges.

Note that if we approximate the sum of such an alternating series by truncating the series after the first m terms (that is, if we only consider the sum of the first m terms) the magnitude of the error involved cannot exceed the next term, that is the $(m+1)$th term. (See Exercise 16.)

For example, if we use the first ten terms to approximate the sum of the series

$$\sum_{n=1}^{\infty}(-1)^{n-1}\frac{1}{n^2} = 1 - \frac{1}{2^2} + \frac{1}{3^2} - \frac{1}{4^2} + \frac{1}{5^2} + \cdots$$

then we are approximating the series by the finite sum

$$\sum_{n=1}^{10}(-1)^{n-1}\frac{1}{n^2} = 1 - \frac{1}{2^2} + \frac{1}{3^2} - \frac{1}{4^2} + \frac{1}{5^2} + \cdots - \frac{1}{10^2}$$

and the magnitude of the error is less than the next term, namely $\frac{1}{11^{\cdot}} \approx 0.008$.

Worked Example 8.4.1 *Show that the series*

$$\sum_{n=1}^{\infty}(-1)^{n+1}\frac{1}{2\sqrt{n}+1}$$

is convergent.
The series is clearly alternating with $a_n = \frac{1}{2\sqrt{n}+1}$.
Since $a_n \to 0$ and $a_{n+1} < a_n$ — that is, a_n is monotonically decreasing to zero — the series must converge. ■

Self-help exercise

Indicate which of the following series are alternating:

1. $\sum(-1)^n\frac{n}{n^2+1}$ 2. $\sum(-1)^{2n+1}\frac{n}{n^2+1}$ 3. $\sum\frac{\cos n\pi}{n^2+1}$.

[1, 3 — note that $\cos n\pi = (-1)^n$]

8.4.1 Absolute Convergence

If a positive term series $\sum a_n$ converges then we shall show that any series obtained from it by changing signs is also convergent. For example, since the positive term series

$$\sum_{n=1}^{\infty} \frac{1}{n^2} = 1 + \frac{1}{2^2} + \frac{1}{3^2} + \frac{1}{4^2} + \frac{1}{5^2} + \cdots$$

is convergent it follows that each of the series

$$1 + \frac{1}{2^2} - \frac{1}{3^2} + \frac{1}{4^2} + \frac{1}{5^2} - \frac{1}{6^2} + \frac{1}{7^2} + \frac{1}{8^2} - \cdots$$

$$1 - \frac{1}{2^2} - \frac{1}{3^2} - \frac{1}{4^2} + \frac{1}{5^2} - \frac{1}{6^2} - \frac{1}{7^2} - \cdots$$

and

$$1 + \frac{1}{2^2} + \frac{1}{3^2} - \frac{1}{4^2} - \frac{1}{5^2} + \frac{1}{6^2} + \frac{1}{7^2} + \cdots$$

also converge.

If $\sum a_n$ is a series of arbitrarily positive/negative terms such that $\sum |a_n|$ is convergent, we then say that $\sum a_n$ is *absolutely convergent.*

Theorem 8.5 (The Absolute Convergence Theorem) *If $\sum a_n$ is absolutely convergent, that is, if $\sum |a_n|$ converges, then the series $\sum a_n$ converges.*

Proof: We shall use the inequality[3] $0 < a_n + |a_n| \leq 2|a_n|$.
But $\sum 2|a_n|$ converges since $\sum |a_n|$ converges.
Thus, by the comparison test, $\sum(a_n + |a_n|)$ converges.
Now,

$$\sum a_n = \sum\{(a_n + |a_n|) - |a_n|\} = \sum(a_n + |a_n|) - \sum |a_n|.$$

Both series on the right converge so that the series on the left $\sum a_n$ must also converge.

- **Note that if $\sum |a_n|$ diverges we cannot deduce that $\sum a_n$ diverges.**

  For example, $\sum \left|\frac{(-1)^n}{n}\right| = \sum \frac{1}{n}$ diverges, but the alternating series $\sum \frac{(-1)^n}{n}$ converges.

---

[3] The inequality follows from $-|a_n| \leq a_n \leq |a_n|$ on adding $|a_n|$ through the inequality.

■ **Worked Example 8.4.2** *Show that the series* $\sum \frac{\sin 2n}{n^2}$ *is convergent.*
First note that the series $\sum \frac{\sin 2n}{n^2}$ is neither a positive term series (why?) nor an alternating series.
Consider the series

$$\sum \left| \frac{\sin 2n}{n^2} \right| = \sum \frac{|\sin 2n|}{n^2}.$$

Now $\sum \frac{|\sin 2n|}{n^2}$ is a positive term series.
We know that $\frac{|\sin 2n|}{n^2} \leq \frac{1}{n^2}$ using $|\sin 2n| \leq 1$.
But the series $\sum \frac{1}{n^2}$ converges.
Thus, by the comparison test, $\sum \frac{|\sin 2n|}{n^2}$ must also converge and so $\sum \frac{\sin 2n}{n^2}$ converges by the absolute convergence theorem. ■

Self-help exercises

1. (a) Does $\sum(-1)^n \frac{2}{n^2}$ converge?
   (b) Is $\sum(-1)^n \frac{2}{n^2}$ absolutely convergent?

   [Yes; Yes]

2. (a) Does $\sum(-1)^n \frac{2}{2n+1}$ converge?
   (b) Is $\sum(-1)^n \frac{2}{2n+1}$ absolutely convergent?

   [Yes; No]

3. (a) Is $\sum(-1)^n \frac{2}{n(n+3)}$ absolutely convergent?
   (b) Does it follow that $\sum(-1)^n \frac{2}{n(n+3)}$ converges?

   [Yes; Yes]

4. The series $\sum \frac{2}{\sqrt{n^2+5}}$ diverges.
   What, if anything, can you deduce about the series $\sum(-1)^n \frac{2}{\sqrt{n^2+5}}$?
   [Nothing! This series does converge but this is established using the alternating series test.]

8.5 Approximating Functions with Polynomials

8.5.1 Introduction

Have you ever wondered how expressions such as $\log 1.2, \arctan 0.4$ and $\sin \frac{\pi}{7}$ were evaluated before the advent of electronic calculators? As we shall soon see, functions such as

$$\log(1 + x), \arctan x \text{ and } \sin x$$

can be approximated by polynomials and polynomials are very easily evaluated.

For example, $\frac{1}{(1+x)^{\cdot}}$ can be approximated by a cubic polynomial as follows:

$$\frac{1}{(1+x)^2} \approx 1 - 2x + 3x^2 - 4x^3.$$

If we denote the error involved in this approximation by $R_3(x)$ we can then write

$$\frac{1}{(1+x)^2} = 1 - 2x + 3x^2 - 4x^3 + R_3(x).$$

It is natural to ask:

1. Can we obtain a better (that is, more accurate) approximation for $\frac{1}{(1+x)^{\cdot}}$ by using a higher degree polynomial? That is, if we use a higher degree polynomial do we reduce the error?

2. How do we obtain the approximating polynomial?

3. Which functions can be approximated by polynomials?

Here we shall be concerned with trying to answer such questions. That is, we shall consider the following problem:

Given a sufficiently well-behaved function[4] $f(x)$ how can we approximate this function using polynomials?

Consider the function $f(x) = \frac{1}{(1+x)^{\cdot}}$ for values of x near to $x = 0$. Let us try to approximate $f(x)$ by a linear function $P_1(x) = a_0 + a_1x$ about $x = 0$.

Note that there are two unknowns a_0 and a_1 to be determined so that it is necessary to impose two conditions upon $P_1(x)$. In order that $P_1(x)$ gives a reasonable approximation to $f(x)$ near $x = 0$ it is natural to insist that the values of both $P_1(x)$ and $f(x)$ agree at $x = 0$ and that the two functions have the same derivative at $x = 0$; that is, $P_1'(x) = f'(x)$ at $x = 0$. Geometrically, this means that the graphs of the two functions both pass through the same point when $x = 0$ and have the same slopes when $x = 0$.

This requires $a_0 = 1$ and $a_1 = -2$ so that $P_1(x) = 1 - 2x$ is a linear approximation for $\frac{1}{(1+x)^{\cdot}}$ near $x = 0$. (See Figure 8.3.)

In a similar way we can approximate the function by a quadratic (a polynomial of degree 2) $P_2(x) = a_0 + a_1x + a_2x^2$ by making the two functions and their first and second derivatives agree at $x = 0$.

These conditions give $a_0 = 1$ and $a_1 = -2$ as before and, in addition we require $f''(x) = \frac{6}{(1+x)^{\cdot}}$ and $P_2''(x) = 2a_2$ to agree at $x = 0$ so that $a_2 = 3$.

---

[4] For our purposes a sufficiently well-behaved function will be a function which can be differentiated as many times as we wish.

Thus $P_2(x) = 1 - 2x + 3x^2$ gives a better approximation than $P_1(x)$ to $f(x) = \frac{1}{(1+x)}$ near $x = 0$.

Better approximations can be obtained by successively increasing the degree of the approximating polynomial, the additional coefficients being obtained by equating higher order derivatives of $f(x)$ and the polynomial at $x = 0$. You should verify that the sequence of approximating polynomials is

$$\begin{gathered} P_1(x) = 1 - 2x \\ P_2(x) = 1 - 2x + 3x^2 \\ P_3(x) = 1 - 2x + 3x^2 - 4x^3 \\ P_4(x) = 1 - 2x + 3x^2 - 4x^3 + 5x^4 \\ \text{and so on.} \end{gathered}$$

The graphs in Figure 8.3 illustrate how these polynomials successfully give better approximations for $f(x) = \frac{1}{(1+x)}$ in successively larger neighbourhoods of $x = 0$ as the degree of the approximating polynomial increases. Note however that the approximations are totally inadequate for values of $x < -1$ or $x > 1$.

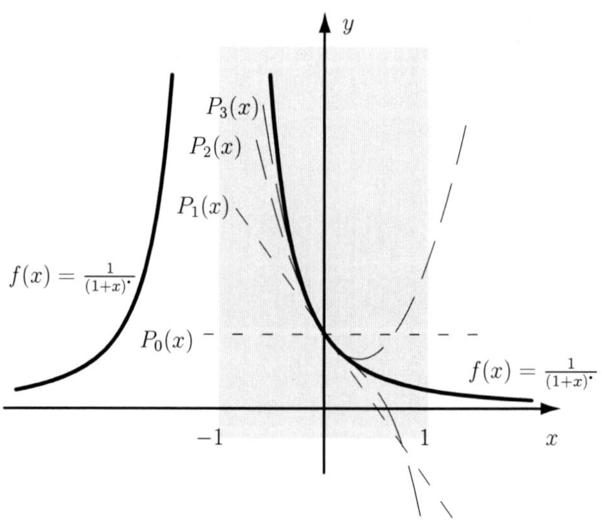

Figure 8.3: Approximating polynomials.

The polynomials used to approximate a function $f(x)$ are called **Taylor polynomials**. For example, the polynomial $P_4(x) = 1-2x+3x^2-4x^3+5x^4$ is called the Taylor polynomial of degree 4 for the function $f(x) = \frac{1}{(1+x)}$ about the point $x = 0$.

Returning to the above example, it can be shown (verify this!) that

$$P_n(x) = 1 - 2x + 3x^2 - 4x^3 + \cdots + (-1)^n(n+1)x^n$$

so that

$$\frac{1}{(1+x)^2} = 1 - 2x + 3x^2 - 4x^3 + \cdots + (-1)^n(n+1)x^n + R_n(x)$$

where $R_n(x)$ is the error involved in approximating the function by the nth degree polynomial.

From Figure 8.3 it is clear that the error term depends not only on the degree of the approximating polynomial, but also on x; for a given polynomial the error term for $x = 0.01$ will be less than the corresponding error at $x = 0.9$.

We can show (see Exercise 22) that

$$\begin{aligned} R_n(x) &= (-1)^{n+1}\frac{(n+2)x^{n+1} + (n+1)x^{n+2}}{(1+x)^2} \\ &= (-1)^{n+1}\left[\frac{(n+1)x^{n+1}}{1+x} + \frac{x^{n+1}}{(1+x)^2}\right]. \end{aligned}$$

We can also show (see the same Exercise 22) that $R_n(x)$ tends to zero as n increases provided $|x| < 1$. Consequently we can say that the infinite series

$$1 - 2x + 3x^2 - 4x^3 + \cdots + (-1)^n(n+1)x^n + \cdots = \sum(-1)^n(n+1)x^n$$

converges to $\frac{1}{(1+x)}$. for $|x| < 1$ and write

$$\begin{aligned} \frac{1}{(1+x)^2} &= 1 - 2x + 3x^2 - 4x^3 + \cdots + (-1)^n(n+1)x^n + \cdots \\ &= \sum_{n=0}^{\infty}(-1)^n(n+1)x^n \text{ if } |x| < 1. \end{aligned}$$

The series on the right is called the **power series** expansion for the function $\frac{1}{(1+x)^2}$.

8.5.2 Taylor Polynomials

In the preceding discussion we considered the problem of approximating the function $\frac{1}{(1+x)}$. near $x = 0$ by polynomials, called Taylor polynomials. Can other functions be approximated by such polynomials at $x = 0$ or, indeed, at other points $x = a$?

Suppose that the function $f(x)$ has derivatives of all orders at $x = a$. The Taylor polynomial, $P_n(x)$, of $f(x)$ at $x = a$ is the polynomial of degree n:

$$\boxed{P_n(x) = a_0 + a_1(x-a) + a_2(x-a)^2 + a_3(x-a)^3 + \cdots + a_n(x-a)^n} \tag{8.8}$$

which satisfies the conditions

$$P_n(x) = f(x), P_n'(x) = f'(x), P_n''(x) = f''(x), \dots, P_n^{(n)}(x) = f^{(n)}(x)$$

at $x = a$; that is, which satisfies

$$\boxed{P_n(a) = f(a), P_n'(a) = f'(a), P_n''(a) = f''(a), \dots, P_n^{(n)}(a) = f^{(n)}(a)}$$

The coefficient a_0 can be obtained from Equation 8.8 by putting $x = a$ so that $a_0 = f(a)$; to obtain a_1, differentiate both sides of Equation 8.8 and then put $x = a$ to obtain $a_1 = f'(a)$. It is left as an exercise (see Exercise 21) to obtain the coefficients $a_2, a_3, \dots, a_n$.

Definition 8.1 *The Taylor polynomial of degree n for $f(x)$ at $x = a$ is*

$$\boxed{\begin{aligned} P_n(x) &= f(a) + f'(a)(x-a) + \frac{f''(a)}{2!}(x-a)^2 + \\ &\quad \frac{f^{(3)}(a)}{3!}(x-a)^3 + \cdots + \frac{f^{(n)}(a)}{n!}(x-a)^n \end{aligned}}$$

It should be noted that Taylor polynomials do not exist for every function. For example, we cannot find Taylor polynomials for $f(x) = \log x$ or $f(x) = \sqrt{x}$ about $x = 0$ (why?) but they do have Taylor polynomials of any degree about $x = 1$.

Worked Example 8.5.1 *Obtain the Taylor polynomial of degree n for*

$$\log(1+x) \text{ about } x = 0.$$

Hence find an approximate value for $\log 1.2$ *using Taylor polynomials of degrees 2, 4 and 8.*
Compare your approximations with the value 0.182322 which is the correct six decimal place representation of $\log 1.2$.
We have
$f(x) = \log(1+x)$ so that $f(0) = 0$
$f'(x) = \frac{1}{1+x} \Rightarrow f'(0) = 1$
$f''(x) = -\frac{1}{(1+x)^{\cdot}} \Rightarrow f''(0) = -1$
$f^{(3)}(x) = \frac{2!}{(1+x)^{\cdot}} \Rightarrow f^{(3)}(0) = 2!$
$f^{(4)}(x) = -\frac{3!}{(1+x)^{\cdot}} \Rightarrow f^{(4)}(0) = -3!$.
Continuing in this way we get:
$f^{(n)}(x) = (-1)^{n-1}\frac{(n-1)!}{(1+x)^n} \Rightarrow f^{(n)}(0) = (-1)^{n-1}(n-1)!$.
Thus the Taylor polynomial of degree n approximating $\log(1+x)$ near $x = 0$ is

$$\begin{aligned} P_n(x) &= f(0) + f'(0)x + \tfrac{f''(0)}{2!}x^2 + \tfrac{f^{\cdots}(0)}{3!}x^3 + \cdots + \tfrac{f^{\cdot n \cdot}(0)}{n!}x^n \\ &= x - \tfrac{x^{\cdot}}{2} + \tfrac{2!}{3!}x^3 - \tfrac{3!}{4!}x^4 + \cdots + (-1)^{n-1}\tfrac{(n-1)!}{n!}x^n \\ &= x - \tfrac{x^{\cdot}}{2} + \tfrac{x^{\cdot}}{3} - \tfrac{x^{\cdot}}{4} + \cdots + (-1)^{n-1}\tfrac{x^n}{n}. \end{aligned}$$

In particular $P_2(x) = x-\frac{1}{2}x^2$ so that $P_2(1.2) = 0.2-\frac{1}{2}0.04 = 0.18$ which agrees with $\log 1.2$ to two decimal places; $P_4(x) = x - \frac{x^{\cdot}}{2} + \frac{x^{\cdot}}{3} - \frac{x^{\cdot}}{4}$ so that $P_4(1.2) = 0.2 - \frac{0.04}{2} + \frac{0.008}{3} - \frac{0.0016}{4} = 0.1823$ which agrees with $\log 1.2$ to four decimal places; finally

$$\begin{aligned} P_8(1.2) &= 0.2 - \frac{0.04}{2} + \frac{0.008}{3} - \frac{0.0016}{4} + \cdots - \frac{(1.2)^8}{8} \\ &= 0.182322 \text{ to six places.} \end{aligned}$$

which agrees with $\log 1.2$ to six decimal places.
It is clear that we obtain more accurate approximations for $\log 1.2$ by approximating the function with higher degree Taylor polynomials. ■

Worked Example 8.5.2 *Obtain a Taylor polynomial of degree 4 for*

$$f(x) = \sin x \textit{ about } x = \frac{\pi}{6}.$$

Hence obtain an approximate value for $\sin 32^\circ$.
Let $T_4(x) = a_0 + a_1(x-\frac{\pi}{6}) + a_2(x-\frac{\pi}{6})^2 + a_3(x-\frac{\pi}{6})^3 + a_4(x-\frac{\pi}{6})^4$.
We now find $a_0, a_1, \ldots, a_4$ by insisting that $T_4(x)$ and $f(x) = \sin x$ agree at $x = \frac{\pi}{6}$, as do their first four derivatives. (Why do we specify exactly four derivatives?)
We have $T_4(\frac{\pi}{6}) = a_0$ and $\sin\frac{\pi}{6} = 0.5 \Rightarrow a_0 = 0.5$.
Now $T_4'(x) = a_1 + 2a_2(x-\frac{\pi}{6}) + 3a_3(x-\frac{\pi}{6})^2 + 4a_4(x-\frac{\pi}{6})^3$; therefore $T_4'(\frac{\pi}{6}) = a_1 = f'(\frac{\pi}{6}) = \cos\frac{\pi}{6} = 0.8660$
$T_4''(x) = 2a_2 + 6a_3(x-\frac{\pi}{6}) + 12a_4(x-\frac{\pi}{6})^2$; therefore $T_4''(\frac{\pi}{6}) = 2a_2 = f''(\frac{\pi}{6}) = -\sin\frac{\pi}{6} = -0.5$. Hence $a_2 = -0.2500$
$T_4^{(3)}(x) = 6a_3 + 24a_4(x-\frac{\pi}{6})$; therefore $T_4^{(3)}(\frac{\pi}{6}) = 6a_3 = f^{(3)}(\frac{\pi}{6}) = -\cos\frac{\pi}{6} = -0.8660$. Hence $a_3 = -0.1443$
$T_4^{(4)}(x) = 24a_4$; therefore $T_4^{(4)}(\frac{\pi}{6}) = 24a_4 = f^{(4)}(\frac{\pi}{6}) = \sin\frac{\pi}{6} = 0.5$. Hence $a_4 = 0.0208$.
Thus we have obtained the Taylor polynomial of degree 4 which approximates $\sin x$ near $x = \frac{\pi}{6}$. This polynomial is

$$\begin{aligned} T_4(x) &= 0.5 + 0.8660(x-\frac{\pi}{6}) - 0.2500(x-\frac{\pi}{6})^2 - \\ &\quad 0.1443(x-\frac{\pi}{6})^3 + 0.0208(x-\frac{\pi}{6})^4. \end{aligned}$$

Using this to approximate $\sin 32^\circ$ we put $x = \frac{32\pi}{180}$. This gives $\sin 32^\circ \approx 0.5299$ which is correct to four decimal places. (Why is it necessary to use radians in this calculation?) ■

Self-help exercises

1. Show that the Taylor polynomial of degree 1 for $f(x) = \sqrt{1+x}$ about $x = 0$ is $T_1(x) = 1 + \frac{x}{2}$.

   Use this to obtain approximate values for $\sqrt{1.2}$ and $\sqrt{4.2}$.

   (Note that $\sqrt{4+y} = 2\sqrt{1+\frac{1}{4}y}$.) [1.1; 2.05]

2. Write down (by inspection) the Taylor polynomials of degrees 0, 2 and 3 for $f(x) = 1 + x^2$.

   $[1;\ 1+x^2; 1+x^2]$

8.6 Taylor Series

In the previous section we saw how to approximate a sufficiently well-behaved function by polynomials, the approximation being improved by using higher degree polynomials. Moreover, we were able to express functions which had derivatives of all orders at $x = a$ in the form

$$f(x) = f(a) + f'(a)(x-a) + \frac{f''(a)}{2!}(x-a)^2 + \cdots + \frac{f^{(n)}(a)}{n!}(x-a)^n + R_n(x). \tag{8.9}$$

where $R_n(x)$ is the error (or remainder) involved in approximating $f(x)$ by the Taylor polynomial of degree n.

For $f(x) = 1/(1+x)^2$ expanded about $x = 0$ we showed that $R_n(x) \to 0$ as $n \to \infty$ provided $|x| < 1$. We thus obtained the power series expansion

$$\frac{1}{(1+x)^2} = 1 - 2x + 3x^2 - 4x^3 + \cdots + (-1)^n(n+1)x^n + \cdots = \sum_{n=0}^{\infty}(-1)^n(n+1)x^n$$

valid for $|x| < 1$.

Returning to the general case in Equation 8.9, if we can show that $R_n(x) \to 0$ as $n \to \infty$ we obtain the Taylor series expansion for $f(x)$ about $x = a$. That is

$$\begin{aligned} f(x) &= f(a) + f'(a)(x-a) + \frac{f''(a)}{2!}(x-a)^2 + \cdots + \frac{f^{n}(a)}{n!}(x-a)^n + \cdots \\ &= \sum_{n=0}^{\infty} \frac{f^{(n)}(a)}{n!}(x-a)^n \end{aligned}$$

When $a = 0$, this power series is usually referred to as simply the Maclaurin series of $f(x)$.

Note that the series expansion for $f(x)$ will only be valid for those values of x for which $R_n(x) \to 0$ as $n \to \infty$; that is, for those x for which the power series converges. For the time being we shall concentrate on the formal processes involved in determining the Taylor series expansions and shall defer the question of convergence to later.

Worked Example 8.6.1 *Determine the Taylor series expansion for*

$$f(x) = \cos x \textit{ about } x = 0.$$

That is, determine the Maclaurin series for $\cos x$.

Method 1

Let $f(x) = \cos x$.
Since the Taylor series expansion for $f(x)$ about $x = 0$ is:

$$f(x) = f(0) + f'(0)x + \frac{f''(0)}{2!}x^2 + \cdots + \frac{f^{(n)}(0)}{n!}x^n + \cdots$$

we need to find $f(0), f'(0), f''(0)$ and so on.

$$\begin{aligned}
f(x) &= \cos x &\Rightarrow\ & f(0) = 1 \\
f'(x) &= -\sin x &\Rightarrow\ & f'(0) = 0 \\
f''(x) &= -\cos x &\Rightarrow\ & f''(0) = -1 \\
f^{(3)}(x) &= \sin x &\Rightarrow\ & f^{(3)}(0) = 0 \\
f^{(4)}(x) &= \cos x &\Rightarrow\ & f^{(4)}(0) = 1
\end{aligned}$$

etc.
Thus

$$\cos x = 1 - \frac{x^2}{2!} + \frac{x^4}{4!} - \frac{x^6}{6!} + \cdots = \sum_{n=0}^{\infty} \frac{(-1)^n x^{2n}}{(2n)!}.$$

In the next section it will be shown that this series converges for all x; that is, this power series expansion is valid for all x.

Method 2

Let $\cos x = a_0 + a_1x + a_2x^2 + a_3x^3 + a_4x^4 + a_5x^5 + \cdots$.
Clearly a_0 can be found by putting $x = 0$; this gives $a_0 = 1$.
Assuming that term-by-term differentiation of power series is valid (see Section 8.8.1) we obtain

$$-\sin x = a_1 + 2a_2x + 3a_3x^2 + 4a_4x^3 + 5a_5x^4 + \cdots$$

Putting $x = 0$ gives $a_1 = 0$.

Successive differentiation yields:

$$\begin{aligned} -\cos x &= 2a_2 + 3.2a_3x + 4.3a_4x^2 + 5.4a_5x^3 + \cdots \\ \sin x &= 3.2a_3 + 4.3.2a_4x + 5.4.3a_5x^2 + 6.5.4a_6x^3 + \cdots \\ \cos x &= 4.3.2.a_4 + 5.4.3.2a_5x + \cdots \end{aligned}$$

Substituting $x = 0$ into these expressions gives

$$\begin{aligned} -1 = 2a_2 &\Rightarrow a_2 = -\tfrac{1}{2!} \\ 0 = 3.2a_3 &\Rightarrow a_3 = 0 \\ 1 = 4.3.2a_4 &\Rightarrow a_4 = \tfrac{1}{4!} \end{aligned}$$

and so on.
This gives, as before

$$\cos x = 1 - \frac{x^2}{2!} + \frac{x^4}{4!} - \frac{x^6}{6!} + \cdots = \sum_{n=0}^{\infty} \frac{(-1)^n x^{2n}}{(2n)!}.$$

(The advantage of this latter method lies in the fact that the precise form of the Taylor series does not need to be remembered.)

■

Note that the Maclaurin series for $\cos x$ only contains even powers of x. This is to be expected when it is remembered that $\cos x$ is an even function; that is, it remains unchanged when x is replaced by $-x$.

Worked Example 8.6.2 *Obtain the Taylor series expansion for $f(x) = \log x$ about $x = 2$.*
(Note that $\log x$ does *not* have a Taylor series expansion about $x = 0$. Why?)
Let

$$\log x = a_0 + a_1(x-2) + a_2(x-2)^2 + a_3(x-2)^3 + a_4(x-2)^4 + \cdots$$

To find a_0 we put $x = 2$, giving $a_0 = \log 2$.
Successive differentiation yields

$$\begin{aligned} \tfrac{1}{x} &= a_1 + 2a_2(x-2) + 3a_3(x-2)^2 + 4a_4(x-2)^3 + \cdots \\ -\tfrac{1}{x^{\cdot}} &= 2a_2 + 3.2a_3(x-2) + 4.3a_4(x-2)^2 + \cdots \\ \tfrac{2}{x^{\cdot}} &= 3.2a_3 + 4.3.2a_4(x-2) + \cdots \end{aligned}$$

and so on.
Substituting $x = 2$ into these expressions gives

$$\begin{aligned} 1/2 &= a_1 \\ -1/4 &= 2a_2 \quad \Rightarrow \quad a_2 = -1/8 \\ 2/8 &= 3.2a_3 \quad \Rightarrow \quad a_3 = 1/24 \end{aligned}$$

and so on.
Thus

$$\log x = \log 2 + \frac{1}{2}(x-2) - \frac{1}{8}(x-2)^2 + \frac{1}{24}(x-2)^3 + \cdots$$

for those values of x for which the series on the right converges. ∎

(See the next section — Section 8.7 — for a discussion of the convergence of this series.)

We have indicated how to obtain power series expansions; in fact, you should be able to derive (see Exercise 23) the following series which are included in Appendix B (Page 489).

$$\begin{aligned} e^x &= 1 + x + \tfrac{x^{\bullet}}{2!} + \tfrac{x^{\bullet}}{3!} + \tfrac{x^{\bullet}}{4!} + \cdots + \tfrac{x^n}{n!} + \cdots \\ \sin x &= x - \tfrac{x^{\bullet}}{3!} + \tfrac{x^{\bullet}}{5!} - \tfrac{x^{\bullet}}{7!} + \cdots + (-1)^{n-1}\tfrac{x^{\bullet n-\bullet}}{(2n-1)!} + \cdots \\ \cos x &= 1 - \tfrac{x^{\bullet}}{2!} + \tfrac{x^{\bullet}}{4!} - \tfrac{x^{\bullet}}{6!} + \cdots + (-1)^n\tfrac{x^{\bullet n}}{(2n)!} + \cdots \\ \log(1+x) &= x - \tfrac{x^{\bullet}}{2} + \tfrac{x^{\bullet}}{3} - \tfrac{x^{\bullet}}{4} + \cdots + (-1)^{n-1}\tfrac{x^n}{n} + \cdots \end{aligned}$$

Before looking at how such series can be manipulated we must first consider the question of convergence.

8.7 Convergence of Power Series

The series

1. $-1 + \frac{1}{2} - \frac{1}{3} + \frac{1}{4} - \frac{1}{5} + \cdots$
2. $1 + \frac{1}{2} + \frac{1}{3} + \frac{1}{4} + \frac{1}{5} + \cdots$
3. $2 + \frac{2^{\bullet}}{2} + \frac{2^{\bullet}}{3} + \frac{2^{\bullet}}{4} + \frac{2^{\bullet}}{5} + \cdots$
4. $(1/2) + \frac{(1/2)^{\bullet}}{2} + \frac{(1/2)^{\bullet}}{3} + \frac{(1/2)^{\bullet}}{4} + \frac{(1/2)^{\bullet}}{5} + \cdots$

are all special cases of the series

$$x + \frac{x^2}{2} + \frac{x^3}{3} + \frac{x^4}{4} + \frac{x^5}{5} + \cdots \tag{8.10}$$

Substituting

- $x = -1$ into the series 8.10 gives the series in 1 which converges by the alternating series test
- $x = 1$ into the series 8.10 gives the series in 2 which is the divergent harmonic series
- $x = 2$ into the series 8.10 gives the series in 3 which diverges since the nth term does not tend to zero
- $x = 1/2$ into the series 8.10 gives the series in 4 which converges by the ratio test.

Thus the series 8.10 converges for some values of x and diverges for other values. In fact, we can show that it converges only if $-1 \leq x < 1$.

To show this we appeal to the absolute convergence theorem; that is, we shall consider the series $\sum |a_n|$ where $a_n = \frac{x^n}{n}$. By the ratio test

$$\begin{aligned}
\lim_{n\to\infty} \frac{|a_{n+1}|}{|a_n|} &= \lim_{n\to\infty} |\frac{a_{n+1}}{a_n}| \\
&= \lim_{n\to\infty} |\frac{x^{n+1}}{n+1}\frac{n}{x^n}| \\
&= \lim_{n\to\infty} \frac{n}{n+1}|x| \\
&= |x| \quad \text{since } \frac{n}{n+1} \to 1 \text{ as } n \to \infty.
\end{aligned}$$

The series is absolutely convergent **and hence convergent** if $|x| < 1$.

For $|x| > 1, \lim_{n\to\infty} |a_n| \neq 0$, so that the nth term of the series does not tend to zero and the series diverges. We have seen that the series diverges when $x = 1$ and converges when $x = -1$. (Note that the ratio test yields no information about the convergence of the series when $x = \pm 1$. Why?)

Thus this power series converges for $-1 \leq x < 1$.

Now consider the general power series about $x = 0$:

$$\sum_{n=0}^{\infty} c_n x^n = c_0 + c_1 x + c_2 x^2 + c_3 x^3 + \cdots + c_n x^n + \cdots$$

This series obviously converges when $x = 0$. To determine its convergence behaviour elsewhere, consider

$$\lim_{n\to\infty} \left|\frac{a_{n+1}}{a_n}\right| = \lim_{n\to\infty} \left|\frac{c_n x^n}{c_{n-1}x^{n-1}}\right| = \lim_{n\to\infty} \left|\frac{c_n}{c_{n-1}}\right| |x|.$$

If we define R by $\lim_{n\to\infty} |\frac{c_n}{c_{n-\cdot}}| = \frac{1}{R}$ then $\lim_{n\to\infty} |\frac{a_{n\cdot\,\cdot}}{a_n}| = \frac{|x|}{R}$.

Thus the power series is absolutely convergent and hence convergent when $|x| < R$. For $|x| > R$ the series diverges since for such x, a_n does not tend to zero. For $|x| = R$, that is, for $x = R$ and $x = -R$, some other test needs to be used to establish convergence or divergence.

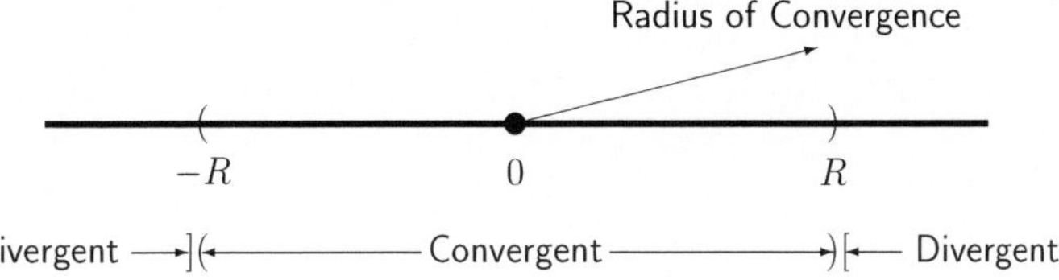

Note that

1. If $R = 0$ the series is convergent **only** for $x = 0$.
2. If R is infinite the series converges for all x.

The set of real numbers for which a power series converges is called the **interval of convergence**, and R is called the **radius of convergence**.

Worked Example 8.7.1 *Determine the radius of convergence of each of the series:*

1. $\sum_{n=1}^{\infty} n!x^n$ *2.* $\sum_{n=1}^{\infty} \frac{x^{n-1}}{(2n)!}$ *3.* $\sum_{n=1}^{\infty} \frac{x^n}{n^2 3^n}$.

1.
$$\begin{aligned}\lim_{n\to\infty}\left|\frac{a_{n+1}}{a_n}\right| &= \lim_{n\to\infty}\left|\frac{(n+1)!x^{n+1}}{n!x^n}\right| \\ &= \lim_{n\to\infty}(n+1)|x| \\ &= \infty \text{ unless } x = 0.\end{aligned}$$

Therefore $\sum_{n=1}^{\infty} n!x^n$ converges only if $x = 0$.

2.
$$\begin{aligned}\lim_{n\to\infty}\left|\frac{a_{n+1}}{a_n}\right| &= \lim_{n\to\infty}\left|\frac{x^n}{(2n+2)!}\frac{(2n)!}{x^{n-1}}\right| \\ &= \lim_{n\to\infty}\frac{|x|}{(2n+2)(2n+1)} \\ &= 0 \text{ for all } x.\end{aligned}$$

Therefore $\sum_{n=1}^{\infty} \frac{x^n}{(2n)!}$ converges for all x; that is, its radius of convergence is infinite.

3.
$$\begin{aligned}\lim_{n\to\infty}\left|\frac{a_{n+1}}{a_n}\right| &= \lim_{n\to\infty}\left|\frac{x^{n+1}}{(n+1)^2 3^{n+1}}\frac{n^2 3^n}{x^n}\right| \\ &= \lim_{n\to\infty}\frac{n^2}{3(n+1)^2}|x| \\ &= \frac{|x|}{3} \text{ since } \frac{n^2}{(n+1)^2} \to 1 \text{ as } n \to \infty.\end{aligned}$$

Therefore $\sum_{n=1}^{\infty} \frac{x^n}{n \cdot 3^n}$ converges for $|x| < 3$ and diverges for $|x| > 3$. The radius of convergence is 3. (Does the series converge for $x = 3$? For $x = -3$?)

■

Worked Example 8.7.2 *Determine the radius of convergence of the series:*

$$2x + \frac{2.4}{1.3}x^2 + \frac{2.4.6}{1.3.5}x^3 + \frac{2.4.6.8}{1.3.5.7}x^4 + \cdots$$

Here

$$a_n = \frac{2.4.6.8\ldots 2n}{1.3.5.7\ldots(2n-1)}x^n$$

and

$$a_{n+1} = \frac{2.4.6.8\ldots 2n(2n+2)}{1.3.5.7\ldots(2n-1)(2n+1)}x^{n+1}$$

so that

$$\begin{aligned}
\lim_{n\to\infty} \left|\frac{a_{n+1}}{a_n}\right| &= \lim_{n\to\infty} \left|\frac{2.4.6...2n(2n+2)}{1.3.5...(2n-1)(2n+1)} \frac{1.3.5...(2n-1)}{2.4.6...2n} x\right| \\
&= \lim_{n\to\infty} \frac{2n+2}{2n+1}|x| \\
&= |x| \text{ since } \frac{2n+2}{2n+1} \to 1 \text{ as } n \to \infty
\end{aligned}$$

Thus the series converges for $|x| < 1$ and diverges for $|x| > 1$ so it has radius of convergence 1. ■

Power series of the form

$$\sum_{n=0}^{\infty} c_n(x-a)^n = c_0 + c_1(x-a) + c_2(x-a)^2 + c_3(x-a)^3 + \cdots + c_n(x-a)^n + \cdots \tag{8.11}$$

can be discussed in a similar manner to those for which $a = 0$ by a simple change of variable $x' = x - a$ so that

$$\sum_{n=0}^{\infty} c_n(x-a)^n = \sum_{n=0}^{\infty} c_n x'^{\,n}.$$

It follows that the series 8.11 converges for $|x - a| < R$ and diverges for $|x - a| > R$ where R is the radius of convergence.

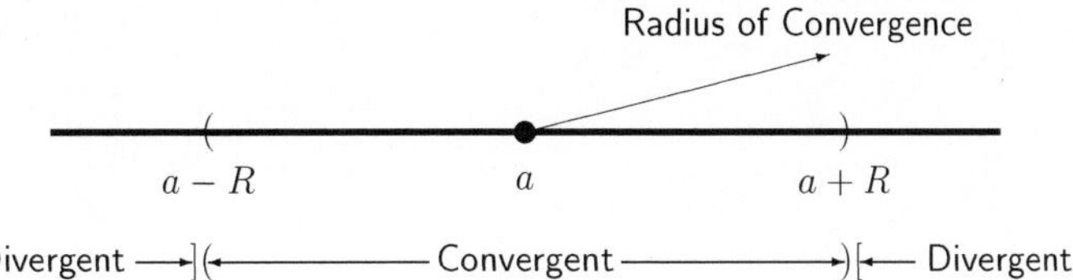

If $R = 0$ the series only converges for $x = a$; if R is infinite the series converges for all x.

Worked Example 8.7.3 *Determine the radius of convergence of the power series*

$$\sum_{n=0}^{\infty} \frac{(-1)^n}{4^n(n+1)}(x-2)^n =$$

$$1 - \tfrac{1}{4.2}(x-2) + \tfrac{1}{4^{\cdot}.3}(x-2)^2 - \cdots + \tfrac{(-1)^n}{4^n(n+1)}(x-2)^n + \cdots$$

We have

$$\begin{aligned}
\lim_{n\to\infty} \left|\frac{a_{n+1}}{a_n}\right| &= \lim_{n\to\infty} \left|\frac{(-1)^{n+1}(x-2)^{n+1}}{4^{n+1}(n+2)} \frac{4^n(n+1)}{(-1)^n(x-2)^n}\right| \\
&= \frac{1}{4}|x-2| \lim_{n\to\infty} \frac{n+1}{n+2} \\
&= \frac{1}{4}|x-2| \text{ since } \frac{n+1}{n+2} \to 1 \text{ as } n \to \infty
\end{aligned}$$

Hence the series converges for $\frac{1}{4}|x-2| < 1$ (that is, $|x-2| < 4$) and diverges for $|x-2| > 4$ so that the radius of convergence of the series is 4. ■

Self-help exercises

1. Write down the nth term a_n of each of the series:

(a) $\dfrac{x}{2.5} + \dfrac{x^2}{5.7} + \dfrac{x^3}{8.9} + \dfrac{x^4}{11.11} + \dfrac{x^5}{14.13} + \cdots$ $\left[\frac{x^n}{(3n-1)(2n+3)}\right]$

(b) $\dfrac{x}{2.5} - \dfrac{x^2}{5.7} + \dfrac{x^3}{8.9} - \dfrac{x^4}{11.11} + \dfrac{x^5}{14.13} - \cdots$ $\left[\frac{(-1)^{n\cdot} \cdot x^n}{(3n-1)(2n+3)}\right]$

(c) $\dfrac{3.3}{2.2}x + \dfrac{4.3^2}{3.2^2}x^2 + \dfrac{5.3^3}{4.2^3}x^3 + \dfrac{6.3^4}{5.2^4}x^4 + \cdots$ $\left[\frac{(n+2)3^n}{(n+1)2^n}x^n\right]$

2. For each of the following series determine

- the nth term T_n
- $\left|\dfrac{T_{n+1}}{T_n}\right|$
- $\lim\limits_{n\to\infty}\left|\dfrac{T_{n+1}}{T_n}\right|$
- the radius of convergence.

(a) $\sum \dfrac{(-1)^n(n+1)x^{2n-1}}{n.3^n}$ $[R=\sqrt{3}]$

(b) $\sum \dfrac{(-1)^n(2n+1)x^n}{n!3^n}$ $[R=\infty]$

(c) $\sum \dfrac{2^n(n+1)}{n^2+1}x^n$ $[R=1/2]$

(d) $\sum \dfrac{1.3.5...(2n-1)}{2.4.6...2n}\dfrac{x^n}{n+1}$ $[R=1]$

(e) $\dfrac{1.2}{2.5}x+\dfrac{2.3}{5.7}x^2+\dfrac{3.4}{8.9}x^3+\dfrac{4.5}{11.11}x^4+\dfrac{5.6}{14.13}x^5+\cdots$ $[R=1]$

8.8 Differentiation and Integration of Power Series

8.8.1 Term-by-term Differentiation of Power Series

A power series $\sum\limits_{n=0}^{\infty} c_n x^n$ defines a function f by

$$f(x)=\sum_{n=0}^{\infty} c_n x^n = c_0+c_1x+c_2x^2+c_3x^3+\cdots \text{ if } |x|<R \tag{8.12}$$

where R is the radius of convergence of the power series. For example, the geometric series

$$\sum_{n=0}^{\infty}(-1)^n x^n = 1-x+x^2-x^3+x^4-\cdots$$

defines the function $f(x)=\dfrac{1}{1+x}$ if $|x|<1$; that is,

$$\frac{1}{1+x}=1-x+x^2-x^3+x^4-\cdots \text{ if } |x|<1.$$

In Section 8.5.1 we showed that

$$\frac{1}{(1+x)^2}=1-2x+3x^2-4x^3+5x^4-\cdots \text{ if } |x|<1.$$

so that the series expansion for the derivative of $\dfrac{1}{1+x}$ is

$$-\frac{1}{(1+x)^2} = -1 + 2x - 3x^2 + 4x^3 - 5x^4 + \cdots \text{ if } |x| < 1.$$

You should note that this series could have been obtained by simply differentiating the series for $\frac{1}{1+x}$ term-by-term. This last observation regarding the series expansion of $\frac{1}{1+x}$ is also true in general.

Suppose $\sum_{n=0}^{\infty} c_n x^n$ is a power series having radius of convergence R. Let

$$\boxed{f(x) = \sum_{n=0}^{\infty} c_n x^n \text{ if } |x| < R}$$

Then the power series $\sum_{n=0}^{\infty} n c_n x^{n-1}$, obtained by differentiating the series 8.12 term-by-term, also has radius of convergence R (verify this!) and

$$\boxed{f'(x) = \sum_{n=0}^{\infty} n c_n x^{n-1} \text{ if } |x| < R}$$

Although both series have the same radius of convergence the two series may exhibit different convergence behaviour at the endpoints $x = R$ and $x = -R$.

This result also extends naturally to power series of the form $\sum c_n (x-a)^n$.

▌**Example:** In Worked Example 8.6.1 we obtained the Maclaurin series expansion for $\cos x$:

$$\cos x = 1 - \frac{x^2}{2!} + \frac{x^4}{4!} - \frac{x^6}{6!} + \cdots = \sum_{n=0}^{\infty} \frac{(-1)^n x^{2n}}{(2n)!}$$

which is valid for all x.
Use this series to determine the Maclaurin series expansion for $\sin x$.
Solution: Differentiating this series term-by-term gives

$$-\sin x = -x + \frac{x^3}{3!} - \frac{x^5}{5!} + \frac{x^7}{7!} + \cdots = \sum_{n=0}^{\infty} \frac{(-1)^{n+1} x^{2n+1}}{(2n+1)!}$$

for all x.
Thus the Maclaurin series for $\sin x$ is

$$\sin x = x - \frac{x^3}{3!} + \frac{x^5}{5!} - \frac{x^7}{7!} + \cdots = \sum_{n=0}^{\infty} \frac{(-1)^n x^{2n+1}}{(2n+1)!}$$

8.8.2 Term-by-term Integration of Power Series

Just as a power series can be differentiated term-by-term within its interval of convergence, it can also be integrated term-by-term.

Suppose the power series $\sum c_n x^n$ has radius of convergence R and that

$$\boxed{f(x) = \sum_{n=0}^{\infty} c_n x^n \text{ if } |x| < R}$$

then the series $\sum_{n=0}^{\infty} \frac{c_n}{n+1} x^{n+1}$ which is obtained by integrating the series 8.12 term-by-term also has radius of convergence R (Verify this!) and

$$\begin{aligned} \int f(x)\,dx &= \sum_{n=0}^{\infty} \frac{c_n}{n+1} x^{n+1} + C \\ &= \left[c_0 x + \tfrac{c_1}{2} x^2 + \tfrac{c_2}{3} x^3 + \tfrac{c_3}{4} x^4 + \cdots\right] + C \text{ if } |x| < R \end{aligned}$$

Worked Example 8.8.1 *Use the geometric series* $\sum_{n=0}^{\infty}(-1)^n x^n$ *to obtain a power series for* $\log(1+x)$.
Hence obtain estimates for $\log 1.1$ *and* $\log 3$.
The series

$$\sum_{n=0}^{\infty} (-1)^n x^n = 1 - x + x^2 - x^3 + x^4 - x^5 + \cdots$$

is a convergent geometric series with sum $\frac{1}{1+x}$ if $|x| < 1$.
That is, $\frac{1}{1+x} = 1 - x + x^2 - x^3 + x^4 - x^5 + \cdots$ if $|x| < 1$.
Integrating both sides yields

$$\log(1+x) = C + x - \frac{x^2}{2} + \frac{x^3}{3} - \frac{x^4}{4} + \frac{x^5}{5} - \frac{x^6}{6} + \cdots$$

if $|x| < 1$.
On substituting $x = 0$ we obtain $C = 0$ so that

$$\log(1+x) = x - \frac{x^2}{2} + \frac{x^3}{3} - \frac{x^4}{4} + \frac{x^5}{5} - \frac{x^6}{6} + \cdots$$

if $|x| < 1$.
Note that this series also converges for $x = 1$. (Why?)
Putting $x = 0.1$ we obtain

$$\begin{aligned} \log 1.1 &= 0.1 - \tfrac{0.01}{2} + \tfrac{0.001}{3} - \tfrac{0.0001}{4} + \cdots \\ &\approx 0.1 - 0.005 + 0.0003 \end{aligned}$$

upon truncating the series after the third term. The resulting error is at most $\frac{1}{4}(0.0001)$ since the error involved in truncating a convergent alternating series does not exceed the magnitude of the next term.

Thus $\log 1.1 \approx 0.0953$ correct to four decimal places.

In order to obtain an estimate for $\log 3$ it is tempting to substitute $x = 2$ in the power series for $\log(1+x)$. **Explain why this is not possible!**

However if we put $x = -\frac{2}{3}$ (which is OK) we obtain

$$\log\frac{1}{3} = -\frac{2}{3} - \frac{(\frac{2}{3})^2}{2} - \frac{(\frac{2}{3})^3}{3} - \frac{(\frac{2}{3})^4}{4} - \cdots$$

and so

$$\log 3 = \frac{2}{3} + \frac{1}{2}(\frac{2}{3})^2 + \frac{1}{3}(\frac{2}{3})^3 + \frac{1}{4}(\frac{2}{3})^4 + \cdots$$

This last series converges very slowly; a better approach is to obtain a power series for $\log\frac{1+x}{1-x} = \log(1+x) - \log(1-x)$ and then put $x = \frac{1}{2}$. (See Exercise 32.) ■

8.9 The Binomial Series

You should be familiar with the binomial theorem

$$\begin{aligned}(1+x)^n &= 1 + nx + \frac{n(n-1)}{2!}x^2 + \frac{n(n-1)(n-2)}{3!}x^3 \\ &\quad + \cdots + \frac{n(n-1)\ldots(n-k+1)}{k!}x^k + \cdots + x^n\end{aligned}$$

where n is a positive integer. This result extends when n is not a positive integer as follows.

If n is not a positive integer the series on the right becomes an infinite series called the **binomial series**. It becomes, on replacing n by α and k by n:

$$\begin{aligned}\sum_{n=0}^{\infty} c_n x^n &= 1 + \alpha x + \frac{\alpha(\alpha-1)}{2!}x^2 + \frac{\alpha(\alpha-1)(\alpha-2)}{3!}x^3 \\ &\quad + \cdots + \frac{\alpha(\alpha-1)\ldots(\alpha-n+1)}{n!}x^n + \cdots\end{aligned}$$

Now, if α is not a positive integer,

$$\begin{aligned}
&\lim_{n\to\infty}\left|\frac{T_{n+1}}{T_n}\right| \\
&= \lim_{n\to\infty}\left|\frac{\alpha(\alpha-1)\dots(\alpha-n+1)(\alpha-n)}{(n+1)!}\frac{n!}{\alpha(\alpha-1)\dots(\alpha-n+1)}x\right| \\
&= \lim_{n\to\infty}\left|\frac{\alpha-n}{n+1}\right||x| \\
&= |x| \text{ since } \left|\frac{\alpha-n}{n+1}\right| \to 1 \text{ as } n\to\infty.
\end{aligned}$$

Therefore the binomial series has radius of convergence 1.

We now show that this series represents the function $(1+x)^\alpha$ if $|x|<1$. If f is the function defined by the series for $|x|<1$, then

$$\begin{aligned}
f(x) &= 1+\alpha x+\frac{\alpha(\alpha-1)}{2!}x^2+\frac{\alpha(\alpha-1)\dots(\alpha-n+1)}{n!}x^n+\cdots \\
&= \sum_{n=0}^{\infty} c_n x^n
\end{aligned}$$

and

$$f'(x)=\sum_{n=1}^{\infty} nc_n x^{n-1} \text{ if } |x|<1.$$

Thus, if $|x|<1$, we have

$$\begin{aligned}
(1+x)f'(x) &= f'(x)+xf'(x) \\
&= \sum_{n=1}^{\infty} nc_n x^{n-1}+\sum_{n=1}^{\infty} nc_n x^n \\
&= c_1+2c_2x+3c_3x^2+\cdots+(n+1)c_{n+1}x^n+\cdots \\
&\quad +c_1x+2c_2x^2+\cdots+nc_nx^n+\cdots \\
&= c_1+\sum_{n=1}^{\infty}\{(n+1)c_{n+1}+nc_n\}x^n \\
&= \alpha+\sum_{n=1}^{\infty}\{(n+1)c_{n+1}+nc_n\}x^n
\end{aligned}$$

Now $c_n = \frac{\alpha(\alpha-1)\ldots(\alpha-n+1)}{n!}$ so that

$$\begin{aligned}
&(n+1)c_{n+1} + nc_n \\
&= (n+1)\frac{\alpha(\alpha-1)\ldots(\alpha-n+1)(\alpha-n)}{(n+1)!} + n\frac{\alpha(\alpha-1)\ldots(\alpha-n+1)}{n!} \\
&= \frac{\alpha(\alpha-1)(\alpha-2)\ldots(\alpha-n+1)}{n!}(\alpha-n+n) \\
&= \alpha c_n.
\end{aligned}$$

Thus

$$(1+x)f'(x) = \alpha + \sum_{n=1}^{\infty} \alpha c_n x^n = \alpha \sum_{n=0}^{\infty} c_n x^n = \alpha f(x) \text{ if } |x| < 1.$$

This is a first order separable differential equation for $f(x)$. You should verify that the unique solution which satisfies the initial condition $f(0) = 1$ is $f(x) = (1+x)^\alpha$.

Hence we have the binomial series

$$\boxed{(1+x)^\alpha = 1 + \alpha x + \frac{\alpha(\alpha-1)}{2!}x^2 + \cdots + \frac{\alpha(\alpha-1)\ldots(\alpha-n+1)}{n!}x^n + \cdots}$$

valid for all real α and all x such that $|x| < 1$. (For an alternative derivation of this result, see Exercise 23e.)

Note the form of the left-hand side; the 1 in the term $(1+x)^\alpha$ is highly significant. In order to obtain expansions for functions such as $(a+x)^\alpha$ the function needs to be modified to $(a[1+\frac{x}{a}])^\alpha$; that is, $b(1+x')^\alpha$ where $b = a^\alpha$ and $x' = x/a$. This is illustrated in some of the following examples. Also note that if α is a positive integer m, this result reduces to the familiar binomial theorem.

Example:

1. $$\begin{aligned}
\frac{1}{1+x} &= (1+x)^{-1} \\
&= 1 + (-1)x + \tfrac{(-1)(-2)}{2!}x^2 + \tfrac{(-1)(-2)(-3)}{3!}x^3 + \cdots \\
&= 1 - x + x^2 - x^3 + \cdots \text{ if } |x| < 1.
\end{aligned}$$

2. $$\begin{aligned}
\frac{1}{\sqrt{1+x^2}} &= (1+x^2)^{-1/2} \\
&= 1 - \tfrac{1}{2}x^2 + \tfrac{(-1/2)(-3/2)}{2!}x^4 + \tfrac{(-1/2)(-3/2)(-5/2)}{3!}x^6 + \cdots \\
&= 1 - \tfrac{1}{2}x^2 + \tfrac{3}{8}x^4 - \tfrac{5}{16}x^6 + \cdots
\end{aligned}$$

   if $|x^2| < 1$; that is, if $|x| < 1$.

$$3.\quad \frac{1}{4+x^2} = \frac{1}{4}\frac{1}{1+\frac{x^{\bullet}}{4}}$$

$$= \tfrac{1}{4}(1+\tfrac{x^{\bullet}}{4})^{-1}$$

$$= \tfrac{1}{4}(1-\tfrac{x^{\bullet}}{4}+(\tfrac{x^{\bullet}}{4})^2-(\tfrac{x^{\bullet}}{4})^3+\cdots) \text{ if } \left|\tfrac{x^{\bullet}}{4}\right|<1$$

$$= \tfrac{1}{4}(1-\tfrac{x^{\bullet}}{4}+\tfrac{x^{\bullet}}{16}-\tfrac{x^{\bullet}}{64}+\cdots) \text{ if } |x|<2.$$

The binomial series provides a powerful tool for finding power series expansions for the inverse circular and inverse hyperbolic functions. This is illustrated in the following worked example.

▌**Worked Example 8.9.1** *Use the binomial series to write down the power series for* $\dfrac{1}{\sqrt{4+x^2}}$ *and deduce a power series for* *arsinh* $\frac{x}{2}$.

We have

$$\frac{1}{\sqrt{4+x^2}} = \frac{1}{2\sqrt{1+\frac{x^{\bullet}}{4}}}$$

$$= \frac{1}{2}(1+\frac{x^2}{4})^{-1/2}$$

$$= \tfrac{1}{2}(1-\tfrac{1}{2}\tfrac{x^{\bullet}}{4}+\tfrac{(-1/2)(-3/2)}{2!}(\tfrac{x^{\bullet}}{4})^2+\cdots) \text{ if } \left|\tfrac{x^{\bullet}}{4}\right|<1$$

$$= \frac{1}{2}\left(1-\frac{x^2}{8}+\frac{3x^4}{128}-\cdots\right) \text{ if } |x|<2.$$

On integrating both sides we obtain

$$\text{arsinh}\,\frac{x}{2} = C+\frac{1}{2}\left(x-\frac{x^3}{3.8}+\frac{3x^5}{5.128}-\cdots\right) \text{ if } |x|<2.$$

Putting $x=0$ yields $C=0$ so that

$$\text{arsinh}\,\frac{x}{2} = \frac{1}{2}\left(x-\frac{x^3}{3.8}+\frac{3x^5}{5.128}-\cdots\right) \text{ if } |x|<2.$$

■

Self-help exercises

1. Determine the first four non-zero terms of the power series expansions for

(a) $\sqrt{1-x}$ $[1 - \frac{x}{2} - \frac{x^{\cdot}}{8} - \frac{x^{\cdot}}{16} + \cdots]$

(b) $\dfrac{1}{(1+x)^2}$ $[1 - 2x + 3x^2 - 4x^3 + \cdots]$

(c) $(1-2x)^{-3}$ $[1 + 6x + 24x^2 + 80x^3 + \cdots]$

(d) $\sqrt{9-x}$ $[3 - \frac{x}{6} - \frac{x^{\cdot}}{216} - \frac{x^{\cdot}}{3888} + \cdots]$

(e) $\dfrac{1}{\sqrt{1-3x}}$. $[1 + \frac{3x}{2} + \frac{27x^{\cdot}}{8} + \frac{135x^{\cdot}}{16} + \cdots]$

2. Write down the radius of convergence of each of the series in Question 1. $[1; 1; \frac{1}{2}; 9; \frac{1}{3}]$

8.10 Formal Manipulation of Power Series

We have already considered the operations of term-by-term differentiation and term-by-term integration of power series. Here we are concerned with the formal operations of addition, multiplication, division and composition of power series but we shall not concern ourselves with the justification of such procedures.

For ease of reference we include the following important series which are given in Appendix B (Page 489).

---

e^x

$$= 1 + x + \frac{x^2}{2!} + \frac{x^3}{3!} + \frac{x^4}{4!} + \cdots + \frac{x^n}{n!} + \cdots \text{ for all } x$$

$\sin x$

$$= x - \frac{x^3}{3!} + \frac{x^5}{5!} - \frac{x^7}{7!} + \cdots + (-1)^{n-1}\frac{x^{2n-1}}{(2n-1)!} + \cdots \text{ for all } x$$

$\cos x$

$$= 1 - \frac{x^2}{2!} + \frac{x^4}{4!} - \frac{x^6}{6!} + \cdots + (-1)^n\frac{x^{2n}}{(2n)!} + \cdots \text{ for all } x$$

$\log(1+x)$

$$= x - \frac{x^2}{2} + \frac{x^3}{3} - \frac{x^4}{4} + \cdots + (-1)^{n-1}\frac{x^n}{n} + \cdots \text{ if } |x| < 1$$

$(1+x)^\alpha$

$$= 1 + \alpha x + \frac{\alpha(\alpha-1)}{2!}x^2 + \cdots + \frac{\alpha(\alpha-1)\ldots(\alpha-n+1)}{n!}x^n + \cdots$$

if $|x| < 1$

8.10.1 Addition of Power Series

Suppose

$$f(x) = \sum_{n=0}^{\infty} a_n x^n = a_0 + a_1 x + a_2 x^2 + \cdots + a_n x^n + \cdots \text{ if } |x| < R_1$$

and

$$g(x) = \sum_{n=0}^{\infty} b_n x^n = b_0 + b_1 x + b_2 x^2 + \cdots + b_n x^n + \cdots \text{ if } |x| < R_2$$

then

$$\begin{aligned} f(x) + g(x) &= \sum_{n=0}^{\infty} (a_n + b_n) x^n \\ &= (a_0 + b_0) + (a_1 + b_1)x + (a_2 + b_2)x^2 + \cdots + \\ &\quad (a_n + b_n)x^n + \cdots \\ &\qquad \text{if } |x| < R \text{ where } R \text{ is the smaller of } R_1 \text{ and } R_2. \end{aligned}$$

Similarly

$$\begin{aligned} f(x) - g(x) &= \sum_{n=0}^{\infty} (a_n - b_n) x^n \\ &= (a_0 - b_0) + (a_1 - b_1)x + (a_2 - b_2)x^2 + \cdots + \\ &\quad (a_n - b_n)x^n + \cdots \\ &\qquad \text{if } |x| < R \text{ where } R \text{ is the smaller of } R_1 \text{ and } R_2. \end{aligned}$$

Thus, power series can be added (or subtracted) term-by-term for those values of x for which **both** series converge.

For example

$$\log(1 + x) = x - \frac{x^2}{2} + \frac{x^3}{3} - \frac{x^4}{4} + \cdots$$

converges for $|x| < 1$ and

$$\cos x = 1 - \frac{x^2}{2!} + \frac{x^4}{4!} - \frac{x^6}{6!} + \cdots$$

converges for all x so that

$$\log(1 + x) + \cos x = 1 + x - x^2 + \frac{x^3}{3} - \frac{5x^4}{24} - \cdots$$

converges if $|x| < 1$.

Moreover, if k is a constant

$$kf(x) = \sum_{n=0}^{\infty} ka_n x^n = ka_0 + ka_1 x + ka_2 x^2 + \cdots + ka_n x^n + \cdots$$

if $|x| < R_1$.

▌**Worked Example 8.10.1** *Using the exponential series determine power series expansions for* $\cosh x$ *and* $\sinh x$.
From Appendix B (Page 489).

$$e^x = 1 + x + \frac{x^2}{2!} + \frac{x^3}{3!} + \frac{x^4}{4!} + \frac{x^5}{5!} + \cdots \text{ for all } x.$$

On replacing x by $-x$ we have

$$e^{-x} = 1 - x + \frac{x^2}{2!} - \frac{x^3}{3!} + \frac{x^4}{4!} - \frac{x^5}{5!} + \cdots \text{ for all } x.$$

Now

$$\begin{aligned}
\cosh x &= \tfrac{1}{2}(e^x + e^{-x}) \\
&= \tfrac{1}{2}\Big\{(1 + x + \tfrac{x^2}{2!} + \tfrac{x^3}{3!} + \tfrac{x^4}{4!} + \cdots) + \\
&\qquad (1 - x + \tfrac{x^2}{2!} - \tfrac{x^3}{3!} + \tfrac{x^4}{4!} - \cdots)\Big\} \\
&= 1 + \tfrac{x^2}{2!} + \tfrac{x^4}{4!} + \cdots \text{ for all } x
\end{aligned}$$

and

$$\begin{aligned}
\sinh x &= \tfrac{1}{2}(e^x - e^{-x}) \\
&= \tfrac{1}{2}\Big\{(1 + x + \tfrac{x^2}{2!} + \tfrac{x^3}{3!} + \tfrac{x^4}{4!} + \cdots) - \\
&\qquad (1 - x + \tfrac{x^2}{2!} - \tfrac{x^3}{3!} + \tfrac{x^4}{4!} - \cdots)\Big\} \\
&= x + \tfrac{x^3}{3!} + \tfrac{x^5}{5!} + \cdots \text{ for all } x.
\end{aligned}$$

(Note that the series for $\cosh x$ only contains even powers of x and the series for $\sinh x$ only contains odd powers of x. Why is this not unexpected?) ■

8.10.2 Composition of Power Series

Suppose

$$f(x) = \sum_{n=0}^{\infty} a_n x^n = a_0 + a_1 x + a_2 x^2 + \cdots + a_n x^n + \cdots$$

converges for $|x| < R$. The power series for $f(g(x))$ can be obtained from

$$f(g(x)) = \sum_{n=0}^{\infty} a_n \{g(x)\}^n = a_0 + a_1 g(x) + a_2 \{g(x)\}^2 + \cdots + a_n \{g(x)\}^n + \cdots$$

which converges for $|g(x)| < R$ and then substituting the power series expansion of $g(x)$ for $g(x)$ in this expression. This technique is illustrated in the following example.

▌ **Worked Example 8.10.2** *Determine the first few terms of the power series for $e^{\cos x}$.*
From Appendix B (Page 489) we have

$$\cos x = 1 - \frac{x^2}{2!} + \frac{x^4}{4!} - \frac{x^6}{6!} + \cdots \text{ for all } x$$

and

$$e^x = 1 + x + \frac{x^2}{2!} + \frac{x^3}{3!} + \frac{x^4}{4!} + \cdots \text{ for all } x$$

Therefore

$$\begin{aligned}
e^{\cos x} &= e^{1+g(x)} \text{ where } g(x) = -\tfrac{x^{\bullet}}{2} + \tfrac{x^{\bullet}}{24} - \tfrac{x^{\bullet}}{720} + \cdots \\
&= e \times e^{g(x)} \\
&= e\left[1 + g(x) + \tfrac{1}{2}\{g(x)\}^2 + \tfrac{1}{6}\{g(x)\}^3 + \cdots\right] \\
&= e\Big\{1 + \left(-\tfrac{x^{\bullet}}{2} + \tfrac{x^{\bullet}}{24} - \tfrac{x^{\bullet}}{720} + \cdots\right) + \\
&\quad \tfrac{1}{2}\left(-\tfrac{x^{\bullet}}{2} + \tfrac{x^{\bullet}}{24} - \tfrac{x^{\bullet}}{720} + \cdots\right)^2 + \\
&\quad \tfrac{1}{6}\left(-\tfrac{x^{\bullet}}{2} + \tfrac{x^{\bullet}}{24} - \tfrac{x^{\bullet}}{720} + \cdots\right)^3 + \cdots\Big\} \\
&= e\left\{1 - \tfrac{x^{\bullet}}{2} + (\tfrac{1}{24} + \tfrac{1}{8})x^4 - (\tfrac{1}{720} + \tfrac{1}{48} + \tfrac{1}{48})x^6 + \cdots\right\} \\
&= e\left(1 - \tfrac{x^{\bullet}}{2} + \tfrac{x^{\bullet}}{6} - \tfrac{31x^{\bullet}}{720} + \cdots\right)
\end{aligned}$$

■

8.10.3 Multiplication of Power Series

Suppose

$$f(x) = \sum_{n=0}^{\infty} a_n x^n = a_0 + a_1 x + a_2 x^2 + \cdots + a_n x^n + \cdots \text{ if } |x| < R_1$$

and

$$g(x) = \sum_{n=0}^{\infty} b_n x^n = b_0 + b_1 x + b_2 x^2 + \cdots + b_n x^n + \cdots \text{ if } |x| < R_2$$

then

$$\begin{aligned} f(x)g(x) &= (a_0 + a_1 x + a_2 x^2 + a_3 x^3 + \cdots)\times \\ &\quad (b_0 + b_1 x + b_2 x^2 + b_3 x^3 + \cdots) \\ &= a_0 b_0 + (a_0 b_1 + a_1 b_0)x + (a_2 b_0 + a_1 b_1 + a_0 b_2)x^2 + \\ &\quad (a_3 b_0 + a_2 b_1 + a_1 b_2 + a_0 b_3)x^3 + \cdots \text{ if } |x| < R \end{aligned}$$

where R is the smaller of R_1 and R_2.

The power series are multiplied just like polynomials and terms are then grouped starting with the lowest powers.

Worked Example 8.10.3 *Determine a power series expansion for* $\dfrac{\log(1+x)}{1+x}$.

From Appendix B (Page 489)

$$\log(1+x) = x - \frac{x^2}{2} + \frac{x^3}{3} - \frac{x^4}{4} + \cdots \text{ if } |x| < 1.$$

Using the binomial series we obtain

$$\frac{1}{1+x} = (1+x)^{-1} = 1 - x + x^2 - x^3 + x^4 - \cdots \text{ if } |x| < 1.$$

Therefore, the power series for $\frac{\log(1+x)}{1+x}$ is

$$\begin{aligned} \frac{\log(1+x)}{1+x} &= (x - \tfrac{x^2}{2} + \tfrac{x^3}{3} - \tfrac{x^4}{4} + \cdots)\times \\ &\quad (1 - x + x^2 - x^3 + x^4 - \cdots) \\ &= x - x^2 + x^3 - x^4 + \cdots \\ &\quad - \tfrac{x^2}{2} + \tfrac{x^3}{2} - \tfrac{x^4}{2} + \cdots \\ &\quad + \tfrac{x^3}{3} - \tfrac{x^4}{3} + \cdots \\ &\quad - \tfrac{x^4}{4} + \cdots \\ &= x - (1 + \tfrac{1}{2})x^2 + (1 + \tfrac{1}{2} + \tfrac{1}{3})x^3 - \\ &\quad (1 + \tfrac{1}{2} + \tfrac{1}{3} + \tfrac{1}{4})x^4 + \cdots \end{aligned}$$

and is valid for $|x| < 1$ since each of the original series converge if $|x| < 1$. ■

8.10.4 Division of Power Series

Suppose

$$f(x) = \sum_{n=0}^{\infty} a_n x^n = a_0 + a_1 x + a_2 x^2 + \cdots + a_n x^n + \cdots \quad \text{if } |x| < R_1$$

and

$$g(x) = \sum_{n=0}^{\infty} b_n x^n = b_0 + b_1 x + b_2 x^2 + \cdots + b_n x^n + \cdots \quad \text{if } |x| < R_2$$

then the quotient

$$\frac{f(x)}{g(x)} = \frac{a_0 + a_1 x + a_2 x^2 + \cdots + a_n x^n + \cdots}{b_0 + b_1 x + b_2 x^2 + \cdots + b_n x^n + \cdots}$$

can be expressed in the form

$$c_0 + c_1 x + c_2 x^2 + c_3 x^3 + \cdots + c_n x^n + \cdots$$

provided $b_0 \neq 0$ when $a_0 \neq 0$. The coefficients $c_0, c_1, c_2, c_3, \ldots$ may be obtained by the procedure of long division (see Worked Example 8.10.4) or by writing

$$\begin{aligned} &a_0 + a_1 x + a_2 x^2 + a_3 x^3 + \cdots + a_n x^n + \cdots \\ &= (c_0 + c_1 x + c_2 x^2 + \cdots + c_n x^n + \cdots)\times \\ &\quad (b_0 + b_1 x + b_2 x^2 + \cdots + b_n x^n + \cdots) \end{aligned}$$

expanding the right side and equating coefficients of like terms.

The resulting power series may not converge for $|x| < R$ where R is the smaller of R_1 and R_2 as there is an added complication. Problems may occur wherever the denominator is zero. This is illustrated in the following example.

Worked Example 8.10.4 *Determine the first few terms of the power series for* $\tan x$.

We have

$$\tan x = \frac{\sin x}{\cos x} = \frac{x - \frac{x^3}{3!} + \frac{x^5}{5!} - \frac{x^7}{7!} + \cdots}{1 - \frac{x^2}{2!} + \frac{x^4}{4!} - \frac{x^6}{6!} + \cdots}$$

Thus, using long division (see Figure 8.4), we have

$$\tan x = x + \frac{1}{3}x^3 + \frac{2}{15}x^5 + \frac{17}{315}x^7 + \cdots$$

$$
\begin{array}{r|ccccccccc}
 & x & + & \frac{1}{3}x^3 & + & \frac{2}{15}x^5 & + & \frac{17}{315}x^7 & + & \cdots \\
\hline
1 - \frac{1}{2}x^2 + \frac{1}{24}x^4 - \frac{1}{720}x^6 + \cdots & x & - & \frac{1}{6}x^3 & + & \frac{1}{120}x^5 & - & \frac{1}{5040}x^7 & + & \cdots \\
 & x & - & \frac{1}{2}x^3 & + & \frac{1}{24}x^5 & - & \frac{1}{720}x^7 & + & \cdots \\
\hline
 & & & \frac{1}{3}x^3 & - & \frac{1}{30}x^5 & + & \frac{1}{840}x^7 & - & \cdots \\
 & & & \frac{1}{3}x^3 & - & \frac{1}{6}x^5 & + & \frac{1}{72}x^7 & - & \cdots \\
\hline
 & & & & & \frac{2}{15}x^5 & - & \frac{4}{315}x^7 & + & \cdots \\
 & & & & & \frac{2}{15}x^5 & - & \frac{1}{15}x^7 & + & \cdots \\
\hline
 & & & & & & & \frac{17}{315}x^7 & - & \cdots
\end{array}
$$

Figure 8.4: Illustration of the long division process.

(It can be shown that the radius of convergence of this series is $\frac{\pi}{2}$ even though the series for $\sin x$ and $\cos x$ are both convergent for all x. This is not surprising when you consider that the function in the denominator, $\cos x$, is zero when $x = \pm\frac{\pi}{2}$.) ■

We conclude this section on series by showing how power series can be used to obtain precise approximations to definite integrals which are either impossible or extremely difficult to evaluate exactly.

▌**Worked Example 8.10.5** *(1990 Examination) Using the series for e^x and $(1+x)^\alpha$ from Appendix B (Page 489):*

1. *Show that* $\dfrac{e^{-x}}{1+2x} = 1 - 3x + \dfrac{13}{2}x^2 - \dfrac{79}{6}x^3 + \cdots$ *if* $|x| < \dfrac{1}{2}$.
2. *Hence obtain an approximation to* $\displaystyle\int_0^{0.1} \frac{e^{-x}}{1+2x}\,dx$.

From Appendix B (Page 489)

$$e^x = 1 + x + \frac{x^2}{2!} + \frac{x^3}{3!} + \frac{x^4}{4!} + \cdots \text{ for all } x$$

and

$$\begin{aligned}(1+x)^\alpha &= 1 + \alpha x + \frac{\alpha(\alpha-1)}{2!}x^2 + \cdots + \\ &\quad \frac{\alpha(\alpha-1)\ldots(\alpha-n+1)}{n!}x^n + \cdots \text{ if } |x| < 1\end{aligned}$$

Thus

$$e^{-x} = 1 - x + \frac{x^2}{2!} - \frac{x^3}{3!} + \frac{x^4}{4!} - \cdots \text{ for all } x$$

and

$$\frac{1}{1+2x} = (1+2x)^{-1} = 1 - 2x + 4x^2 - 8x^3 + 16x^4 + \cdots \text{ if } |2x| < 1$$

so that

$$\begin{aligned}\frac{e^{-x}}{1+2x} &= (1 - x + \tfrac{1}{2}x^2 - \tfrac{1}{6}x^3 + \cdots)\times \\ &\quad (1 - 2x + 4x^2 - 8x^3 + \cdots) \text{ if } |x| < \tfrac{1}{2} \\ &= 1 + (-2-1)x + (4 + 2 + \tfrac{1}{2})x^2 + \\ &\quad (-8 - 4 - 1 - \tfrac{1}{6})x^3 + \cdots \\ &= 1 - 3x + \tfrac{13}{2}x^2 - \tfrac{79}{6}x^3 + \cdots\end{aligned}$$

Using this result we have

$$\begin{aligned}
\int_0^{0.1} \frac{e^{-x}}{1+2x}\,dx &= \int_0^{0.1} [1 - 3x + \frac{13}{2}x^2 - \frac{79}{6}x^3 + \cdots]\,dx \\
&= [x - \tfrac{3x^{\cdot}}{2} + \tfrac{13x^{\cdot}}{6} - \tfrac{79x^{\cdot}}{24} + \cdots]_0^{0.1} \\
&\approx 0.1 - 0.015 + 0.00217 - 0.00033 \\
&= 0.0868 \text{ to four decimal places.}
\end{aligned}$$

The series for $\frac{e^{-x}}{1+2x}$ is alternating (try to prove it!) so if the first four terms of this series are used to evaluate the integral (as in the above example) the magnitude of the error would be no greater than $\frac{211}{8}(0.1)^5 \approx 3.0 \times 10^{-4}$. (Verify this!) ■

Self-help exercises

1. Use the series expansions for $\sin x$ and $\cos x$ to obtain the first three non-zero terms of their product.

   Write down the first three terms of the series expansion for $\sin 2x$.

   What do you notice? $[x - \frac{2}{3}x^3 + \frac{2}{15}x^5;\ 2x - \frac{4}{3}x^3 + \frac{4}{15}x^5]$

2. Find the first three terms of the power series expansion of $f(x) = \frac{e^{-x}}{\sqrt{1-2x}}$.

   $[1 + x^2 + \frac{4}{3}x^3]$

In this topic we have

- Introduced infinite series and the concept of convergence.
- Examined the harmonic and general "p-series".
- Discussed partial sums for telescoping and geometric series.
- Introduced the ratio, comparison and alternating series tests.
- Discussed Taylor polynomials.
- Examined Maclaurin and Taylor series expansions.
- Introduced the concept of a radius of convergence.
- Manipulated power series using addition, multiplication, substitution and term-by-term differentiation and integration.

8.11 Quick Test Number 8

Introduction to Infinite Series

Question **Selection**

True or false quiz:

1. The infinite series $\sum_{n=0}^{\infty} a_n$ is just a list of terms, i.e. $a_0, a_1, a_2, \ldots$

   (a) True (b) False

2. The infinite series $\sum_{n=0}^{\infty} a_n$ is just a sum of terms i.e. $a_0 + a_1 + a_2 + \cdots$

   (a) True (b) False

3. The first three terms of the infinite series $\sum_{n=1}^{\infty} \frac{1}{n^2}$ are 1, $\frac{1}{4}$ and $\frac{1}{9}$.

   (a) True (b) False

4. The sum of the first three terms of the infinite series $\sum_{n=1}^{\infty} \frac{1}{n^2}$ is $1 + \frac{1}{4} + \frac{1}{9}$.

   (a) True (b) False

5. The sum of the first three terms of the infinite series $\sum_{n=1}^{\infty} \frac{1}{n^2}$ is known as the third partial sum S_3 of the series.

   (a) True (b) False

6. For the infinite series $\sum_{n=1}^{\infty} \frac{1}{n^2}$ the fourth partial sum is $\frac{1}{16}$.

   (a) True (b) False

7. The infinite series $\sum_{n=1}^{\infty} \frac{1}{n^2}$ is a geometric series. (a) True (b) False

8. 3^n is the same as $3n$. (a) True (b) False

9. $\sum_{n=0}^{\infty} \frac{5}{3^n} = \frac{15}{2}$. (a) True (b) False

10. "$\lim_{n\to\infty} \frac{1}{n^2} \to 0$." This is the correct notation. (a) True (b) False

11. "$\frac{1}{n^2} = 0$ as $n \to \infty$." This is the correct notation.

(a) True (b) False

12. $\lim_{n\to\infty} \frac{1}{n^2} = 0$, therefore $\sum_{n=1}^{\infty} \frac{1}{n^2} = 0$. (a) True (b) False

13. Any series $\sum_{n=1}^{\infty} a_n$ will converge if $a_n \to 0$ as $n \to \infty$.

(a) True (b) False

14. The series $\sum_{n=1}^{\infty} \frac{1}{n}$ must converge because the terms get smaller and smaller as $n \to \infty$. (a) True (b) False

15. $\sum_{n=0}^{\infty} \frac{n+3}{3n+5} = \frac{1}{3}$. (a) True (b) False

---

Tests for Convergence

Question **Selection**

Which test would you use to test the convergence of:

1. $\sum_{k=1}^{\infty} \frac{k^2+2k+3}{2k^4-1}$ (a) comparison (b) alternating series (c) ratio

2. $\sum_{k=1}^{\infty} \frac{(-1)^k k!}{5^k}$ (a) comparison (b) alternating series (c) ratio

3. $\sum_{k=1}^{\infty} \frac{2+\sin k}{k}$ (a) comparison (b) alternating series (c) ratio

4. $\sum_{k=1}^{\infty} \frac{(-1)^{k+1}}{k}$ (a) comparison (b) alternating series (c) ratio

5. $\sum_{k=1}^{\infty} \frac{\log k}{k^3+1}$ (a) comparison (b) alternating series (c) ratio

6. $\sum_{k=1}^{\infty} \frac{(k^2+1)2^k}{k!}$ (a) comparison (b) alternating series (c) ratio

7. $\sum_{k=1}^{\infty} \frac{(-1)^k \log k}{k^3+1}$ (a) comparison (b) alternating series (c) ratio

Additional questions:

8. $\sum_{k=1}^{\infty} \frac{1}{k}$ diverges even though the terms get successively smaller.

(a) True (b) False

9. $\sum_{k=1}^{\infty} \frac{(-1)^k}{k!}$ is an alternating series. (a) True (b) False

10. $\sum_{k=1}^{\infty} \frac{(-1)^k}{k}$ converges but is not absolutely convergent.

(a) True (b) False

Ratio test:

11. If $a_k = \frac{2k+1}{2^{3k}k!}$, what is a_{k+1}? (a) $\frac{2k+2}{2^{3k+1}(k+1)!}$ (b) $\frac{2k+3}{2^{3(k+1)}(k+1)!}$

12. $\frac{(2(k+1))!}{(2k)!}$ equals (a) $k+1$ (b) $\frac{k+1}{k}$ (c) $(2k+2)(2k+1)$

13. For the series $\sum_{k=1}^{\infty} \frac{(-1)^k}{k+2}$ the ratio test says that $\left|\frac{a_{k+1}}{a_k}\right| \to 1$ as $k \to \infty$.

This means that

(a) the series converges to 1 (b) the series diverges

(c) nothing can be said using this test

Power Series

Question **Selection**

---

1. $\sum_{k=1}^{\infty} x^k$ is a power series. (a) True (b) False

Can you recognize the following frequently occurring series?

2. $\sum_{k=0}^{\infty} x^k = 1 + x + x^2 + x^3 + x^4 + \cdots$

(a) $\cos x$ (b) e^x (c) $\dfrac{1}{1-x}$ (d) $\sin x$

3. $\sum_{k=0}^{\infty} \dfrac{x^k}{k!} = 1 + x + \dfrac{1}{2}x^2 + \dfrac{1}{6}x^3 + \dfrac{1}{24}x^4 + \cdots$

(a) $\cos x$ (b) e^x (c) $\dfrac{1}{1-x}$ (d) $\sin x$

4. $\sum_{k=0}^{\infty} \dfrac{(-1)^k x^{2k+1}}{(2k+1)!} = x - \dfrac{1}{6}x^3 + \dfrac{1}{120}x^5 - \cdots$

(a) $\cos x$ (b) e^x (c) $\dfrac{1}{1-x}$ (d) $\sin x$

5. $\sum_{k=0}^{\infty} \dfrac{(-1)^k x^{2k}}{(2k)!} = 1 - \dfrac{1}{2}x^2 + \dfrac{1}{24}x^4 - \cdots$

(a) $\cos x$ (b) e^x (c) $\dfrac{1}{1-x}$ (d) $\sin x$

Ratio test:

For the series $\sum_{k=0}^{\infty}(k+1)\,x^k$, $\left|\dfrac{a_{k+1}}{a_k}\right| = \left|\dfrac{(k+2)x^{k+1}}{(k+1)x^k}\right|$.

6. $\left|\dfrac{a_{k+1}}{a_k}\right|$ simplifies to give (a) $\dfrac{k+2}{k+1}\dfrac{1}{|x|}$ (b) $\dfrac{k+2}{k+1}|x|$ (c) $\dfrac{k+2}{k+1}$

7. As $k \to \infty$, $\left|\dfrac{a_{k+1}}{a_k}\right|$ approaches

(a) 1 (b) $|x|$ (c) $\dfrac{1}{|x|}$ (d) ∞ (e) $\dfrac{k+2}{k+1}$ (f) 0

8. The ratio test thus tells us that the series converges

(a) for all values of x (b) for no values of x

(c) for $|x| < 1$ (d) for $|x| > 1$

Manipulation of series:

$\dfrac{1}{1-x} = 1 + x + x^2 + x^3 + \cdots$ is valid for $|x| < 1$.

9. The first three terms of the series for $\dfrac{1}{1-3x}$ are (choose one):

(a) $1 - 3x - 3x^2$ (b) $1 - 3x + 9x^2$ (c) $1 + 3x + 9x^2$ (d) $1 + 3x + 3x^2$

10. This series is valid for (a) $|x| < 1$ (b) $|x| < 3$ (c) $|x| < \frac{1}{3}$

11. What is the **radius of convergence** of this series?

    (a) 1 (b) 3 (c) $|x| < \frac{1}{3}$ (d) $\frac{1}{3}$

8.12 Exercises

Before attempting any of these miscellaneous exercises, make sure that you have successfully answered all the self-help exercises appearing throughout the chapter.

1. If S_n denotes the nth partial sum of
$$\sum_{m=1}^{\infty} \frac{3}{(3m-2)(3m+1)}$$
show that
$$S_n = 1 - \frac{1}{3n+1}.$$
Deduce that the series converges.
What is its sum?

2. If S_n denotes the nth partial sum of
$$\sum_{m=1}^{\infty} \frac{3m-2}{m(m+1)(m+2)}$$
show that
$$S_n = 1 + \frac{1}{n+1} - \frac{4}{n+2}.$$
Deduce that the series converges.
What is its sum?

3. Determine the nth partial sum of each of the following series and hence determine whether the series converges.

(a) $\sum_{m=1}^{\infty} \log(1 + \frac{1}{m})$

(b) $\sum_{m=1}^{\infty} \frac{(-1)^{m+1}}{m(m+2)}$

(c) $\sum_{n=0}^{\infty} (\frac{3}{4})^{n+1}$

(d) $2 - \frac{4}{3} + \frac{8}{9} - \frac{16}{27} + \frac{32}{81} + \cdots$

(e) $0.1\dot{4}$

(f) $\sum_{m=1}^{\infty} (-1)^{m+1} \frac{2^{m-1}}{5^{m+1}}$

(g) $\sum_{m=1}^{\infty} (-1)^{m+1} \frac{3}{2^m}.$

4. For the series
$$\sum_{m=1}^{\infty} \log \frac{(m+1)^2}{m(m+2)}$$
show that the nth partial sum is
$$S_n = \log 2 + \log(n+1) - \log(n+2) = \log 2 + \log \frac{n+1}{n+2}.$$
Deduce that the series converges.

What is its sum?

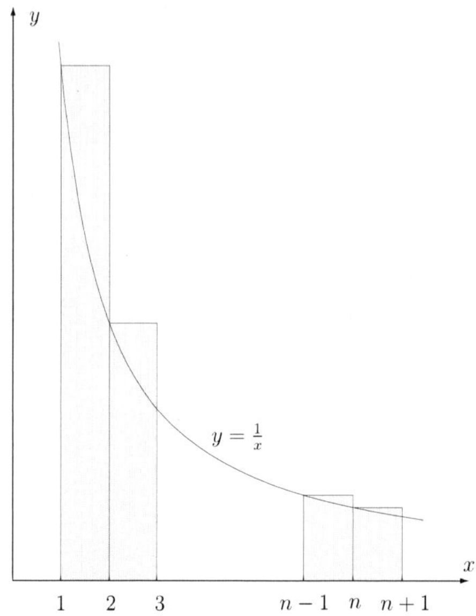

Figure 8.5: Graph of $y = \frac{1}{x}$.

5. Consider the area of the region under the graph of $y = \frac{1}{x}$ between $x = 1$ and $x = n + 1$, and the areas of the shaded rectangles. (See Figure 8.5.) By comparing these areas show that

$$1 + \frac{1}{2} + \frac{1}{3} + \frac{1}{4} + \cdots + \frac{1}{n} > \log(n+1)$$

Deduce that the series $\sum_{m=1}^{\infty} \frac{1}{m}$ diverges by showing that its sequence of partial sums is unbounded.

6. Consider the area of the region under the graph of $y = \frac{1}{x^{\cdot}}$ between $x = 1$ and $x = n$, and the areas of the shaded rectangles. (See Figure 8.6.) By comparing these areas show that

$$\frac{1}{2^2} + \frac{1}{3^2} + \frac{1}{4^2} + \frac{1}{4^2} + \cdots + \frac{1}{n^2} < 1 - \frac{1}{n}$$

Deduce that the series $\sum_{m=1}^{\infty} \frac{1}{m^{\cdot}}$ converges by showing that its sequence of partial sums is bounded above by 1.

7. Try to generalize Exercise 5 to show that $\sum_{m=1}^{\infty} \frac{1}{m^p}$ diverges if $p < 1$. [Hint: Consider the area under the graph of $y = \frac{1}{x^p}$ where $p < 1$.]

8. Try to generalize Exercise 6 to show that $\sum_{m=1}^{\infty} \frac{1}{m^p}$ converges if $p > 1$. [Hint: Consider the area under the graph of $y = \frac{1}{x^p}$ where $p > 1$.]

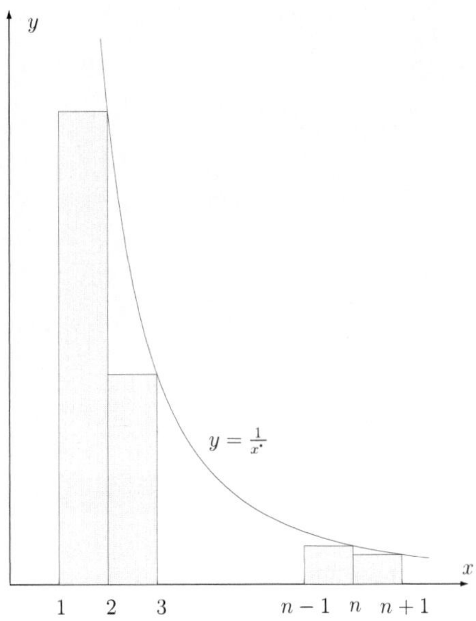

Figure 8.6: Graph of $y = \frac{1}{x^{\cdot}}$.

9. Use the comparison test to determine which of the following series converge.

   (a) $\sum \frac{n}{100n^2 + 1}$

   (b) $\sum \frac{1}{n\sqrt{n^3 + 1}}$

   (c) $\sum \frac{m}{(m+2)(m+1)}$

   (d) $\sum \frac{1}{n + n^{\div}}$

   (e) $\frac{1}{2.3.5} + \frac{2}{3.4.6} + \frac{3}{4.5.7} + \cdots$

   (f) $\frac{1}{1.1} + \frac{1}{2.2} + \frac{1}{3.4} + \frac{1}{4.8} + \cdots$

   (g) $\sum \frac{k+2}{\sqrt{k}(k+1)^2}$

   (h) $\sum \frac{2k+3}{k^2+5}$.

10. Using one of the inequalities

    - $\sin x \leq 1$ for all x
    - $\sin x < x$ for all $x > 0$

    show that each of the following series converges

    (a) $\sum \frac{\sin^2 m}{m^2}$

    (b) $\sum \frac{1}{n} \sin \frac{1}{n}$

    (c) $\sum \sin^2 \frac{1}{n}$

(d) $\sum \frac{2+\sin n}{n^2}$ (e) $\sum \frac{\sin^2 m}{2^m}$.

Does the comparison test apply to $\sum \frac{\sin m}{m^2}$? If so, does it converge? If not, explain why it does not.

The comparison test does not apply to $\sum \frac{1+2\sin m}{m^2}$. Why?

11. Use the inequality

$$\frac{\pi}{4} \leq \arctan m < \frac{\pi}{2}$$

for positive integers m to determine the convergence behaviour of

(a) $\sum \frac{\arctan n}{n}$

(b) $\sum \frac{\arctan n}{n^2}$.

12. Determine which of the following series converge:

(a) $\sum_{n=2}^{\infty} \frac{\log n}{n^2}$

(b) $\sum_{n=2}^{\infty} \frac{\log n}{n}$.

[Hint: You may find the following inequalities useful:

- $\log n < n^\alpha$ for each positive constant α
- $\log n > 1$ for all $n > 2$.]

13. Determine which of the following series converge.

(a) $\sum \frac{n^2}{2^n}$

(b) $\sum \frac{3^n}{n^2}$

(c) $\sum \frac{1}{5^n}$

(d) $\sum \frac{2^n}{n!}$

(e) $\sum \frac{n10^n}{(n+1)!}$

(f) $\sum \frac{n^3}{5^n}$

(g) $\sum \frac{n!n!}{(2n)!}$

(h) $\sum \frac{n!}{n^n}$.

[Hint: The result

$$\lim_{n\to\infty} \left(1+\frac{1}{n}\right)^n = e$$

can be assumed.]

14. *Raabe's test for a positive term series $\sum a_n$ states that if*

$$\lim_{n\to\infty} n\left(\frac{a_n}{a_{n+1}} - 1\right) = L$$

then the series converges if $L > 1$ and diverges if $L < 1$.

Show that the ratio test breaks down for the series

$$\sum_{n=1}^{\infty} \frac{1.3.5\ldots(2n-1)}{2.4.6\ldots 2n} \frac{1}{2n+1}.$$

Using Raabe's test determine whether this series converges or diverges.

15. Determine which of the following series converge.

(a) $\sum(-1)^n \frac{1}{\sqrt{n}}$

(b) $\sum(-1)^n \frac{n}{3n+1}$

(c) $\sum \frac{(-1)^n}{n^2+1}$

(d) $\sum \frac{(-1)^m}{\sqrt{m^3+1}}$

(e) $\sum(-1)^n \sin\frac{1}{n}$

(f) $\sum(-1)^k \cos\frac{1}{k}$

(g) $\sum(-1)^n \frac{1}{\log n}$.

16. Suppose that the series $\sum_{n=1}^{\infty}(-1)^{n+1}a_n$, where each $a_n > 0$, satisfies the conditions of the alternating series test; that is, $a_{n+1} < a_n$ for all n and $a_n \to 0$ as $n \to \infty$.

If the series is truncated after the mth term show that the magnitude of the error involved is

$$R_m = a_{m+1} - (a_{m+2} - a_{m+3}) - (a_{m+4} - a_{m+5}) - (a_{m+6} - a_{m+7}) - \cdots$$

Deduce that the magnitude of the error does not exceed a_{m+1}.

17. Show that the series $\sum \frac{\sin n\theta}{n^{\cdot}}$ and $\sum \frac{\cos n\theta}{n^{\cdot}}$ are absolutely convergent for all θ.

18. **Miscellaneous exercises**

Determine, with reasons, which of the following series converge.

(a) $\sum \frac{1}{n(n+1)}$

(b) $\sum(-1)^n \frac{3^{n-2}}{4^{n+1}}$

(c) $\sum(-1)^n \frac{2}{\sqrt{4n+1}}$

(d) $\sum \frac{4m^2}{2^m}$

(e) $\sum \frac{1}{\sqrt{4n^3+\log n}}$

(f) $\sum \frac{2}{\sqrt{4k+1}}$

(g) $\sum \frac{4+\cos^2 n}{\sqrt{n}}$

(h) $\sum \frac{e^n}{n!}$

(i) $\sum \frac{n+2}{n+1}(\frac{5}{6})^n$

(j) $\sum \frac{(m+2)!}{m!m^2}$

(k) $\sum(-1)^{n-1}\frac{1}{n\sqrt{n}}$

(l) $\sum \frac{k}{2k+1}$

(m) $\sum \frac{1.3.5\ldots(2n-1)}{3^n}$

(n) $\sum \frac{1.3.5\ldots(2n-1)}{1.4.7\ldots(3n-2)}$.

19. **More miscellaneous exercises**

Determine, with reasons, which of the following series converge.

(a) $\sum(-1)^n \frac{1}{n\log n}$

(b) $\sum \frac{2^m}{(2m)!}$

(c) $\sum \frac{1.2.3\ldots n}{2.5.8\ldots(3n-1)}2^n$

(d) $\sum \dfrac{2+\sin n}{n^2}$

(e) $\sum \dfrac{1+2\sin n}{n^2}$

(f) $\sum \cos \dfrac{1}{k}$

(g) $\sum \dfrac{1}{\sqrt{n(n+1)(n+2)}}$

(h) $\sum \dfrac{n!}{1.4.7\ldots(3n-2)}$

(i) $\sum \dfrac{2^m}{3^m+m}$

(j) $\sum (-1)^n \dfrac{1}{\sqrt[3]{n}}$

(k) $\sum \dfrac{\sqrt{n+1}-\sqrt{n}}{n}$

(l) $\frac{1}{2\sqrt{1}}+\frac{1}{5\sqrt{4}}+\frac{1}{8\sqrt{7}}+\frac{1}{11\sqrt{10}}+\cdots$

(m) $\frac{1}{2\sqrt{1}}+\frac{2}{5\sqrt{4}}+\frac{3}{8\sqrt{7}}+\frac{4}{11\sqrt{10}}+\cdots$

(n) $\sum \dfrac{3^k k!}{k^k}$

(o) $\sum \dfrac{n}{2^n}$.

20. Determine Taylor polynomials of degree 3 for each of the following functions about the specified point.

    (a) $f(x)=\sqrt{x}$ about $x=4$

    (b) $f(x) = \sin x$ about $x = \pi/3$

    (c) $f(x) = \cos x$ about $x = \pi/4$.

    Use your answers to obtain estimates of $\sqrt{4.2}$, $\sin 58°$ and $\cos 44°$.

21. Suppose $f(x)$ is a sufficiently well-behaved function and

$$P_n(x) = a_0 + a_1(x-a) + a_2(x-a)^2 + a_3(x-a)^3 + \cdots + a_n(x-a)^n$$

is a polynomial which satisfies

$$P_n(a) = f(a), P_n'(a) = f'(a), P_n''(a) = f''(a), \ldots, P_n^{(n)}(a) = f^{(n)}(a)$$

Show that

$$a_0 = f(a), a_1 = f'(a), a_2 = \frac{f''(a)}{2!}, a_3 = \frac{f^{(3)}(a)}{3!}, \ldots a_n = \frac{f^{(n)}(a)}{n!}.$$

22. By differentiating the identity

$$1 - x + x^2 - x^3 + x^4 - \cdots + (-1)^{n+1}x^{n+1} = \frac{1-(-x)^{n+2}}{1+x} \quad \text{for } x \neq -1$$

show that

$$\frac{1}{(1+x)^2} = 1 - 2x + 3x^2 - 4x^3 + 5x^4 + \cdots + (-1)^n(n+1)x^n + (-1)^{n+1}\left\{\frac{(n+1)x^{n+1}}{1+x} + \frac{x^{n+1}}{(1+x)^2}\right\}.$$

Deduce that if $|x| < 1$ then

$$R_n(x) = (-1)^{n+1}\left\{\frac{(n+1)x^{n+1}}{1+x} + \frac{x^{n+1}}{(1+x)^2}\right\} \to 0$$

as $n \to \infty$.

23. Obtain the following Maclaurin series:

    (a) $\arctan x = x - \frac{x^{\cdot}}{3} + \frac{x^{\cdot}}{5} - \frac{x^{\cdot}}{7} + \cdots$

    (b) $e^x = 1 + x + \frac{x^{\cdot}}{2!} + \frac{x^{\cdot}}{3!} + \frac{x^{\cdot}}{4!} + \cdots$

    (c) $\sin x = x - \frac{x^{\cdot}}{3!} + \frac{x^{\cdot}}{5!} - \frac{x^{\cdot}}{7!} + \cdots$

    (d) $\log(1+x) = x - \frac{x^{\cdot}}{2} + \frac{x^{\cdot}}{3} - \frac{x^{\cdot}}{4} + \cdots$

    (e) $(1+x)^\alpha = 1 + \alpha x + \frac{\alpha(\alpha-1)}{2!}x^2 + \cdots + \frac{\alpha(\alpha-1)\ldots(\alpha-n+1)}{n!}x^n + \cdots$.

24. Determine the Taylor series of each of the following functions about the specified point $x = a$.

    (a) $f(x) = \frac{1}{x}, \quad a = -1$

    (b) $f(x) = \sin x, \quad a = \pi/3$

    (c) $f(x) = \cos x, \quad a = \pi/3$.

    Use your answers to (b) and (c) to obtain estimates of $\sin 62°$ and $\cos 58°$.

25. Obtain a Taylor series for $\tan x$ in the form

$$\tan x = a_0 + a_1(x - \frac{\pi}{4}) + a_2(x - \frac{\pi}{4})^2 + \cdots$$

26. Determine the radius of convergence of each of the following power series.

    (a) $\sum \frac{nx^n}{2^{n+1}}$

    (b) $\sum \frac{(x-1)^{n+1}}{n3^{n-1}}$

    (c) $\sum (-1)^n \frac{x^n}{\sqrt{n+1}}$

    (d) $\sum \frac{2^{n-1}x^n}{n^3}$

    (e) $\sum \frac{(-1)^n(x-1)^{n+1}}{n3^n}$

    (f) $\sum \frac{n!}{1.4.7\ldots(3n-2)}x^n$

    (g) $\sum \frac{(n!)^2}{(2n)!}x^n$

    (h) $\sum \frac{(-1)^n(2x-1)^{n+1}}{n4^{n-1}}$

    (i) $\sum \frac{(-1)^n n}{n^2+1}x^n$

    (j) $\sum \frac{(-1)^n(x-3)^n}{n!3^n}$

    (k) $\sum \frac{(x+1)^{n+1}}{n3^{2n-1}}$

    (l) $\sum (-1)^n \frac{n+1}{n+2}x^n$

(m) $\sum \frac{1.3.5 \ldots (2n-1)}{2.4.6 \ldots 2n} \frac{x^n}{n}$

(n) $\sum \frac{1.3.5 \ldots (2n-3)}{n!} \frac{(-x)^n}{2^n}$.

27. (a) Use the geometric series

$$\sum_{n=0}^{\infty} (-x^2)^n$$

to obtain the expansion

$$\arctan x = x - \frac{x^3}{3} + \frac{x^5}{5} - \cdots$$

for $|x| < 1$.

(b) Use (a) to show that

$$\lim_{x \to 0} \frac{\arctan x}{x} = 1.$$

(c) Find a power series for

$$\int_0^x \frac{\arctan t}{t}\, dt.$$

28. Assuming that the power series for $\sin x$ and $\cos x$ hold for all complex numbers show that $\sin i\theta = i \sinh \theta$ and $\cos i\theta = \cosh \theta$.

29. Using the binomial series, write down power series expansions for each of the following functions. In each case write down the radius of convergence.

(a) $\sqrt{1-x}$

(b) $\sqrt{2-x}$

(c) $(1+2x)^{-3}$

(d) $\frac{1}{(1+x)^{\cdot}}$

(e) $\frac{1}{\sqrt{1+3x^{\cdot}}}$

(f) $\frac{1}{\sqrt{4-x^{\cdot}}}$

(g) $(1+2x^2)^{-2}$.

30. Show that

$$\frac{7}{3x^2+5x-2} = \frac{3}{3x-1} - \frac{1}{x+2}.$$

Hence obtain a power series expansion for $f(x) = \frac{7}{3x^{\cdot}+5x-2}$ using the binomial series.

[Hint: Write the expression $\frac{3}{3x-1}$ in the form $-3(1-3x)^{-1}$ and the expression $\frac{1}{x+2}$ in the form $\frac{1}{2}(1+\frac{x}{2})^{-1}$.]

31. After using partial fractions use the binomial series to determine the Maclaurin series of $\frac{1}{1-x-2x^2}$. What is its radius of convergence?

32. Use the power series expansion for $\log(1+x)$ to determine a power series expansion for $\log\left(\frac{1+x}{1-x}\right)$. Hence obtain an estimate for $\log 3$.

33. Use the binomial series for $\frac{1}{\sqrt{1-x^{\cdot}}}$ to determine a power series for $\arcsin x$. Using a similar method find a power series expansion for $\operatorname{arsinh} x$.

34. Using the standard series from Appendix B (Page 489), determine the first few terms of the power series for:

(a) $e^{-\frac{1}{2}x^2}$ (b) $e^x \cos x$ (c) $\log(1+e^x)$.

35. Using the series for $\log(1+x)$, $\cos x$ and $\sin x$ show that

(a) $\log(1+\sin x) = x - \dfrac{x^2}{2} + \dfrac{x^3}{6} - \dfrac{x^4}{12} + \dfrac{x^5}{24} + \cdots$

(b) $\log(\cos x) = -\dfrac{x^2}{2} - \dfrac{x^4}{12} - \dfrac{x^6}{45} + \cdots$.

36. Use the power series for $\log(1+x)$ to show that

$$\{\log(1+x)\}^2 = x^2 - (1+\frac{1}{2})\frac{2x^3}{3} + (1+\frac{1}{2}+\frac{1}{3})\frac{2x^4}{4} + \cdots$$

if $|x| < 1$.

37. Use the first four terms of appropriate power series to obtain estimates of

(a) $\displaystyle\int_0^1 \frac{dx}{\sqrt{1-\frac{1}{4}x^4}}$

(b) $\displaystyle\int_0^{0.1} \frac{e^x - 1}{x}\, dx$

(c) $\displaystyle\int_0^{0.5} \sqrt{1+\frac{1}{2}x^3}\, dx$

(d) $\displaystyle\int_0^{0.2} \frac{\sin x}{x}\, dx$.

In (c) and (d) give an estimate of the error.

38. Using the binomial series and the exponential series show that

$$\frac{e^{-x}}{\sqrt{1-2x}} = 1 + x^2 + \frac{4}{3}x^3 + \cdots$$

if $|x| < \frac{1}{2}$.

Hence obtain an approximate value for $\displaystyle\int_0^{0.2} \frac{e^{-x}}{\sqrt{1-2x}}\, dx$.

39. Use the series

$$\cos x = 1 - \frac{x^2}{2!} + \frac{x^4}{4!} - \frac{x^6}{6!} + \cdots$$

for all x and

$$(1+x)^\alpha = 1 + \alpha x + \frac{\alpha(\alpha-1)}{2!}x^2 + \frac{\alpha(\alpha-1)(\alpha-2)}{3!}x^3 + \cdots$$

if $|x| < 1$ to find the first four terms of the power series for $\sqrt{1-x}\cos x$.

Hence obtain an approximate value for $\displaystyle\int_0^{0.1} \sqrt{1-x}\cos x\, dx$.

Topic 9

Matrices

9.1 Systems of Linear Equations

In order to introduce some of the ideas involved in solving systems of linear equations, we shall begin with a system of three equations in three unknowns.

Consider the system

$$\begin{aligned} x &- 5y + 2z = -5 \\ 3x &- 14y + 3z = -8 \\ 4x &- 18y + 3z = -8. \end{aligned}$$

Using the first equation to eliminate x from the second and third equations we obtain

$$\begin{aligned} x - 5y + 2x &= -5 \\ y - 3z &= 7 \\ 2y - 5z &= 12. \end{aligned}$$

Using the second equation to eliminate y from the third equation we obtain

$$\begin{aligned} x - 5y + 2x &= -5 \\ y - 3z &= 7 \\ z &= -2. \end{aligned}$$

The third equation now yields $z = -2$ and "back substitution" into the second equation yields $y + 6 = 7 \Rightarrow y = 1$. Finally, back substitution into the first equation yields $x - 5 - 4 = -5 \Rightarrow x = 4$.

In the above solution it is clear that the symbols used for the unknowns are unimportant. The same solution would have been obtained for the system

$$\begin{array}{rcrcrcr} a & - & 5b & + & 2c & = & -5 \\ 3a & - & 14b & + & 3c & = & -8 \\ 4a & - & 18b & + & 3c & = & -8. \end{array}$$

The coefficients of the unknowns together with the right-hand side constants are all that is needed when determining the solution. That is, the solution could have been obtained by considering the array (matrix) of numbers

$$\left[\begin{array}{rrr|r} 1 & -5 & 2 & -5 \\ 3 & -14 & 3 & -8 \\ 4 & -18 & 3 & -8 \end{array}\right]$$

called the augmented matrix of the system. Each row of this matrix corresponds to the coefficients and corresponding right-hand sides of an equation in the original system. The successive elimination of unknowns is equivalent to reducing this matrix to

$$\left[\begin{array}{rrr|r} 1 & -5 & 2 & -5 \\ 0 & 1 & -3 & 7 \\ 0 & 0 & 1 & -2 \end{array}\right]$$

by adding (subtracting) multiples of one row to another row. We compare the two approaches as follows.

As a System of Equations

Working with the system of equations as such we have

$$\begin{array}{rcrcrcrl} x & - & 5y & + & 2z & = & -5 & \quad (1) \\ 3x & - & 14y & + & 3z & = & -8 & \quad (2) \\ 4x & - & 18y & + & 3z & = & -8. & \quad (3) \end{array}$$

Step 1.

$$\begin{array}{rcrl} x - 5y + 2z & = & -5 & \\ y - 3z & = & 7 & \text{Equation (2)} - 3 \times \text{Equation (1)} \\ 2y - 5z & = & 12 & \text{Equation (3)} - 4 \times \text{Equation (1)} \end{array}$$

Step 2.

$$\begin{aligned} x - 5y + 2z &= -5 \\ y - 3z &= 7 \\ z &= -2 \quad \text{Equation (3)} - 2 \times \text{Equation (2)}. \end{aligned}$$

As an Augmented Matrix

Using the equivalent augmented matrix instead we have

$$\left[\begin{array}{rrr|r} 1 & -5 & 2 & -5 \\ 3 & -14 & 3 & -8 \\ 4 & -18 & 3 & -8 \end{array}\right]$$

$$\sim \left[\begin{array}{rrr|r} 1 & -5 & 2 & -5 \\ 0 & 1 & -3 & 7 \\ 0 & 2 & -5 & 12 \end{array}\right] \begin{array}{l} \\ R_2 - 3R_1 \\ R_3 - 4R_1 \end{array}$$

$$\sim \left[\begin{array}{rrr|r} 1 & -5 & 2 & -5 \\ 0 & 1 & -3 & 7 \\ 0 & 0 & 1 & -2 \end{array}\right] \begin{array}{l} \\ \\ R_3 - 2R_2. \end{array}$$

The last matrix has zeros everywhere below the main diagonal and is said to be in row-echelon form. Generally, given any system of linear equations we can begin with the augmented matrix of the system and then reduce this matrix to row-echelon form by using one or more of the following elementary row operations:

- interchanging any two rows (equivalent to interchanging two equations)
- dividing (or multiplying) a row by a scalar (equivalent to dividing — or multiplying — an equation by a scalar)
- adding or subtracting a multiple of one row to another row (equivalent to adding or subtracting a multiple of one equation to another equation).

Once the matrix of the system has been reduced to row-echelon form the solution of the system is readily obtained by a process of back substitution.

Worked Example 9.1.1 *Determine the solution of the system of equations*

$$\begin{array}{rcrcrcrcr} 2x_1 & - & x_2 & + & 2x_3 & + & x_4 & = & 0 \\ x_1 & + & 2x_2 & - & x_3 & + & 3x_4 & = & 9 \\ x_1 & + & x_2 & & & + & 2x_4 & = & 5 \\ 3x_1 & - & 4x_2 & + & 3x_3 & + & x_4 & = & -1. \end{array}$$

We reduce the matrix of this system to row-echelon form using elementary row operations as follows:

$$\left[\begin{array}{rrrr|r} 2 & -1 & 2 & 1 & 0 \\ 1 & 2 & -1 & 3 & 9 \\ 1 & 1 & 0 & 2 & 5 \\ 3 & -4 & 3 & 1 & -1 \end{array}\right]$$

$$\sim \left[\begin{array}{rrrr|r} 1 & 2 & -1 & 3 & 9 \\ 2 & -1 & 2 & 1 & 0 \\ 1 & 1 & 0 & 2 & 5 \\ 3 & -4 & 3 & 1 & -1 \end{array}\right] \begin{array}{l} R_1 \to R_2 \\ R_2 \to R_1 \\ \\ \\ \end{array}$$

$$\sim \left[\begin{array}{rrrr|r} 1 & 2 & -1 & 3 & 9 \\ 0 & -5 & 4 & -5 & -18 \\ 0 & -1 & 1 & -1 & -4 \\ 0 & -10 & 6 & -8 & -28 \end{array}\right] \begin{array}{l} \\ R_2 - 2R_1 \\ R_3 - R_1 \\ R_3 - 3R_1 \end{array}$$

$$\sim \left[\begin{array}{rrrr|r} 1 & 2 & -1 & 3 & 9 \\ 0 & 1 & -1 & 1 & 4 \\ 0 & -5 & 4 & -5 & -18 \\ 0 & -10 & 6 & -8 & -28 \end{array}\right] \begin{array}{l} \\ R_2 \to R_3 \\ R_3 \to R_2 \\ \\ \end{array}$$

Therefore

$$\left[\begin{array}{cccc|c} 2 & -1 & 2 & 1 & 0 \\ 1 & 2 & -1 & 3 & 9 \\ 1 & 1 & 0 & 2 & 5 \\ 3 & -4 & 3 & 1 & -1 \end{array}\right]$$

$$\sim \left[\begin{array}{cccc|c} 1 & 2 & -1 & 3 & 9 \\ 0 & 1 & -1 & 1 & 4 \\ 0 & 0 & -1 & 0 & 2 \\ 0 & 0 & -4 & 2 & 12 \end{array}\right] \begin{array}{l} \\ \\ R_3 + 5R_2 \\ R_4 + 10R_2 \end{array}$$

$$\sim \left[\begin{array}{cccc|c} 1 & 2 & -1 & 3 & 9 \\ 0 & 1 & -1 & 1 & 4 \\ 0 & 0 & 1 & 0 & -2 \\ 0 & 0 & 0 & 2 & 4 \end{array}\right] \begin{array}{l} \\ \\ R_3 \times -1 \\ R_4 - 4R_3. \end{array}$$

From this last matrix we can read off the solutions as follows:

$$x_1 = 1, x_2 = 0, x_3 = -2 \text{ and } x_4 = 2.$$

■

Worked Example 9.1.2 *Show that the system of equations*

$$\begin{array}{rcrcrcl} x & - & 2y & + & 3z & = & 2 \\ 2x & + & 3y & - & z & = & 4 \\ x & + & 12y & - & 11z & = & 6 \end{array}$$

has no solution.

Consider the matrix of the system

$$\left[\begin{array}{ccc|c} 1 & -2 & 3 & 2 \\ 2 & 3 & -1 & 4 \\ 1 & 12 & -11 & 6 \end{array}\right].$$

Reducing this to row-echelon form we obtain

$$\left[\begin{array}{ccc|c} 1 & -2 & 3 & 2 \\ 0 & 7 & -7 & 0 \\ 0 & 14 & -14 & 4 \end{array}\right] \begin{array}{l} \\ R_2 - 2R_1 \\ R_3 - R_1 \end{array}$$

$$\sim \left[\begin{array}{ccc|c} 1 & -2 & 3 & 2 \\ 0 & 1 & -1 & 0 \\ 0 & 0 & 0 & 4 \end{array}\right] \begin{array}{l} \\ \frac{1}{7}R_2 \\ R_3 - 2R_2. \end{array}$$

The last row of this matrix implies $0x + 0y + 0z = 4$ which is clearly impossible. Thus this system has no solution; such systems of equations are said to be *inconsistent*. ■

Worked Example 9.1.3 *Show that the system*

$$\begin{array}{rcrcrcrcl} x_1 & + & 2x_2 & - & 3x_3 & + & x_4 & = & 0 \\ x_1 & + & 3x_2 & + & x_3 & + & x_4 & = & 2 \\ 2x_1 & - & x_2 & + & x_3 & + & 3x_4 & = & 3 \\ 4x_1 & + & 4x_2 & - & x_3 & + & 5x_4 & = & 5 \end{array}$$

has infinitely many solutions and find all solutions.

Reducing the matrix of the system to row-echelon form we obtain:

$$\left[\begin{array}{cccc|c} 1 & 2 & -3 & 1 & 0 \\ 1 & 3 & 1 & 1 & 2 \\ 2 & -1 & 1 & 3 & 3 \\ 4 & 4 & -1 & 5 & 5 \end{array}\right]$$

$$\sim \left[\begin{array}{cccc|c} 1 & 2 & -3 & 1 & 0 \\ 0 & 1 & 4 & 0 & 2 \\ 0 & -5 & 7 & 1 & 3 \\ 0 & -4 & 11 & 1 & 5 \end{array}\right] \begin{array}{l} \\ R_2 - R_1 \\ R_3 - 2R_1 \\ R_4 - 4R_1. \end{array}$$

Therefore

$$\left[\begin{array}{rrrr|r} 1 & 2 & -3 & 1 & 0 \\ 1 & 3 & 1 & 1 & 2 \\ 2 & -1 & 1 & 3 & 3 \\ 4 & 4 & -1 & 5 & 5 \end{array}\right]$$

$$\sim \left[\begin{array}{rrrr|r} 1 & 2 & -3 & 1 & 0 \\ 0 & 1 & 4 & 0 & 2 \\ 0 & 0 & 27 & 1 & 13 \\ 0 & 0 & 27 & 1 & 13 \end{array}\right] \begin{array}{l} \\ \\ R_3 + 5R_2 \\ R_4 + 4R_2 \end{array}$$

$$\sim \left[\begin{array}{rrrr|r} 1 & 2 & -3 & 1 & 0 \\ 0 & 1 & 4 & 0 & 2 \\ 0 & 0 & 27 & 1 & 13 \\ 0 & 0 & 0 & 0 & 0 \end{array}\right] \begin{array}{l} \\ \\ \\ R_4 - R_3. \end{array}$$

The final row of zeros indicates that the fourth equation in the system is redundant.
From row 3 we have $27x_3 + x_4 = 13 \Rightarrow x_4 = 13 - 27x_3$. (Alternatively, we could have obtained x_3 in terms of x_4.)
Back substitution into the equation represented by row 2 gives $x_2 + 4x_3 = 2 \Rightarrow x_2 = 2 - 4x_3$.
Further back substitution into row 1 yields

$$\begin{aligned} & x_1 + 2x_2 - 3x_3 + x_4 = 0 \\ \Rightarrow \quad & x_1 + 2(2 - 4x_3) - 3x_3 + (13 - 27x_3) = 0 \\ \Rightarrow \quad & x_1 = -17 + 38x_3. \end{aligned}$$

Thus the general solution is

$$\begin{array}{rcrcr} x_1 & = & -\ 17 & + & 38x_3 \\ x_2 & = & 2 & - & 4x_3 \\ x_4 & = & 13 & - & 27x_3 \end{array}$$

where x_3 can take any value.
Alternatively, the general solution could be represented as

$$x_1 = -17 + 38k, x_2 = 2 - 4k, x_3 = k, x_4 = 13 - 27k,$$

where k is any real number.
(This solution is also expressible as

$$\mathbf{x} = (-17, 2, 0, 13) + k(38, -4, 1, -27).$$

You should observe that the four-dimensional vector (38, −4, 1, −27) is a solution of the *homogeneous* system of equations and that the four-dimensional vector $(-17, 2, 0, 13)$ is a *particular* solution of the non-homogeneous equations. Note the similarity of this form of the answer to that of the solution of constant coefficient higher order ordinary differential equations.) ■

Self-help exercises

1. Write down the general solution of the following equations:

   (a) $x_1 + 2x_2 = 0$ [$x_1 = -2k, x_2 = k$; k any real number]

   (b) $x_1 + 2x_2 = 3$ [$x_1 = 3 - 2k, x_2 = k$; k any real number]

   (c) $x_1 - 2x_2 + x_3 = 0.$
   [$x_1 = 2k - h, x_2 = k, x_3 = h$; h and k any real numbers]

2. Determine the general solution of the system of equations:

$$\begin{array}{rcrcrcl} x_1 & - & 2x_2 & + & x_3 & = & 2 \\ 2x_1 & + & x_2 & - & x_3 & = & 0. \end{array}$$

[$x_1 = k, x_2 = 3k - 2, x_3 = 5k - 2$; k any real number]

9.2 Matrix Algebra

Although matrix notation provides us with an economical method of solving systems of linear equations, matrices play a far more important role than merely a notational convenience. Matrices and their algebra are important in numerous areas of applied mathematics such as linear programming, transportation problems, electrical circuit theory, the numerical approximation of solutions to partial differential equations describing such physical and engineering applications as fluid flow, elasticity etc.

9.2.1 Review of Matrix Algebra

A matrix is a rectangular array of numbers containing m rows and n columns; such a matrix is called an $m \times n$ matrix and we then say that the matrix has order $m \times n$. If $m = n$ the matrix is said to be a square matrix. Matrices are usually denoted by upper-case letters A, B, X, Y etc. For example,

$$A = \begin{bmatrix} -1 & 23 \\ 17 & 0 \\ 4 & 12 \end{bmatrix} \text{ is a } 3 \times 2 \text{ matrix} \qquad \text{and } B = \begin{bmatrix} 2 \\ 12 \\ -3 \\ 7 \end{bmatrix} \text{ is a } 4 \times 1 \text{ matrix.}$$

Note:

An $m \times 1$ matrix is usually referred to as a *column vector*

and a $1 \times n$ matrix is usually referred to as a *row vector*.

Frequently we use the notation $\boxed{A = (a_{ij})_{m \times n}}$ to denote an $m \times n$ matrix whose element in the ith row and jth column is a_{ij}. For example,

$$A = \begin{bmatrix} 2 & -1 & 13 \\ 0 & 17 & 23 \end{bmatrix}$$

could be represented as $A = (a_{ij})_{2\times 3}$ where $a_{11} = 2, a_{12} = -1, a_{13} = 13, a_{21} = 0, a_{22} = 17$ and $a_{23} = 23$.

Equality of Matrices

Two matrices are said to be *equal* if they have the same order and all corresponding elements are equal; that is, for two matrices $A = (a_{ij})_{m\times n}$ and $B = (b_{ij})_{r\times s}$, we have $A = B$ if $m = r, n = s$ and $a_{ij} = b_{ij}$ for all i and j.

Addition of Matrices

If $A = (a_{ij})$ and $B = (b_{ij})$ both have the same order we define their sum $A+B$ as the matrix $(a_{ij}+b_{ij})$. That is, provided two matrices have the same order we add them by adding the corresponding elements. For example,

$$\begin{bmatrix} 1 & -2 & 7 \\ -3 & 0 & 5 \end{bmatrix} + \begin{bmatrix} 8 & 5 & -2 \\ 4 & -4 & 6 \end{bmatrix} = \begin{bmatrix} 9 & 3 & 5 \\ 1 & -4 & 11 \end{bmatrix}.$$

It follows from the definition of addition that

$$\begin{aligned} A + B &= B + A \\ A + (B + C) &= (A + B) + C \end{aligned}$$

provided the matrices A, B and C have the same orders.

Multiplication by a Scalar

If A is a matrix and k is a scalar (usually a real or complex number) we define the scalar multiple of A as $kA = (ka_{ij})$; that is, kA is obtained from A by multiplying each element of A by k.

For example, if

$$A = \begin{bmatrix} 1 & -12 & 2 \\ -3 & 4 & -3 \end{bmatrix} \text{ and } B = \begin{bmatrix} 2 & -15 & -7 \\ 0 & -3 & 4 \end{bmatrix}$$

then

$$2A = \begin{bmatrix} 2 & -24 & 4 \\ -6 & 8 & -6 \end{bmatrix} \text{ and } 2A - 3B = \begin{bmatrix} -4 & 21 & 25 \\ -6 & 17 & -18 \end{bmatrix}.$$

Matrix Multiplication

If $A = (a_{ij})$ is an $m \times n$ matrix and $B = (b_{ij})$ is an $r \times s$ matrix then the product $C = AB$ is defined only if $n = r$; that is, if *the number of columns in A equals the number of rows in B*. Provided $n = r$, the order of the product is $m \times s$.

The product $AB = C = (c_{ij})$ is defined as follows:

$$\begin{aligned} c_{ij} &= \text{the element in the } i\text{th row and } j\text{th column of } C \\ &= \text{the scalar (inner) product of the } i\text{th row of } A \text{ with the } j\text{th column of } B \\ &= a_{i1}b_{1j} + a_{i2}b_{2j} + \cdots + a_{in}b_{nj}. \end{aligned}$$

For example,

$$\begin{aligned} c_{11} &= \text{the element in the first row and first column of } AB \\ &= \text{the scalar product of the first row of } A \text{ with the first column of } B \\ &= a_{11}b_{11} + a_{12}b_{21} + \ldots + a_{1n}b_{n1} \text{ and} \\ c_{23} &= \text{the element in the second row and third column of } AB \\ &= \text{the scalar product of the second row of } A \text{ with the third column of } B \\ &= a_{21}b_{13} + a_{22}b_{23} + \ldots + a_{2n}b_{n3}. \end{aligned}$$

For example, if

$$A = \begin{bmatrix} 3 & 4 & -7 \\ 2 & 0 & 1 \\ -3 & 1 & 5 \end{bmatrix} \text{ and } B = \begin{bmatrix} 5 & 1 & 11 & -2 & 3 \\ -1 & 6 & 9 & 0 & 12 \\ 1 & -1 & 0 & 7 & 5 \end{bmatrix}$$

then

$$AB = \begin{bmatrix} 4 & 34 & 69 & -55 & 22 \\ 11 & 1 & 22 & 3 & 11 \\ -11 & -2 & -24 & 41 & 28 \end{bmatrix}$$

where, for example, $c_{11} = 3(5)+4(-1)+(-7)1 = 4$ and $c_{23} = 2(11)+0(9)+1(0) = 22$.

Worked Example 9.2.1 *If*

$$A = \begin{bmatrix} -1 & 2 & -4 \\ 0 & 2 & 7 \\ -5 & 3 & 1 \end{bmatrix}, B = \begin{bmatrix} -2 & 7 & 3 \end{bmatrix},$$

$$C = \begin{bmatrix} -4 \\ 12 \\ 1 \end{bmatrix} \text{ and } D = \begin{bmatrix} 8 & -6 \\ 1 & 0 \\ -4 & 3 \end{bmatrix}$$

determine which of the following products exist:

$$AB, AC, BC, CB, A(B+D), DC \text{ and } CD.$$

1. Since A is a 3×3 matrix and B is a 1×3 matrix the product AB is not defined.
2. AC is defined since the number of columns of A equals the number of rows of C and
$$AC = \begin{bmatrix} 24 \\ 31 \\ 57 \end{bmatrix}.$$
3. BC is defined (why?) and equals $[8 + 84 + 3] = [95] = 95$.
4. CB is defined and equals
$$\begin{bmatrix} 8 & -28 & -12 \\ -24 & 84 & 36 \\ -2 & 7 & 3 \end{bmatrix}.$$
5. $A(B + D)$ is not defined since $B + D$ is not defined (B and D have different orders).
6. DC is not defined. (Why?)
7. CD is not defined. (Why?)

■

If AB is defined there is no guarantee that BA is defined so that in general $AB \neq BA$. For example, if A has order 2×3 and B has order 3×4, AB is defined but not BA.

Even if both AB and BA are defined, AB does not generally equal BA.

Example: If

$$A = \begin{bmatrix} -1 & 2 \\ 3 & 4 \end{bmatrix} \text{ and } B = \begin{bmatrix} 2 & -3 \\ 1 & 7 \end{bmatrix}$$

then

$$AB = \begin{bmatrix} 0 & 17 \\ 10 & 19 \end{bmatrix} \text{ while } BA = \begin{bmatrix} -11 & -8 \\ 20 & 30 \end{bmatrix}$$

so that $AB \neq BA$.

The following theorem is easily proven.

Theorem 9.1 *If all operations are defined*

$$\begin{aligned} A(B+C) &= AB+AC \\ (B+C)A &= BA+CA \\ A(BC) &= (AB)C. \end{aligned}$$

Since $AB \neq BA$, in general, we have

$$\begin{aligned} (A+B)^2 &= (A+B)(A+B) \neq A^2+2AB+B^2 \\ A^2-B^2 &\neq (A-B)(A+B) \\ (AB)^2 &\neq A^2B^2. \end{aligned}$$

In fact, as you should verify,

$$\begin{aligned} (A+B)^2 &= A^2+AB+BA+B^2 \text{ and} \\ (A-B)(A+B) &= A^2+AB-BA-B^2. \end{aligned}$$

Transpose of a Matrix

If A is an $m \times n$ matrix (a_{ij}) then the transpose of A, denoted by A^T, is the $n \times m$ matrix (a_{ji}) obtained by interchanging the rows and columns of A; for example, if

$$A = \begin{bmatrix} -2 & 6 & -7 \\ 0 & -3 & 4 \end{bmatrix} \text{ then } A^T = \begin{bmatrix} -2 & 0 \\ 6 & -3 \\ -7 & 4 \end{bmatrix}.$$

That is, the ith row of A^T is the ith column of A and the jth column of A^T is the jth row of A. The following theorem is easily proven.

Theorem 9.2

1. $(A+B)^T = A^T + B^T$ *if* A *and* B *have the same order and*

2. $(A^T)^T = A$.

Moreover we will prove the following theorem.

Theorem 9.3 *Provided AB is defined then $(AB)^T = B^T A^T$.*

The theorem holds because the element in the ith row and jth column of the matrix product $(AB)^T$

= the element in the jth row and ith column of AB

= the scalar product of the jth row of A with the ith column of B

= the scalar product of the ith row of B^T with the jth column of A^T

= the element in the ith row and jth column of $B^T A^T$.

Worked Example 9.2.2 *If*

$$A = \begin{bmatrix} -2 & 3 & 7 \\ 1 & -4 & 0 \end{bmatrix} \text{ and } B = \begin{bmatrix} 2 & -8 & 0 \\ 5 & -1 & 6 \\ 0 & -9 & 1 \end{bmatrix}$$

verify that $(AB)^T = B^T A^T$.

$$\begin{aligned} AB &= \begin{bmatrix} -2 & 3 & 7 \\ 1 & -4 & 0 \end{bmatrix} \begin{bmatrix} 2 & -8 & 0 \\ 5 & -1 & 6 \\ 0 & -9 & 1 \end{bmatrix} \\ &= \begin{bmatrix} 11 & -50 & 25 \\ -18 & -4 & -24 \end{bmatrix} \end{aligned}$$

so that

$$(AB)^T = \begin{bmatrix} 11 & -18 \\ -50 & -4 \\ 25 & -24 \end{bmatrix}.$$

Also $B^T A^T = \begin{bmatrix} 2 & 5 & 0 \\ -8 & -1 & -9 \\ 0 & 6 & 1 \end{bmatrix} \begin{bmatrix} -2 & 1 \\ 3 & -4 \\ 7 & 0 \end{bmatrix} = \begin{bmatrix} 11 & -18 \\ -50 & -4 \\ 25 & -24 \end{bmatrix}.$

Thus $(AB)^T = B^T A^T$. ■

Self-help exercises

1. $A = \begin{bmatrix} 1 & 2 \\ 3 & 4 \\ -1 & 2 \end{bmatrix}$, $B = \begin{bmatrix} 1 & -2 & 1 \\ -3 & 1 & 2 \end{bmatrix}$ and $C = \begin{bmatrix} -1 & 2 \\ 1 & 4 \end{bmatrix}$.

   (a) Evaluate each of the products AB, BA, AC, BC, A^2 which exist.

   $$[AB = \begin{bmatrix} -5 & 0 & 5 \\ -9 & -2 & 11 \\ -7 & 4 & 3 \end{bmatrix}; BA = \begin{bmatrix} -6 & -4 \\ -2 & 2 \end{bmatrix};$$

   $$AC = \begin{bmatrix} 1 & 10 \\ 1 & 22 \\ 3 & 6 \end{bmatrix}]$$

   (b) Verify that $(AB)^T = B^T A^T$.

2. If $A = \begin{bmatrix} 4 & 6 & -1 \\ 3 & 0 & 2 \\ 1 & -2 & 5 \end{bmatrix}$ and $B = \begin{bmatrix} -2 & 1 & 1 \\ 1 & 2 & -2 \\ 2 & -1 & -4 \end{bmatrix}$ find AB and BA.

   $$[AB = \begin{bmatrix} -4 & 17 & -4 \\ -2 & 1 & -5 \\ 6 & -8 & -15 \end{bmatrix}; BA = \begin{bmatrix} -4 & -14 & 9 \\ 8 & 10 & -7 \\ 1 & 20 & -24 \end{bmatrix}]$$

3. If $A = \begin{bmatrix} 4 & 0 \\ 6 & 0 \end{bmatrix}$ and $B = \begin{bmatrix} 0 & 0 \\ 2 & -3 \end{bmatrix}$ show that $AB = O$.

Some Special Matrices

- **The zero matrix** of any order • is a matrix which has all its elements equal to zero.

Note that

$$\begin{aligned} A+O &= A = O+A \\ AO &= OA = O. \end{aligned}$$

Note also that, unlike ordinary algebra, the equation $AB = O$ does *not* imply $A = O$ or $B = O$. (See Self-help Exercise 3 above.)

- **The unit (identity) matrix** • is a square matrix which has entries of 1s down the main diagonal and entries of 0s everywhere else. The unit matrix of order $n \times n$ is denoted by I_n (or simply I when the order is clear from the context).

  For example,

$$I_2 = \begin{bmatrix} 1 & 0 \\ 0 & 1 \end{bmatrix} \text{ and } I_3 = \begin{bmatrix} 1 & 0 & 0 \\ 0 & 1 & 0 \\ 0 & 0 & 1 \end{bmatrix}.$$

  You should verify that for any square matrix A

$$AI = IA = A.$$

- **Diagonal matrices**

  The unit matrix is a special case of a more general type of matrix called a diagonal matrix — a square matrix in which all elements off the main diagonal are zero. That is, a square matrix is said to be a diagonal matrix if all elements above and below the main diagonal are zero. Diagonal matrices are extremely easy to work with — for example, if A and B are diagonal matrices of the same order then $A + B$ and AB are both diagonal; moreover, for diagonal matrices $AB = BA$.

- **Triangular matrices**

  A square matrix is said to be lower triangular if all elements above the main diagonal are zero, and upper triangular if all elements below the main diagonal are zero. For example,

$$\begin{bmatrix} 3 & 0 & 0 \\ -2 & 7 & 0 \\ 1 & 4 & 3 \end{bmatrix} \quad \text{is a lower triangular matrix, and}$$

$$\begin{bmatrix} -2 & 1 & 6 \\ 0 & 5 & 3 \\ 0 & 0 & -1 \end{bmatrix} \quad \text{is an upper triangular matrix.}$$

- **Symmetric matrices**

  A matrix A is said to be symmetric if $A^T = A$; that is, symmetric matrices are square and display symmetry about the main diagonal. (That is, $a_{ij} = a_{ji}$.)

- **Skew-symmetric matrices**

  A matrix A is said to be skew-symmetric if $A^T = -A$; that is skew-symmetric matrices are square and display anti-symmetry about the main diagonal. (That is, $a_{ij} = -a_{ji}$. You should note that the main diagonal of a skew-symmetric matrix will consist solely of zeros.)

Self-help exercise

Show that *any* square matrix A can be written as the sum of a symmetric and a skew-symmetric matrix.

- **Orthogonal matrices**

  A square matrix P is said to be an orthogonal matrix if $PP^T = P^TP = I$. It is readily shown that the columns (rows) of an orthogonal matrix are mutually orthogonal unit vectors. For example,

$$P = \begin{bmatrix} 1/3 & -2/3 & -2/3 \\ 2/3 & 2/3 & -1/3 \\ 2/3 & -1/3 & 2/3 \end{bmatrix}$$

is an orthogonal matrix.

- **Invertible matrices**

A square matrix A is said to invertible (or non-singular) if there exists a matrix B such that $AB = BA = I$. The matrix B is then called the inverse of A and is denoted by A^{-1}; if A is an invertible matrix, then its inverse is *unique*.

For example, for the matrix A defined by

$$A = \begin{bmatrix} 3 & 5 \\ 1 & 2 \end{bmatrix}$$

we have

$$A^{-1} = \begin{bmatrix} 2 & -5 \\ -1 & 3 \end{bmatrix} \text{ since } AB = \begin{bmatrix} 1 & 0 \\ 0 & 1 \end{bmatrix}.$$

However not every square matrix has an inverse. For example, the matrix

$$A = \begin{bmatrix} 1 & 2 \\ 2 & 4 \end{bmatrix}$$

has no inverse because

$$\begin{bmatrix} 1 & 2 \\ 2 & 4 \end{bmatrix}\begin{bmatrix} a & b \\ c & d \end{bmatrix} = \begin{bmatrix} 1 & 0 \\ 0 & 1 \end{bmatrix} \Rightarrow \begin{bmatrix} a+2c & b+2d \\ 2a+4c & 2b+4d \end{bmatrix} = \begin{bmatrix} 1 & 0 \\ 0 & 1 \end{bmatrix}.$$

But this last (matrix) equation is impossible to satisfy. (Why?)

Recall that $\det AB = \det A \times \det B$ (see Appendix A) so that if $\det A = 0$ it is impossible for AB to equal I since

$$\det AB = \det A \times \det B = 0 \text{ while } \det I = 1.$$

The following theorem is easily proven.

Theorem 9.4 *If det $A = 0$ then A^{-1} cannot exist.*

In the following theorem we collect together some of the important properties of the matrix inverse.

Theorem 9.5 *If A and B are non-singular matrices of the same order, then*

$$\begin{aligned} (A^{-1})^{-1} &= A \\ (A^T)^{-1} &= (A^{-1})^T \\ (AB)^{-1} &= B^{-1}A^{-1}. \end{aligned}$$

Self-help exercises

Let A and B be non-singular matrices of the same order.

1. By considering
$$(AA^{-1})^T = I^T$$
show that
$$(A^T)^{-1} = (A^{-1})^T.$$

2. By considering
$$AB(AB)^{-1} = I$$
show that
$$(AB)^{-1} = B^{-1}A^{-1}.$$

9.3 The Inverse of a Matrix

Suppose A is a 3×3 matrix which has an inverse

$$A^{-1} = \begin{bmatrix} a_1 & b_1 & c_1 \\ a_2 & b_2 & c_2 \\ a_3 & b_3 & c_3 \end{bmatrix}.$$

Then the system of equations

$$A \begin{bmatrix} x \\ y \\ z \end{bmatrix} = \begin{bmatrix} 1 \\ 0 \\ 0 \end{bmatrix}$$

has the solution

$$\begin{bmatrix} x \\ y \\ z \end{bmatrix} = A^{-1} \begin{bmatrix} 1 \\ 0 \\ 0 \end{bmatrix} = \begin{bmatrix} a_1 & b_1 & c_1 \\ a_2 & b_2 & c_2 \\ a_3 & b_3 & c_3 \end{bmatrix} \begin{bmatrix} 1 \\ 0 \\ 0 \end{bmatrix} = \begin{bmatrix} a_1 \\ a_2 \\ a_3 \end{bmatrix}.$$

Similarly the solution of

$$AX = \begin{bmatrix} 0 \\ 1 \\ 0 \end{bmatrix} \text{ is } X = \begin{bmatrix} b_1 \\ b_2 \\ b_3 \end{bmatrix}$$

and the solution of

$$AX = \begin{bmatrix} 0 \\ 0 \\ 1 \end{bmatrix} \text{ is } X = \begin{bmatrix} c_1 \\ c_2 \\ c_3 \end{bmatrix}.$$

The three systems of equations can be solved at the one time by considering the "augmented" matrix of the *system*; that is, the matrix

$$\left[\begin{array}{c|ccc} & 1 & 0 & 0 \\ A & 0 & 1 & 0 \\ & 0 & 0 & 1 \end{array}\right].$$

Applying elementary row operations we can reduce this to the equivalent form

$$\left[\begin{array}{ccc|ccc} 1 & 0 & 0 & a_1 & b_1 & c_1 \\ 0 & 1 & 0 & a_2 & b_2 & c_2 \\ 0 & 0 & 1 & a_3 & b_3 & c_3 \end{array}\right]$$

where the inverse A^{-1} of A will "appear" in the "right-hand half" of this augmented matrix.

Worked Example 9.3.1 *Find the inverse of*

$$A = \begin{bmatrix} 3 & -2 \\ -1 & 2 \end{bmatrix}.$$

We start with

$$\left[\begin{array}{cc|cc} 3 & -2 & 1 & 0 \\ -1 & 2 & 0 & 1 \end{array}\right]$$

and reduce this to the form $[I|B]$ using elementary row operations as follows:

$$\left[\begin{array}{rr|rr} 3 & -2 & 1 & 0 \\ -1 & 2 & 0 & 1 \end{array}\right]$$
$$\sim \left[\begin{array}{rr|rr} 1 & -2/3 & 1/3 & 0 \\ 0 & 4/3 & 1/3 & 1 \end{array}\right] \begin{array}{l} \frac{1}{3}R_1 \\ R_2 + \frac{1}{3}R_1 \end{array}$$
$$\sim \left[\begin{array}{rr|rr} 1 & -2/3 & 1/3 & 0 \\ 0 & 1 & 1/4 & 3/4 \end{array}\right] \begin{array}{l} \\ \frac{3}{4}R_2 \end{array}$$
$$\sim \left[\begin{array}{rr|rr} 1 & 0 & 1/2 & 1/2 \\ 0 & 1 & 1/4 & 3/4 \end{array}\right] \begin{array}{l} R_1 + \frac{2}{3}R_2 \\ \end{array}$$

so that

$$A^{-1} = \begin{bmatrix} 1/2 & 1/2 \\ 1/4 & 3/4 \end{bmatrix} = \frac{1}{4}\begin{bmatrix} 2 & 2 \\ 1 & 3 \end{bmatrix}.$$

(Alternatively, we could have begun by interchanging rows 1 and 2 of the augmented matrix in order to obtain 1 in the top left-hand corner — this has the advantage of deferring the need for fractions until a later step.) ■

Note that for invertible 2×2 matrices (see Exercise 10)

$$\boxed{A = \begin{bmatrix} a & b \\ c & d \end{bmatrix} \quad \Rightarrow \quad A^{-1} = \frac{1}{ad - bc}\begin{bmatrix} d & -b \\ -c & a \end{bmatrix}}$$

That is, interchange the main diagonal elements, change the signs of the off-diagonal elements and divide by the determinant.

Self-help exercises

Write down the inverse of each of the following 2×2 matrices:

1. $\begin{bmatrix} 2 & -5 \\ -1 & -3 \end{bmatrix}$ $[\frac{1}{11}\begin{bmatrix} 3 & -5 \\ -1 & -2 \end{bmatrix}]$ 2. $\begin{bmatrix} 1 & 3 \\ 2 & 7 \end{bmatrix}$ $[\begin{bmatrix} 7 & -3 \\ -2 & 1 \end{bmatrix}]$

Worked Example 9.3.2 *Find the inverse of*

$$A = \begin{bmatrix} 0 & 1 & -1 \\ 1 & -1 & 0 \\ 1 & 1 & 2 \end{bmatrix}.$$

We have

$$[A|I] = \left[\begin{array}{ccc|ccc} 0 & 1 & -1 & 1 & 0 & 0 \\ 1 & -1 & 0 & 0 & 1 & 0 \\ 1 & 1 & 2 & 0 & 0 & 1 \end{array}\right]$$

$$\sim \left[\begin{array}{ccc|ccc} 1 & -1 & 0 & 0 & 1 & 0 \\ 0 & 1 & -1 & 1 & 0 & 0 \\ 1 & 1 & 2 & 0 & 0 & 1 \end{array}\right] \quad \begin{array}{l} R_2 \to R_1 \\ R_1 \to R_2 \\ \end{array}$$

$$\sim \left[\begin{array}{ccc|ccc} 1 & -1 & 0 & 0 & 1 & 0 \\ 0 & 1 & -1 & 1 & 0 & 0 \\ 0 & 2 & 2 & 0 & -1 & 1 \end{array}\right] \quad \begin{array}{l} \\ \\ R_3 - R_1 \end{array}$$

$$\sim \left[\begin{array}{ccc|ccc} 1 & 0 & -1 & 1 & 1 & 0 \\ 0 & 1 & -1 & 1 & 0 & 0 \\ 0 & 0 & 4 & -2 & -1 & 1 \end{array}\right] \quad \begin{array}{l} R_1 + R_2 \\ \\ R_3 - 2R_2 \end{array}$$

$$\sim \left[\begin{array}{ccc|ccc} 1 & 0 & 0 & 1/2 & 3/4 & 1/4 \\ 0 & 1 & 0 & 1/2 & -1/4 & 1/4 \\ 0 & 0 & 1 & -1/2 & -1/4 & 1/4 \end{array}\right] \quad \begin{array}{l} R_1 + \frac{1}{4}R_3 \\ R_2 + \frac{1}{4}R_3. \\ \end{array}$$

(In the above manipulations, why was it necessary to first interchange two rows, one of which is the first row?)

Thus

$$A^{-1} = \frac{1}{4}\begin{bmatrix} 2 & 3 & 1 \\ 2 & -1 & 1 \\ -2 & -1 & 1 \end{bmatrix}.$$

■

The above method of determining A^{-1} will in fact indicate the existence or otherwise of an inverse. If A cannot be reduced to I using elementary row operations (which means that a row of zeros occurs in the reduction of A) then A is singular (that is, it does not have an inverse).

9.4 Linear Mappings

Consider the 2×2 matrix

$$A = \begin{bmatrix} 1 & -2 \\ 3 & 4 \end{bmatrix}.$$

This can be interpreted as a mapping T from two-dimensional space $\mathbf{R}^2$ into itself defined by $T(X) = AX$ where

$$X = \begin{bmatrix} x \\ y \end{bmatrix} \text{ and } T(X) = \begin{bmatrix} x' \\ y' \end{bmatrix}.$$

That is, T can be defined by

$$T(X) = \begin{bmatrix} x' \\ y' \end{bmatrix} = \begin{bmatrix} 1 & -2 \\ 3 & 4 \end{bmatrix}\begin{bmatrix} x \\ y \end{bmatrix}$$

or, equivalently

$$\begin{aligned} x' &= x - 2y \\ y' &= 3x + 4y \end{aligned}$$

which maps

- points into points
- curves into curves and
- regions into regions.

Example: With reference to the transformation T described above

1. The point $(1, 1)$ or the vector $[1\ 1]^T$ in the xy-plane maps into the point $(-1, 7)$ or the vector $[-1\ 7]^T$ in the $x'y'$-plane.
2. The line $y = x$ maps into the line $y' = -7x'$. (Verify this!)

If X_1 and X_2 are any two vectors in the xy-plane it is easily shown that

$$\begin{aligned} T(X_1 + X_2) &= T(X_1) + T(X_2) \\ T(aX_1) &= aT(X_1) \end{aligned}$$

where a is any scalar quantity (that is, any real constant).

This means that T maps any linear combination of X_1 and X_2 into the same linear combination of their images; that is, T preserves linear combinations. Such mappings are called *linear transformations.*

In a similar manner, if A is a 3×3 matrix then the mapping defined by $T(X) = AX$ can be interpreted as a mapping from three-dimensional space into itself.

9.4.1 Some Important Linear Mappings

1. **Dilations**

   The matrix

$$A = \begin{bmatrix} k & 0 & 0 \\ 0 & 1 & 0 \\ 0 & 0 & 1 \end{bmatrix}$$

   represents a stretching in the x-direction since the point (x, y, z) maps into the point (kx, y, z). In fact, all distances parallel to the x-axis are stretched by a factor of k.

   Similarly, the matrices

$$\begin{bmatrix} 1 & 0 & 0 \\ 0 & k & 0 \\ 0 & 0 & 1 \end{bmatrix} \quad \text{and} \quad \begin{bmatrix} 1 & 0 & 0 \\ 0 & 1 & 0 \\ 0 & 0 & k \end{bmatrix}$$

   represent "stretchings" in the y-direction and z-direction respectively.

   Such mappings are often called *dilations*.

2. **Reflections**

It is a simple matter to verify that

- $\begin{bmatrix} -1 & 0 \\ 0 & 1 \end{bmatrix}$ represents a reflection in the y-axis in two dimensions
- $\begin{bmatrix} 1 & 0 & 0 \\ 0 & -1 & 0 \\ 0 & 0 & 1 \end{bmatrix}$ represents a reflection in the xz-plane in three dimensions
- $\begin{bmatrix} 0 & 1 \\ 1 & 0 \end{bmatrix}$ represents a reflection in the line $y = x$ in two dimensions.

It is left as an exercise (see Exercise 15) to verify that a reflection in the line $y = mx$ where $m = \tan\theta$ in two dimensions can be represented by the matrix

$$\begin{bmatrix} \cos 2\theta & \sin 2\theta \\ \sin 2\theta & -\cos 2\theta \end{bmatrix}.$$

3. **Rotations**

Consider an anti-clockwise rotation of angle θ in two dimensions. Suppose the point (x, y) maps into the point (x', y').

In particular, consider the points $(1, 0)$ and $(0, 1)$. From Figure 9.1 it is clear that these points map onto the points $(\cos\theta, \sin\theta)$ and $(-\sin\theta, \cos\theta)$ respectively. Thus, if

$$A = \begin{bmatrix} a & b \\ c & d \end{bmatrix}$$

is the matrix of this rotation, then we have

$$\begin{bmatrix} a & b \\ c & d \end{bmatrix}\begin{bmatrix} 1 \\ 0 \end{bmatrix} = \begin{bmatrix} \cos\theta \\ \sin\theta \end{bmatrix} \text{ and } \begin{bmatrix} a & b \\ c & d \end{bmatrix}\begin{bmatrix} 0 \\ 1 \end{bmatrix} = \begin{bmatrix} -\sin\theta \\ \cos\theta \end{bmatrix}.$$

The above equations yield

$$a = \cos\theta, c = \sin\theta, b = -\sin\theta \text{ and } d = \cos\theta.$$

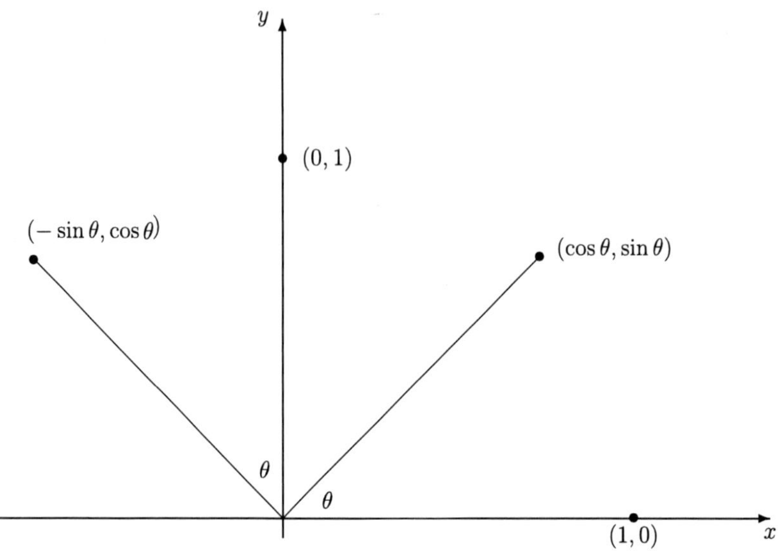

Figure 9.1: Rotation in the plane.

In general, the point (x, y) will map to the point (x', y') where

$$x' = x\cos\theta - y\sin\theta \text{ and } y' = x\sin\theta + y\cos\theta.$$

Thus the matrix

$$\begin{bmatrix} \cos\theta & -\sin\theta \\ \sin\theta & \cos\theta \end{bmatrix}$$

represents an anti-clockwise rotation through an angle θ.

▮**Example:** The matrix

$$\begin{bmatrix} 1/2 & -\sqrt{3}/2 \\ \sqrt{3}/2 & 1/2 \end{bmatrix}$$

represents an anti-clockwise rotation through 60^0.

In *three* dimensions the matrix

$$A = \begin{bmatrix} 1 & 0 & 0 \\ 0 & 1/2 & -\sqrt{3}/2 \\ 0 & \sqrt{3}/2 & 1/2 \end{bmatrix}$$

represents an anti-clockwise rotation through 60^0 about the x-axis.

(Note that under this mapping the x-co-ordinate of each point does not change.)

Similarly, the matrix

$$B = \begin{bmatrix} 1/2 & -\sqrt{3}/2 & 0 \\ \sqrt{3}/2 & 1/2 & 0 \\ 0 & 0 & 1 \end{bmatrix}$$

represents an anti-clockwise rotation of 60^0 about the z-axis.

Note that the two 3×3 matrices above have the following properties:

- each column (and row) of the matrices is a unit vector
- the columns (rows) of the matrices are mutually orthogonal vectors and
- the determinant of each matrix is $+1$.

Thus, each of the matrices A and B which represent rotations in three dimensions is an orthogonal matrix having determinant 1. This is generally true; that is, any orthogonal 3×3 matrix which has determinant 1 represents a rotation.

Worked Example 9.4.1 *Show that the matrix*

$$A = \frac{1}{7}\begin{bmatrix} 6 & 2 & 3 \\ 2 & 3 & -6 \\ -3 & 6 & 2 \end{bmatrix}$$

is an orthogonal matrix having determinant 1.
Given that A represents a rotation, determine both the axis of rotation and the angle of rotation.
To show that A is orthogonal we may verify either

- the columns of A are mutually orthogonal unit vectors

or, equivalently,

- $A^T A = AA^T = I$.

You should also verify that $\det A = 1$.
Since A is an orthogonal matrix with determinant 1, it represents a rotation.

1. **The axis of rotation**

   First note that any vector along the axis of rotation maps into itself; that is, it undergoes no change. All other vectors will change in direction.

   If

   $$Y = \begin{bmatrix} a \\ b \\ c \end{bmatrix}$$

   is any vector along the axis of rotation, we need to find a, b and c such that $AY = Y$; that is, we need to solve

   $$\begin{array}{rcrcrcl} 6a & + & 2b & + & 3c & = & 7a \\ 2a & + & 3b & - & 6c & = & 7b \\ -3a & + & 6b & + & 2c & = & 7c. \end{array}$$

   Solving this system gives $a = 2b, c = 0$ so that the axis of rotation lies along the line joining the origin to the point $(2, 1, 0)$.

   Thus all vectors (or points) along this line map into themselves.

2. **The angle of rotation**

   To determine the angle of rotation we consider a vector perpendicular to the axis of rotation, say the vector $Y = [0\ 0\ 1]^T$.

   This maps into the vector

   $$AY = A \begin{bmatrix} 0 \\ 0 \\ 1 \end{bmatrix} = \begin{bmatrix} 3/7 \\ -6/7 \\ 2/7 \end{bmatrix}.$$

   The angle of rotation is simply the angle between Y and AY obtained as follows using the normal formula for the dot product of two vectors:

   $$Y \cdot AY = |Y||AY| \cos\theta \Rightarrow \theta = \arccos 2/7.$$

   Thus A represents a rotation through an angle of $\arccos 2/7$ about the line joining the origin to the point $(2, 1, 0)$.

■

9.5 Eigenvalues and Eigenvectors

In many applications of matrices to physical problems it is sometimes important to know which vectors undergo no change in direction under a mapping defined by a matrix A. (See Worked Example 9.4.1.)

For example, suppose A represents the mapping in two dimensions defined as follows:

> *Each vector is projected onto the line $y = 2x$ as indicated in Figure 9.2.*

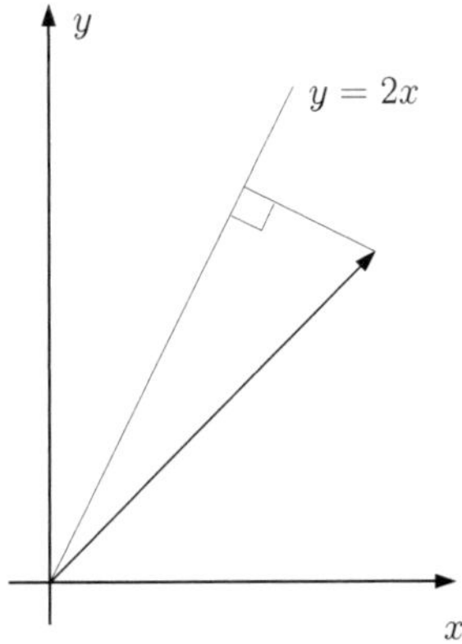

Figure 9.2: Projection onto the line $y = 2x$.

Clearly, as a result of this mapping every vector undergoes a change in direction except those vectors which lie along the line $y = 2x$ (these vectors map onto themselves; that is $X \to X$) and those which are perpendicular to the line $y = 2x$; that is, those which lie along the line $y = -x/2$ (these vectors map onto the zero vector; that is, $X \to 0X = O$).

Vectors such as these which undergo no change in direction under a linear mapping are called *eigenvectors* of the mapping.

Thus an eigenvector of a matrix (linear mapping) is simply a vector which maps onto a scalar multiple of itself. Since the zero vector always maps onto itself, we are not interested in such vectors.

Definition 9.1 *If A is a matrix (of a linear mapping) then a* non-zero vector X *is called an eigenvector of A if $AX = \lambda X$ for some scalar λ.*

The scalar λ which gives a measure of how X is "stretched" is called a corresponding *eigenvalue*. We usually speak of X as an eigenvector corresponding to the eigenvalue λ.

Suppose A is a 3×3 matrix which represents a linear mapping of three-dimensional space into itself. Then X is an eigenvector of A if

$$AX = \lambda X \Rightarrow (A - \lambda I)X = O.$$

This matrix equation represents a homogeneous system of three equations in three unknowns. Trivially $X = O$ is a solution but we are seeking non-zero solutions. The system will have non-trivial solutions if and only if the coefficient matrix $A - \lambda I$ is *singular*; that is, if $\det(A - \lambda I) = 0$. For this 3×3 matrix, we get a cubic equation for λ which can be solved to obtain three eigenvalues.

Definition 9.2 *The equation*

$$\boxed{\det(A - \lambda I) = 0}$$

is called the characteristic equation *of the matrix A.*

In general if A is an $n \times n$ matrix the characteristic equation is a polynomial equation of degree n in λ.

Corresponding to each eigenvalue we can then determine the eigenvectors by solving the system of equations $AX = \lambda X$ as illustrated in the following worked examples.

Worked Example 9.5.1 *Determine the eigenvalues of the matrix*

$$A = \begin{bmatrix} 0 & -3 & 0 \\ -1 & 2 & 0 \\ 1 & 1 & 1 \end{bmatrix}$$

and determine all the eigenvectors of A.

The characteristic equation is

$$\det(A - \lambda I) = \begin{vmatrix} -\lambda & -3 & 0 \\ -1 & 2-\lambda & 0 \\ 1 & 1 & 1-\lambda \end{vmatrix} = 0$$

$$\Rightarrow (1 - \lambda)(\lambda^2 - 2\lambda - 3) = 0,$$

where the determinant has been expanded about its last column. Thus the eigenvalues are $-1, 1$ and 3.

We now determine the eigenvectors corresponding to each of the eigenvalues.

- $\lambda = -1$

  We need to find all non-zero vectors $X = [a\ b\ c]^T$ which satisfy $AX = -X$; that is, we need to solve

$$\begin{bmatrix} 0 & -3 & 0 \\ -1 & 2 & 0 \\ 1 & 1 & 1 \end{bmatrix} \begin{bmatrix} a \\ b \\ c \end{bmatrix} = - \begin{bmatrix} a \\ b \\ c \end{bmatrix} \Rightarrow \begin{aligned} a - 3b &= 0 \\ -a + 3b &= 0 \\ a + b + 2c &= 0 \end{aligned}$$

  The solution of this set of equations is clearly $a = 3b, c = -2b$ where b is *any* real number.

  That is, all eigenvectors corresponding to the eigenvalue $\lambda = -1$ have the form

$$\begin{bmatrix} 3b \\ b \\ -2b \end{bmatrix} = b \begin{bmatrix} 3 \\ 1 \\ -2 \end{bmatrix}.$$

  Thus a possible eigenvector is $[3 \quad 1 \ -2]^T$.

- $\lambda = 1$

  We need to solve $AX = X$ for $X = [a\ b\ c]^T$.

  This leads to the system of equations:

$$\begin{aligned} -a - 3b &= 0 \\ -a + b &= 0 \\ a + b &= 0 \end{aligned}$$

  which has general solution $a = b = 0$ with c any real value. Thus each eigenvector corresponding to $\lambda = 1$ has the form

$$\begin{bmatrix} 0 \\ 0 \\ c \end{bmatrix} = c \begin{bmatrix} 0 \\ 0 \\ 1 \end{bmatrix}.$$

  A possible eigenvector is therefore $[0\ 0\ 1]^T$.

- $\lambda = 3$

  A possible eigenvector corresponding to this eigenvalue is $X = [1 \ -1 \ \ 0]^T$. (Verify this!)

■

In the preceding example suppose we construct a matrix P whose columns are eigenvectors corresponding to the eigenvalues $-1, 1$ and 3 respectively. For example we could take

$$P = \begin{bmatrix} 3 & 0 & 1 \\ 1 & 0 & -1 \\ -2 & 1 & 0 \end{bmatrix}.$$

It then turns out that $P^{-1}AP$ is a diagonal matrix D (you should verify this!) where

$$D = \begin{bmatrix} -1 & 0 & 0 \\ 0 & 1 & 0 \\ 0 & 0 & 3 \end{bmatrix}.$$

Note that the terms in the main diagonal of D are the eigenvalues of A and they appear in *the same order* along the diagonal as the order of the eigenvectors in the columns of P. This is always the case if A has three distinct eigenvalues; in such cases the matrix P whose columns are eigenvectors corresponding to the distinct eigenvalues is non-singular. This is not always the case if A does not have distinct eigenvalues.

Worked Example 9.5.2 *Determine the eigenvalues of the matrix*

$$A = \begin{bmatrix} 1 & 2 & -4 \\ 0 & -1 & 6 \\ 0 & -1 & 4 \end{bmatrix}$$

and show that there does not exist a set of three linearly independent eigenvectors.

We first find the eigenvalues by solving the equation

$$\begin{vmatrix} 1-\lambda & 2 & -4 \\ 0 & -1-\lambda & 6 \\ 0 & -1 & 4-\lambda \end{vmatrix} = 0$$

$$\Rightarrow \quad (1-\lambda)(\lambda^2 - 3\lambda + 2) = 0$$

$$\Rightarrow \quad \lambda = 1, 1 \text{ and } 2.$$

The eigenvectors corresponding to $\lambda = 2$ are vectors $X = [a\ b\ c]^T$ which satisfy the equation $AX = 2X$; that is, they satisfy

$$\begin{bmatrix} 1 & 2 & -4 \\ 0 & -1 & 6 \\ 0 & -1 & 4 \end{bmatrix} \begin{bmatrix} a \\ b \\ c \end{bmatrix} = 2 \begin{bmatrix} a \\ b \\ c \end{bmatrix}.$$

Thus

$$\begin{aligned} -a + 2b - 4c &= 0 \\ -3b + 6c &= 0 \\ -b + 2c &= 0. \end{aligned}$$

Therefore $b = 2c$ and $a = 0$.
Thus each eigenvector corresponding to $\lambda = 2$ has the form

$$\begin{bmatrix} 0 \\ 2c \\ c \end{bmatrix} = c \begin{bmatrix} 0 \\ 2 \\ 1 \end{bmatrix}.$$

Corresponding to $\lambda = 1$ the eigenvectors $X = [a\ b\ c]^T$ satisfy

$$\begin{bmatrix} 1 & 2 & -4 \\ 0 & -1 & 6 \\ 0 & -1 & 4 \end{bmatrix} \begin{bmatrix} a \\ b \\ c \end{bmatrix} = \begin{bmatrix} a \\ b \\ c \end{bmatrix}.$$

which implies

$$\begin{aligned} 2b - 4c &= 0 \\ -2b + 6c &= 0 \\ -b + 3c &= 0 \end{aligned}$$

giving $b = c = 0$ so that each eigenvector corresponding to $\lambda = 1$ has the form

$$\begin{bmatrix} a \\ 0 \\ 0 \end{bmatrix} = a \begin{bmatrix} 1 \\ 0 \\ 0 \end{bmatrix}.$$

Note that all eigenvectors corresponding to the eigenvalue $\lambda = 1$ are scalar multiples of $[1\ 0\ 0]^T$.
Consequently there do not exist two linearly independent eigenvectors corresponding to $\lambda = 1$ so that we cannot find three linearly independent eigenvectors of the matrix A.
Thus there does not exist a matrix P such that $P^{-1}AP$ is diagonal. ■

Self-help exercises

1. $X = [1\ 1\ 1]^T$ is an eigenvector of the matrix

$$A = \begin{bmatrix} -1 & 1 & 2 \\ 1 & 2 & -1 \\ 0 & 1 & 1 \end{bmatrix}.$$

What is the corresponding eigenvalue?

(Hint: Calculate AX.) [$\lambda = 2$]

2. Which of the following vectors are eigenvectors of

$$A = \begin{bmatrix} 0 & 1 & 1 \\ 1 & 0 & 1 \\ 1 & 1 & 0 \end{bmatrix}?$$

(a) $[0\ 0\ 0]^T$ (b) $[1\ 1\ 1]^T$ (c) $[2\ 2\ 3]^T$ (d) $[1\ 0\ -1]^T$

[(b), (d)]

In this topic we have

- Reviewed basic matrix algebra.
- Evaluated inverses using elementary row operations.
- Obtained solution(s) of systems of linear equations using row operations.
- Interpreted matrices as representing mappings: reflections, dilations and rotations.
- Reviewed determinants.
- Discussed eigenvalues and eigenvectors.
- Used exercises to explore applications to systems of differential equations and quadratic forms.

9.6 Quick Test Number 9

Elementary Row Operations

Question **Selection**

---

1. The next (sensible) elementary row operation to perform on the tableau

$$\begin{array}{rrr|r} 1 & -2 & 3 & 7 \\ 0 & -1 & 4 & 6 \\ -2 & 1 & -5 & 2 \end{array}$$

is (a) $R_3 - 2R_1 \to R_3$ (b) $R_2 + R_3 \to R_3$ (c) $R_3 + 2R_1 \to R_3$ (d) $R_3/2 \to R_3$

2. The next (sensible) elementary row operation to perform on the tableau

$$\begin{array}{rrrr|r} 3 & -2 & 4 & -1 & 7 \\ 0 & 0 & 2 & 3 & 5 \\ 0 & -1 & 4 & 2 & 1 \\ 0 & 0 & -3 & 7 & 2 \end{array}$$

is

(a) $2R_2 - R_1 \to R_2$ (b) $R_2 \leftrightarrow R_3$
(c) $R_3 - 2R_2 \to R_3$ (d) $R_1/3 \to R_1$

3. A system of three equations in three unknowns x, y and z reduces to the final tableau

$$\begin{array}{ccc|c} 1 & 2 & 3 & 7 \\ 0 & 1 & 2 & 5 \\ 0 & 0 & 0 & 0. \end{array}$$

The general solution is $(x, y, z) =$

(a) $(-3, 5, 0)$ (b) $(-3, 5, 0) + t(1, -2, 1)$
(c) There is no solution! (d) $(0, 0, 0)$

4. A system of three equations in three unknowns x, y and z reduces to the tableau

$$\begin{array}{ccc|c} 1 & 0 & -4 & 0 \\ 0 & 0 & 0 & 0 \\ -4 & 0 & 1 & 0. \end{array}$$

The general solution is $(x, y, z) =$

(a) $t(1, 0, 4)$ (b) $t(4, 0, 1)$ (c) $t(0, 1, 0)$

Finding an Inverse

Question **Selection**

Reduction to the identity:

1. The next (sensible) row operation to perform on the tableau

$$\begin{array}{ccc|c} 1 & -3 & 4 & 7 \\ 0 & 1 & 2 & 3 \\ 0 & 0 & 1 & 1 \end{array}$$

is (circle the correct answer(s))

(a) $2R_2 - R_1 \to R_2$ (b) $3R_2 + R_1 \to R_2$
(c) $R_2 - 2R_3 \to R_2$ (d) $4R_3 - R_1 \to R_3$

2. The next (sensible) elementary row operation to perform on the tableau

$$\begin{array}{ccc|c} 1 & 1 & 2 & 5 \\ 0 & 1 & 2 & 3 \\ 0 & 0 & 1 & 1 \end{array}$$

is (circle the correct answer(s))

(a) $R_2 - 2R_3 \to R_2$ (b) $R_2 - R_1 \to R_2$

(c) $R_1 - R_2 \to R_1$ (d) $2R_3 - R_1 \to R_3$

Finding the inverse of a matrix:

3. On applying a sequence of elementary row operations a matrix A reduces to

$$\begin{bmatrix} 1 & -2 & 5 \\ 0 & 7 & 4 \\ 0 & 0 & -3 \end{bmatrix}.$$

Does A have an inverse A^{-1}? (a) Yes (b) No

4. On applying a sequence of elementary row operations a matrix A reduces to

$$\begin{bmatrix} 1 & 0 & 0 \\ 0 & 1 & 0 \\ 0 & 0 & 0 \end{bmatrix}.$$

Does A have an inverse A^{-1}? (a) Yes (b) No

Eigenvalues and Eigenvectors

Question **Selection**

---

1. The determinant $\begin{vmatrix} 1 & 2 & -5 \\ 7 & 1 & 3 \\ 8 & 9 & -1 \end{vmatrix}$ is

(a) a matrix (b) a vector (c) a scalar

2. The determinant $\begin{vmatrix} 1-\lambda & 0 & 0 \\ 0 & -\lambda & 0 \\ 0 & 0 & 3-\lambda \end{vmatrix}$ is

(a) a matrix (b) a cubic in λ (c) a scalar

3. Evaluate $\begin{vmatrix} 3 & 2 & 4 & 1 \\ 0 & 2 & 1 & 0 \\ 0 & 0 & -7 & 2 \\ 0 & 0 & 0 & 4 \end{vmatrix}$.

Let $A = \begin{bmatrix} 1 & 2 \\ 2 & -2 \end{bmatrix}$.

4. $\begin{bmatrix} 1 & 2 \\ 2 & -2 \end{bmatrix}\begin{bmatrix} 6 \\ 3 \end{bmatrix} = \begin{bmatrix} 12 \\ 6 \end{bmatrix}$ Is $\begin{bmatrix} 6 \\ 3 \end{bmatrix}$ an eigenvector of A?

(a) Yes (b) No

5. $\begin{bmatrix} 1 & 2 \\ 2 & -2 \end{bmatrix}\begin{bmatrix} 1 \\ 1 \end{bmatrix} = \begin{bmatrix} 3 \\ 0 \end{bmatrix}$ Is $\begin{bmatrix} 1 \\ 1 \end{bmatrix}$ an eigenvector of A?

(a) Yes (b) No

6. $\begin{bmatrix} 1 & 2 \\ 2 & -2 \end{bmatrix}\begin{bmatrix} 2 \\ -4 \end{bmatrix} = \begin{bmatrix} -6 \\ 12 \end{bmatrix}$ Is $\begin{bmatrix} 2 \\ -4 \end{bmatrix}$ an eigenvector of A?

(a) Yes (b) No

7. The eigenvalues of A are

(a) $\lambda_1 = 12, \lambda_2 = -1$ (b) $\lambda_1 = 2, \lambda_2 = -3$ (c) $\lambda_1 = 6, \lambda_2 = 3$

8. $\det(A - \lambda I) = 0$, i.e. $\begin{vmatrix} 1-\lambda & 2 \\ 2 & -2-\lambda \end{vmatrix} = 0$ is

(a) a matrix (b) a system of equations

(c) a number (d) a quadratic equation for λ

9. The condition $\det(A - \lambda I) = 0$ means that

   (a) $A\mathbf{x} = \lambda\mathbf{x}$ has non-trivial solutions $\mathbf{x}$ called eigenvectors

   (b) the eigenvalues of A are all zero

   (c) $A = \lambda I$

   (d) $A\mathbf{x} = \lambda\mathbf{x}$ has only the trivial solution $\mathbf{x} = \begin{bmatrix} 0 \\ 0 \end{bmatrix}$.

10. It is claimed that $\begin{bmatrix} -1 \\ 2 \end{bmatrix}$ and $\begin{bmatrix} 2 \\ 1 \end{bmatrix}$ are also eigenvectors of A. Is this correct?

    (a) No, we already have two distinct eigenvectors.

    (b) Yes, every vector is an eigenvector of A.

    (c) Yes, any multiple of an eigenvector is also an eigenvector.

    (d) No, they're in the wrong order.

11. The matrix $P = \dfrac{1}{\sqrt{5}}\begin{bmatrix} -1 & 2 \\ 2 & 1 \end{bmatrix}$

    (a) is symmetric (b) is its own inverse

    (c) has distinct eigenvectors of A as columns

12. Evaluate $P^T AP = \dfrac{1}{5}\begin{bmatrix} -1 & 2 \\ 2 & 1 \end{bmatrix}\begin{bmatrix} 1 & 2 \\ 2 & -2 \end{bmatrix}\begin{bmatrix} -1 & 2 \\ 2 & 1 \end{bmatrix}$.

9.7 Exercises

Before attempting any of these miscellaneous exercises, make sure that you have successfully answered all the self-help exercises appearing throughout the chapter.

1. Determine the solutions of the systems of equations

(a)
$$\begin{aligned} x_1 - 2x_2 + 3x_3 &= 5 \\ 2x_1 + x_2 - 5x_3 &= -7 \\ 4x_1 - 3x_2 + 2x_3 &= 5 \end{aligned}$$

(b)
$$\begin{aligned} 2x_1 - x_2 + 7x_3 &= 18 \\ x_1 + x_2 + x_3 &= 3 \\ 5x_1 + 2x_2 + 3x_3 &= 6. \end{aligned}$$

2. Determine the solutions (if they exist) of the systems of equations

(a)
$$\begin{aligned} x_1 + 2x_2 + 3x_3 &= 2 \\ 4x_1 + 5x_2 + 6x_3 &= 8 \\ 7x_1 + 8x_2 + 9x_3 &= 13 \end{aligned}$$

(b)
$$\begin{aligned} x_1 + 2x_2 + 3x_3 &= 2 \\ 4x_1 + 5x_2 + 6x_3 &= 8 \\ 7x_1 + 8x_2 + 9x_3 &= 14. \end{aligned}$$

3. Find the general solution of the systems of equations

(a)
$$\begin{aligned} x_1 + x_2 + x_3 + 2x_4 - x_5 &= 0 \\ 2x_1 - x_2 - x_3 + x_4 + 2x_5 &= 0 \\ x_1 + 3x_2 - 2x_3 + x_4 + x_5 &= 0 \end{aligned}$$

(b)
$$\begin{aligned} x_1 - 2x_2 + x_3 - x_4 &= 0 \\ 2x_1 + 4x_2 - 3x_3 &= 0 \\ 3x_1 + 2x_2 + 2x_3 - x_4 &= 0. \end{aligned}$$

4. Determine all solutions $\mathbf{x}$ of $A\mathbf{x} = \lambda\mathbf{x}$ where

$$A = \begin{bmatrix} 0 & 1 & 1 \\ 1 & 0 & 1 \\ 1 & 1 & 0 \end{bmatrix}$$

and

$$\mathbf{x} = \begin{bmatrix} a \\ b \\ c \end{bmatrix}$$

when

(a) $\lambda = 2$
(b) $\lambda = -1$ and
(c) $\lambda = 3$.

5. Given

$$A = \begin{bmatrix} 1 & 2 \\ 3 & 4 \end{bmatrix},$$

$$B = \begin{bmatrix} 0 & -1 \\ 2 & 1 \end{bmatrix}$$

and

$$C = \begin{bmatrix} 5 & 1 \\ 1 & 3 \end{bmatrix}$$

show that

(a) $(A + B)C = AC + BC$
(b) $(AB)C = A(BC)$ and
(c) $(AB)^T = B^T A^T$.

6. If

$$A = \begin{bmatrix} 3 & 2 & -2 \\ -1 & -4 & 1 \\ 2 & -4 & -1 \end{bmatrix}$$

show that

$$A^3 + 2A^2 - A - 2I = O.$$

7. Given

$$A = \begin{bmatrix} 2 & -3 & -5 \\ -1 & 4 & 5 \\ 1 & -3 & -4 \end{bmatrix},$$

$$B = \begin{bmatrix} -1 & 3 & 5 \\ 1 & -3 & -5 \\ -1 & 3 & 5 \end{bmatrix}$$

and

$$C = \begin{bmatrix} 2 & -2 & -4 \\ -1 & 3 & 4 \\ 1 & -2 & -3 \end{bmatrix}$$

show that

(a) $AB = O$ even though $A \neq O$ and $B \neq O$ and
(b) $AC = A$ even though $C \neq I$.

8. Determine the inverse (if it exists) of each of the following matrices:

(a) $$\begin{bmatrix} 2 & 6 \\ 1 & 3 \end{bmatrix}$$

(b) $$\begin{bmatrix} 2 & -5 \\ -1 & -3 \end{bmatrix}$$

(c) $$\begin{bmatrix} 0 & 1 & 1 \\ 1 & 0 & 1 \\ 1 & 1 & 0 \end{bmatrix}$$

(d) $$\begin{bmatrix} 1 & 2 & 1 \\ 2 & 5 & 2 \\ 1 & 3 & 3 \end{bmatrix}$$

(e) $$\begin{bmatrix} 1 & 2 & 3 \\ 4 & 5 & 6 \\ 7 & 8 & 9 \end{bmatrix}$$ and

(f) $$\begin{bmatrix} -1 & 1 & 1 & 1 \\ 1 & -1 & 1 & 1 \\ 1 & 1 & -1 & 1 \\ 1 & 1 & 1 & -1 \end{bmatrix}.$$

9. If

$$A = \begin{bmatrix} 0 & 1 & 1 \\ 1 & 0 & 1 \\ 1 & 1 & 0 \end{bmatrix}$$

and

$$B = \begin{bmatrix} 1 & 2 & 1 \\ 2 & 5 & 2 \\ 1 & 3 & 3 \end{bmatrix}$$

verify that $(AB)^{-1} = B^{-1}A^{-1}$.

10. By solving

$$\begin{bmatrix} a & b \\ c & d \end{bmatrix}\begin{bmatrix} x & y \\ z & w \end{bmatrix} = \begin{bmatrix} 1 & 0 \\ 0 & 1 \end{bmatrix}$$

(where $ad - bc \neq 0$) for x, y, z and w show that

$$\begin{bmatrix} a & b \\ c & d \end{bmatrix}^{-1} = \frac{1}{ad - bc}\begin{bmatrix} d & -b \\ -c & a \end{bmatrix}.$$

Hence write down the inverse of

$$A = \begin{bmatrix} -3 & -7 \\ 2 & 8 \end{bmatrix}.$$

11. Prove that

$$(P^{-1}AP)^n = P^{-1}A^nP.$$

12. A square matrix P is said to be orthogonal if

$$PP^T = P^T P = I.$$

Prove that

(a) the columns of an orthogonal matrix are mutually orthogonal unit vectors

(b) the transpose of an orthogonal matrix is orthogonal and

(c) the product of two orthogonal matrices is orthogonal.

13. A square matrix A is said to be skew-symmetric if $A^T = -A$. For example

$$A = \begin{bmatrix} 0 & -1 & 2 \\ 1 & 0 & 4 \\ -2 & -4 & 0 \end{bmatrix}$$

is a skew-symmetric matrix.

Verify that $(I - A)(I + A)^{-1}$ is orthogonal.

14. (a) Describe (in geometric terms) the linear mappings defined by

i. $A = \begin{bmatrix} 1/2 & -\sqrt{3}/2 \\ \sqrt{3}/2 & 1/2 \end{bmatrix}$

ii. $B = \begin{bmatrix} -\sqrt{3}/2 & 1/2 \\ -1/2 & -\sqrt{3}/2 \end{bmatrix}$ and

iii. $C = \begin{bmatrix} 1/\sqrt{2} & -1/\sqrt{2} \\ 1/\sqrt{2} & 1/\sqrt{2} \end{bmatrix}.$

(b) Write down the matrix of the linear transformations defined by

$$AB, BA \text{ and } C^{-1}$$

without any calculations.

15. Suppose the point $P(x, y)$ is reflected in the line $y = mx$ where $m = \tan\theta$.

Let the point $Q(X, Y)$ be the image of the point P and N be the point of intersection of the line segment PQ with the line $y = mx$. (See Figure 9.3.)

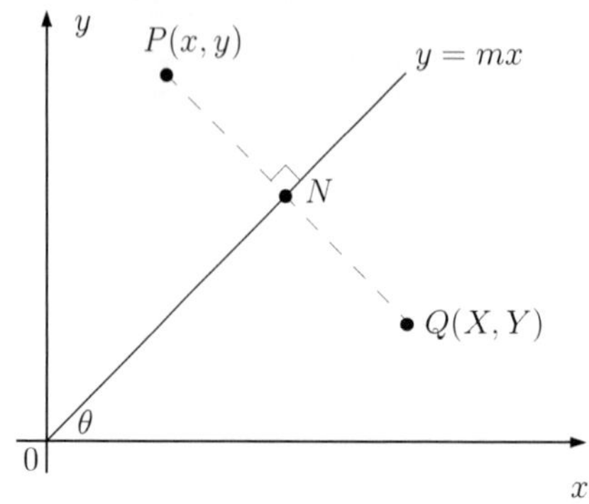

Figure 9.3: Reflection in the line $y = mx$.

(a) Write down the co-ordinates of the point N.

(b) By considering the gradients of the lines ON and PQ show that

i. $y + Y = m(x + X)$ and

ii. $m(y - Y) = -x + X$.

(c) By substituting $m = \tan\theta$, show that

$$\begin{bmatrix} X \\ Y \end{bmatrix} = \begin{bmatrix} \cos 2\theta & \sin 2\theta \\ \sin 2\theta & -\cos 2\theta \end{bmatrix} \begin{bmatrix} x \\ y \end{bmatrix}.$$

16. Describe (in geometric terms) the linear transformations defined by the following matrices:

(a) $\begin{bmatrix} 0 & 1 \\ 1 & 0 \end{bmatrix}$ (b) $\begin{bmatrix} 2 & 0 \\ 0 & 1 \end{bmatrix}$ and (c) $\begin{bmatrix} -1 & 0 \\ 0 & 1 \end{bmatrix}$.

17. Describe (in geometric terms) the linear transformations defined by

(a) $A = \begin{bmatrix} 1 & 0 & 0 \\ 0 & 1/\sqrt{2} & -1/\sqrt{2} \\ 0 & 1/\sqrt{2} & 1/\sqrt{2} \end{bmatrix}$

(b) $B = \begin{bmatrix} 1/\sqrt{2} & 0 & -1/\sqrt{2} \\ 0 & 1 & 0 \\ 1/\sqrt{2} & 0 & 1/\sqrt{2} \end{bmatrix}$

(c) Without performing any further calculations write down A^{-1}.

18. The linear transformation defined by

$$P = \frac{1}{3}\begin{bmatrix} 1 & -2 & 2 \\ 2 & 2 & 1 \\ -2 & 1 & 2 \end{bmatrix}$$

represents a rotation.

Determine the axis of rotation.

Describe how you find the angle of rotation.

19. For the linear transformation defined by

$$A = \begin{bmatrix} 1 & 0 & 1 \\ 0 & 1 & 0 \\ 1 & 0 & 1 \end{bmatrix}$$

determine which of the following vectors are eigenvectors.

$$\begin{bmatrix} 1 \\ 1 \\ 1 \end{bmatrix}, \begin{bmatrix} 1 \\ 0 \\ -1 \end{bmatrix}, \begin{bmatrix} 1 \\ 0 \\ 1 \end{bmatrix} \text{ and } \begin{bmatrix} 0 \\ 1 \\ 0 \end{bmatrix}$$

What are the eigenvalues of A?

20. Determine the eigenvalues and a set of three linearly independent eigenvectors for each of the following matrices A.

(a) $\begin{bmatrix} 1 & -2 & 7 \\ 0 & -1 & 3 \\ 0 & 0 & 2 \end{bmatrix}$ (b) $\begin{bmatrix} 1 & 1 & 1 \\ 1 & 1 & 1 \\ 1 & 1 & 1 \end{bmatrix}$

(c) $\begin{bmatrix} 1 & 1 & 1 \\ 1 & 0 & 1 \\ 1 & 1 & 0 \end{bmatrix}$ (d) $\begin{bmatrix} 3 & 2 & 4 \\ 2 & 0 & 2 \\ 4 & 2 & 3 \end{bmatrix}$.

In each case determine a matrix P such that $P^{-1}AP$ is a diagonal matrix D and write down D.

21. Show that the expression

$$3x^2 + 2xy + 3y^2$$

can be expressed in matrix form as

$$X^T AX$$

where

$$X = \begin{bmatrix} x \\ y \end{bmatrix} \text{ and } A = \begin{bmatrix} 3 & 1 \\ 1 & 3 \end{bmatrix}.$$

(a) Determine the eigenvalues of A and hence find an *orthogonal* matrix P such that

$$P^{-1}AP = P^T AP$$

is a diagonal matrix.
(Note: In order to make P orthogonal you will need to take eigenvectors which are *unit* vectors.)

(b) By making the substitution

$$X = PY$$

where

$$Y = \begin{bmatrix} \xi \\ \eta \end{bmatrix}$$

express the equation of the curve

$$3x^2 + 2xy + 3y^2 = 10$$

relative to the $\xi, \eta-$axes and identify the curve.
Describe the transformation from the x, y-plane to the ξ, η-plane.

22. Show that under an anti-clockwise rotation through an angle θ, that is

$$\begin{bmatrix} X \\ Y \end{bmatrix} = \begin{bmatrix} \cos\theta & \sin\theta \\ -\sin\theta & \cos\theta \end{bmatrix} \begin{bmatrix} x \\ y \end{bmatrix}$$

the equation of the curve

$$ax^2 + 2hxy + by^2 = 1$$

becomes

$$\begin{aligned} &(a\cos^2\theta + h\sin 2\theta + b\sin^2\theta)X^2 \\ &+(2h\cos 2\theta - a\sin 2\theta + b\sin 2\theta)XY \\ &+(a\sin^2\theta - h\sin 2\theta + b\cos^2\theta)Y^2 \\ &= 1. \end{aligned}$$

Deduce that if the angle θ of rotation is suitably chosen the equation of the curve becomes

$$\alpha X^2 + \beta Y^2 = 1$$

relative to the new axes.

Identify the curve

$$3x^2 + 2xy + 3y^2 = 10$$

by carrying out a suitable rotation of axes and compare your answer with that of Exercise 21.

23. Consider the following system of two linked differential equations

$$\begin{aligned} \ddot{y}_1 &= -5y_1 + 2y_2 \\ \ddot{y}_2 &= 2y_1 - 2y_2 \end{aligned}$$

that arises in the modelling of two masses coupled by springs and performing small vibrations about an equilibrium position.

By introducing the matrices

$$Y = \begin{bmatrix} y_1 \\ y_2 \end{bmatrix}$$

and

$$A = \begin{bmatrix} -5 & 2 \\ 2 & -2 \end{bmatrix}$$

show that the above system of equations can be expressed in matrix form as

$$\ddot{Y} = AY.$$

Determine a matrix P such that $P^{-1}AP$ is a diagonal matrix D.

Show that the matrix form of the system of differential equations is expressible as

$$\ddot{Y} = PDP^{-1}Y.$$

Hence by defining the new matrix Z through

$$Y = PZ$$

show that this system is equivalent to

$$\ddot{Z} = DZ.$$

Solve this system of equations for

$$Z = \begin{bmatrix} z_1 \\ z_2 \end{bmatrix}.$$

Deduce the solution

$$Y = PZ$$

of the original system of equations.

(Note that this approach does *not* require the evaluation of the inverse of any matrices and only requires solving the simple differential equation

$$\ddot{z} = kz$$

for some constant k — which are related to the eigenvalues of the matrix of coefficients A.)

Topic 10

Vector-valued Functions

In this chapter we consider vector-valued functions of one real variable. The simplest example of this type of function is exemplified by the parametric representation of a straight line. Here, for each value of the parameter t, we associate the vector $\mathbf{r}(t)$ that represents an arbitrary point on the line. We will find that all curves in space can be represented in this form.

10.1 Curves in Space

In an earlier section we saw that to uniquely specify a line we needed a point on the line and the line's direction. For example, the equation of the straight line passing through the point $P(-1, 2, 4)$ and parallel to the direction $(-3, 4, 8) = -3\,\mathbf{i} + 4\,\mathbf{j} + 8\,\mathbf{k}$, in vector form, is given by

$$\begin{aligned}\mathbf{r}(t) &= (x, y, z) = (-1, 2, 4) + t(-3, 4, 8) \\ &= (-1 - 3t)\,\mathbf{i} + (2 + 4t)\,\mathbf{j} + (4 + 8t)\,\mathbf{k} \text{ for all real } t.\end{aligned}$$

The vector quantity $\mathbf{r}(t)$ is called a *vector-valued function of the real variable* t.

Similarly, all points on the ellipse

$$\frac{x^2}{a^2} + \frac{y^2}{b^2} = 1$$

can be expressed in parametric form as

$$\mathbf{r}(t) = (a\cos t, b\sin t, 0) = a\cos t\,\mathbf{i} + b\sin t\,\mathbf{j}; \quad t \in [0, 2\pi].$$

(The values of a and b are taken to be positive.) In these notes we will often use the notation

$$\mathbf{r}(t) = \begin{cases} x &= a\cos t \\ y &= b\sin t \end{cases} \quad \text{if } 0 \le t \le 2\pi$$

to represent such vector functions.

There exist many possible parametric representations of these points on the ellipse. The "best" parameterization for practical applications is often the one that uses a known trigonometric or hyperbolic identity to satisfy the equation.

▌**Example:** The parameterization

$$\mathbf{r}(t) = \begin{cases} x = at \\ y = b\sqrt{1-t^2} \end{cases} \quad \text{if } -1 \le t \le 1$$

represents the *top* half of the above ellipse — here a and b are again taken to be positive. (Determine a similar parameterization for the *bottom* half of this ellipse.)

Any curve in space can be represented in the vector parametric form $\mathbf{r}(t)$. For each value of the parameter t we associate a point in three-dimensional space. The locus of all such points, formed as t assumes all its values, traces the curve in question. In many applications the parameter t is simply the time. For example, the position of a satellite in space at each instance in time is defined once we know (or have calculated) the function $\mathbf{r}(t)$.

▌ **Worked Example 10.1.1** *Determine a parametric representation of the curve*

$$\frac{x^2}{a^2} - \frac{y^2}{b^2} = 1.$$

(The equation is of the form

$$(\,)^2 - (\,)^2 = 1$$

so use a hyperbolic representation.)

Define

$$\mathbf{r}(t) = \begin{cases} x = a\cosh t \\ y = b\sinh t \end{cases} \quad \text{for all real } t$$

or

$$\mathbf{r}(t) = (a\cosh t, b\sinh t, 0) = a\cosh t\,\mathbf{i} + b\sinh t\,\mathbf{j}.$$

(Here a and b are again taken to be positive.) Note that this parameterization represents only the *right-hand half* of the hyperbola. (Why?) ■

Self-help exercises

1. Determine a parameterization of the *left-hand half* of the curve defined in Worked Example 10.1.1. $[x = -a\cosh t, y = b\sinh t]$

2. Determine a parameterization of the curve

$$x^2 + 4y^2 = 9.$$

$[x = 3\cos t, y = \frac{3}{2}\sin t]$

Each of the examples discussed above has taken a curve defined in Cartesian form and represented the same curve in parametric form. In solving many mechanics problems for the path of a particle moving under specified forces using vector methods, for example, the final solution will be expressed in parametric form. We then face the task of describing the resulting curve. If the curve is planar then we may be able to recognize the resulting curve when written in its Cartesian form. Many such curves may lead to extremely complicated Cartesian formulations and this method is not appropriate for curves defined in three dimensions. One possible approach in this case is to adopt the procedure followed in the next example.

Worked Example 10.1.2 *Describe the space curve*

$$\mathbf{r}(t) = a\cos t\,\mathbf{i} + a\sin t\,\mathbf{j} + t\,\mathbf{k} \textit{ for all } t \geq 0.$$

In parametric form the curve is given by

$$\left.\begin{aligned} x &= a\cos t \\ y &= a\sin t \\ z &= t \end{aligned}\right\} \text{ for all } t \geq 0.$$

The first two of these equations imply that

$$x^2 + y^2 = a^2 \text{ for all } z$$

which represents the surface of a *circular cylinder* in three-dimensional space. The curve (being one-dimensional) must lie on this cylindrical surface.

The curve has its starting point defined by $t = 0$; that is, at the point $(a, 0, 0)$. As t increases, the z-co-ordinate of points on the curve also increase linearly with t but with all points on the curve restricted to lying on the surface of the cylinder. (Note that as t increases by 2π, both the x- and y-co-ordinates return to their previous values.)

The curve is depicted in Figure 10.1 and is referred to as a circular *helix*. ■

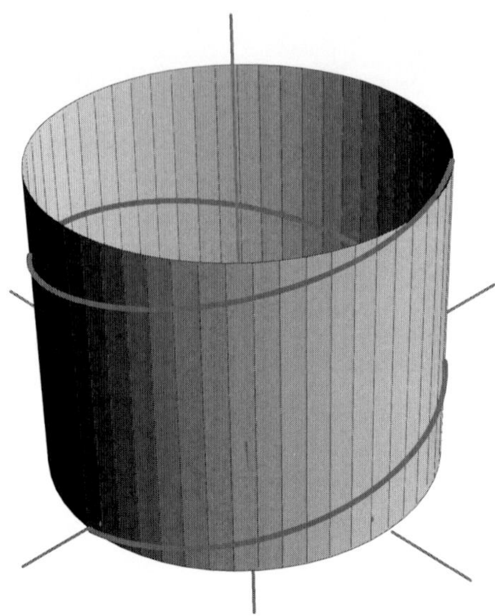

Figure 10.1: Circular helix of Worked Example 10.1.2.

Worked Example 10.1.3 *Describe the curve*

$$\mathbf{r}(t) = t\cos t\,\mathbf{i} + t\sin t\,\mathbf{j} + t\,\mathbf{k} \textit{ for all } t \geq 0.$$

In parametric form the curve is given by

$$\left.\begin{array}{rcl} x & = & t\cos t \\ y & = & t\sin t \\ z & = & t \end{array}\right\} \text{for all } t \geq 0.$$

Eliminating the parameter t between these three equations gives as the surface on which the curve lies

$$x^2 + y^2 = t^2 = z^2.$$

This is the equation of a double (circular) cone in three-dimensional space. (But note that because $t \geq 0$ we have $z \geq 0$ and thus the curve only lies on the *top half* of this cone.)
The curve starts at the origin $(0, 0, 0)$, corresponding to the value of $t = 0$, and, as t increases, the z-co-ordinate of points on the curve also increase linearly with t but with all points on the curve restricted to lie on the surface of the cone.
The curve is depicted in the diagram in Figure 10.2 and is described as a spiral lying on the surface of the cone. ■

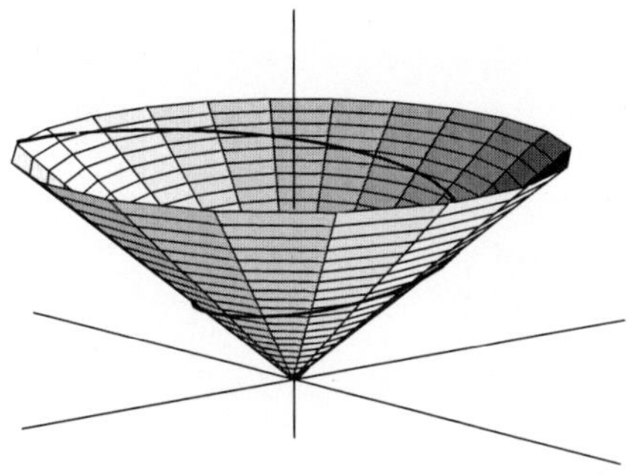

Figure 10.2: Spiral on a cone of Worked Example 10.1.3.

Self-help exercise

Describe the curve

$$\mathbf{r}(t) = t\cos 4\pi t\,\mathbf{i} + t\sin 4\pi t\,\mathbf{j} + \sqrt{1-t^2}\,\mathbf{k} \text{ for all } t \in [0,1].$$

[See Figure 10.3.]

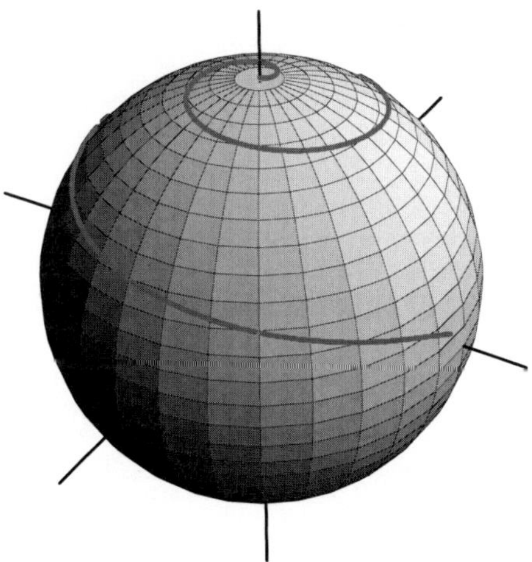

Figure 10.3: Answer to the self-help exercise above.

10.2 Arc Length

Consider the curve C represented in parametric form by

$$C : \mathbf{r}(t) = x(t)\,\mathbf{i} + y(t)\,\mathbf{j} + z(t)\,\mathbf{k} \text{ for } t \in [a,b].$$

A parameterization that is often found to be useful is that using the *arc length* parameter s for the curve C. This arc length parameter measures the distance along the curve from some convenient point on C (usually the value of $t = a$ corresponding to the start of the curve).

The relationship between the arc length parameter s and any other parameter t is obtained as follows. Since

$$\left(\frac{ds}{dt}\right)^2 = \left(\frac{dx}{dt}\right)^2 + \left(\frac{dy}{dt}\right)^2 + \left(\frac{dz}{dt}\right)^2 = \frac{d\mathbf{r}}{dt} \cdot \frac{d\mathbf{r}}{dt},$$

and if we measure the sense of s increasing in the same sense as t is increasing then

$$\boxed{\frac{ds}{dt} = \left|\frac{d\mathbf{r}}{dt}\right|}$$

The parameter s is thus given by

$$\boxed{s = \int_0^t \left|\frac{d\mathbf{r}}{dt}\right| \, du}$$

(In the above integral, the integrand $|d\mathbf{r}/dt|$ is to be treated as a function of the integration variable u. The value of the integral represents the area under the speed-time graph from the initial time ($t = 0$) to general time t.)

Worked Example 10.2.1 *Find the length of the portion of the circular helix defined by*

$$\mathbf{r}(t) = (a\cos t, a\sin t, t) \textit{ for } t \in [0, 2\pi].$$

(You should sketch the portion of the helix being defined by this range of the parameter t.)
The length ℓ of the helix is given by

$$\begin{aligned}
\ell &= \int_{s_a}^{s_b} ds \quad \text{(where } s_a \text{ and } s_b \text{ correspond to } t = 0 \text{ and } 2\pi\text{)} \\
&= \int_0^{2\pi} \frac{ds}{dt}\, dt \\
&= \int_0^{2\pi} \sqrt{\dot{x}^2 + \dot{y}^2 + \dot{z}^2}\, dt \\
&= \int_0^{2\pi} \sqrt{(-a\sin t)^2 + (a\cos t)^2 + (1)^2}\, dt \\
&= \int_0^{2\pi} \sqrt{a^2 + 1}\, dt = 2\pi\sqrt{a^2 + 1}.
\end{aligned}$$

■

Self-help exercise

Obtain an integral expression that represents the length of the spiral

$$\mathbf{r}(t) = t\cos t\,\mathbf{i} + t\sin t\,\mathbf{j} + t^2\,\mathbf{k}, \quad t \in [0, 1].$$

Use the tables appearing in Appendix C (Pages 493–495) to evaluate this integral.

$$\left[\int_0^1 \sqrt{1+5t^2}\,dt,\ \tfrac{1}{2}\sqrt{6} + \tfrac{1}{2\sqrt{5}}\log(\sqrt{5}+\sqrt{6})\right]$$

10.3 Tangent Vector

Consider the space curve C represented in parametric form by

$$C : \mathbf{r}(t) = x(t)\,\mathbf{i} + y(t)\,\mathbf{j} + z(t)\,\mathbf{k} \text{ for } t \in [a, b].$$

With reference to Figure 10.4, we have

$$\overrightarrow{PQ} = \mathbf{r}(t+\delta t) - \mathbf{r}(t).$$

In the limit of δt tending to zero, the "direction" of this vector is parallel

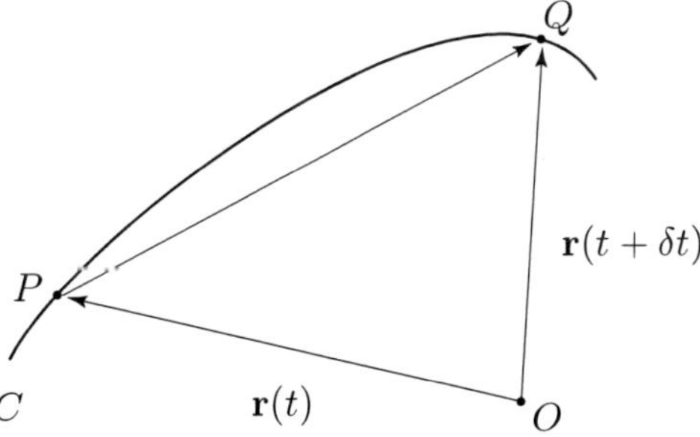

Figure 10.4: Calculation of tangent vector.

to the direction of the tangent to the curve at the point P. To obtain a vector that represents this tangential direction we wish to construct a vector parallel to $\overrightarrow{PQ}$ which has a non-vanishing limit as δt tends to zero.

For this purpose, consider

$$\lim_{\delta t \to 0} \frac{\mathbf{r}(t+\delta t) - \mathbf{r}(t)}{\delta t},$$

which (if the limit exists and is finite) will represent the tangential direction to the curve at the point P. Following the same terminology as used in the real variable calculus, we define such a limit to be the derivative of the vector function $\mathbf{r}(t)$ with respect to the parameter t and denote the limit by

$$\frac{d\mathbf{r}(t)}{dt} = \lim_{\delta t \to 0} \frac{\mathbf{r}(t+\delta t) - \mathbf{r}(t)}{\delta t}.$$

As for functions of one real variable, we often use the notation $\dot{\mathbf{r}}(t)$ to denote the derivative $\frac{d\mathbf{r}(t)}{dt}$.

It can be easily shown (for example by considering each component in turn) that the following rules are satisfied by such derivatives.

Theorem 10.1 (Rules of Differentiation)

1. $\dfrac{d}{dt}(\mathbf{r}(t) \pm \mathbf{s}(t)) = \dfrac{d\mathbf{r}(t)}{dt} \pm \dfrac{d\mathbf{s}(t)}{dt}$.

2. $\dfrac{d}{dt}(\mathbf{r}(t) \cdot \mathbf{s}(t)) = \mathbf{r}(t) \cdot \dfrac{d\mathbf{s}(t)}{dt} + \dfrac{d\mathbf{r}(t)}{dt} \cdot \mathbf{s}(t)$.

3. $\dfrac{d}{dt}(\mathbf{r}(t) \times \mathbf{s}(t)) = \mathbf{r}(t) \times \dfrac{d\mathbf{s}(t)}{dt} + \dfrac{d\mathbf{r}(t)}{dt} \times \mathbf{s}(t)$.

Note carefully that the terms on the right-hand side of the formulae for the derivative of a cross product *preserve the sense of the terms appearing in the cross product*. (Also note the similarity of items 2 and 3 with the ordinary product rule for differentiation of a product.)

The *unit tangent vector* to the curve C at the point P is thus given by

$$\boxed{\hat{\boldsymbol{\tau}} = \frac{d\mathbf{r}(t)/dt}{|d\mathbf{r}(t)/dt|} = \frac{d\mathbf{r}(t)/dt}{ds/dt}}$$

If the curve is parameterized with respect to the arc length parameter s, then the formula for the unit tangent vector simplifies to

$$\hat{\boldsymbol{\tau}} = \frac{d\mathbf{r}(s)}{ds}.$$

Worked Example 10.3.1 *Find a unit tangent vector and the equation of the tangent line to the circular helix*

$$\mathbf{r}(t) = a\cos t\,\mathbf{i} + a\sin t\,\mathbf{j} + ct\,\mathbf{k} \text{ for } t \in [0, 3\pi]$$

at the point

1. $t = 2\pi$

2. $(-a, 0, c\pi)$.

For each value of t, the tangent vector is parallel to

$$\frac{d\mathbf{r}}{dt} = (-a\sin t, a\cos t, c).$$

1. Thus at the nominated point $t = 2\pi$, $\hat{\boldsymbol{\tau}}$ is parallel to the vector $(0, a, c) = a\,\mathbf{j} + c\,\mathbf{k}$.

   Therefore $\hat{\boldsymbol{\tau}} = \dfrac{(0, a, c)}{\sqrt{a^2 + c^2}}$.

   For the tangent *line*, we want the equation of the straight line passing through the point corresponding to $t = 2\pi$ with the same *direction* as the tangent vector there. In vector parametric form the equation is

$$\mathbf{r}(s) = (x, y, z) = (a, 0, 2\pi c) + s(0, a, c) \text{ for all real } s.$$

2. The point $(-a, 0, c\pi)$ corresponds to the value $t = \pi$. Therefore $\hat{\boldsymbol{\tau}}$ is parallel to the vector $(0, -a, c) = -a\,\mathbf{j} + c\,\mathbf{k}$.

   Thus $\hat{\boldsymbol{\tau}} = \dfrac{(0, -a, c)}{\sqrt{a^2 + c^2}}$.

   The tangent line is given by

$$\mathbf{r}(s) = \begin{cases} x = -a \\ y = -as \\ z = c\pi + cs \end{cases} \quad \text{for all } s.$$

■

Self-help exercise

Determine the direction of the tangent line to the curve

$$\mathbf{r}(t) = t \sin t\,\mathbf{i} + 3t \cos t\,\mathbf{j} + 4t^2\,\mathbf{k}$$

at the point $t = \frac{1}{2}\pi$. [$\mathbf{i} - \frac{3}{2}\pi\,\mathbf{j} + 4\pi\,\mathbf{k}$)]

10.4 Velocity and Acceleration

Let $\mathbf{r}(t)$ represent the position vector of a particle P in space where t is the time. The path described by P will be some curve C. Using results obtained above we have that

$$\frac{d\mathbf{r}}{dt} \text{ is tangent to } C$$

and

$$\left|\frac{d\mathbf{r}}{dt}\right| = \frac{ds}{dt} \text{ is the speed of } P.$$

Therefore we call $d\mathbf{r}/dt$ the *velocity* $\mathbf{v}$ of P.

The acceleration $\mathbf{a}$ is then given by

$$\boxed{\mathbf{a} = \frac{d\mathbf{v}}{dt} = \frac{d^2\mathbf{r}}{dt^2}}$$

Worked Example 10.4.1 *Determine the velocity and acceleration of a particle P moving along the helical path*

$$\mathbf{r}(t) = (a\cos\omega t, a\sin\omega t, bt) \textit{ for } t \in [0, 4\pi/\omega]$$

at the time corresponding to $t = 2\pi/\omega$.
What is the distance d travelled by the particle in time $t = \frac{4\pi}{\omega}$?
We have

$$\mathbf{v} = \frac{d\mathbf{r}}{dt} = (-a\omega\sin\omega t, a\omega\cos\omega t, b)$$

$$\mathbf{a} = \frac{d\mathbf{v}}{dt} = (-a\omega^2\cos\omega t, -a\omega^2\sin\omega t, 0).$$

Thus at time $t = 2\pi/\omega$

$$\mathbf{v} = (0, a\omega, b) \text{ and } \mathbf{a} = (-a\omega^2, 0, 0).$$

The distance d travelled in time $\frac{4\pi}{\omega}$ is given by

$$\begin{aligned} d &= \int_0^{\frac{\bullet\pi}{\omega}} |\dot{\mathbf{r}}(t)|\,dt \\ &= \int_0^{\frac{\bullet\pi}{\omega}} \sqrt{a^2\omega^2 + b^2}\,dt \\ &= \frac{4\pi}{\omega}\sqrt{a^2\omega^2 + b^2}. \end{aligned}$$

(Note that the acceleration vector in this example is

1. parallel to the x, y-plane
2. perpendicular to the tangent to the curve at each point and
3. always directed towards the axis of the cylinder.

Furthermore the above statements will also hold for the force that is producing this particular motion since by Newton's second law

$$\mathbf{F} = \frac{d}{dt}(m\mathbf{v}).)$$

■

Self-help exercise

Determine the velocity and acceleration of a particle P moving along the spiral

$$\mathbf{r}(t) = t\cos 4\pi t\,\mathbf{i} + t\sin 4\pi t\,\mathbf{j} + \sqrt{1-t^2}\,\mathbf{k}$$

at the time $t = \frac{1}{2}$. $[(1, 2\pi, -\frac{1}{\sqrt{3}}), (-8\pi^2, 8\pi, -\frac{8}{\sqrt{27}})]$

In this topic we have

- Examined curves in space.
- Determined tangent vectors and arc-lengths.
- Discussed the velocity and acceleration of particles.

10.5 Quick Test Number 10

Vector-valued Functions

Question **Selection**

Space curves:

The position of a moving particle at time t, $t \geq 0$, is given by

$$\mathbf{r}(t) = (t^2 + 2t + 5)\,\mathbf{i} + (t^3 + 1)\,\mathbf{j} + (5t - 3)\,\mathbf{k}.$$

1. What is the position of the particle at $t = 0$?
   (a) $5\,\mathbf{i} + \mathbf{j} - 3\,\mathbf{k}$ (b) $5\,\mathbf{i} + \mathbf{j} + 2\,\mathbf{k}$ (c) 3
2. At what time does the particle reach the point $(13, 9, 7)$?
   (a) $t = 1$ (b) $t = 2$ (c) never
3. Does the particle ever pass through the point $(8, 3, 2)$?
   (a) Yes (b) No
4. Find $\dot{\mathbf{r}}(t)$. (a) $(2t + 2)\,\mathbf{i} + 3t^2\,\mathbf{j} + 5\,\mathbf{k}$ (b) $2t + 2 + 3t^2 + 5$
5. In what direction is the particle travelling at $t = 1$?
   (a) $4\,\mathbf{i} + 3\,\mathbf{j} + 5\,\mathbf{k}$ (b) 12
6. What is the speed of the particle at $t = 1$? (a) $\sqrt{50}$ (b) 12

10.6 Exercises

> Before attempting any of these miscellaneous exercises, make sure that you have successfully answered all the self-help exercises appearing throughout the chapter.

1. A plane curve C is defined by the parametric equations

$$\left.\begin{array}{rcl} x & = & t\cos t \\ y & = & t\sin t \end{array}\right\} \quad t \in [0, \pi].$$

   (a) Identify and sketch the curve C.

   (b) Determine a tangent vector to C at the point on C corresponding to the value of $t = \frac{\pi}{2}$.

   (c) Find the length of the curve C.

2. A space curve C is defined by the parametric equations

$$\left.\begin{array}{rcl} x & = & 3\cos t \\ y & = & 2\sin t \\ z & = & t \end{array}\right\} \quad t \in [0, 2\pi].$$

   (a) Determine a *surface* S on which the curve C lies and sketch both the surface S and curve C.

   (b) Find a tangent vector to the curve C at the point $(-3, 0, \pi)$.

3. A space curve C is defined by the parametric equations

$$\left.\begin{array}{rcl} x & = & 1+t \\ y & = & (1+t)^{\div} \\ z & = & t \end{array}\right\} \quad t \in [0, 2].$$

   (a) Determine a unit tangent vector to the curve C at the point P on C corresponding to the value $t = 1$.

   Derive a vector equation for the tangent line to the curve C at this point P.

   (b) Find the length of the curve C.

4. A space curve C is defined by the vector equation

$$C: \quad \mathbf{r}(t) = t\cos t\,\mathbf{i} + t\sin t\,\mathbf{j} + t^2\,\mathbf{k}, \quad t \geq 0.$$

   (a) Find a surface S on which the curve C lies and sketch both S and C.

   (b) Determine a vector equation of the tangent line to the curve C at the point on C corresponding to the value of $t = \frac{\pi}{2}$.

   (c) Express, as a definite integral, the arc-length of C between the points on C corresponding to the values of $t = 0$ and $t = \frac{\pi}{2}$.

5. A particle of mass $m = 3$ moves along a space curve C under the action of a force $\mathbf{F}$ so that, at time t, its position vector $\mathbf{r} = \mathbf{r}(t)$ relative to an origin O is given by

$$C: \quad \mathbf{r}(t) = 3t^2\,\mathbf{i} + 6t\,\mathbf{j} + (1 - t^2)\,\mathbf{k}$$

   (a) Determine the particle's acceleration $\mathbf{a}$.

   Using $\mathbf{F} = m\mathbf{a}$, show that $\mathbf{F} = 18\,\mathbf{i} - 6\,\mathbf{k}$.

   (b) The moment $\mathbf{L}$ of this force about the origin O is given by $\mathbf{L} = \mathbf{r}\times\mathbf{F}$.

   Find $\mathbf{L}$.

   (c) The angular momentum $\mathbf{H}$ of the particle about O is given by $\mathbf{H} = \mathbf{r}\times m\mathbf{v}$ where $\mathbf{v}$ is the particle's velocity.

   Determine $\mathbf{H}$ and verify that

$$\mathbf{L} = \frac{d\mathbf{H}}{dt}.$$

6. The position vector of a particle at time t is given by

$$\mathbf{r}(t) = \cos 2t\,\mathbf{i} + \sin 2t\,\mathbf{j} + 2\cosh t\,\mathbf{k}.$$

   (a) Find the velocity and acceleration of the particle at time t.

   (b) Show that the distance travelled by the particle between the times $t = 0$ and $t = 2$ is $2\sinh 2$.

7. The space curve

$$\mathbf{r}(t) = \sin t\,\mathbf{i} + \cos t\,\mathbf{j} + t^3\,\mathbf{k}$$

represents the position of a particle in space at time t.

   (a) Determine the velocity and acceleration vectors.

   (b) Find the time $t = T$ (if any such time exists) at which the velocity and acceleration vectors are perpendicular to one another.

8. The space curve

$$\mathbf{r}(t) = 3\cos 2t\,\mathbf{i} + 3\sin 2t\,\mathbf{j} + t\,\mathbf{k}$$

represents the position vector of a moving particle at time t seconds.

Determine

(a) the velocity and speed of the particle at time $t = 2$

(b) the distance the particle travels in the first three seconds

(c) a unit tangent vector at time $t = 3$.

9. At time t the position vector of a particle moving along a circular helix is given by

$$\mathbf{r}(t) = 4\cos t\,\mathbf{i} + 4\sin t\,\mathbf{j} + 3t\,\mathbf{k}.$$

(a) Show that the particle travels with constant speed.

(b) Determine the particle's acceleration at time t.

(c) Show that the acceleration of the particle is always perpendicular to the particle's direction of motion.

10. Given the space curve

$$C: \quad \mathbf{r}(u) = u^2\,\mathbf{i} + (1-u)\,\mathbf{j} + 2\sqrt{u}\,\mathbf{k}, \quad u \geq 0,$$

find

(a) a unit tangent vector to the curve C at the point $(1, 0, 2)$

(b) parametric equations of the tangent line to the curve C at the point $(1, 0, 2)$

(c) an integral that represents the arc-length of the curve between the points $(1, 0, 2)$ and $(16, -3, 4)$.

11. Show that the space curve C

$$C: \quad \mathbf{r}(t) = t^2\,\mathbf{i} + 2t^2\,\mathbf{j} + 4t\,\mathbf{k}, \quad t \geq 0$$

intersects the plane Π

$$\Pi: \quad -x + y + 4z = 17$$

at the point $P(1, 2, 4)$.

What is the angle between the curve C and the plane Π at this point P of intersection?

12. The path of a particle moving in a region of static magnetic induction $\mathbf{B}$ is given as the solution of the vector differential equation

$$m\ddot{\mathbf{r}} = q\dot{\mathbf{r}} \times \mathbf{B},$$

where m is the particle's mass and q is the charge on the particle.

In the special case of $\mathbf{B} = B\mathbf{k}$, show that the system of scalar equations defining the path is given by

$$\begin{aligned} m\ddot{x} &= qB\dot{y} \\ m\ddot{y} &= -qB\dot{x} \\ \ddot{z} &= 0. \end{aligned}$$

(The solution of these equations shows the resulting path is a helix.)

Topic 11

Functions of Several Variables

11.1 Introduction

In this chapter we discuss real-valued functions of several real variables. Such functions occur frequently in applications of mathematics to physical and engineering problems. Some examples of such functions are given below.

Example:

$r = \sqrt{x^2 + y^2 + z^2}$	defines the distance of a point $P(x, y, z)$ from the origin.
$V = \frac{1}{3}\pi r^2 h$	defines the volume of a cone in terms of its height and the radius of its base.
$P = \frac{\rho RT}{M}$	defines the pressure in a perfect gas in terms of its density and temperature through the gas constant R and the molar mass M.

With reference to the above examples and as for functions of one real variable, we call

x, y, z	the independent variables and	r	the dependent variable,
h, r	the independent variables and	V	the dependent variable,
ρ, T	the independent variables and	P	the dependent variable.

Furthermore,

since to each point $\begin{cases} (x,y,z) & \in & \mathbf{R}^3 \\ (h,r) & \in & \mathbf{R}^2 \\ (\rho,T) & \in & \mathbf{R}^2 \end{cases}$

there is a unique number $\begin{cases} r \\ V \\ P \end{cases}$

we can write $\begin{cases} r & = & f(x,y,z) & = & \sqrt{x^2+y^2+z^2} \\ V & = & g(h,r) & = & \frac{1}{3}\pi r^2 h \\ P & = & h(\rho,T) & = & \frac{\rho RT}{M} \end{cases}$

and we call f, g, h (real-valued) *functions of several variables.* Values of these functions are specified in the same manner as for the single variable case.

▮Example: With reference to the above functions, for example, we have

$$g(1,2) = \tfrac{4}{3}\pi$$
$$h(4,3) = 12R/M.$$

11.2 Domains

The domain of functions of several real variables is often not stated explicitly but is implied from the formula that defines the functions' value. As for functions of one real variable, the domain is the set of possible values that can be taken by the independent variables so that the function is *defined.*

We illustrate this idea in the following examples.

▮Example: Consider the function defined by

$$z = f(x,y) = \sqrt{25 - x^2 - y^2}.$$

The domain is the set of points in the x, y-plane for which values of z exist. In this case it is the set of points defined by

$$25 - x^2 - y^2 \geq 0;$$

that is, the points inside and on the circle centred at the origin and of radius 5.
Clearly the range of the function is the interval $[0, 5]$.

Example: Consider the function

$$z = g(x, y) = \frac{\sqrt{x^2 + y^2 - 25}}{x}.$$

The domain is the set of points for which

$$x^2 + y^2 \geq 25 \text{ and } x \neq 0.$$

The range of g is all $\mathbf{R}$.

Example: Consider the function

$$z = h(x, y) = \frac{1}{x\sqrt{25 - x^2 - y^2}}.$$

The domain is the set of all points (x, y) satisfying

$$x^2 + y^2 < 25 \text{ and } x \neq 0;$$

that is, all points inside the circle centred at the origin and of radius 5 but *excluding* that portion of the y-axis between $y = -5$ and $y = 5$.
What is the range of the function h?

Self-help exercises

Determine the domain of each of the functions.

1. $z = x\sqrt{x^2 + y^2 - 16}$ [$x^2 + y^2 \geq 16$]
2. $z = \dfrac{x}{\sqrt{x^2 + y^2 - 16}}$ [$x^2 + y^2 > 16$]
3. $z = \dfrac{4}{x^2 + y^2}$ [all x and y except $x = y = 0$]
4. $z = e^{x-2y}$ [all x and y]
5. $z = \log(2x - y)$. [$2x - y > 0$]

11.3 Graphs and Level Curves

To represent a surface of the form $z = f(x, y)$ we can attempt to sketch the three-dimensional surface defined by the equation or generate some level curves for the function. For example, the surface

$$z = \sqrt{25 - x^2 - y^2}$$

is clearly the upper hemisphere[1] defined by

$$x^2 + y^2 + z^2 = 25, \quad z \geq 0.$$

For other surfaces its geometric description may not always be so obvious. For this reason we draw instead the function's **level curves**. This latter approach is one you are already familiar with. Contour lines for lines of altitude, isobars for lines of constant pressure and isotherms for lines of constant temperature are all examples of level curves. For a general function

$$z = f(x, y)$$

they are defined as the curves

$$f(x, y) = k$$

for constant values of k. These curves arise as the intersection of the planes $z = k$ and the surface $z = f(x, y)$ but projected onto the x, y-plane. From our experience of viewing physical level curves such as contour lines, we can mentally reconstitute the surface from the information contained in the graph of the level curves. For example, we can readily distinguish where a given path will be steepest or horizontal and what path would be a convenient one to follow in navigating from one valley to another.

Three-dimensional graphing software that displays an image of such three-dimensional surfaces on the simplest of computing devices is becoming more readily available. Such surfaces can be viewed from any specified angle, rotated, magnified, etc. so that a complete appreciation of the surface is immediately available. Some examples of such software include Derive, Maple and Mathematica. For example, consider the surface

$$z = f(x, y) = xye^{-(x^{\bullet} + y^{\bullet})}.$$

It is not immediately obvious what this surface might look like; however, we can obtain the sketch of both the level curves and the surface using these packages. A sketch of the surface is given in Figure 11.1 and a sketch of some of the level curves is given in Figure 11.2; the darker shaded portions of Figure 11.2 correspond to points on the surface that lie below the x, y-plane and the lighter shaded portions to those lying above the x, y-plane.

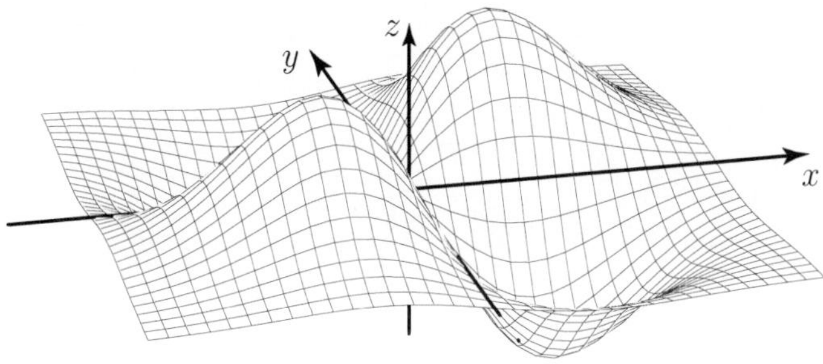

Figure 11.1: The surface $z = f(x, y) = xye^{-(x^2+y^2)}$.

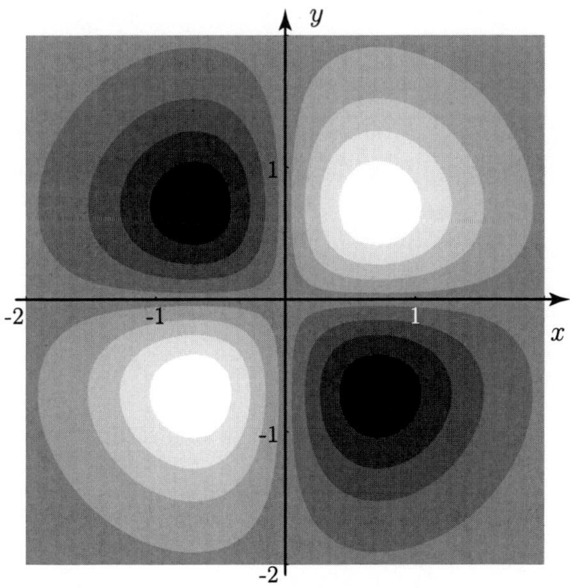

Figure 11.2: Level curves of the surface $f(x, y) = xye^{-(x^2+y^2)}$.

Worked Example 11.3.1 *Sketch some of the level curves for the surface*

$$z = f(x, y) = 10 - x^2 - y^2.$$

The level curves are defined by

$$f(x, y) = 10 - x^2 - y^2 = k.$$

For each value of k these are the circles

$$x^2 + y^2 = 10 - k$$

having centre $(0, 0)$ and radius $\sqrt{10 - k}$ (for $k \leq 10$). Some examples of these level curves are given in Figure 11.3.
The complete surface is an inverted paraboloid — see Figure 11.4.

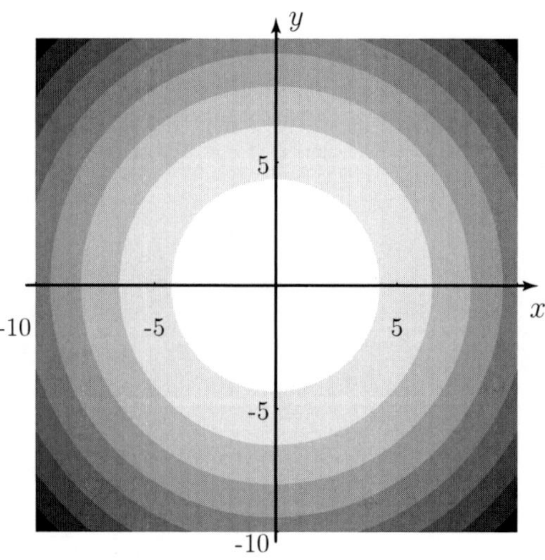

Figure 11.3: Level curves of the inverted paraboloid.

■

Worked Example 11.3.2 *Sketch some level curves of the surface*

$$z = f(x, y) = 9 - 2x^2 - 8y.$$

---

* You should be able to recognize the equations of the surfaces of a cone, a paraboloid, cylinders and ellipsoids including spheres. These basic quadric surfaces were used to help identify space curves earlier. Cross-sections of each of these surfaces need not be circular.

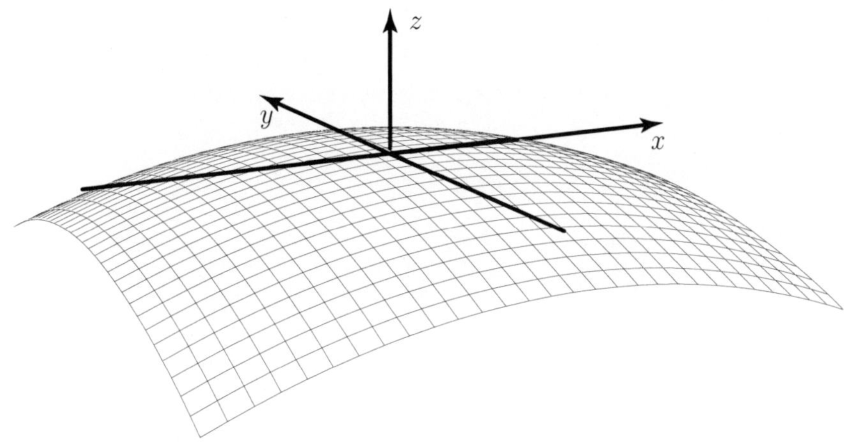

Figure 11.4: The surface of the inverted paraboloid.

The level curves are defined by

$$f(x, y) = 9 - 2x^2 - 8y = k.$$

For each value of k these are parabolas symmetric about the x-axis. Some examples of these curves are given in Figure 11.5. The complete surface is given in Figure 11.6. ■

Self-help exercises

1. Draw some level curves defined by the following functions:

   (a) $f(x, y) = x^2 + y^2 - 12$

   (b) $f(x, y) = \sqrt{x^2 + y^2 - 10}$

   (c) $f(x, y) = \sqrt{9 - 4x^2 - 6y^2}$.

   $[x^2 + y^2 = k;\ x^2 + y^2 = k, k \geq 10;\ 4x^2 + 6y^2 = k, k \leq 9]$

2. Describe and identify the surfaces defined by the equations $z = f(x, y)$. (If you have access to a surface graphics plotting programme, sketch the actual surface to confirm the conclusions you have drawn from your sketch of the level curves.)

   [paraboloid; upper surface of cone; upper surface of ellipsoid]

11.4 Partial Derivatives

Let $z = f(x, y)$. The partial derivative of f with respect to x is the function

$$\frac{\partial f}{\partial x} = f_x = \frac{\partial z}{\partial x} = z_x$$

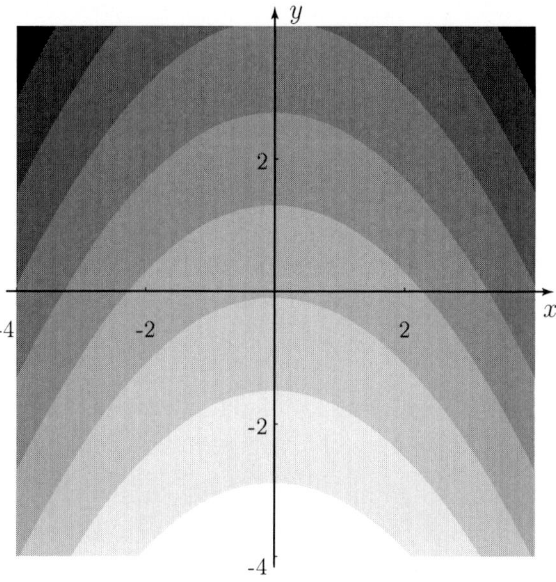

Figure 11.5: Level curves of the surface $f(x, y) = 9 - 2x^2 - 8y$.

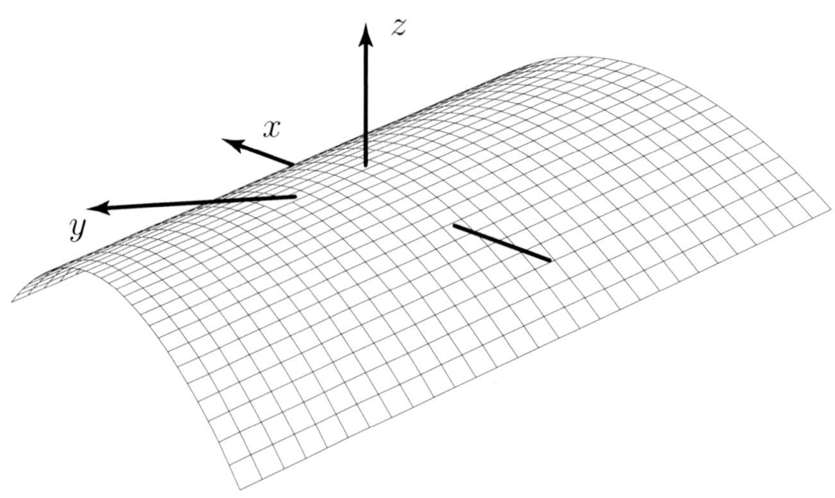

Figure 11.6: The surface $f(x, y) = 9 - 2x^2 - 8y$.

whose value at the point (x, y) (in the domain of f) is given by

$$\lim_{h \to 0} \frac{f(x+h, y) - f(x, y)}{h}$$

if this limit exists. (In the expression f_x, for example, the subscript x refers to partial differentiation with respect to that variable. This convention is used quite extensively in practice and is followed in these notes.) In Figure 11.7, the partial derivative $f_x(x_0, y_0)$ is shown to represent the slope of the tangent line to the curve $z = f(x, y_0)$ at the point (x_0, y_0).

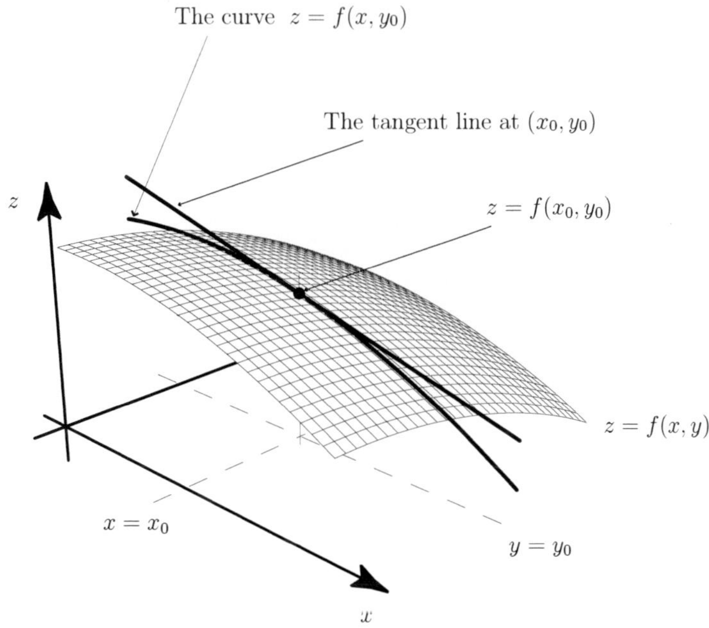

Figure 11.7: The x partial derivative.

This partial derivative is evaluated by holding the variable y fixed and differentiating in the normal way with respect to x. Consequently, f_x represents the rate of change of the function $f(x, y)$ in the direction of the x-axis. Similarly

$$\frac{\partial f}{\partial y} = f_y = \frac{\partial z}{\partial y} = z_y = \lim_{k \to 0} \frac{f(x, y+k) - f(x, y)}{k}$$

if this limit exists. This partial derivative is evaluated by holding the variable x fixed and differentiating in the normal way with respect to y and represents the rate of change of f in the direction of the y-axis.

The same ideas extend naturally to functions of more than two independent variables. For example, for the function $f(x, y, z, w, u, v)$ we can define the six partial derivatives f_x, f_y, f_z, f_w, f_u and f_v. They are all defined as the limit of the appropriate difference quotient. For example

$$f_u = \lim_{h \to 0} \frac{f(x, y, z, w, u+h, v) - f(x, y, z, w, u, v)}{h},$$

whenever this limit exists. In calculating f_u we again perform the ordinary derivative of f with respect to u, holding all the other independent variables x, y, z, w and v constant.

It follows immediately from the definition of the partial derivative that the product and quotient rules of ordinary differentiation carry over to partial differentiation. Thus if f and g are functions of several variables that include x, for example, as an independent variable then

$$\frac{\partial}{\partial x}(fg) = f\frac{\partial g}{\partial x} + g\frac{\partial f}{\partial x} \qquad \text{Product rule}$$

$$\frac{\partial}{\partial x}\left(\frac{f}{g}\right) = \frac{g\frac{\partial f}{\partial x} - f\frac{\partial g}{\partial x}}{g^2} \qquad \text{Quotient rule.}$$

Also, if $f = f(w)$ where $w = w(x, y, z)$ then the chain rule for functions of one real variable can be used to determine all of the first partial derivatives f_x, f_y and f_z. For example

$$f_z = \frac{\partial f}{\partial z} = \frac{df}{dw} \cdot \frac{\partial w}{\partial z} = f'(w)w_z.$$

(This is a limited form of the chain rule for functions of this type. The complete chain rule is discussed in Section 11.7.)

▌**Worked Example 11.4.1** *Determine both the x and y partial derivatives of the functions f and g defined below. Also determine $f_x(3, -2)$ and $g_y(2, -1)$.*

1. $f(x, y) = 3x^2 - 2xy + y^2$ *2.* $g(x, y) = 5\sqrt{28 - 2x^2 - 4y^2}$.

1. $\dfrac{\partial f}{\partial x} = 6x - 2y$ and $\dfrac{\partial f}{\partial y} = -2x + 2y$.
   Hence $f_x(3, -2) = 22$.
2. $\dfrac{\partial g}{\partial x} = 5\dfrac{1}{2}(28 - 2x^2 - 4y^2)^{-1/2}(-4x)$ and
   $\dfrac{\partial g}{\partial y} = 5\dfrac{1}{2}(28 - 2x^2 - 4y^2)^{-1/2}(-8y)$.
   Hence $g_y(2, -1) = 5\frac{1}{2}(28 - 8 - 4)^{-1/2}(8) = 5$.
   (An alternative procedure for determining partial derivatives of g is to use the ideas of implicit differentiation borrowed from functions of one real variable. That is, if $z = g(x, y)$ then

$$\frac{1}{25}z^2 = 28 - 2x^2 - 4y^2$$

which upon differentiating implicitly with respect to x gives

$$\tfrac{2}{25} z \frac{\partial z}{\partial x} = -4x.$$

Solving for z_x gives the previous result for the partial derivative.)

■

Worked Example 11.4.2 *Find all first partial derivatives of the function z where*

1. $z(x, y) = x \log(2x + 3y)$
2. $z(x, y) = e^{\frac{x}{y}}$
3. $z(x, y) = \frac{x+2y}{3x+5y}$
4. $z(x, y) = \arctan \frac{x}{y}$.

1. Using the product and chain rules we have

$$z_x = x \frac{2}{2x + 3y} + \log(2x + 3y) \text{ and } z_y = x \frac{3}{2x + 3y}.$$

2. The chain rule gives

$$z_x = e^{\frac{x}{y}} \left(\frac{1}{y} \right) \text{ and } z_y = e^{\frac{x}{y}} \left(\frac{-x}{y^2} \right).$$

3. The quotient rule gives

$$z_x = \frac{(3x + 5y) - (x + 2y)3}{(3x + 5y)^2} = -\frac{y}{(3x + 5y)^2}$$

and

$$z_y = \frac{(3x + 5y)2 - (x + 2y)5}{(3x + 5y)^2} = \frac{x}{(3x + 5y)^2}.$$

4. The chain and quotient rules give

$$z_x = \frac{1}{1 + \left(\frac{x}{y} \right)^2} \left(\frac{1}{y} \right) = \frac{y}{y^2 + x^2}$$

and

$$z_y = \frac{1}{1 + \left(\frac{x}{y} \right)^2} \left(\frac{-x}{y^2} \right) = -\frac{x}{y^2 + x^2}.$$

■

Self-help exercises

Determine all first partial derivatives for the given functions.

1. $f(x, y) = x^2y + xy^2$ $[2xy + y^2;\ x^2 + 2xy]$

2. $g(x, y) = \frac{xy}{x+y}$ $[\left(\frac{y}{x+y}\right)^2;\ \left(\frac{x}{x+y}\right)^2]$

3. $h(x, y, z) = \sinh xyz.$ $[yz\cosh xyz;\ xz\cosh xyz;\ xy\cosh xyz]$

11.4.1 Cylindrical and Spherical Polar Co-ordinates

For regions that exhibit either cylindrical or spherical symmetries the rectangular Cartesian co-ordinate system of representing points in three-dimensional space is no longer the most suitable. Two co-ordinate systems that exploit these respective symmetries are given below.

1. **Cylindrical co-ordinate system**

   With reference to Figure 11.8, the cylindrical co-ordinates are the co-ordinates (r, θ, z).

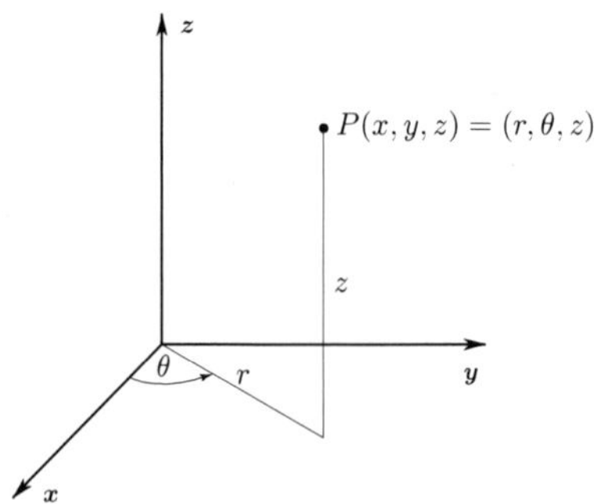

Figure 11.8: Cylindrical polar co-ordinates r, θ, z.

Conversion from the cylindrical to Cartesian co-ordinate systems and vice versa is given as follows:

Cylindrical to Cartesian	Cartesian to Cylindrical
$x = r\cos\theta$	$r = \sqrt{x^2 + y^2}$
$y = r\sin\theta$	$\theta = \tan^{-1}\frac{y}{x}$
$z = z$	$z = z.$

The cylindrical co-ordinates r and θ are simply the plane polar co-ordinates; the value of θ may need to be adjusted by the amount $\pm\pi$

to correspond to which octant the point P lies in — in exactly the same way as for plane polar co-ordinates.

2. **Spherical polar co-ordinates**

   With reference to Figure 11.9, the spherical co-ordinates are the co-ordinates (ρ, θ, ϕ).

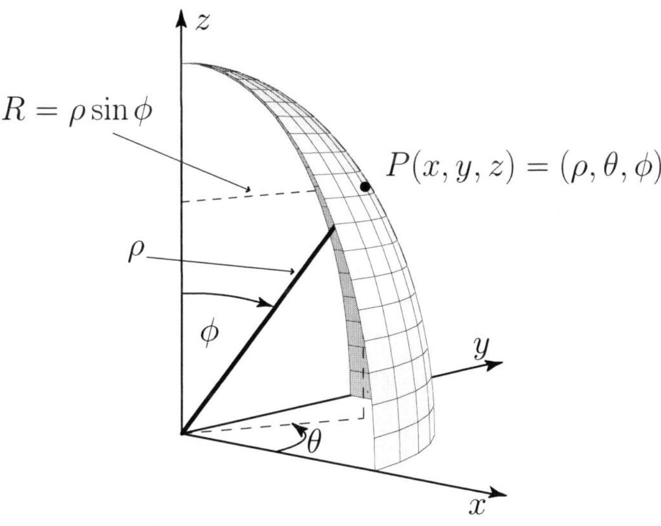

Figure 11.9: Spherical polar co-ordinates ρ, θ, ϕ.

Conversion from the spherical to Cartesian co-ordinates systems and vice versa is given as follows:

Spherical to Cartesian	**Cartesian to Spherical**
$x = R\cos\theta = \rho\sin\phi\cos\theta$	$\rho = \sqrt{x^2+y^2+z^2}$
$y = R\sin\theta = \rho\sin\phi\sin\theta$	$\theta = \arctan\frac{y}{x}$
$z = \rho\cos\phi$	$\phi = \arctan\frac{\sqrt{x^2+y^2}}{z}$.

The values of θ and ϕ may need to be adjusted by the factor $\pm\pi$ for the same reasons outlined above in the cylindrical co-ordinate case. Also note that there are several other options for defining the variable ϕ; for example, we could instead have used $\phi = \arcsin\frac{\sqrt{x^2+y^2}}{\sqrt{x^2+y^2+z^2}}$.

In each of these transformation equations, it is important to recognize the dependent variables and the independent variables in each set. For example, in converting from Cartesian co-ordinates to cylindrical co-ordinates the independent variables are x, y, z and the dependent variables are r, θ, z; in converting from cylindrical to Cartesian the independent variables are r, θ, z and the dependent variables are x, y, z. Thus, for example, in evaluating the partial derivative $\frac{\partial r}{\partial x}$ we treat x as the independent variable (keeping y fixed) whereas in evaluating the partial derivative $\frac{\partial x}{\partial r}$ we treat x as the dependent variable (keeping θ fixed).

Worked Example 11.4.3 *In the notation of Cartesian and cylindrical co-ordinates, evaluate the partial derivatives*

1. $\frac{\partial r}{\partial x}$ *2.* $\frac{\partial x}{\partial r}$ *3.* $\frac{\partial \theta}{\partial y}$ *4.* $\frac{\partial y}{\partial \theta}$.

1. Here we have $r = \sqrt{x^2 + y^2}$. Thus $\frac{\partial r}{\partial x} = \frac{x}{\sqrt{x^{\cdot} + y^{\cdot}}}$.

   (Why is it *incorrect* to use the equation $r = \frac{x}{\cos\theta}$ to obtain $\frac{\partial r}{\partial x} = \frac{1}{\cos\theta}$?)
2. Here we have $x = r\cos\theta$. Thus $\frac{\partial x}{\partial r} = \cos\theta$.
3. Here we have $\theta = \tan^{-1}\frac{y}{x}$.

   Thus $\frac{\partial \theta}{\partial y} = \frac{1}{1+\left(\frac{y}{x}\right)^{\cdot}}\left(\frac{1}{x}\right) = \frac{x}{x^{\cdot}+y^{\cdot}}$.
4. Here we have $y = r\sin\theta$. Thus $\frac{\partial y}{\partial \theta} = r\cos\theta$.

The results of items 1 and 2 illustrate the important result that, unlike for ordinary derivatives,

$$\boxed{\frac{\partial r}{\partial x} \neq \frac{1}{\frac{\partial x}{\partial r}}}$$

■

Self-help exercises

Determine each of the following partial derivatives where the notation is that for either cylindrical or spherical polar co-ordinates.

1. $\frac{\partial z}{\partial z}$ 2. $\frac{\partial \rho}{\partial z}$ 3. $\frac{\partial \phi}{\partial z}$ 4. $\frac{\partial z}{\partial \phi}$.

$[1; \frac{z}{\sqrt{x^{\cdot}+y^{\cdot}+z^{\cdot}}}; -\frac{\sqrt{x^{\cdot}+y^{\cdot}}}{x^{\cdot}+y^{\cdot}+z^{\cdot}}; -\rho\sin\phi]$

11.5 Higher Order Partial Derivatives

For a given function $z = f(x, y)$, the first partial derivatives f_x and f_y are themselves functions of the independent variables x and y. Consequently, we can consider their partial derivatives with respect to both x and y. Thus we have the four new second partial derivatives

$$\frac{\partial^2 f}{\partial x^2} = \frac{\partial}{\partial x}\left(\frac{\partial f}{\partial x}\right) \qquad \frac{\partial^2 f}{\partial y^2} = \frac{\partial}{\partial y}\left(\frac{\partial f}{\partial y}\right)$$
$$\frac{\partial^2 f}{\partial x \partial y} = \frac{\partial}{\partial x}\left(\frac{\partial f}{\partial y}\right) \qquad \frac{\partial^2 f}{\partial y \partial x} = \frac{\partial}{\partial y}\left(\frac{\partial f}{\partial x}\right).$$

As was the case for the first partial derivatives, we will often use subscript notation to represent these partial derivatives. For example

$$f_{xx} = \frac{\partial^2 f}{\partial x^2} \text{ and } f_{xy} = \frac{\partial^2 f}{\partial y \partial x}.$$

Similarly each of these derivatives possess partial derivatives such as

$$\frac{\partial^3 f}{\partial x^3}, \frac{\partial^3 f}{\partial y^3}, \frac{\partial^3 f}{\partial x^2 \partial y} \text{ and } \frac{\partial^3 f}{\partial x \partial y^2},$$

with the corresponding subscript notation

$$f_{xxx}, \ f_{yyy}, \ f_{yxx} \text{ and } f_{yyx}.$$

Worked Example 11.5.1 *Given*

$$f(x, y) = e^x \cos y + x^2 y^2,$$

determine

1. $f_{xx} = \dfrac{\partial^2 f}{\partial x^2}$ 2. $f_{xy} = \dfrac{\partial^2 f}{\partial y \partial x}$ 3. $f_{xyy} = \dfrac{\partial^3 f}{\partial y^2 \partial x}$.

1. We have $f_x = e^x \cos y + 2xy^2$.
   Therefore $f_{xx} = e^x \cos y + 2y^2$.
2. Using the above we have $f_{xy} = -e^x \sin y + 4xy$.
3. Finally we have $f_{xyy} = -e^x \cos y + 4x$.

■

In the above worked example, note carefully the order in which the partial derivatives are taken depending on the notation used. For example, in the case $\frac{\partial^2 f}{\partial x \partial y}$ or f_{yx}, the y partial derivative is taken first.

The partial derivatives described above for functions of two independent variables extend naturally to functions of three and more independent variables. (See the next worked example as an illustration of this.)

Worked Example 11.5.2 *Given* $f(x, y, z) = \cos(x^2y^2 + 5z)$, *determine the partial derivative* f_{xzy}.
We have $f_x = -2xy^2 \sin(x^2y^2 + 5z)$.
Then $f_{xz} = -2xy^2 \cos(x^2y^2 + 5z)\ 5$ and finally

$$\begin{aligned} &f_{xzy} \\ &\quad = -10xy^2\{-2x^2y \sin(x^2y^2 + 5z)\} - 20xy \cos(x^2y^2 + 5z) \\ &\quad = 20x^3y^3 \sin(x^2y^2 + 5z) - 20xy \cos(x^2y^2 + 5z). \end{aligned}$$

■

Worked Example 11.5.3 *Given*

$$f(x, y) = x^3y^2 + x \sinh xy^2$$

show that $f_{xy} = f_{yx}$.
We have

$$\begin{aligned} &f_{xy} \\ &\quad = \frac{\partial}{\partial y}(3x^2y^2 + xy^2 \cosh xy^2 + \sinh xy^2) \\ &\quad = 6x^2y + xy^2(\sinh xy^2)\ 2xy + 2xy \cosh xy^2 + 2xy \cosh xy^2. \end{aligned}$$

Similarly

$$\begin{aligned} f_{yx} &= \frac{\partial}{\partial x}(2x^3y + 2x^2y \cosh xy^2) \\ &= 6x^2y + 2x^2y^3 \sinh xy^2 + 4xy \cosh xy^2. \end{aligned}$$

The required result follows immediately. ■

The result obtained in the above example holds for a wide class of functions including those that arise in practice. We refer to the result by saying that "the order of mixed partial derivatives is not important". That is, in the special case of second order partial derivatives

$$\boxed{f_{xy} = f_{yx}}$$

In the case of higher order derivatives we have, for example,

$$f_{xxyx} = f_{xxxy} = f_{yxxx} = \cdots$$

This result regarding the order of mixed partial derivatives can be made precise but requires a thorough understanding of the continuity of functions of many variables; we will not go into this here but rely on the fact that most functions arising as the solution of practical problems will have the required continuity to guarantee that the result holds true.

Self-help exercises

1. For each of the functions

(a) $f(x, y, z) = x^3y^2z$
(b) $h(x, y, z) = x^2 + y^2 + z^2$
(c) $p(x, y, z) = xyz^2$

determine their xy, yxz and $zzxy$ partial derivatives.

$[6x^2yz, 6x^2y, 0;\ 0, 0, 0;\ z^2, 2z, 2]$

In the case of the function p, show further that

$$p_{xyz} = p_{yzx} = p_{zxy} = p_{xzy} = p_{zyx} = p_{yxz}.$$

2. Verify that the function $\phi(x, y) = 4x^3y - 4xy^3$ is a solution of Laplace's equation

$$\phi_{xx} + \phi_{yy} = 0.$$

11.6 Approximation to $\Delta f(x_0, y_0)$

In this section we seek an analytic approximation to the change produced in a function $f(x, y)$ when both x and y change by a small amount; similar arguments provide approximation formulae for functions of more independent variables. Such formulae are useful in generating error estimates for physical quantities that rely on measurements that may themselves be in error and form the basis of the derivation of many numerical methods procedures.

That is, we seek an approximation to the value of

$$\Delta f(x_0, y_0) = f(x_0 + \Delta x, y_0 + \Delta y) - f(x_0, y_0).$$

Geometrically, if $z = f(x, y)$ measures the height above sea level then the value $\Delta f(x_0, y_0)$ measures the difference in altitude for neighbouring points (x_0, y_0) and $(x_0 + \Delta x, y_0 + \Delta y)$ on the surface $z = f(x, y)$; see Figure 11.10. The approach is the same as that used for functions of one real variable; that is, we seek to obtain a linear approximation with respect to the variables Δx and Δy. The procedure can easily be extended to functions of more than two independent variables.

We have

$$\begin{aligned}
\Delta f(x_0, y_0) &= f(x_0 + \Delta x, y_0 + \Delta y) - f(x_0, y_0) \\
&= \{f(x_0 + \Delta x, y_0 + \Delta y) - f(x_0, y_0 + \Delta y)\} \\
&\quad + \{f(x_0, y_0 + \Delta y) - f(x_0, y_0)\} \\
&= \frac{\{f(x_0 + \Delta x, y_0 + \Delta y) - f(x_0, y_0 + \Delta y)\}}{\Delta x}\Delta x \\
&\quad + \frac{\{f(x_0, y_0 + \Delta y) - f(x_0, y_0)\}}{\Delta y}\Delta y \\
&\approx \frac{\partial f}{\partial x}(x_0, y_0)\Delta x + \frac{\partial f}{\partial y}(x_0, y_0)\Delta y.
\end{aligned} \tag{11.1}$$

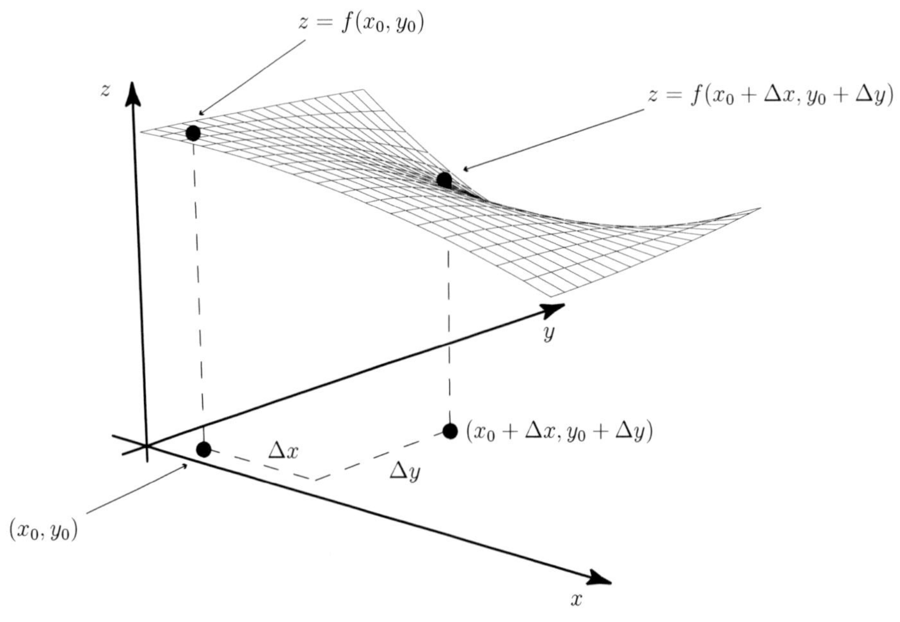

Figure 11.10: Approximation to $\Delta f(x_0, y_0)$.

In words the change in f consists of the sum of two terms representing the change of f with respect to x multiplied by the change in x together with the change of f with respect to y multiplied by the change in y. (In this equation the partial derivatives are all to be evaluated at the point (x_0, y_0); also compare this with the formula from the one-dimensional real variable case $\Delta y \approx \frac{dy}{dx}\Delta x$.)

The result used to simplify the expressions between the brackets { and

} is the one variable result that

$$\frac{g(x+h) - g(x)}{h} \approx g'(x)$$

using the observation that the tangent line to the curve $y = g(x)$ can provide an estimate for $g(x+h)$ for small values of h; see Figure 11.11.

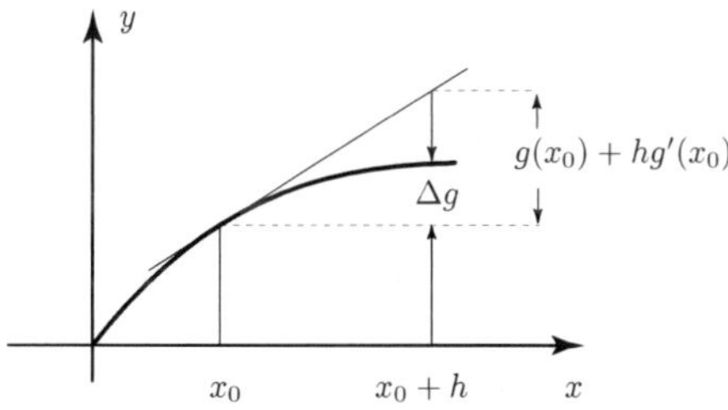

Figure 11.11: One-dimensional approximation to $\Delta g = g(x_0 + h) - g(x_0)$.

We are usually interested in comparing the values of a given function f at nearby points so that both Δx and Δy are generally small. The expression 11.1 for Δf then provides a useful approximation for many physical calculations.

An immediate consequence of Equation 11.1 is that

$$\boxed{f(x_0 + \Delta x, y_0 + \Delta y) \approx f(x_0, y_0) + \frac{\partial f}{\partial x}\Delta x + \frac{\partial f}{\partial y}\Delta y}$$

If we write $x = x_0 + \Delta x$ and $y = y_0 + \Delta y$ then this equation becomes

$$\boxed{f(x, y) \approx f(x_0, y_0) + \frac{\partial f}{\partial x}(x - x_0) + \frac{\partial f}{\partial y}(y - y_0)}$$

It will be referred to as the **total increment theorem**. This result provides a way of estimating the value of a function of two real variables near to a point where f and its two partial derivatives are known. It is a *linear approximation* in the variables Δx and Δy and represents the tangent plane approximation to the value of $f(x_0+\Delta x, y_0+\Delta y)$. (This should be compared with the tangent line approximation used to approximate the value of $g(x_0+h)$ using information about the function g and its derivative at a nearby point x illustrated in Figure 11.11.)

A similar result holds for functions of more than two real variables. For example, the linearization of the function $f(x, y, z, u, v)$ near to the point

$(x_0, y_0, z_0, u_0, v_0)$ is given by

$$\begin{aligned} f(x_0, y_0, z_0, u_0, v_0) + \frac{\partial f}{\partial x}(x - x_0) + \frac{\partial f}{\partial y}(y - y_0) \\ + \frac{\partial f}{\partial z}(z - z_0) + \frac{\partial f}{\partial u}(u - u_0) + \frac{\partial f}{\partial v}(v - v_0). \end{aligned}$$

▮ **Worked Example 11.6.1** *Estimates for the values of a function $f(x, y)$ and its partial derivatives f_x and f_y at the point $(1, 2)$ are given as $f(1, 2) = 0.1348 \times 10^{-1}$, $f_x(1, 2) = -0.1348 \times 10^{-1}$ and $f_y(1, 2) = -0.4717 \times 10^{-1}$. Use the total increment theorem to obtain estimates for*

1. $f(1.1, 2.1)$ *2.* $f(1.2, 1.9)$ *3.* $f(1.4, 1.6)$.

Given that the function generating these values is in fact

$$f(x, y) = xye^{-(x^{\bullet} + y^{\bullet})},$$

determine the error involved in each of these approximations.
Using the total increment theorem, we have

$$f(x + h, y + k) \approx f(x, y) + hf_x(x, y) + kf_y(x, y).$$

Thus

1. $$\begin{aligned} f(1.1, 2.1) &\approx f(1, 2) + 0.1f_x(1, 2) + 0.1f_y(1, 2) \\ &= 0.1348 \times 10^{-1} + 0.1 \times (-0.1348 \times 10^{-1}) \\ &\quad + 0.1 \times (-0.4717 \times 10^{-1}) \\ &= 0.7415 \times 10^{-2} \end{aligned}$$

2. $$\begin{aligned} f(1.2, 1.9) &\approx 0.1348 \times 10^{-1} + 0.2 \times (-0.1348 \times 10^{-1}) \\ &\quad + (-0.1) \times (-0.4717 \times 10^{-1}) \\ &= 0.1550 \times 10^{-1} \end{aligned}$$

3. $$\begin{aligned} f(1.4, 1.6) &\approx 0.1348 \times 10^{-1} + 0.4 \times (-0.1348 \times 10^{-1}) \\ &\quad + (-0.4) \times (-0.4717 \times 10^{-1}) \\ &= 0.2696 \times 10^{-1}. \end{aligned}$$

(Note that in the last item above, since h and k are not "small", the estimate obtained there cannot be expected to be very accurate.)
Using the given formula for the function f we have $f(1.1, 2.1) = 0.8373 \times 10^{-2}$, $f(1.2, 1.9) = 0.1461 \times 10^{-1}$, and $f(1.4, 1.6) = 0.2439 \times 10^{-1}$. Thus the error in each case is 0.9580×10^{-3}, 0.8900×10^{-3} and 0.2569×10^{-2}. ■

Worked Example 11.6.2 *A closed metal can in the shape of a right circular cylinder is to have an inside height of 20 mm, an inside radius of 5 mm and a thickness of 0.4 mm. If the cost of metal to be used is 0.5 cents per cubic millimetre, estimate the cost of metal used to manufacture the can.*
The volume (V) of a cylinder of height h and radius r is given by

$$V = \pi r^2 h = V(r, h).$$

The exact volume of metal required is

$$\begin{aligned}
\Delta V &= V(5.4, 20.8) - V(5, 20) \\
&\approx \frac{\partial V}{\partial r}(5, 20)\Delta r + \frac{\partial V}{\partial h}(5, 20)\Delta h \\
&= 2\pi\ (5)\ (20)\ (0.4) + \pi\ (5^2)\ (0.8) \\
&= 100\pi.
\end{aligned}$$

Thus the approximate cost to manufacture the can is $100\pi \times 0.5 \approx 157$ cents or \$1.57 per can. ■

Self-help exercise

The dimensions of a rectangular box are measured as 40 mm, 50 mm and 60 mm with all measurements correct to 0.05mm.

Estimate the largest error in calculating the volume of the box using these measurements.

[370 mm]

11.7 The Chain Rule

For functions of one real variable the chain rule takes the following form. If $w = f(x)$ and $x = g(t)$ then $w = f(g(t))$ and

$$\frac{dw}{dt} = \frac{dw}{dx}\frac{dx}{dt} = f'(x)g'(t) = f'(g(t))g'(t).$$

We now extend this rule in two stages. First to the case

$$w = f(x, y) \text{ where } x = g(t) \text{ and } y = h(t)$$

and then to the more general case

$$w = f(x, y) \text{ where } x = g(u, v) \text{ and } y = h(u, v).$$

The first of these arises in the case where w is a temperature function expressed in Cartesian co-ordinates and we wish to determine the rate of change of the temperature along a specified space curve defined by $\mathbf{r}(t) = g(t)\ \mathbf{i} + h(t)\ \mathbf{j}$. The second of these arises when we wish to determine rates of change in circumferential and radial directions when converting from a Cartesian co-ordinate system to plane polar co-ordinates.

[Note: The main use of the general chain rule is in converting derivatives expressed in one co-ordinate system to equivalent derivatives in another co-ordinate system; see Worked Example 11.7.6.]

11.7.1 The Simple Case

Here $w = f(x, y)$ where $x = g(t)$ and $y = h(t)$. We have

$$\begin{aligned} \Delta w &= f(x + \Delta x, y + \Delta y) - f(x, y) \\ &\approx \frac{\partial f}{\partial x}\Delta x + \frac{\partial f}{\partial y}\Delta y. \end{aligned}$$

Therefore, upon dividing both sides by Δt,

$$\frac{\Delta w}{\Delta t} \approx \frac{\partial f}{\partial x}\frac{\Delta x}{\Delta t} + \frac{\partial f}{\partial y}\frac{\Delta y}{\Delta t}.$$

Taking the limit of this equation as $\Delta t \to 0$ leads to

$$\frac{dw}{dt} = \frac{\partial f}{\partial x}\frac{dx}{dt} + \frac{\partial f}{\partial y}\frac{dy}{dt}. \tag{11.2}$$

We often write this equation in the more obvious notation

$$\boxed{\frac{dw}{dt} = \frac{\partial w}{\partial x}\frac{dx}{dt} + \frac{\partial w}{\partial y}\frac{dy}{dt}}$$

[Note that

- each term on the right-hand side "looks like" the derivative $\frac{dw}{dt}$ on the left-hand side and

- the right-hand side consists of the sum of two terms corresponding to the fact that the function f is expressed in terms of *two* independent variables: the sum of the rate of change in the x direction times the change in x together with the change in the y direction times the change in y.]

For functions of more independent variables a similar chain rule expansion holds. For example, if $f = f(x, y, z, u, v)$ where $x = x(t)$, $y = y(t)$, $z = z(t)$, $u = u(t)$ and $v = v(t)$ so that $f = f(x(t), y(t), z(t), u(t), v(t)) = f(t)$, then

$$\boxed{\frac{df}{dt} = \frac{\partial f}{\partial x}\frac{dx}{dt} + \frac{\partial f}{\partial y}\frac{dy}{dt} + \frac{\partial f}{\partial z}\frac{dz}{dt} + \frac{\partial f}{\partial u}\frac{du}{dt} + \frac{\partial f}{\partial v}\frac{dv}{dt}}$$

Again, each term on the right-hand side "looks like" the derivative $\frac{df}{dt}$ on the left-hand side and the right-hand side consists of the sum of five terms corresponding to the fact that the function f is expressed in terms of *five* independent variables.

Worked Example 11.7.1 *Using the chain rule, determine* $\frac{dw}{dt}$ *where*

$$w = x^2 + y^2 + xy \text{ and } \begin{cases} x = t\cos t \\ y = t\sin t. \end{cases}$$

We have

$$\begin{aligned} \frac{dw}{dt} &= \frac{\partial w}{\partial x}\frac{dx}{dt} + \frac{\partial w}{\partial y}\frac{dy}{dt} \\ &= (2x + y)(-t\sin t + \cos t) + (2y + x)(t\cos t + \sin t). \end{aligned}$$

The last expression can finally be written as a function only of the variable t if required. However, in most cases, if the value of $\frac{dw}{dt}$ is required at any specified value of t then we use the given formulae to determine values of x and y first and then substitute to determine the corresponding value of $\frac{dw}{dt}$. ■

Worked Example 11.7.2 *Using the chain rule, determine* $\frac{du}{dt}$ *where*

$$u = xy + yz + xz \text{ and } \begin{cases} x = \cos t \\ y = \sin t \\ z = t. \end{cases}$$

Here

$$\begin{aligned}\frac{du}{dt} &= \frac{\partial u}{\partial x}\frac{dx}{dt} + \frac{\partial u}{\partial y}\frac{dy}{dt} + \frac{\partial u}{\partial z}\frac{dz}{dt}\\ &= (y+z)(-\sin t) + (x+z)(\cos t) + (y+x)(1).\end{aligned}$$

■

In the previous two examples the required t derivative can be found directly by substituting the given functions of t to obtain explicitly $w = w(t)$ and $u = u(t)$ and then differentiating these functions of t using ordinary differentiation. You should verify that the result obtained is the same as that given using the chain rule. In the next example we have no choice but to use the chain rule.

Worked Example 11.7.3 *Using the chain rule, determine* $\frac{du}{dt}$ *where*

$$u(x,y) = x^3 y \text{ and } \begin{cases} x^5 + y = t \\ x^2 + y^3 = t^2. \end{cases}$$

We have

$$\begin{aligned}\frac{du}{dt} &= \frac{\partial u}{\partial x}\frac{dx}{dt} + \frac{\partial u}{\partial y}\frac{dy}{dt}\\ &= 3x^2 y\dot{x} + x^3\dot{y}.\end{aligned}$$

But from the constraint equations we have, upon differentiating implicitly with respect to t

$$5x^4\dot{x} + \dot{y} = 1 \text{ and } 2x\dot{x} + 3y^2\dot{y} = 2t.$$

The above two equations are a pair of simultaneous equations for $\dot{x}$ and $\dot{y}$. The solutions are

$$\dot{x} = \frac{3y^2 - 2t}{15x^4y^2 - 2x} \text{ and } \dot{y} = \frac{10x^4t - 2x}{15x^4y^2 - 2x}.$$

Thus $\displaystyle\frac{du}{dt} = 3x^2 y\frac{3y^2 - 2t}{15x^4y^2 - 2x} + x^3\frac{10x^4t - 2x}{15x^4y^2 - 2x}.$ ■

Worked Example 11.7.4 *Show that the tangent vector for the curve* $f(x,y) = k$ *in the* x, y*-plane, where* k *is a constant, is parallel to the vector*

$$(f_y, -f_x) = f_y\,\mathbf{i} - f_x\,\mathbf{j}.$$

Let the space curve be denoted by $\mathbf{r}(t)$.

Differentiating both sides of the equation defining the curve with respect to t and using the chain rule gives

$$f_x\dot{x} + f_y\dot{y} = 0.$$

Therefore

$$\dot{\mathbf{r}}(t) = (\dot{x}, \dot{y}) = \dot{x}\,\mathbf{i} + \dot{y}\,\mathbf{j} = \dot{x}\,\mathbf{i} - \frac{f_x}{f_y}\dot{x}\,\mathbf{j},$$

assuming that $f_y \neq 0$. Hence the direction of the tangent vector is parallel to $f_y\,\mathbf{i} - f_x\,\mathbf{j}$. ■

Self-help exercises

Using the chain rule, determine $\frac{dw}{dt}$ if

1. $w = x^2 + y^2 + z^2$ and $x = t\cos t,\ y = t\sin t,\ z = t^2$ [$2t + 4t^3$]

2. $w = xyz$ and $x = t,\ y = t^2,\ z = t^3$. [$6t^5$]

11.7.2 The General Case

Here $w = f(x, y)$ but both the variables x and y are themselves functions of two independent variables u and v; that is, substitution of these formulae for x and y gives $w = f(x(u, v), y(u, v)) = f(u, v)$. Consequently, we need to consider partial derivatives of w (or f) with respect to u and v. Both these derivatives will expand using the previous simple case of the chain rule. Hence

$$\boxed{\begin{aligned} \frac{\partial w}{\partial u} &= \frac{\partial w}{\partial x}\frac{\partial x}{\partial u} + \frac{\partial w}{\partial y}\frac{\partial y}{\partial u} \\ \frac{\partial w}{\partial v} &= \frac{\partial w}{\partial x}\frac{\partial x}{\partial v} + \frac{\partial w}{\partial y}\frac{\partial y}{\partial v} \end{aligned}}$$

[Note that there are

- two equations corresponding to the *two* new independent variables u and v and

- two terms on the right-hand side of each equation corresponding to the *two* original independent variables x and y and each term on the right-hand side "looks like" the corresponding derivative on the left.]

This result generalizes in the obvious way for functions of more independent variables. For example, if

$$f = f(x, y, z, p) \text{ where } \begin{cases} x = x(u, v, w) \\ y = y(u, v, w) \\ z = z(u, v, w) \\ p = p(u, v, w) \end{cases}$$

then upon substitution we get $f = f(u, v, w)$ so there will be three equations for the three partial derivatives $\frac{\partial f}{\partial u}$, $\frac{\partial f}{\partial v}$ and $\frac{\partial f}{\partial w}$ with each equation having four terms on the right-hand side. For example:

$$\boxed{\frac{\partial f}{\partial v} = \frac{\partial f}{\partial x}\frac{\partial x}{\partial v} + \frac{\partial f}{\partial y}\frac{\partial y}{\partial v} + \frac{\partial f}{\partial z}\frac{\partial z}{\partial v} + \frac{\partial f}{\partial p}\frac{\partial p}{\partial v}}$$

Worked Example 11.7.5 *Using the chain rule, determine* $\frac{\partial w}{\partial u}$ *and* $\frac{\partial w}{\partial v}$ *where*

$$w = x^2y + y^2x \text{ and } \begin{cases} x = u\cos v \\ y = u\sin v. \end{cases}$$

Here

$$\begin{aligned} \frac{\partial w}{\partial u} &= \frac{\partial w}{\partial x}\frac{\partial x}{\partial u} + \frac{\partial w}{\partial y}\frac{\partial y}{\partial u} \\ &= (2xy + y^2)(\cos v) + (x^2 + 2yx)(\sin v) \\ \frac{\partial w}{\partial v} &= \frac{\partial w}{\partial x}\frac{\partial x}{\partial v} + \frac{\partial w}{\partial y}\frac{\partial y}{\partial v} \\ &= (2xy + y^2)(-u\sin v) + (x^2 + 2yx)(u\cos v). \end{aligned}$$

■

In the remaining examples in this section, we do not give an explicit formula for the function f. That is, we treat f as arbitrary. Consequently, any derivatives of f cannot be explicitly calculated but rather are left as general expressions such as f_x etc. Once a specific formula has been given for f, *then* we can specify the value of the derivatives of f.

Worked Example 11.7.6 *Using the chain rule, show that the expression*

$$\left(\frac{\partial f}{\partial r}\right)^2 + \frac{1}{r^2}\left(\frac{\partial f}{\partial \theta}\right)^2$$

in terms of the plane polar co-ordinates r and θ, transforms to the expression

$$\left(\frac{\partial f}{\partial x}\right)^2 + \left(\frac{\partial f}{\partial y}\right)^2$$

in terms of the Cartesian co-ordinates x and y.

We have $x = r\cos\theta$ and $y = r\sin\theta$.

Hence, using the chain rule

$$\begin{aligned}\frac{\partial f}{\partial r} &= \frac{\partial f}{\partial x}\frac{\partial x}{\partial r} + \frac{\partial f}{\partial y}\frac{\partial y}{\partial r} \\ &= \frac{\partial f}{\partial x}\cos\theta + \frac{\partial f}{\partial y}\sin\theta\end{aligned}$$

and

$$\begin{aligned}\frac{\partial f}{\partial \theta} &= \frac{\partial f}{\partial x}\frac{\partial x}{\partial \theta} + \frac{\partial f}{\partial y}\frac{\partial y}{\partial \theta} \\ &= -\frac{\partial f}{\partial x} r\sin\theta + \frac{\partial f}{\partial y} r\cos\theta.\end{aligned}$$

Therefore

$$\begin{aligned}&\left(\frac{\partial f}{\partial r}\right)^2 + \frac{1}{r^2}\left(\frac{\partial f}{\partial \theta}\right)^2 \\ &= \left(\frac{\partial f}{\partial x}\cos\theta + \frac{\partial f}{\partial y}\sin\theta\right)^2 + \frac{1}{r^2}\left(-\frac{\partial f}{\partial x} r\sin\theta + \frac{\partial f}{\partial y} r\cos\theta\right)^2 \\ &= (f_x^2\cos^2\theta + 2f_xf_y\cos\theta\sin\theta + f_y^2\sin^2\theta) \\ &\quad + \quad (f_x^2\sin^2\theta - 2f_xf_y\cos\theta\sin\theta + f_y^2\cos^2\theta) \\ &= \left(\frac{\partial f}{\partial x}\right)^2 + \left(\frac{\partial f}{\partial y}\right)^2.\end{aligned}$$

■

In the above worked example note carefully the distinction that needs to be made between the expressions

$$\frac{\partial^2 f}{\partial x^2} \text{ and } f_x^2,$$

and between the expressions

$$\left(\frac{\partial f}{\partial x}\right)^2 \text{ and } f_{xx}.$$

Also note that $f_xf_y \neq f_{xy}$.

▌**Worked Example 11.7.7** *If $w = f(u)$ and $u = \alpha x + \beta y$ where α and β are constants, show that*

$$\alpha\frac{\partial w}{\partial y} = \beta\frac{\partial w}{\partial x}.$$

Upon substitution of u in the formula for w we get $w = w(x, y)$. Therefore, using the chain rule

$$\begin{aligned}\frac{\partial w}{\partial x} &= \frac{dw}{du}\frac{\partial u}{\partial x} = f'(u)\alpha \\ \frac{\partial w}{\partial y} &= \frac{dw}{du}\frac{\partial u}{\partial y} = f'(u)\beta.\end{aligned}$$

Hence

$$\alpha\frac{\partial w}{\partial y} - \beta\frac{\partial w}{\partial x} = \alpha\beta f'(u) - \beta\alpha f'(u) = 0.$$

■

Self-help exercise

If $w = xyz$ and $x = uv$, $y = u^2v^2$ and $z = u^3v^3$, use the chain rule to evaluate the partial derivatives $\frac{\partial w}{\partial u}$ and $\frac{\partial w}{\partial v}$.

$[yz \cdot v + xz \cdot 2uv^2 + xy \cdot 3u^2v^3;\ yz \cdot u + xz \cdot 2u^2v + xy \cdot 3u^3v^2]$

11.8 Directional Derivative

The directional derivative of a function provides the rate of change of that function (with respect to distance) in any specified direction — not just the direction of the axes as generated by the partial derivatives.

Let $f = f(x, y, z)$ be a given function and P a given point (a, b, c). Suppose $\mathbf{u} = (u_1, u_2, u_3)$ is a *unit vector* in the direction in which we wish to determine the rate of change of f at the point P. If Q represents a point on the line passing through P in the direction of $\mathbf{u}$, then the directional derivative of f in the direction of $\mathbf{u}$ at the point P is given by

$$\boxed{\frac{df}{du} = \lim_{PQ\to 0}\frac{f(Q) - f(P)}{PQ}}$$

(Note that the symbol denoting the directional derivative is not standard. Other notations that are often used include

$$\frac{df}{dn}, \frac{df}{ds}, \frac{\partial f}{\partial u} \text{ and } D_u.)$$

But the equation of the straight line passing through P in the direction $\mathbf{u}$ is given by

$$x = a + hu_1$$
$$y = b + hu_2$$
$$z = c + hu_3.$$

(Normally we would use t as the parameter of the path but here it is more usual to use h.) Hence, the directional derivative of f in the direction of $\mathbf{u}$ at the point P is then given by

$$\boxed{\frac{df}{du} = \lim_{h\to 0} \frac{f(a+hu_1, b+hu_2, c+hu_3) - f(a,b,c)}{h}}$$

if this limit exists. (The situation relevant to the two-dimensional case is given in Figure 11.12.)

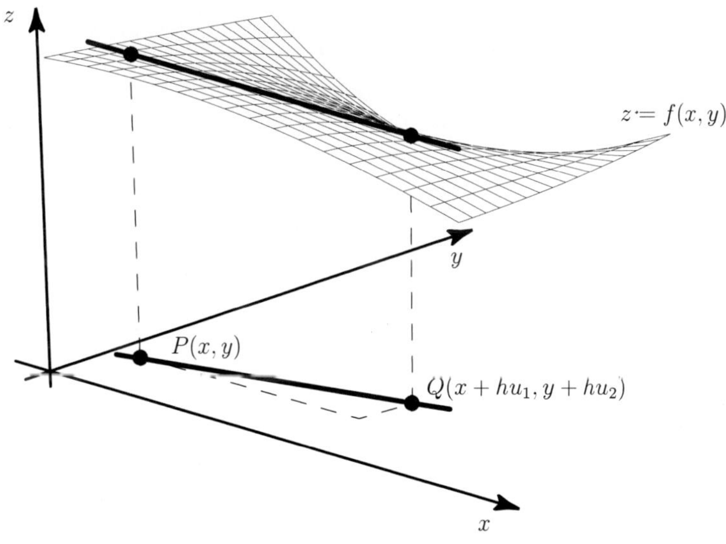

Figure 11.12: The directional derivative in the two-dimensional case.

Example: We note the special cases:

$$\text{if } \mathbf{u} = \mathbf{i} \text{ then } \frac{df}{du} = \frac{\partial f}{\partial x}, \quad (u_2 = u_3 = 0)$$
$$\text{if } \mathbf{u} = \mathbf{j} \text{ then } \frac{df}{du} = \frac{\partial f}{\partial y}, \quad (u_1 = u_3 = 0)$$
$$\text{if } \mathbf{u} = \mathbf{k} \text{ then } \frac{df}{du} = \frac{\partial f}{\partial z} \quad (u_1 = u_2 = 0).$$

(Because of the above results for the special cases when the specified direction is parallel to one of the standard axes, we could evaluate the directional derivative in *any* other direction $\mathbf{u}$ by first rotating the axes so that the new x-axis, for example, is in the direction of the vector $\mathbf{u}$. The required directional derivative is then the ordinary partial derivative with respect to the new x-co-ordinate.)

Using the result 11.1 we have

$$f(a+hu_1, b+hu_2, c+hu_3) - f(a,b,c) \approx \frac{\partial f}{\partial x}hu_1 + \frac{\partial f}{\partial y}hu_2 + \frac{\partial f}{\partial z}hu_3.$$

That is

$$\frac{f(a+hu_1, b+hu_2, c+hu_3) - f(a,b,c)}{h} \approx \frac{\partial f}{\partial x}u_1 + \frac{\partial f}{\partial y}u_2 + \frac{\partial f}{\partial z}u_3$$

for small h so that, taking the limit as $h \to 0$, we get

$$\frac{df}{du} = \frac{\partial f}{\partial x}u_1 + \frac{\partial f}{\partial y}u_2 + \frac{\partial f}{\partial z}u_3.$$

It is convenient to identify this result as the following dot product:

$$\begin{aligned}\frac{df}{du} &= \left(\frac{\partial f}{\partial x}, \frac{\partial f}{\partial y}, \frac{\partial f}{\partial z}\right) \cdot (u_1, u_2, u_3) \\ &= \left(\frac{\partial f}{\partial x}, \frac{\partial f}{\partial y}, \frac{\partial f}{\partial z}\right) \cdot \mathbf{u}.\end{aligned}$$

We often use the notation

$$\boxed{\nabla f = \left(\frac{\partial f}{\partial x}, \frac{\partial f}{\partial y}, \frac{\partial f}{\partial z}\right) = \frac{\partial f}{\partial x}\mathbf{i} + \frac{\partial f}{\partial y}\mathbf{j} + \frac{\partial f}{\partial z}\mathbf{k}}$$

so that the directional derivative of f in the direction of the **unit vector u** is given by

$$\boxed{\frac{df}{du} = \nabla f \cdot \mathbf{u}}$$

(Note that the operator ∇ transforms a scalar function of position into a vector with components $\frac{\partial f}{\partial x}$, $\frac{\partial f}{\partial y}$ and $\frac{\partial f}{\partial z}$. For example, if $f(x,y,z) = xyz$ then

$$\nabla f = (f_x, f_y, f_z) = (yz, xz, xy) = yz\,\mathbf{i} + xz\,\mathbf{j} + xy\,\mathbf{k}.$$

In particular, at the point $P = (1,2,3)$ we have $\nabla f = (6,3,2) = 6\,\mathbf{i} + 3\,\mathbf{j} + 2\,\mathbf{k}$.)

Thus

$$\boxed{\frac{df}{du} = \nabla f \cdot \mathbf{u} = |\nabla f| \cos\phi}$$

where ϕ is the angle between the direction of the unit vector $\mathbf{u}$ and the vector ∇f. Note that the directional derivative $\frac{df}{du}$ is a maximum when $\cos\phi = 1$; that is, $\mathbf{u}$ is parallel to the vector $\nabla f = f_x\,\mathbf{i} + f_y\,\mathbf{j} + f_z\,\mathbf{k}$. In this case

$$\frac{df}{du} = |\nabla f| = \sqrt{f_x^2 + f_y^2 + f_z^2}.$$

That is,

> The gradient ∇f of a function f is in the direction in which f has its **maximum rate of increase**. Also the value of this maximum rate of increase is $|\nabla f|$.
>
> The direction in which f decreases most rapidly is in the direction of the vector $-\nabla f$.

The vector ∇f is called the **gradient** of the function f and called grad$f$. (For a motivation of this term see Worked Example 11.8.3.)

Worked Example 11.8.1 *If $f(x, y, z) = x^2 + y^2 + z^2$, determine the directional derivative of f in the direction of the vector $\mathbf{v} = \mathbf{i} - \mathbf{j} + 2\,\mathbf{k}$ at the point $P(2, -1, 3)$.*
Here $\nabla f = (2x, 2y, 2z) = 2x\,\mathbf{i} + 2y\,\mathbf{j} + 2z\,\mathbf{k}$.
Therefore, the required directional derivative is given by

$$\begin{aligned} \frac{df}{du} &= \nabla f \cdot \hat{\mathbf{v}}|_P \\ &= (2x, 2y, 2z) \cdot \frac{(1,-1,2)}{\sqrt{6}}\Big|_P \\ &= \tfrac{1}{\sqrt{6}}(4, -2, 6) \cdot (1, -1, 2) \quad = \quad \tfrac{1}{\sqrt{6}}18 = 3\sqrt{6}. \end{aligned}$$

■

Worked Example 11.8.2 *The temperature T at each point of a rectangular plate is given by*

$$T(x, y) = 3x^2 + 7y^2.$$

1. *Determine the rate of change of temperature at the point $P(2, 5)$ in the direction*

   (a) $\mathbf{v} = (1, -1) = \mathbf{i} - \mathbf{j}$ and

   (b) making an angle $\pi/4$ with the positive x-axis.

2. *Find the direction for which the rate of change of temperature at P is a maximum and give the maximum rate of change of T at this point.*

1. (a) The directional derivative is given by

$$\begin{aligned}\frac{df}{du} &= \nabla f \cdot \hat{\mathbf{v}}|_P \\ &= (6x, 14y) \cdot \left.\frac{(1,-1)}{\sqrt{2}}\right|_P \\ &= \tfrac{1}{\sqrt{2}}(12, 70) \cdot (1, -1) \\ &= -\tfrac{1}{\sqrt{2}}58.\end{aligned}$$

(The negative sign in this answer indicates that the temperature is decreasing at this point in this direction.)

(b) In this case the required direction is given by the vector $(\cos\frac{\pi}{4}, \sin\frac{\pi}{4})$. Note that this is a *unit* vector in the required direction. Therefore

$$\begin{aligned}\frac{df}{du} &= \nabla f \cdot \mathbf{u}|_P \\ &= (6x, 14y) \cdot \left.\tfrac{1}{\sqrt{2}}(1, 1)\right|_P \\ &= \tfrac{1}{\sqrt{2}}(12, 70) \cdot (1, 1) \\ &= \tfrac{1}{\sqrt{2}}82.\end{aligned}$$

2. The maximum rate of change of T is in the direction of ∇T, which at the point P is given by

$$\nabla T = 12\,\mathbf{i} + 70\,\mathbf{j}.$$

The maximum rate of change of T at this point is then

$$|\nabla T| = \sqrt{144 + 4900} = \sqrt{5044} = 2\sqrt{1261}.$$

■

▌**Worked Example 11.8.3** *The level curves of the surface $z = f(x, y)$ are given by $f(x, y) = k$ where k is scalar constant. Show that the vector* grad$f$ *is perpendicular to the level curves.*

Using Worked Example 11.7.4, the unit tangent vector $\hat{\boldsymbol{\tau}}$ to these curves is parallel to the vector

$$(f_y, -f_x) = f_y\,\mathbf{i} - f_x\,\mathbf{j}.$$

But

$$\nabla f = f_x\,\mathbf{i} + f_y\,\mathbf{j}$$

so that

$$\nabla f \cdot (f_y, -f_x) = f_x f_y - f_y f_x = 0.$$

Therefore the vector $\nabla f = f_x\,\mathbf{i} + f_y\,\mathbf{j}$ is perpendicular to the level curves. ■

(Because the direction perpendicular to the contour lines gives the direction in which the "slope" or "gradient" of the surface is greatest, it is natural to use the term "gradient" as a synonym for the vector ∇f.)

Worked Example 11.8.4 *The electric potential V at a point $P(x, y, z)$ in three-dimensional space is given by*

$$V = \frac{1}{\sqrt{x^2 + y^2 + z^2}}.$$

Determine

1. *the rate of change of V at the point $P(2, -1, 2)$ in the direction of the vector $(3, -1, 4) = 3\,\mathbf{i} - \mathbf{j} + 4\,\mathbf{k}$*
2. *the direction of the greatest rate of increase of V at the point P and the value of this greatest rate of increase.*

It is convenient to introduce the notation

$$r = \sqrt{x^2 + y^2 + z^2}$$

and treat $V = V(r)$. Thus

1. $$\begin{aligned} V_x &= \frac{dV}{dr}\frac{\partial r}{\partial x} = -\frac{1}{r^2}\frac{x}{r} &= -\frac{x}{r^3} \\ V_y &= \qquad \cdots &= -\frac{y}{r^3} \\ V_z &= \qquad \cdots &= -\frac{z}{r^3}. \end{aligned}$$

   Thus the required directional derivative is given by

   $$\begin{aligned} \frac{dV}{du} &= \nabla V \cdot \mathbf{u}|_P \\ &= \left(-\frac{x}{r^3}, -\frac{y}{r^3}, -\frac{z}{r^3}\right) \cdot \frac{(3, -1, 4)}{\sqrt{26}}\bigg|_P \\ &= \left(-\frac{2}{27}, \frac{1}{27}, -\frac{2}{27}\right) \cdot \frac{(3, -1, 4)}{\sqrt{26}} \\ &= -\frac{1}{\sqrt{26}}\frac{15}{27}. \end{aligned}$$

   Therefore at the point $P(2, -1, 2)$ the electric potential is decreasing at the rate of $-\frac{1}{\sqrt{26}}\frac{15}{27}$ in the direction of $(3, -1, 4) = 3\,\mathbf{i} - \mathbf{j} + 4\,\mathbf{k}$.

2. Here $\nabla V = \left(-\frac{2}{27}, \frac{1}{27}, -\frac{2}{27}\right)$ is parallel to the vector

$$(-2, 1, -2) = -2\,\mathbf{i} + \mathbf{j} - 2\,\mathbf{k}$$

which is the direction of the greatest increase of V at the point P. The value of this greatest increase is

$$|\nabla V| = \sqrt{\frac{4}{27^2} + \frac{1}{27^2} + \frac{4}{27^2}} = \frac{3}{27} = \frac{1}{9}.$$

■

11.9 Tangent Planes and Normals to Surfaces

Suppose S is a surface defined (implicitly) by the equation

$$F(x, y, z) = 0.$$

▌**Example:** $F(x, y, z) = (x-1)^2 + y^2 + (z+2)^2 - 16 = 0$ is the surface of a sphere centred on the point $(1, 0, -2)$ of radius 4.

(Note that a surface given explicitly, for example

$$z = 10 - x^2 - (y-2)^2$$

which represents the surface of an inverted paraboloid, can be simply written in the required form as

$$F(x, y, z) = z - 10 + x^2 + (y-2)^2.)$$

If the point $P_0(x_0, y_0, z_0)$ is on the surface then

$$F(x_0, y_0, z_0) = 0.$$

Suppose now that the curve C_0 lies on the surface S and passes through the point P_0; see Figure 11.13.

Let the curve C_0 be defined parametrically by

$$\mathbf{r}(t) = (x(t), y(t), z(t)),$$

where $r(t_0)$ represents the point P_0. Then all points on this curve lie on S providing

$$F(x(t), y(t), z(t)) = 0 \text{ for all } t.$$

Thus using the chain rule and differentiating with respect to t gives

$$\begin{aligned} 0 &= \frac{dF}{dt} \\ &= \frac{\partial F}{\partial x}\frac{dx}{dt} + \frac{\partial F}{\partial y}\frac{dy}{dt} + \frac{\partial F}{\partial z}\frac{dz}{dt} \\ &= \nabla F \cdot \frac{d\mathbf{r}}{dt}. \end{aligned}$$

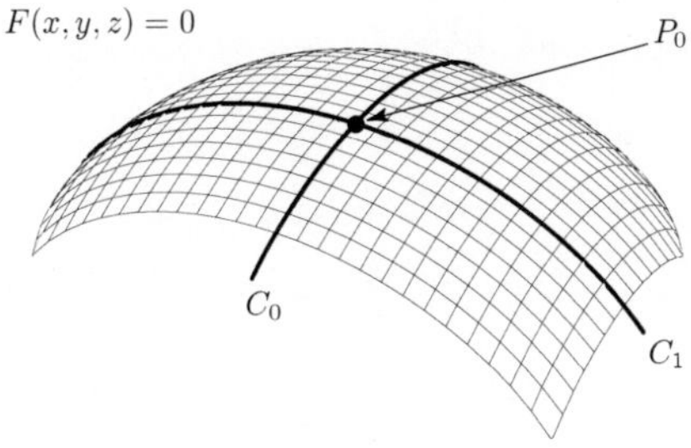

Figure 11.13: Curves on a surface.

But $\frac{d\mathbf{r}}{dt}$ is parallel to the tangent to the curve C_0, so that ∇F is perpendicular to this tangent to the curve C_0. This is true, in particular, for the point P_0.

The above argument applies to any other curve, such as C_1, that lies on the surface S that passes through the point P_0. Consequently, see Figure 11.14.

> For the surface S defined by $F(x, y, z) = 0$
>
> a **normal vector to S** is
>
> $$\nabla F = F_x\,\mathbf{i} + F_y\,\mathbf{j} + F_z\,\mathbf{k}.$$

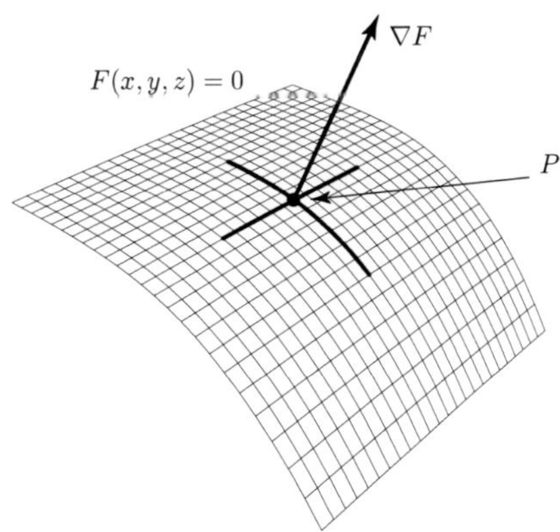

Figure 11.14: Normal to a surface.

The *normal line* to a surface S at a point P is the straight line passing through P in the direction parallel to the normal to S. The *tangent plane* to a surface S at a point P is the plane passing through P with normal parallel to the normal of S.

▌ **Worked Example 11.9.1** *Determine the equation of the normal line and tangent planes to the elliptical paraboloid*

$$z = 4x^2 + y^2 - 26$$

at the point $P(2, 4, 6)$.

The surface is given by

$$F(x, y, z) = 4x^2 + y^2 - z - 26 = 0.$$

Thus the normal to S is parallel to the vector

$$\nabla F = 8x\,\mathbf{i} + 2y\,\mathbf{j} - \mathbf{k} = 16\,\mathbf{i} + 8\,\mathbf{j} - \mathbf{k} \text{ at the point } P.$$

Therefore the equation of the normal line (in parametric form) is

$$\begin{aligned} x &= 2 + 16t \\ y &= 4 + 8t \\ z &= 6 - t. \end{aligned}$$

The equation of the tangent plane is

$$16x + 8y - z = d = 16(2) + 8(4) - (6) = 58.$$

■

11.10 Double Integrals

Consider a function f which is defined on a closed bounded region R of the x, y-plane. Select any partition of R into n sub-regions having area ΔA_i (for $1 \le i \le n$) and let (x_i, y_i) be an arbitrary point in the ith sub-region.

If the Riemann sum $\sum_{i=1}^{n} f(x_i, y_i)\Delta A_i$ approaches a finite limit as the dimensions of the sub-region tend to 0 (and $n \to \infty$), independently of how R is partitioned and how the points (x_i, y_i) are chosen, the limit is called the double integral of $f(x, y)$ over the region R. We write this double integral as

$$\iint_R f(x, y)\, dA.$$

(Note the similarity with the definition of the Riemann integral in, for example, [4, Topic 5].)

It can be shown that the double integral exists if $f(x, y)$ is continuous on R.

If R is partitioned into rectangles using lines parallel to the x- and y-axis (see Figure 11.15), then

$$\Delta A_i = \Delta x_i \Delta y_i$$

and we denote the double integral by

$$\iint_R f(x,y)\,dA = \iint_R f(x,y)\,dx\;dy.$$

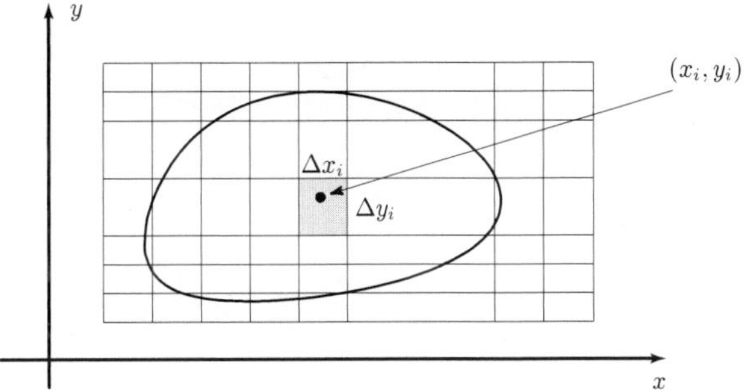

Figure 11.15: Partition of R into rectangles parallel to the axis.

Note that if

- $f(x,y) = 1$ then $\iint_R f(x,y)\,dx\;dy$ gives the area of the region R
- $f(x,y)$ represents the density of the region R then $\iint_R f(x,y)\,dx\;dy$ gives the mass of the region R
- $f(x,y) \geq 0$ for all (x,y) in R then $\iint_R f(x,y)\,dx\;dy$ gives the volume of the solid beneath the surface $z = f(x,y)$ and above the region R. (See Figure 11.16 where $f(x_i, y_i)\Delta A_i$ gives the volume of the rectangular parallelepiped whose base area is ΔA_i and whose altitude is $f(x_i, y_i)$.)

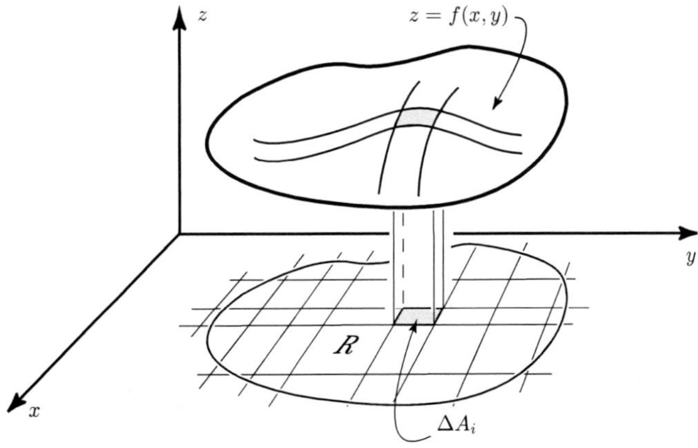

Figure 11.16: Volume of solid beneath $z = f(x,y)$ and above the region R.

In practice we evaluate double integrals by writing them as iterated integrals of the form

1. $\displaystyle\int \left(\int f(x,y)\,dy \right) dx$ with suitable limits

2. $\displaystyle\int \left(\int f(x,y)\,dx \right) dy$ with suitable limits.

The inner integrals in general will have limits which depend on the outer integration variable in each case. We consider these two cases in more detail.

Type 1 Region: Here R is described by

$$R = \{(x,y) : g_1(x) \leq y \leq g_2(x), a \leq x \leq b\}.$$

With reference to Figure 11.17, the summation over R is first carried out over strips parallel to the y-axis, followed by a summation over all such strips.

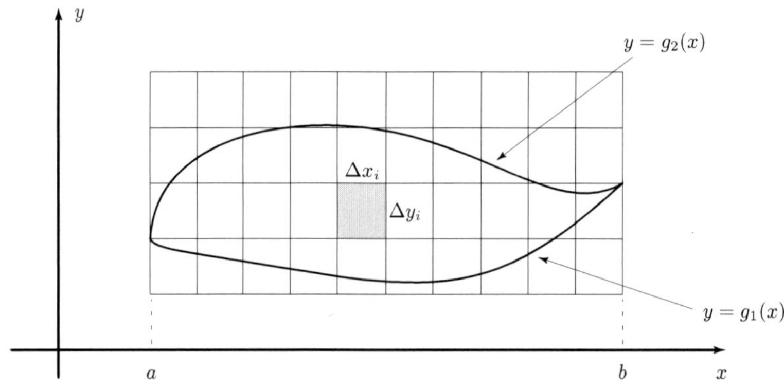

Figure 11.17: A region R of the first type.

That is

$$\int_{y=g_{\cdot}(x)}^{y=g_{\cdot}(x)} f(x,y)\,dy$$

is evaluated first, treating x as a constant. The resulting function of x is then integrated with respect to x over $[a,b]$. This gives the repeated or iterated integral

$$\int_{x=a}^{x=b} \left[\int_{y=g_{\cdot}(x)}^{y=g_{\cdot}(x)} f(x,y)\,dy \right] dx$$

which is often written as

$$\int_a^b \int_{g_{\cdot}(x)}^{g_{\cdot}(x)} f(x,y)\,dy\,dx = \int_a^b dx \int_{g_{\cdot}(x)}^{g_{\cdot}(x)} f(x,y)\,dy.$$

For example, with reference to Figure 11.18

$$\begin{aligned}\int_1^2\int_x^{x^{\cdot}}(2x+y)\,dy\,dx &= \int_{x=1}^{x=2}\left[\int_{y=x}^{y=x^{\cdot}}(2x+y)\,dy\right]dx\\ &= \int_{x=1}^{x=2}\left[2xy+\frac{1}{2}y^2\right]_{y=x}^{y=x^{\cdot}}dx\\ &= \int_1^2\left(2x^3+\frac{1}{2}x^4-2x^2-\frac{1}{2}x^2\right)dx\\ &= \tfrac{143}{30}.\end{aligned}$$

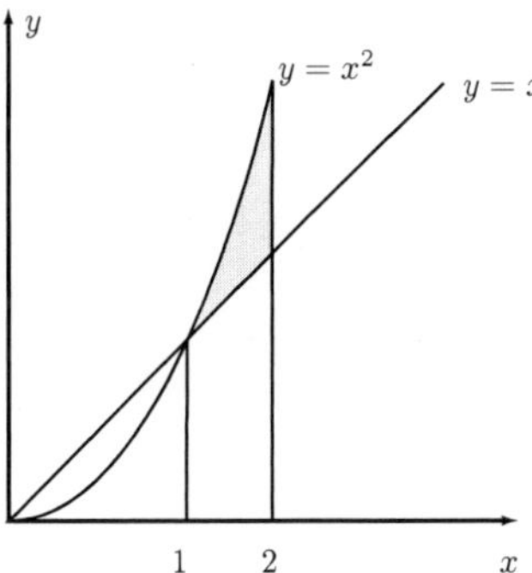

Figure 11.18: Example of a first type region.

Self-help exercise

Evaluate the double integral

$$\iint_R (1+x+y)\,dx\,dy$$

where R is the region in the first octant bounded by the straight lines $y=\frac{1}{2}x$, $y=x$ and $x=2$. $[\frac{10}{3}]$

Type 2 Region: Here R is described by

$$R=\{(x,y): h_1(y)\le x\le h_2(y), c\le y\le d\}.$$

With reference to Figure 11.19, the summation over R is first carried out over strips parallel to the x-axis, followed by a summation over all such strips.

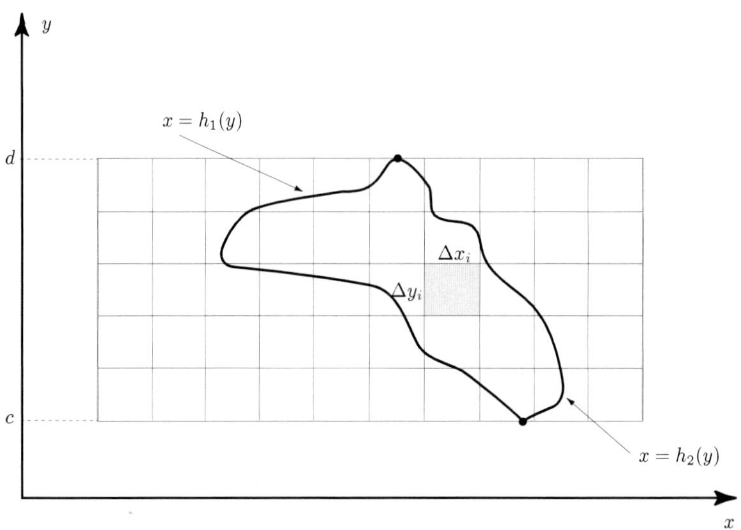

Figure 11.19: A region R of the second type.

That is

$$\int_{x=h_{\bullet}(y)}^{x=h_{\bullet}(y)} f(x,y)\,dx$$

is evaluated first, treating y as a constant. The resulting function of y is then integrated over $[c,d]$. This gives the repeated or iterated integral

$$\int_{y=c}^{y=d}\left[\int_{x=h_{\bullet}(y)}^{x=h_{\bullet}(y)} f(x,y)\,dx\right]dy$$

which is often written as

$$\int_c^d\int_{h_{\bullet}(y)}^{h_{\bullet}(y)} f(x,y)\,dx\,dy = \int_c^d dy\int_{h_{\bullet}(y)}^{h_{\bullet}(y)} f(x,y)\,dx.$$

For example, with reference to Figure 11.20,

$$\begin{aligned}\int_0^1\int_{-1}^{1+y}(2x+y)\,dx\,dy &= \int_{y=0}^{y=1}\left[\int_{x=-1}^{x=1+y}(2x+y)\,dx\right]dy\\ &= \int_0^1\left[x^2+xy\right]_{x=-1}^{x=1+y}dy\\ &= \int_{y=0}^{y=1}\left[(1+y)^2+y(1+y)-1+y\right]dy\\ &= \tfrac{8}{3}.\end{aligned}$$

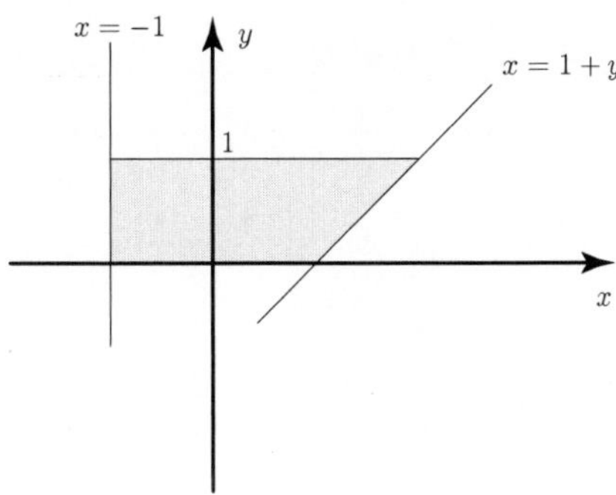

Figure 11.20: Example of a second type region.

Self-help exercise

Evaluate the double integral

$$\iint_R (1 + x + y)\, dx\, dy$$

where R is the region in the first octant bounded by the curves $y = x-1$ and $y = 3$.

$[\frac{63}{2}]$

(What would you need to do if you treated this as a region of type 1?)

[Express as the sum of two integrals]

Frequently the region R is neither one of the above types but can be expressed as the union of such regions. For example, in Figure 11.21, $R = R_1 \cup R_2$ and

$$\iint_R f(x, y)\, dx\, dy = \iint_{R_\bullet} f(x, y)\, dx\, dy + \iint_{R_\bullet} f(x, y)\, dx\, dy.$$

When R is a simple region, the double integral $\iint_R f(x, y)\, dA$ can often be evaluated using either of the two possible forms of the iterated integral

$$\int_a^b \int_{g_\bullet(x)}^{g_\bullet(x)} f(x, y)\, dy\, dx \text{ or } \int_c^d \int_{h_\bullet(y)}^{h_\bullet(y)} f(x, y)\, dx\, dy.$$

Which iterated integral is "best" to use depends on the region R and the integrand $f(x, y)$; often one of them is more easily evaluated than the other — see the following worked examples.

▮ **Worked Example 11.10.1** *Evaluate*

$$I = \int_0^1 \int_x^{\sqrt{x}} (x^2 + y^2)\, dy\, dx.$$

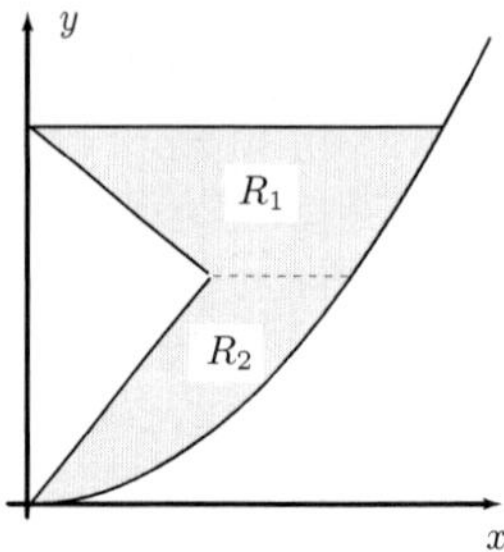

Figure 11.21: A non-simple region expressible as the union of simple regions.

Sketch the region of integration and evaluate the iterated integral with the order of the integration reversed.

$$\begin{aligned}
\int_0^1 \int_x^{\sqrt{x}} (x^2 + y^2)\, dy\, dx &= \int_0^1 \left[\int_x^{\sqrt{x}} (x^2 + y^2)\, dy \right] dx \\
&= \int_0^1 \left[x^2 y + \frac{1}{3} y^3 \right]_{y=x}^{y=\sqrt{x}} dx \\
&= \int_0^1 \left(x^{\div} + \frac{1}{3} x^{\div} - x^3 - \frac{1}{3} x^3 \right) \\
&= \frac{2}{7} + \frac{2}{15} - \frac{1}{3} \quad = \quad \frac{3}{35}.
\end{aligned}$$

The sketch of the region is given in Figure 11.22.

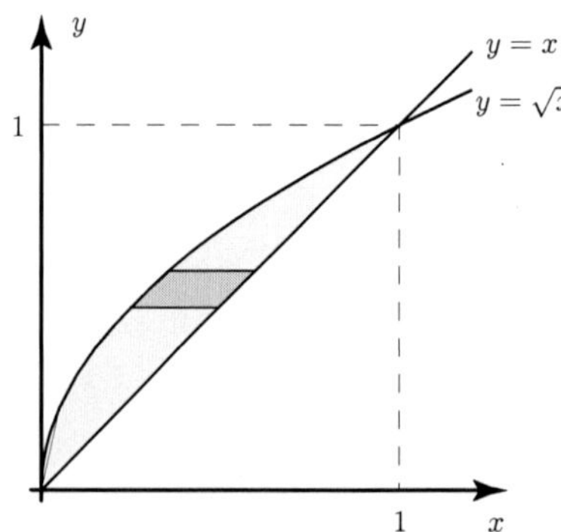

Figure 11.22: The region R for Worked Example 11.10.1.

Reversing the order of integration gives

$$I = \int_0^1 \int_x^{\sqrt{x}} (x^2 + y^2)\, dy\, dx = \int_0^1 \int_{y^{\cdot}}^{y} (x^2 + y^2)\, dx\, dy$$

where on the right-hand side x varies from $x = y^2$ to $x = y$ for each value of y between 0 and 1. Thus

$$\begin{aligned} I &= \int_0^1 \int_x^{\sqrt{x}} (x^2 + y^2)\, dy\, dx \\ &= \int_0^1 \int_{y^{\cdot}}^{y} (x^2 + y^2)\, dx\, dy \\ &= \int_{y=0}^{y=1} \left[\int_{x=y^{\cdot}}^{x=y} (x^2 + y^2)\, dx \right] dy \\ &= \int_0^1 \left[\frac{1}{3}x^3 + xy^2 \right]_{x=y^{\cdot}}^{x=y} dy \\ &= \int_0^1 \left(\frac{y^3}{3} + y^3 - \frac{y^6}{3} - y^4 \right) dy \\ &= \tfrac{3}{35}. \end{aligned}$$

■

▮ **Worked Example 11.10.2** *By first reversing the order of integration, evaluate*

$$J = \int_0^1 \int_y^1 e^{x^{\cdot}}\, dx\, dy.$$

(Since $e^{x^{\cdot}}$ does not have an antiderivative in the class of elementary functions, it seems worthwhile to experiment with writing the integral with the order of integration reversed. Here the resulting integral can be completely evaluated without recourse to higher transcendental functions.) Reversing the order of integration (see Figure 11.23) gives

$$\begin{aligned} J &= \int_0^1 \int_y^1 e^{x^{\cdot}}\, dx\, dy \\ &= \int_{x=0}^{x=1} \left[\int_{y=0}^{y=x} e^{x^{\cdot}}\, dy \right] dx \\ &= \int_{x=0}^{x=1} \left[y e^{x^{\cdot}} \right]_{y=0}^{y=x} dx \\ &= \int_{x=0}^{x=1} x e^{x^{\cdot}}\, dx \\ &= \left[\frac{1}{2} e^{x^{\cdot}} \right]_0^1 \;=\; \frac{1}{2}(e - 1). \end{aligned}$$

■

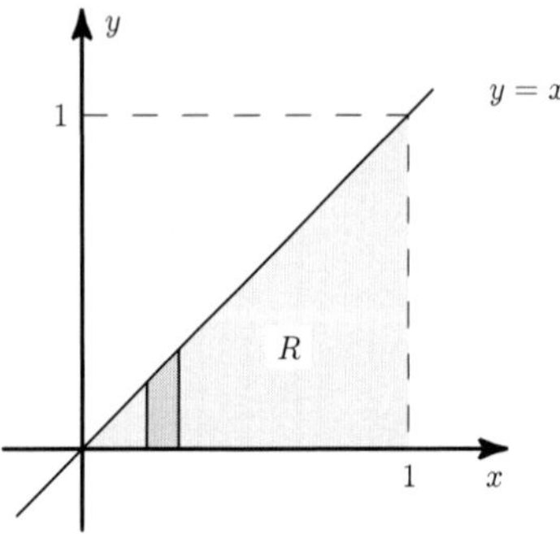

Figure 11.23: The region R for Worked Example 11.10.2.

Worked Example 11.10.3 *Determine the volume of the region in the first octant bounded by the elliptic cylinder* $16x^2 + 9y^2 = 144$*, the plane* $16x + 9y - 12z = 0$ *and the co-ordinate planes.*

With reference to Figure 11.24 the volume V of the region is given by

$$V = \iint_R z\,dx\,dy = \iint_R \frac{1}{12}(16x + 9y)\,dx\,dy$$

where the region R is shown in Figure 11.25.

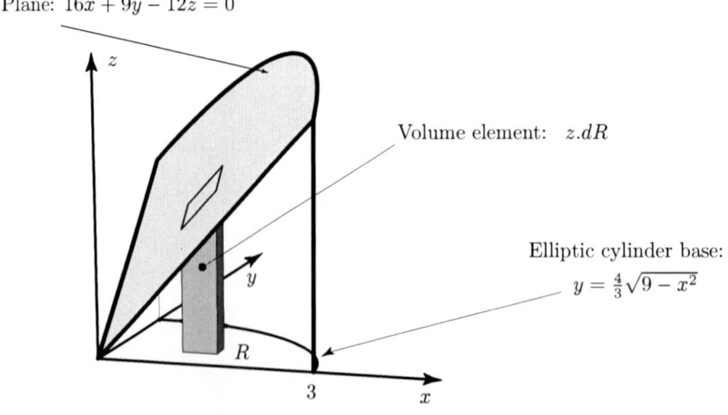

Figure 11.24: The volume V for Worked Example 11.10.3.

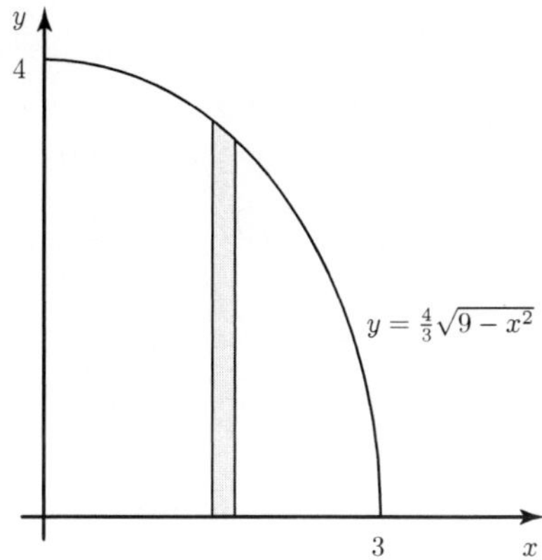

Figure 11.25: The region R for Worked Example 11.10.3.

Writing the volume as an iterated integral gives

$$\begin{aligned}
V &= \frac{1}{12}\int_{x=0}^{x=3}\int_{y=0}^{y=\frac{4}{3}\sqrt{9-x^2}} (16x+9y)\,dy\,dx \\
&= \frac{1}{12}\int_0^3 \left[16xy + \frac{9}{2}y^2\right]_{y=0}^{y=\frac{4}{3}\sqrt{9-x^2}} dx \\
&= \frac{1}{12}\int_0^3 \left[\frac{1}{3}64x\sqrt{9-x^2} + \frac{9}{2}\cdot\frac{16}{9}(9-x^2)\right] dx \\
&= \frac{1}{12}\left[-\frac{64}{9}(9-x^2)^{\frac{3}{2}} + 8(9x - \frac{1}{3}x^3)\right]_0^3 \\
&= 28.
\end{aligned}$$

■

11.11 Extreme Values

In this section we discuss the determination of extreme values of functions of two real variables. That is, we determine the points where a function of two real variables takes its maximum and minimum values.

11.11.1 Introduction

As for a function of one real variable, the point (a, b) is a local **maximum** of $f(x, y)$ if

$$f(x,y) \le f(a,b) \text{ for all points } (x,y) \text{ near } (a,b).$$

The point (a, b) is a local **minimum** if

$$f(x, y) \geq f(a, b) \text{ for all points } (x, y) \text{ near } (a, b).$$

▌Example: The point $(1, -3)$ is a minimum for the function

$$f(x, y) = x^2 - 2x + y^2 + 6y + 10$$

because

$$f(x, y) = (x - 1)^2 + (y + 3)^2 \geq 0 = f(1, -3).$$

11.11.2 Local Extreme Values

In this course we consider only those functions $f(x, y)$ of two variables that have a tangent plane at all points on their surface $z = f(x, y)$. Thus both f_x and f_y exist at all points in their domain. For such functions, local extremes will occur at points on the surface where the tangent plane is horizontal.

To obtain a simple test for such points along the lines available in the one-dimensional case, consider the surface $z = f(x, y)$ corresponding to the function $f(x, y)$ whose extreme points are sought. The tangent plane at a point (a, b) has a normal parallel to the vector $(f_x(a, b), f_y(a, b), -1)$. Consequently, we have a horizontal tangent plane providing both f_x and f_y are zero at the point (a, b). That is,

Extreme points of the function $f(x, y)$ occur where $f_x = f_y = 0$.

Geometrically, extreme points can be either local maxima, local minima or saddle points. For example, for the function $f(x, y) = xye^{-(x^2+y^2)}$ (see Figure 11.1), a local maximum occurs at the point $(\frac{1}{\sqrt{2}}, \frac{1}{\sqrt{2}})$ (and at $(-\frac{1}{\sqrt{2}}, -\frac{1}{\sqrt{2}})$), a local minimum at the point $(\frac{1}{\sqrt{2}}, -\frac{1}{\sqrt{2}})$ (and at $(-\frac{1}{\sqrt{2}}, \frac{1}{\sqrt{2}})$) and a saddle point at $(0, 0)$.

▌Worked Example 11.11.1 *Determine the extreme points for the function*

$$f(x, y) = 27x - 3y - x^3 + y^3.$$

The extreme points are given by $f_x = f_y = 0$.
That is, $27 - 3x^2 = 0$ and $-3 + 3y^2 = 0$.
Thus $x = \pm 3$ and $y = \pm 1$.
Therefore the extreme points are $(-3, -1)$, $(-3, 1)$, $(3, -1)$ and $(3, 1)$.
(It can be shown that $(-3, 1)$ is a minimum, $(3, -1)$ is a maximum and the two points $(-3, -1)$ and $(3, 1)$ are saddle points; such points generalize the one-dimensional case of a point of inflection.) ■

Worked Example 11.11.2 *Determine all extreme points for the function*

$$f(x,y) = x^2 + 2xy + y^2.$$

Extreme points occur where $f_x = f_y = 0$; that is, where $2x+2y = 0$ and $2x + 2y = 0$.
Hence all points satisfying $y = -x$ are extreme points for $f(x,y)$. (In fact $f(x,y) = (x+y)^2$ so that all points satisfying $y = -x$ are points that minimize $f(x,y)$.) ∎

Worked Example 11.11.3 *Determine all extreme points for the function*

$$f(x,y) = x^2 + 3x^2y - y^3.$$

Extreme points occur where $f_x = f_y = 0$; that is, where $2x + 6xy = 0$ and $3x^2 - 3y^2 = 0$.
The solutions of these two equations are $x = 0, y = 0$ and $x = \pm\frac{1}{3}, y = -\frac{1}{3}$.
Therefore, the extreme points are $(0,0)$ and $(\pm\frac{1}{3}, -\frac{1}{3})$.
(In fact all these points are saddle points for $f(x,y)$.) ∎

Although a test involving the second partial derivatives does exist to help decide whether a given extreme point is a maximum, minimum or saddle point, we do not derive this result here. Usually the region on which an extreme value is sought is a finite bounded region and the method developed below decides at which points the maximum and minimum are attained.

11.11.3 Global Extreme Values

In most applications the extreme values of a function are required in some specified region of the x,y-plane. Such extreme values can only occur at local extreme points (as given by $f_x = f_y = 0$) or at points on the boundary of the region.

(This is exactly the same situation that occurs for functions of one real variable — global extrema on a closed interval $[a,b]$ occur either at points of local extremum or at the endpoints of the interval.)

Worked Example 11.11.4 *Determine both the global maximum and the global minimum for the function*

$$f(x,y) = 2 + 2x + 2y - x^2 - y^2$$

for all (x,y) in the square $0 \leq x \leq 9$ and $0 \leq y \leq 9$.
Local extreme points occur where $f_x = f_y = 0$; that is, where $2 - 2x = 0$ and $2 - 2y = 0$. Hence the only local extremum occurs at $(1,1)$. Also $f(1,1) = 4$.

The boundary consists of the four portions:

$$x = 0 \quad 0 \le y \le 9 \qquad x = 9 \quad 0 \le y \le 9$$
$$y = 9 \quad 0 \le x \le 9 \qquad y = 0 \quad 0 \le x \le 9.$$

On the first portion we have $g_1(y) = f(0, y) = 2 + 2y - y^2$. For the function g_1 we require the global maximum on the closed interval $[0, 9]$. The local extremum will occur where $g_1'(y) = 2 - 2y = 0$. That is, $y = 1$. By the second derivative test, $y = 1$ is a local maximum with $g_1(1) = f(0, 1) = 3$. Also, at the endpoints of the interval $y = 0$ and $y = 9$, we have $g_1(0) = f(0, 0) = 2$ and $g_1(9) = f(0, 9) = -61$.
On the second portion we have $g_2(y) = f(9, y) = -61 + 2y - y^2$. For the function g_2 we require the global maximum on the closed interval $[0, 9]$. The local extremum will occur where $g_2'(y) = 2 - 2y = 0$. That is, $y = 1$. By the second derivative test, $y = 1$ is a local maximum with $g_2(1) = f(9, 1) = -60$. Also, at the endpoints of the interval $y = 0$ and $y = 9$, we have $g_2(0) = f(9, 0) = -61$ and $g_2(9) = f(9, 9) = -124$.
On the third portion we have $g_3(x) = f(x, 9) = -61 + 2x - x^2$. For the function g_3 we require the global maximum on the closed interval $[0, 9]$. The local extremum will occur where $g_3'(x) = 2 - 2x = 0$. That is, $x = 1$. By the second derivative test, $x = 1$ is a local maximum with $g_3(1) = f(1, 9)) = -60$. Also, at the endpoints of the interval $x = 0$ and $x = 9$, we have $g_3(0) = f(0, 9) = -61$ and $g_3(9) = f(9, 9) = -124$.
On the fourth portion we have $g_4(x) = f(x, 0) = 2 + 2x - x^2$. For the function g_4 we require the global maximum on the closed interval $[0, 9]$. The local extremum will occur where $g_4'(x) = 2 - 2x = 0$. That is, $x = 1$. By the second derivative test, $x = 1$ is a local maximum with $g_4(1) = f(1, 0) = 3$. Also, at the endpoints of the interval $x = 0$ and $x = 9$, we have $g_4(0) = f(0, 0) = 2$ and $g_4(9) = f(9, 0) = -61$.

Thus, in summary, the global maximum occurs at $(1, 1)$ with a value of 4 and a global minimum of -124 at $(9, 9)$. ■

In this topic we have

- Defined level curves and discussed how they allow vizualisation of surfaces.
- Introduced partial derivatives and their geometrical interpretation.

- Discussed the various forms of the chain rule.
- Evaluated directional derivative.
- Introduced the gradient, normal direction to a surface, tangent planes and normal lines.
- Discussed cylindrical and spherical polar coordinates.
- Developed linear approximations and applied these to estimate errors.
- Introduced double integrals and interpreted these as representing areas, masses and volumes.
- Discussed local and global extreme values.

11.12 Quick Test Number 11

Surfaces

Question	Selection
Identify the surfaces defined by the following equations:	
1. $z = x^2 + y^2$ is a	(a) cone (b) cylinder (c) plane (d) paraboloid (e) circle
2. $z = \sqrt{x^2 + y^2}$ is a	(a) cone (b) cylinder (c) plane (d) paraboloid (e) circle
3. $x^2 + y^2 = 4,\ 0 \leq z \leq 3$ is a	(a) cone (b) cylinder (c) plane (d) paraboloid (e) circle
4. $z = 2x + 3y$ is a	(a) cone (b) cylinder (c) plane (d) paraboloid (e) circle

Partial Derivatives

Question	**Selection**
Find $\dfrac{\partial z}{\partial x}$ for the following functions:	
1. $z = 2x + 4y + 9$	(a) $2 + 4y' + 9$ (b) $2 + 4$ (c) 2
2. $z = 2x^2 - 3y^2 - 5x + 7$	(a) $4x - 6y - 5$ (b) $4x - 5$ (c) $4x - 5 + 7$
3. $z = xy^2 - 3x + 8$	(a) $2xy^2 - 3$ (b) $y^2 - 3$ (c) $y^2 + 2xyy' - 3$
4. $z = \mathrm{e}^{(x^{\cdot} + y^{\cdot})}$	(a) $2x\,\mathrm{e}^{(x^{\cdot} + y^{\cdot})}$ (b) $\mathrm{e}^{(x^{\cdot} + y^{\cdot})}$ (c) $(2x + 2y)\,\mathrm{e}^{(x^{\cdot} + y^{\cdot})}$
5. $z = \sin xy$	(a) $x \sin xy$ (b) $y \sin xy$ (c) $x \cos xy$ (d) $y \cos xy$
6. $z = y/x$	(a) y (b) $-y/x^2$ (c) $1/x$ (d) $-1/x^2$
7. $z = 1/(xy)$	(a) $-1/(x^2y)$ (b) $1/(y)$ (c) $-1/(xy)^2$

Gradient and Directional Derivative

Question	**Selection**
In the following, $F(x, y, z) = x^2 + 2y^2 + 3z^2 - 19$.	
1. ∇F is a	(a) scalar (b) vector
2. ∇F is defined by	(a) $\dfrac{\partial F}{\partial x} + \dfrac{\partial F}{\partial y} + \dfrac{\partial F}{\partial z}$ (b) $\dfrac{\partial F}{\partial x}\mathbf{i} + \dfrac{\partial F}{\partial y}\mathbf{j} + \dfrac{\partial F}{\partial z}\mathbf{k}$
3. ∇F equals	(a) $2x + 4y + 6z$ (b) $2x\mathbf{i} + 4y\mathbf{j} + 6z\mathbf{k}$ (c) $2\mathbf{i} + 4\mathbf{j} + 6\mathbf{k}$
4. What is ∇F at the point $(1, 2, 3)$?	(a) 28 (b) $2\mathbf{i} + 8\mathbf{j} + 18\mathbf{k}$ (c) $2\mathbf{i} + 4\mathbf{j} + 6\mathbf{k}$
5. The directional derivative $\dfrac{dF}{du}$ is a	(a) scalar (b) vector
6. The directional derivative $\dfrac{dF}{du}$ at $(1, 2, 3)$ in the direction of $\mathbf{i} + 2\mathbf{j} + 2\mathbf{k}$ is	(a) 18 (b) 54 (c) 22 (d) $28\mathbf{i} + 56\mathbf{j} + 56\mathbf{k}$

Normal and Tangent Plane

Question **Selection**

In the following, S is the surface $z = 4x^2 + 3y^2 - 11$.

1. This means that $S(x, y, z) = 4x^2 + 3y^2 - 11 - z$ (a) True (b) False

[The correct answer to Q1 is (b) False. S is just a name for the surface!]

Let $F(x, y, z) = 4x^2 + 3y^2 - 11 - z$.

2. ∇F is a (a) scalar (b) vector
3. ∇F equals (a) $8x + 6y - 1$ (b) $8x\mathbf{i} + 6y\mathbf{j} - \mathbf{k}$ (c) $8\mathbf{i} + 6\mathbf{j} - \mathbf{k}$
4. ∇F is normal to S. (a) True (b) False
5. A normal to S at the point $(1, 2, 5)$ is

   (a) $8\mathbf{i} + 12\mathbf{j} - 5\mathbf{k}$ (b) $8\mathbf{i} + 6\mathbf{j} - \mathbf{k}$ (c) $8\mathbf{i} + 12\mathbf{j} - \mathbf{k}$
6. The equation of the tangent plane to S at $(1, 2, 5)$ is

   (a) $8x + 12y - 5z = 7$ (b) $8x + 6y - z = 15$

   (c) $8x + 12y - z = 27$ (d) $8\mathbf{i} + 12\mathbf{j} - \mathbf{k} = 27$

Double Integrals

Question **Selection**

1. A region R is bounded by the lines $y = 0$, $x = 1$ and $y = x$.

   $\int\int_R f(x, y)\, dA$ equals

   (a) $\int_0^1 \left\{ \int_0^1 f(x, y)\, dy \right\} dx$ (b) $\int_0^1 \left\{ \int_0^x f(x, y)\, dy \right\} dx$

   (c) $\int_y^1 \left\{ \int_0^1 f(x, y)\, dy \right\} dx$
2. Sketch the region R of integration for the double integral

   $$\int_0^1 \left\{ \int_{2x}^2 f(x, y)\, dy \right\} dx.$$

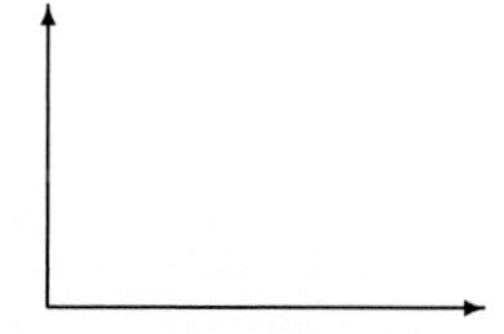

3. On reversing the order of integration, that is by integrating first with respect to x, the integral becomes

(a) $\int_{2x}^{2}\left\{\int_0^1 f(x,y)\,dx\right\}dy$ (b) $\int_0^2\left\{\int_0^{y/2} f(x,y)\,dx\right\}dy$

(c) $\int_0^1\left\{\int_0^2 f(x,y)\,dx\right\}dy$

4. $\int_0^1\left\{\int_0^x (2xy+1)\,dy\right\}dx$ equals

(a) $\int_0^1 \Big[2xy+1\Big]_0^x\,dx$ (b) $\int_0^1 \Big[xy^2+y\Big]_0^x\,dx$

(c) $\int_0^1 \Big[xy^2+1\Big]_0^x\,dx$ (d) $\int_0^x \Big[x^2y+x\Big]_0^1\,dy$

11.13 Exercises

Before attempting any of these miscellaneous exercises, make sure that you have successfully answered all the self-help exercises appearing throughout the chapter.

1. Find $\frac{\partial z}{\partial x}$ and $\frac{\partial z}{\partial y}$ when

   (a) $z = 3x^2 + y^2$
   (b) $z = 2xy - 3y + x^2$
   (c) $z = x^2 \cos xy$
   (d) $z = \frac{x-y}{x+y}$
   (e) $z = \sin(x + 2y)$
   (f) $z = \arctan \frac{y}{x}$
   (g) $z = \frac{xy}{x+y}$
   (h) $z = e^{xy} \sin x$
   (i) $z = x^y$.

2. Find $\frac{\partial^2 z}{\partial x^2}$, $\frac{\partial^2 z}{\partial y^2}$ and verify that

$$\frac{\partial^2 z}{\partial x \partial y} = \frac{\partial^2 z}{\partial y \partial x}$$

   when

   (a) $z = x^3 + 2x^2y + 3xy^2 + 4y^3$
   (b) $z = xy \log(x + 2y)$
   (c) $z = e^x \sin y$.

3. (a) If $u = \frac{xy}{x+y}$ show that

$$x\frac{\partial u}{\partial x} + y\frac{\partial u}{\partial y} = u.$$

   (b) The molecular concentration of a liquid is given by

$$c(x,t) = \frac{1}{\sqrt{t}} e^{-x^2/(4t)}.$$

   Verify that $c(x,t)$ satisfies the diffusion equation

$$\frac{\partial^2 c}{\partial x^2} = \frac{\partial c}{\partial t}.$$

4. If $V = \frac{\cos\theta}{r^2}$ evaluate the determinant

$$\begin{vmatrix} r\dfrac{\partial V}{\partial r} & -\dfrac{\partial V}{\partial \theta} \\ 2\sin\theta & \cos\theta \end{vmatrix}$$

5. Show that the function

$$z = e^y(y\cos x - x\sin x)$$

satisfies Laplace's equation

$$\frac{\partial^2 z}{\partial x^2} + \frac{\partial^2 z}{\partial y^2} = 0.$$

6. If

$$z = \frac{\cos\theta}{r^2}$$

find

(a) $\dfrac{\partial z}{\partial \theta}$ (b) $\dfrac{\partial z}{\partial r}$.

Hence verify that

(c) $\dfrac{\partial}{\partial r}\left(r^2\dfrac{\partial z}{\partial r}\right) + \dfrac{1}{\sin\theta}\dfrac{\partial}{\partial\theta}\left(\sin\theta\dfrac{\partial z}{\partial\theta}\right) = 0.$

7. If $z = \log(x - 3y) + e^{x+3y}$ show that

$$9\frac{\partial^2 z}{\partial x^2} = \frac{\partial^2 z}{\partial y^2}.$$

8. The voltage transient E in a transmission line satisfies

$$\frac{\partial E}{\partial t} = k\frac{\partial^2 E}{\partial x^2}.$$

Show that the function $E = ae^{-\omega^{\bullet} kt}\cos\omega x$ is a possible solution. Here a, ω and k are constants.

9. Show that the function $u = (x - 2t)\sin(x - 2t)$ satisfies the wave equation

$$4\frac{\partial^2 u}{\partial x^2} = \frac{\partial^2 u}{\partial t^2}.$$

10. Show that $z = x\cos x\cosh y + y\sin x\sinh y$ is a solution of Laplace's equation

$$\frac{\partial^2 z}{\partial x^2} + \frac{\partial^2 z}{\partial y^2} = 0.$$

11. If $z = \frac{x^{\bullet}+y^{\bullet}}{x+y}$ verify that

$$x^2 z_{xx} + 2xyz_{xy} + y^2 z_{yy} = 0.$$

12. If $w = x^2 + y^2$ where $x = \sin t$ and $y = \cos t$, find $\frac{dw}{dt}$ using the chain rule.

    Check your answer by first expressing w as a function of t.

13. Using the chain rule, find $\frac{dw}{dt}$ (as a function of t) where
$$w = xy^2 + 3z^2x - y^2$$
    with $x = e^t \cos t, y = e^t \sin t$ and $z = e^t$.

    Check your answer by first expressing w as a function of t.

14. If $f(x, y) = e^x + xy^2$ where $x = u + 2v$ and $y = u - 2v$
    (a) Write down $\frac{\partial f}{\partial x}$ and $\frac{\partial f}{\partial y}$.
    (b) Use the chain rule to determine $\frac{\partial f}{\partial v}$.
    (c) Check your answer to (b) by first expressing f in terms of u and v.

15. Given that u and v are functions of x and y defined by
$$u = 2xy \text{ and } v = x^2 + y^2$$
    and that $f(u, v) = uv^2$, express $\frac{\partial f}{\partial y}$ in terms of x, y, u and v.

16. Let $w = f(x, y)$ where $x = e^u \cos v$ and $y = e^u \sin v$. Using the chain rule show that
$$w_u^2 + w_v^2 = (x^2 + y^2)(w_x^2 + w_y^2).$$

17. If $z = f(x, y)$ and
$$x = u \cos \alpha + v \sin \alpha$$
$$y = -u \sin \alpha + v \cos \alpha$$
    where α is constant, show that
$$\left(\frac{\partial f}{\partial u}\right)^2 + \left(\frac{\partial f}{\partial v}\right)^2 = \left(\frac{\partial f}{\partial x}\right)^2 + \left(\frac{\partial f}{\partial y}\right)^2.$$

18. Let $f(x, y, z) = 2xy^2 + xz^2$.
    (a) Find ∇f at the point $(1, -2, 4)$.
    (b) Determine the directional derivative of $f(x, y, z)$ at the point $P(1, -2, 4)$ in the direction of the vector $\mathbf{i} - 2\mathbf{j} + 2\mathbf{k}$.

19. The equation of a surface S is $z = \sqrt{x^2 + y^2 + 3}$.
    (a) Determine a vector normal to the surface at the point $P(2, 3, 4)$.
    (b) Hence find the equation of the tangent plane to the surface S at the point P.

20. The equation of a surface S is $z = 8 - x^2 - 2y^2$.

    (a) Find a vector normal to S at the point $P(-1, 1, 5)$.

    (b) Hence find the equation of the

        i. tangent plane

        ii. normal line

    to the surface S at the point P.

21. (a) Determine the directional derivative of $f(x, y, z) = xy + yz + zx$ at the point $(1, -1, 2)$ in the direction of the vector $2\,\mathbf{i} - 2\,\mathbf{j} + \mathbf{k}$. Explain the significance of the negative value obtained.

    (b) In what direction does f increase most rapidly at the point P.

22. The equation of a surface S is $z = 14 - 2x^2 - y^2$.

    (a) Find a vector normal to S at the point $P(-1, 1, 11)$.

    (b) Hence find the equation of the tangent plane to the surface S at the point P.

23. (a) Determine the directional derivative of

$$f(x, y, z) = x^2 + 2xyz + y^2 \log z$$

    at the point $P(2, 1, 1)$ in the direction of the vector $\mathbf{i} + 2\,\mathbf{j} - 2\,\mathbf{k}$.

    (b) In what direction does f increase most rapidly at this point?

24. Find the directional derivative of the function $f(x, y, z) = 2xy + z^2$ at the point $P(1, 2, 1)$ in the direction of the vector $7\mathbf{i} - 4\,\mathbf{j} + 4\,\mathbf{k}$.

25. Determine the equation of the tangent plane to the surface $z = x^2 + y^2$ at the point $(1, 1, 2)$.

26. For the function $f(x, y, z) = xy + y^2 + z$ find

    (a) the gradient vector ∇f at the point $A(1, 1, 1)$

    (b) the rate of increase of f in the direction of the vector $\mathbf{n} = -\,\mathbf{i} + 2\,\mathbf{j} + 2\,\mathbf{k}$ at the point A.

27. The equation of the surface S is $z = \sqrt{4xy + 1}$.

    (a) Find a vector normal to S at the point $P(2, 1, 3)$.

    (b) Hence find the equation of the tangent plane to the surface S at P.

28. Let $f(x, y, z) = xy^2 + yz^2 - 4xz + 1$.

    (a) Determine the vector ∇f at the point $P(1, -2, 1)$.

    (b) Find the directional derivative of f at the point P in the direction of the vector $3\,\mathbf{i} + 2\,\mathbf{j} - 6\,\mathbf{k}$.

    (c) In what direction does f *decrease* most rapidly at P?

29. The equation of a surface S is given by $z = \sqrt{25 - 3x^2 - 3y^2}$.

    (a) Find a vector normal to S at the point $P(2, 2, 1)$.

    (b) Hence find the equation of the tangent plane to the surface S at the point P.

30. Let $f(x, y, z) = x + y \cos z + xye^z$.

    (a) Determine the vector ∇f at the point $P(1, 1, 0)$.

    (b) Find the directional derivative of f at the point $P(1, 1, 0)$ in the direction of the vector $\mathbf{i} + 2\,\mathbf{j} - 2\,\mathbf{k}$.

31. If $\mathbf{r} = x\,\mathbf{i} + y\,\mathbf{j} + z\,\mathbf{k}$ and $r = |\mathbf{r}| = \sqrt{x^2 + y^2 + z^2}$ show that $\nabla r^2 = 2\,\mathbf{r}$.

32. The volume of a frustum of a cone is given by $V = \frac{1}{3}\pi h(a^2 + ab + b^2)$, where a and b are the radii of its ends and h is its height. If the radius of each end is increased by 4% and the height is decreased by 1%, determine the approximate percentage change in the volume.

33. The moment of inertia I of a cylinder of mass M, radius r and length h about a line through its centre perpendicular to its axis is given by

    $$I = \frac{1}{12} M(3r^2 + h^2).$$

    Determine the approximate percentage increase in the moment of inertia I if r and h are increased by 1% without any change in the mass.

34. The gravitational attraction F between two uniform thin parallel rods each of mass M and length a at a distance b apart is given by

    $$F = \frac{2\mu M^2}{a^2 b}\left(\sqrt{a^2 + b^2} - b\right)$$

    where μ is a constant.

    If $a = 12$ and $b = 5$, find the greatest possible percentage error in F if a and b may each be in error by as much as 1%.

35. Evaluate the double integral

    $$\iint_R (x + 2y)\, dx\, dy$$

    where R is the region bounded by the graphs of $y = x$ and $y = x^2 - 2x$.

36. (a) Evaluate $\displaystyle\int_{x=0}^{1}\left[\int_{y=x^2}^{x} 2xy\,dy\right]dx.$

    (b) Check your answer to (a) by evaluating the double integral obtained by reversing the order of integration.

37. Evaluate the double integral

$$\iint_R (2x + xy)\,dx\,dy$$

where R is the triangular region having vertices at the points $(0,0)$, $(2,0)$ and $(0,1)$.

38. Evaluate the double integral

$$\iint_R (1 + 2xy)\,dx\,dy$$

where R is the region bounded by the graphs of $y + x = 2$ and $y = x^2$.

39. (a) Evaluate $\displaystyle\int_{x=0}^{2}\left\{\int_{y=0}^{x}(2x + 2y)\,dy\right\}dx.$

    (b) Sketch the region of integration R defined by the given double integral.

    (c) Hence confirm your answer to (a) by evaluating the double integral obtained by reversing the order of integration.

40. Evaluate the double integral

$$\iint_R (x^2 + y^2)\,dx\,dy$$

where R is the region bounded by the graphs of $y = x$ and $y = 2x$ and the line $x = 2$.

41. (a) Sketch the region of integration R defined by the double integral

$$\int_{x=0}^{1}\left[\int_{y=x}^{1} 2x\frac{e^y}{y^2}\,dy\right]dx.$$

    (b) Hence evaluate the integral by reversing the order of integration.

42. Evaluate the double integral

$$\iint_R (x + 2y + 1)\,dx\,dy$$

where R is the region bounded by the graphs of $y = x^2$ and $y = 2 - x^2$.

43. Evaluate $\iint_R 2xy\,dx\,dy$ where R is the triangle having vertices at $(0,0)$, $(1,0)$ and $(0,1)$.

44. The temperature $T(x,y)$ at the points $P(x,y)$ on a flat plate is given by

$$T(x,y) = 19 + 10x + 24y - x^2 - 4y^2.$$

    (a) Determine the hottest point on the plate.

    (b) Find the maximum temperature gradient at the point $(1,2)$ and the direction in which this occurs.

45. The temperature $T(x,y)$ at the points $P(x,y)$ on a flat circular plate occupying the region $x^2 + y^2 \leq 1$ is

$$T(x,y) = 2x^2 + 2y^2 - x + 100.$$

    Determine where the temperature is

    (a) least (b) greatest.

    (Hint: Extreme points may occur on the boundary of the plate.)

46. Show that the distance D from the origin to a point $P(x,y,z)$ on the plane $2x + y + z = 6$ is given by

$$D = \sqrt{x^2 + y^2 + (6 - 2x - y)^2}.$$

    Show that D is a minimum when $x = 2$ and $y = 1$.

    Deduce that the shortest distance from the origin to the plane is $\sqrt{6}$.

Part III

Appendices

Appendix A

Determinants

A.1 What is a Determinant?

To each square matrix A there is associated a number called the determinant of A and is denoted by $\det A$.

If A is the 2×2 matrix $A = \begin{bmatrix} a & b \\ c & d \end{bmatrix}$ then $\det A$ is defined to be $ad - bc$ and is called a 2×2 determinant.

Sometimes we write

$$\det A = \begin{vmatrix} a & b \\ c & d \end{vmatrix} = ad - bc.$$

For example $\begin{vmatrix} 2 & -3 \\ 5 & 4 \end{vmatrix} = 8 - (-15) = 23.$

Note the distinction between the notations for a matrix and a determinant.

A.1.1 Determinants of 3×3 Matrices

A 3×3 determinant is defined in terms of 2×2 determinants as follows:

$$\begin{vmatrix} a_1 & b_1 & c_1 \\ a_2 & b_2 & c_2 \\ a_3 & b_3 & c_3 \end{vmatrix} = a_1 \begin{vmatrix} b_2 & c_2 \\ b_3 & c_3 \end{vmatrix} - b_1 \begin{vmatrix} a_2 & c_2 \\ a_3 & c_3 \end{vmatrix} + c_1 \begin{vmatrix} a_2 & b_2 \\ a_3 & b_3 \end{vmatrix}.$$

The determinant is said to have been expanded across the first row.

In this definition note:

- the determinant $\begin{vmatrix} b_2 & c_2 \\ b_3 & c_3 \end{vmatrix}$ is called the **minor** of the element a_1 and is obtained from the original determinant by deleting the row and column which contains a_1
- the determinant $\begin{vmatrix} a_2 & c_2 \\ a_3 & c_3 \end{vmatrix}$ is called the **minor** of the element b_1 and is obtained from the original determinant by deleting the row and column which contains b_1
- the determinant $\begin{vmatrix} a_2 & b_2 \\ a_3 & b_3 \end{vmatrix}$ is called the **minor** of the element c_1 and is obtained from the original determinant by deleting the row and column which contains c_1
- the sequence of signs $+-+$ associated with the coefficients a_1, b_1, c_1 respectively.

Alternatively we can expand the determinant across the second or third rows or even down any of the columns as follows:

$$\begin{vmatrix} a_1 & b_1 & c_1 \\ a_2 & b_2 & c_2 \\ a_3 & b_3 & c_3 \end{vmatrix} = -a_2 \begin{vmatrix} b_1 & c_1 \\ b_3 & c_3 \end{vmatrix} + b_2 \begin{vmatrix} a_1 & c_1 \\ a_3 & c_3 \end{vmatrix} - c_2 \begin{vmatrix} a_1 & b_1 \\ a_3 & b_3 \end{vmatrix}$$

or

$$\begin{vmatrix} a_1 & b_1 & c_1 \\ a_2 & b_2 & c_2 \\ a_3 & b_3 & c_3 \end{vmatrix} = c_1 \begin{vmatrix} a_2 & b_2 \\ a_3 & b_3 \end{vmatrix} - c_2 \begin{vmatrix} a_1 & b_1 \\ a_3 & b_3 \end{vmatrix} + c_3 \begin{vmatrix} a_1 & b_1 \\ a_2 & b_2 \end{vmatrix}.$$

For example

$$\begin{aligned} \begin{vmatrix} 2 & -5 & 1 \\ -3 & 6 & 1 \\ 4 & -2 & 3 \end{vmatrix} &= 2 \begin{vmatrix} 6 & 1 \\ -2 & 3 \end{vmatrix} - (-5) \begin{vmatrix} -3 & 1 \\ 4 & 3 \end{vmatrix} + \begin{vmatrix} -3 & 6 \\ 4 & -2 \end{vmatrix} \\ &= 2(18+2) + 5(-9-4) + (6-24) = -43 \end{aligned}$$

by expanding across the first row; or

$$\begin{vmatrix} 2 & -5 & 1 \\ -3 & 6 & 1 \\ 4 & -2 & 3 \end{vmatrix} = 2\begin{vmatrix} 6 & 1 \\ -2 & 3 \end{vmatrix} - (-3)\begin{vmatrix} -5 & 1 \\ -2 & 3 \end{vmatrix} + 4\begin{vmatrix} -5 & 1 \\ 6 & 1 \end{vmatrix}$$

$$= 2(18+2) + 3(-15+2) + 4(-5-6) = -43$$

by expanding down the first column.

You may notice the following pattern of signs

$$\begin{matrix} + & - & + \\ - & + & - \\ + & - & + \end{matrix}$$

when expanding a 3×3 determinant. That is, when expanding across the first row the signs $+ \; - \; +$ are associated with the minors, whereas if it is expanded down the second column the appropriate signs are $- \; + \; -$.

A.1.2 Determinants of Higher Order Matrices

Higher order determinants can be expanded in terms of lower order determinants in a similar fashion to 3×3 determinants where the pattern of signs extends naturally as follows:

$$\begin{matrix} + & - & + & - & \cdots \\ - & + & - & + & \cdots \\ + & - & + & - & \cdots \\ - & + & - & + & \cdots \\ \vdots & \vdots & \vdots & \vdots & \ddots \end{matrix}$$

For example

$$\begin{vmatrix} 2 & -5 & 1 & 2 \\ -3 & 6 & 1 & 7 \\ 4 & -2 & 3 & -5 \\ 1 & 2 & 8 & 0 \end{vmatrix}$$

$$= 2\begin{vmatrix} 6 & 1 & 7 \\ -2 & 3 & -5 \\ 2 & 8 & 0 \end{vmatrix} - (-5)\begin{vmatrix} -3 & 1 & 7 \\ 4 & 3 & -5 \\ 1 & 8 & 0 \end{vmatrix} + \begin{vmatrix} -3 & 6 & 7 \\ 4 & -2 & -5 \\ 1 & 2 & 0 \end{vmatrix}$$

$$- 2\begin{vmatrix} -3 & 6 & 1 \\ 4 & -2 & 3 \\ 1 & 2 & 8 \end{vmatrix}$$

by expanding across the first row.

This is not a very good method (especially for even larger determinants); for example, to evaluate a 4×4 determinant by this method requires the evaluation of 12 2×2 determinants. In many applications very large determinants must be evaluated but *not* by this method. The evaluation of large determinants is the subject of Section A.3.

A.2 Properties of Determinants

The following properties of determinants (which are left for the reader to prove for 3×3 determinants — see Exercises 5 to 7) will enable higher order determinants to be evaluated.

1. If any row (or column) of a determinant contains a common factor, that factor may be taken outside the determinant; that is

$$\begin{vmatrix} ka_1 & kb_1 & kc_1 \\ a_2 & b_2 & c_2 \\ a_3 & b_3 & c_3 \end{vmatrix} = k\begin{vmatrix} a_1 & b_1 & c_1 \\ a_2 & b_2 & c_2 \\ a_3 & b_3 & c_3 \end{vmatrix}.$$

For example,

$$\begin{vmatrix} 6 & -15 & 3 \\ -3 & 6 & 1 \\ 4 & -2 & 3 \end{vmatrix} = 3 \begin{vmatrix} 2 & -5 & 1 \\ -3 & 6 & 1 \\ 4 & -2 & 3 \end{vmatrix}.$$

2. The value of a determinant is unaltered if a scalar multiple of one row is added to another row; and similarly for columns.

   That is,

$$\begin{vmatrix} a_1 & b_1 & c_1 \\ a_2 & b_2 & c_2 \\ a_3 & b_3 & c_3 \end{vmatrix} = \begin{vmatrix} a_1 + ka_2 & b_1 + kb_2 & c_1 + kc_2 \\ a_2 & b_2 & c_2 \\ a_3 & b_3 & c_3 \end{vmatrix}$$

   where k times the second row has been added to the first row.

   For example

$$\begin{vmatrix} 1 & -5 & 3 \\ -3 & 6 & 1 \\ 4 & -2 & 3 \end{vmatrix} = \begin{vmatrix} 1 & -5 & 3 \\ 0 & -9 & 10 \\ 4 & -2 & 3 \end{vmatrix}$$

   where three times row 1 has been added to row 2.

   Furthermore

$$\begin{vmatrix} 1 & -5 & 3 \\ 0 & -9 & 10 \\ 4 & -2 & 3 \end{vmatrix} = \begin{vmatrix} 1 & -5 & 3 \\ 0 & -9 & 10 \\ 0 & 18 & -9 \end{vmatrix}$$

   where (-4) times row 1 has been added to row 3; that is, 4 times row 1 has been subtracted from row 3. [Note that Property 2 has been used to obtain as many zeros as possible down the first column. This technique is used in the evaluation of large determinants in Section A.3.]

   Do you notice that the value of the original determinant can now just be written down by inspection. (Why? Expand down the first column.) This is essentially the idea behind the evaluation of larger determinants, such as 4×4 determinants.

3. A determinant which has two identical rows, or two identical columns, is zero.

   For example

$$\begin{vmatrix} 1 & -5 & 3 \\ -3 & 15 & -9 \\ 4 & -2 & 3 \end{vmatrix} = -3 \begin{vmatrix} 1 & -5 & 3 \\ 1 & -5 & 3 \\ 4 & -2 & 3 \end{vmatrix} = 0.$$

4. Interchanging two rows, or two columns, of a determinant changes the sign of the determinant. That is,

$$\begin{vmatrix} a_1 & b_1 & c_1 \\ a_2 & b_2 & c_2 \\ a_3 & b_3 & c_3 \end{vmatrix} = - \begin{vmatrix} a_3 & b_3 & c_3 \\ a_2 & b_2 & c_2 \\ a_1 & b_1 & c_1 \end{vmatrix}$$

   where rows 1 and 3 have been interchanged. (See Exercise 6.)

5. Interchanging the rows and columns of a determinant does not alter its value. That is, if A is a square matrix then $\det A = \det A^T$. (See Exercise 7.)

   For example

$$\begin{vmatrix} a_1 & b_1 & c_1 \\ a_2 & b_2 & c_2 \\ a_3 & b_3 & c_3 \end{vmatrix} = \begin{vmatrix} a_1 & a_2 & a_3 \\ b_1 & b_2 & b_3 \\ c_1 & c_2 & c_3 \end{vmatrix}$$

6. If A and B are square matrices of the same order then $\det AB = \det A \det B$. (See Exercise 8.)

A.3 Evaluation of Large Determinants

The above properties of determinants make it relatively easy to evaluate determinants of large square matrices by reducing these matrices to row-echelon form; that is, to the form

$$R = \begin{bmatrix} \beta_1 & d_{12} & d_{13} & \cdots & \cdots & d_{1n} \\ 0 & \beta_2 & d_{23} & \cdots & \cdots & d_{2n} \\ 0 & 0 & \beta_3 & \cdots & \cdots & \cdot \\ \vdots & \vdots & \vdots & \vdots & \vdots & \cdot \\ 0 & 0 & 0 & \cdots & \cdots & \cdot \end{bmatrix}$$

and determinants of matrices of this form are easily obtained.

For example,

$$\begin{vmatrix} 1 & 3 & -2 & 4 & 1 \\ 0 & 1 & 0 & 3 & 2 \\ 0 & 0 & 3 & 1 & -1 \\ 0 & 0 & 0 & 5 & 2 \\ 0 & 0 & 0 & 0 & 1 \end{vmatrix}$$

$$= \begin{vmatrix} 1 & 0 & 3 & 2 \\ 0 & 3 & 1 & -1 \\ 0 & 0 & 5 & 2 \\ 0 & 0 & 0 & 1 \end{vmatrix} = \begin{vmatrix} 3 & 1 & -1 \\ 0 & 5 & 2 \\ 0 & 0 & 1 \end{vmatrix} = 3.5.1 = 15$$

by continually reducing the order of the determinant by repeatedly expanding down the first column.

This type of reduction is carried out using three types of operations:

- multiplying a row by a scalar; this affects the value of the determinant as described in Property 1
- interchanging two rows; this changes the sign of the determinant. In order to cancel this effect, whenever two rows are interchanged it is often worthwhile interchanging another pair of rows to avoid changing the sign of the determinant.
- adding a scalar multiple of one row to another row; this leaves the value of the determinant unchanged, as described in Property 2 above.

Thus determinants of large matrices can be found by reducing the matrix to row-echelon form, being careful to take account of any row interchanges and any common factors in a row.

We illustrate this in the following example.

Worked Example A.3.1 *Evaluate the determinant of the matrix A given by*

$$A = \begin{bmatrix} 0 & 1 & 3 & 5 \\ 2 & 2 & -6 & 4 \\ -1 & 2 & 0 & -3 \\ 3 & 1 & -2 & 2 \end{bmatrix}.$$

In order to make the reduction to row-echelon form easier to manage, we try to make the element in the top left-hand corner equal to 1. One way of doing this is to interchange the first two rows and then remove a common factor of 2 to give

$$\det A = \begin{vmatrix} 0 & 1 & 3 & 5 \\ 2 & 2 & -6 & 4 \\ -1 & 2 & 0 & -3 \\ 3 & 1 & -2 & 2 \end{vmatrix} = -2 \begin{vmatrix} 1 & 1 & -3 & 2 \\ 0 & 1 & 3 & 5 \\ -1 & 2 & 0 & -3 \\ 3 & 1 & -2 & 2 \end{vmatrix}.$$

Also by

- adding row 1 to row 3
- subtracting three times row 1 from row 4

and then expanding down the first column we get

$$\det A = -2 \begin{vmatrix} 1 & 1 & -3 & 2 \\ 0 & 1 & 3 & 5 \\ 0 & 3 & -3 & -1 \\ 0 & -2 & 7 & -4 \end{vmatrix} = -2.1 \begin{vmatrix} 1 & 3 & 5 \\ 3 & -3 & -1 \\ -2 & 7 & -4 \end{vmatrix}.$$

Finally, on subtracting three times row 1 from row 2, adding two times row 1 to row 3 and expanding down the first column, we get

$$\det A = -2 \begin{vmatrix} 1 & 3 & 5 \\ 0 & -12 & -16 \\ 0 & 13 & 6 \end{vmatrix}.$$

$$= -2 \begin{vmatrix} -12 & -16 \\ 13 & 6 \end{vmatrix} = -2(-72 + 208) = -272.$$

■

A.4 Cramer's Rule

Determinants provide us with a compact way of writing down the solution of a square system of linear equations. For example the simultaneous equations

$$\begin{aligned} ax &+ by &= f \\ cx &+ dy &= g \end{aligned}$$

have solutions $x = \dfrac{df - bg}{ad - bc}$ and $y = \dfrac{ag - cf}{ad - bc}$. We can write these solutions in determinant form as

$$x = \frac{\begin{vmatrix} f & b \\ g & d \end{vmatrix}}{\begin{vmatrix} a & b \\ c & d \end{vmatrix}} \text{ and } y = \frac{\begin{vmatrix} a & f \\ c & g \end{vmatrix}}{\begin{vmatrix} a & b \\ c & d \end{vmatrix}}.$$

For example, the solution of the linear system

$$\begin{aligned} 2x &- 3y &= 9 \\ 3x &+ 7y &= -4 \end{aligned}$$

is

$$x = \frac{\begin{vmatrix} 9 & -3 \\ -4 & 7 \end{vmatrix}}{\begin{vmatrix} 2 & -3 \\ 3 & 7 \end{vmatrix}} = \frac{51}{23} \text{ and } y = \frac{\begin{vmatrix} 2 & 9 \\ 3 & -4 \end{vmatrix}}{\begin{vmatrix} 2 & -3 \\ 3 & 7 \end{vmatrix}} = -\frac{35}{23}.$$

Cramer's rule extends to a system of three equations in three unknowns (in fact, it extends to any square system of linear equations but is impractical for larger systems) as follows.

The simultaneous equations

$$\begin{array}{ccccccc} a_1x & + & b_1y & + & c_1z & = & d_1 \\ a_2x & + & b_2y & + & c_2z & = & d_2 \\ a_3x & + & b_3y & + & c_3z & = & d_3 \end{array}$$

have the solution

$$x = \frac{\begin{vmatrix} d_1 & b_1 & c_1 \\ d_2 & b_2 & c_2 \\ d_3 & b_3 & c_3 \end{vmatrix}}{\begin{vmatrix} a_1 & b_1 & c_1 \\ a_2 & b_2 & c_2 \\ a_3 & b_3 & c_3 \end{vmatrix}}, y = \frac{\begin{vmatrix} a_1 & d_1 & c_1 \\ a_2 & d_2 & c_2 \\ a_3 & d_3 & c_3 \end{vmatrix}}{\begin{vmatrix} a_1 & b_1 & c_1 \\ a_2 & b_2 & c_2 \\ a_3 & b_3 & c_3 \end{vmatrix}} \text{ and } z = \frac{\begin{vmatrix} a_1 & b_1 & d_1 \\ a_2 & b_2 & d_2 \\ a_3 & b_3 & d_3 \end{vmatrix}}{\begin{vmatrix} a_1 & b_1 & c_1 \\ a_2 & b_2 & c_2 \\ a_3 & b_3 & c_3 \end{vmatrix}}.$$

Note that, in each case, the denominator is just the determinant of the matrix of coefficients. In the solution for x, the numerator is simply the determinant in the denominator where the x-column (that is, the first column) is replaced by the column of right-hand sides, and similarly for the unknowns y and z.

▍ **Worked Example A.4.1** *Determine the solutions of the simultaneous equations:*

$$\begin{array}{ccccccc} 2x & - & 3y & + & z & = & 8 \\ 3x & + & y & - & 2z & = & -4 \\ x & - & 7y & + & z & = & 11. \end{array}$$

Using Cramer's rule the solution is given by

$$x = \frac{\begin{vmatrix} 8 & -3 & 1 \\ -4 & 1 & -2 \\ 11 & -7 & 1 \end{vmatrix}}{\begin{vmatrix} 2 & -3 & 1 \\ 3 & 1 & -2 \\ 1 & -7 & 1 \end{vmatrix}} = \frac{-33}{-33} = 1$$

$$y = \frac{\begin{vmatrix} 2 & 8 & 1 \\ 3 & -4 & -2 \\ 1 & 11 & 1 \end{vmatrix}}{\begin{vmatrix} 2 & -3 & 1 \\ 3 & 1 & -2 \\ 1 & -7 & 1 \end{vmatrix}} = \frac{33}{-33} = -1$$

and

$$z = \frac{\begin{vmatrix} 2 & -3 & 8 \\ 3 & 1 & -4 \\ 1 & -7 & 11 \end{vmatrix}}{\begin{vmatrix} 2 & -3 & 1 \\ 3 & 1 & -2 \\ 1 & -7 & 1 \end{vmatrix}} = \frac{-99}{-33} = 3.$$

■

A.5 Exercises

> Before attempting any of these miscellaneous exercises, make sure that you have successfully answered all the self-help exercises appearing throughout the chapter.

1. Evaluate each of the following 2×2 determinants:

(a) $\begin{vmatrix} 2 & -3 \\ -4 & 3 \end{vmatrix}$ (b) $\begin{vmatrix} -2 & -4 \\ -2 & 3 \end{vmatrix}$ (c) $\begin{vmatrix} -2 & 3 \\ -4 & 6 \end{vmatrix}$

(d) $\begin{vmatrix} -2-\lambda & 3 \\ -4 & 6-\lambda \end{vmatrix}$ (as a function of λ)

2. Evaluate each of the following determinants by expanding
   - across the first row
   - down the first column
   - across the second row.

(a) $\begin{vmatrix} 0 & 1 & 1 \\ 1 & 0 & 1 \\ 1 & 1 & 0 \end{vmatrix}$ (b) $\begin{vmatrix} 1 & 2 & 3 \\ 4 & 5 & 6 \\ 7 & 8 & 9 \end{vmatrix}$

(c) $\begin{vmatrix} 1 & -2 & 5 \\ 2 & 0 & -3 \\ -2 & 1 & 7 \end{vmatrix}$ (d) $\begin{vmatrix} -1 & 3 & 1 \\ 2 & 5 & 0 \\ 3 & 2 & -1 \end{vmatrix}$

3. Evaluate the following determinants as functions of λ.

(a) $\begin{vmatrix} 1-\lambda & -2 & 5 \\ 2 & -\lambda & -3 \\ -2 & 1 & 7-\lambda \end{vmatrix}$ (b) $\begin{vmatrix} 1-\lambda & -2 & 5 \\ 0 & -2-\lambda & -3 \\ 0 & 0 & 7-\lambda \end{vmatrix}$

4. Determine all values of λ which satisfy:

(a) $\begin{vmatrix} 1-\lambda & -2 & 4 \\ 0 & 3-\lambda & 3 \\ 0 & 0 & -2-\lambda \end{vmatrix} = 0$

(b) $\begin{vmatrix} -\lambda & 1 & 1 \\ 1 & -\lambda & 1 \\ 1 & 1 & -\lambda \end{vmatrix} = 0$

(c) $\begin{vmatrix} 5-\lambda & 2 & 2 \\ 2 & 2-\lambda & 1 \\ 2 & 1 & 2-\lambda \end{vmatrix} = 0.$

5. By evaluating both sides verify that

$$\begin{vmatrix} a_1 & b_1 & c_1 \\ a_2 & b_2 & c_2 \\ a_3 & b_3 & c_3 \end{vmatrix}$$

$$= \begin{vmatrix} a_1 + ka_2 & b_1 + kb_2 & c_1 + kc_2 \\ a_2 & b_2 & c_2 \\ a_3 & b_3 & c_3 \end{vmatrix}.$$

6. Verify that

$$\begin{vmatrix} a_1 & b_1 & c_1 \\ a_2 & b_2 & c_2 \\ a_3 & b_3 & c_3 \end{vmatrix} = - \begin{vmatrix} a_3 & b_3 & c_3 \\ a_2 & b_2 & c_2 \\ a_1 & b_1 & c_1 \end{vmatrix}.$$

7. If

$$A = \begin{bmatrix} a_1 & b_1 & c_1 \\ a_2 & b_2 & c_2 \\ a_3 & b_3 & c_3 \end{bmatrix}$$

verify that $\det A^T = \det A$.

8. If

$$A = \begin{bmatrix} 2 & 3 & -3 \\ 2 & -2 & 0 \\ -5 & 2 & -1 \end{bmatrix}$$

and

$$B = \begin{bmatrix} -2 & 7 & 3 \\ 0 & -1 & 2 \\ -3 & 4 & 1 \end{bmatrix}$$

find AB and verify that

$$\det AB = \det A \ \det B.$$

Also show that $\det AB = \det BA$.

9. For each of the following pair of vectors **a** and **b**, use determinants to find $\mathbf{a}\times\mathbf{b}$.

   (a) $\mathbf{a} = 2\,\mathbf{i} + \mathbf{j} + 3\,\mathbf{k}$, $\mathbf{b} = \mathbf{i} - \mathbf{j} + 2\,\mathbf{k}$
   (b) $\mathbf{a} = 2\,\mathbf{i} - \mathbf{k}$, $\mathbf{b} = \mathbf{i} + \mathbf{j}$
   (c) $\mathbf{a} = \mathbf{i} - \mathbf{j} + 3\,\mathbf{k}$, $\mathbf{b} = 2\,\mathbf{i} + \mathbf{j}$.

10. Use the properties of determinants to evaluate each of the following determinants.

(a) $\begin{vmatrix} 1 & -1 & 0 & 2 \\ 1 & 2 & -2 & 4 \\ -1 & 12 & 0 & -3 \\ 2 & 1 & -3 & 4 \end{vmatrix}$ (b) $\begin{vmatrix} 0 & 1 & 1 & 1 \\ 1 & 0 & 1 & 1 \\ 1 & 1 & 0 & 1 \\ 1 & 1 & 1 & 0 \end{vmatrix}$

(c) $\begin{vmatrix} -1 & 1 & 1 & 1 \\ 1 & -1 & 1 & 1 \\ 1 & 1 & -1 & 1 \\ 1 & 1 & 1 & -1 \end{vmatrix}$ (d) $\begin{vmatrix} 1 & 0 & 2 & -1 \\ -1 & 2 & 1 & 3 \\ 2 & 3 & 1 & 4 \\ 0 & 1 & 3 & 5 \end{vmatrix}$

11. Use Cramer's rule to find the solutions of:

(a)
$$\begin{aligned} x - 4y + 3z &= 0 \\ 13x + y - 12z &= 2 \\ 2x + 7y + z &= 10. \end{aligned}$$

(b)
$$\begin{aligned} 2x - 3y + 5z &= 2 \\ x + 2y - 3z &= -1 \\ 3x - 7y + 9z &= 2. \end{aligned}$$

(c)
$$\begin{aligned} 3x - 3y + z &= 0 \\ 2x + y - 2z &= -2 \\ 2y + z &= 7. \end{aligned}$$

A.6 Answers to Exercises

1. (a) -6
   (b) -14
   (c) 0
   (d) $\lambda^2 - 4\lambda$
2. (a) 2
   (b) 0
   (c) 29
   (d) 0
3. (a) $-\lambda^3 + 8\lambda^2 - 24\lambda + 29$
   (b) $(1-\lambda)(-2-\lambda)(7-\lambda)$
4. (a) $1, -2, 3$
   (b) $-1, 2$
   (c) 1,7
9. (a) $5\,\mathbf{i} - \mathbf{j} - 3\,\mathbf{k}$.
   (b) $\mathbf{i} - \mathbf{j} + 2\,\mathbf{k}$.
   (c) $-3\,\mathbf{i} + 6\,\mathbf{j} + 3\,\mathbf{k}$.
10. (a) -69
    (b) -3
    (c) -16
    (d) -69
11. (a) $x = 1, y = 1, z = 1$
    (b) $x = 0, y = 1, z = 1$
    (c) $x = 1, y = 2, z = 3$

Appendix B

A Collection of Formulae

B.1 Basic Formulae

Algebra

---

(1) $a^n - b^n = (a-b)(a^{n-1} + a^{n-2}b + \cdots + ab^{n-2} + b^{n-1})$

(2) $(a+x)^n = a^n + na^{n-1}x + \frac{n(n-1)}{2!}a^{n-2}x^2 + \cdots + x^n$

(3) Roots of $ax^2 + bx + c = 0$ are:

$$x = \frac{-b \pm \sqrt{b^2 - 4ac}}{2a}$$

Logarithms and Exponents

---

(1) $a^{n+m} = a^n a^m$, $(a^m)^n = a^{mn}$, $a^m/a^n = a^{m-n}$, $(ab)^n = a^n b^n$

(2) $\log(xy) = \log x + \log y$

(3) $\log(x/y) = \log x - \log y$

(4) $\log(x^n) = n \log x$

(5) $\log_b a = \dfrac{\log a}{\log b}$

Geometry

(1) Area of a triangle is $\frac{1}{2}bc\sin A = \frac{1}{2}$ base length $\times \perp$ height

(2) Circumference of a circle is $2\pi r$

(3) Area of a circle is πr^2

(4) Length of a circular arc is $r\theta$

(5) Area of a circular sector is $\frac{1}{2}r^2\theta$

(6) Volume of a sphere is $\frac{4}{3}\pi r^3$

(7) Surface area of a sphere is $4\pi r^2$

(8) Volume of a right circular cylinder is $\pi r^2 h$

(9) Curved surface area of a right circular cylinder is $2\pi rh$

(10) Volume of a right circular cone is $\frac{1}{3}\pi r^2 h$

(11) Sine rule: $\dfrac{a}{\sin A} = \dfrac{b}{\sin B} = \dfrac{c}{\sin C}$

(12) Cosine rule: $a^2 = b^2 + c^2 - 2bc\cos A$

(13) Distance between points in three dimensions

$$\sqrt{(x_2 - x_1)^2 + (y_2 - y_1)^2 + (z_2 - z_1)^2}$$

(14) Ellipse: $\dfrac{x^2}{a^2} + \dfrac{y^2}{b^2} = 1$ (Area of ellipse is πab)

(15) Hyperbola: $\dfrac{x^2}{a^2} - \dfrac{y^2}{b^2} = 1$

(16) Parabola: $y^2 = 4ax$

Trigonometry

(1) π radians equals 180°

(2) 1 radian equals $(\frac{180}{\pi})^\circ$

(3) 1° equals $\frac{\pi}{180}$ radians

Trigonometry (continued)

(4)

	sin	cos	tan
30°	$\frac{1}{2}$	$\frac{\sqrt{3}}{2}$	$\frac{1}{\sqrt{3}}$
45°	$\frac{1}{\sqrt{2}}$	$\frac{1}{\sqrt{2}}$	1
60°	$\frac{\sqrt{3}}{2}$	$\frac{1}{2}$	$\sqrt{3}$

(5) $$\tan\theta = \frac{\sin\theta}{\cos\theta}$$

(6) $$\operatorname{cosec}\theta = \frac{1}{\sin\theta}, \quad \sec\theta = \frac{1}{\cos\theta}, \quad \cot\theta = \frac{1}{\tan\theta}$$

---

(7) $$\sin^2\theta + \cos^2\theta = 1, \quad \tan^2\theta + 1 = \sec^2\theta, \quad \cot^2\theta + 1 = \operatorname{cosec}^2\theta$$

---

(8) $$\sin(x+y) = \sin x\cos y + \cos x\sin y$$

(9) $$\sin(x-y) = \sin x\cos y - \cos x\sin y$$

(10) $$\cos(x+y) = \cos x\cos y - \sin x\sin y$$

(11) $$\cos(x-y) = \cos x\cos y + \sin x\sin y$$

---

(12) $$\tan(x+y) = \frac{\tan x + \tan y}{1 - \tan x\tan y}$$

(13) $$\tan(x-y) = \frac{\tan x - \tan y}{1 + \tan x\tan y}$$

---

(14) $$\sin x + \sin y = 2\sin\tfrac{1}{2}(x+y)\cos\tfrac{1}{2}(x-y)$$

(15) $$\sin x - \sin y = 2\cos\tfrac{1}{2}(x+y)\sin\tfrac{1}{2}(x-y)$$

(16) $$\cos x + \cos y = 2\cos\tfrac{1}{2}(x+y)\cos\tfrac{1}{2}(x-y)$$

(17) $$\cos x - \cos y = -2\sin\tfrac{1}{2}(x+y)\sin\tfrac{1}{2}(x-y)$$

---

(18) $$2\sin x\cos y = \sin(x+y) + \sin(x-y)$$

(19) $$2\cos x\sin y = \sin(x+y) - \sin(x-y)$$

(20) $$2\cos x\cos y = \cos(x+y) + \cos(x-y)$$

(21) $$2\sin x\sin y = \cos(x-y) - \cos(x+y)$$

Trigonometry (continued)

(22) $\sin 2x = 2\sin x\cos x$

(23) $\cos 2x = \cos^2 x - \sin^2 x = 2\cos^2 x - 1 = 1 - 2\sin^2 x$

$\cos^2 x = \frac{1}{2}(1+\cos 2x)$, $\sin^2 x = \frac{1}{2}(1-\cos 2x)$

(24) $\tan 2x = \dfrac{2\tan x}{1-\tan^2 x}$, $\sin 2x = \dfrac{2\tan x}{1+\tan^2 x}$, $\cos 2x = \dfrac{1-\tan^2 x}{1+\tan^2 x}$

(25) $\sin \frac{1}{2}x = \pm\sqrt{\dfrac{1-\cos x}{2}}$

(26) $\cos \frac{1}{2}x = \pm\sqrt{\dfrac{1+\cos x}{2}}$

(27) $\tan \frac{1}{2}x = \pm\sqrt{\dfrac{1-\cos x}{1+\cos x}} = \dfrac{\sin x}{1+\cos x} = \dfrac{1-\cos x}{\sin x}$

Hyperbolic functions

(1) $\sinh x = \frac{1}{2}(e^x - e^{-x})$

(2) $\cosh x = \frac{1}{2}(e^x + e^{-x})$

(3) $\tanh x = \dfrac{\sinh x}{\cosh x}$

(4) $\operatorname{cosech} x = \dfrac{1}{\sinh x}$, $\operatorname{sech} x = \dfrac{1}{\cosh x}$, $\coth x = \dfrac{1}{\tanh x}$

(5) $\cosh x + \sinh x = e^x$

(6) $\cosh x - \sinh x = e^{-x}$

(7) $\cosh^2 x - \sinh^2 x = 1$

(8) $\tanh^2 x + \operatorname{sech}^2 x = 1$

(9) $\coth^2 x - \operatorname{cosech}^2 x = 1$

(10) $\cosh(x \pm y) = \cosh x\cosh y \pm \sinh x\sinh y$

(11) $\sinh(x \pm y) = \sinh x\cosh y \pm \cosh x\sinh y$

(12) $\tanh(x \pm y) = \dfrac{\tanh x \pm \tanh y}{1 \pm \tanh x\tanh y}$

Hyperbolic functions (continued)

(13) $\sinh 2x = 2\sinh x \cosh x$

(14) $\cosh 2x = \cosh^2 x + \sinh^2 x = 2\cosh^2 x - 1 = 2\sinh^2 x + 1$

$\cosh^2 x = \frac{1}{2}(\cosh 2x + 1)$, $\sinh^2 x = \frac{1}{2}(\cosh 2x - 1)$

(15) $\tanh 2x = \dfrac{2\tanh x}{1 + \tanh^2 x}$

(16) $\sinh \frac{1}{2}x = \pm\sqrt{\dfrac{\cosh x - 1}{2}}$

(17) $\cosh \frac{1}{2}x = \sqrt{\dfrac{\cosh x + 1}{2}}$

(18) $\tanh \frac{1}{2}x = \pm\sqrt{\dfrac{\cosh x - 1}{\cosh x + 1}}$

Logarithmic equivalents

(1) $\operatorname{arsinh} x = \log[x + \sqrt{x^2 + 1}]$

(2) $\operatorname{arcosh} x = \log[x + \sqrt{x^2 - 1}]$

(3) $\operatorname{artanh} x = \dfrac{1}{2}\log\dfrac{1 + x}{1 - x}$

Quadric surfaces

(1) Sphere: $(x - a)^2 + (y - b)^2 + (z - c)^2 = r^2$

(2) Ellipsoid: $\dfrac{x^2}{a^2} + \dfrac{y^2}{b^2} + \dfrac{z^2}{c^2} = 1$

(3) Cone: $a^2x^2 + b^2y^2 - c^2z^2 = 0$

(4) Elliptic paraboloid: $\dfrac{x^2}{a^2} + \dfrac{y^2}{b^2} = z$

Complex numbers

(1) $z = x + iy$ where x and y are real and $i = \sqrt{-1}$

(2) $i^2 = -1$, $i^3 = -i$, $i^4 = 1$

Complex numbers (continued)

(3) $\Re\text{e}\ \{z\} = x$ and $\Im\text{m}\ \{z\} = y$

(4) $|z| = \sqrt{x^2 + y^2}$

(5) $\overline{z} = x - iy$

(6) $e^{i\theta} = \operatorname{cis}\theta = \cos\theta + i\sin\theta$

(7) If $z = x + iy = re^{i\theta}$ then $r = |z|$ and $\theta = \arg z$

(8) $|z^n| = |z|^n$ and $\arg z^n = n \arg z \pm 2k\pi$

(9) $\sinh iz = i\sin z$

(10) $\sin iz = i\sinh z$

(11) $\cosh iz = \cos z$

(12) $\cos iz = \cosh z$

Differentiation and integration rules

(1) $\dfrac{d}{dx}(u \cdot v) = u'v + uv'$

(2) $\dfrac{d}{dx}\left(\dfrac{u}{v}\right) = \dfrac{u'v - uv'}{v^2}$

(3) $\dfrac{d}{dx}(f(g(x))) = f'(g(x))g'(x)$

(4) If $y = f(x)$ then $\Delta y \approx f'(x)\Delta x$

(5) $\int f(g(x))g'(x)\,dx = \int f(u)\,du$ by setting $u = g(x)$

(6) $\int u\,dv = uv - \int v\,du$

Series

(1) **Arithmetic:**

$$a + (a + d) + \cdots + (a + (n - 1)d) = \tfrac{1}{2}n(2a + (n - 1)d)$$

Series (continued)

(2) **Geometric:**

$$a + ar + ar^2 + \cdots + ar^{n-1} = \frac{a(1-r^n)}{1-r}, \quad r \neq 1,$$

$$S_\infty = \frac{a}{1-r}, \quad \text{if } |r| < 1$$

(3) $$\sin z = z - \frac{z^3}{3!} + \frac{z^5}{5!} - \frac{z^7}{7!} + \cdots \quad \text{for all } z$$

(4) $$\cos z = 1 - \frac{z^2}{2!} + \frac{z^4}{4!} - \frac{z^6}{6!} + \cdots \quad \text{for all } z$$

(5) $$\tan z = z + \frac{z^3}{3} + \frac{2z^5}{15} + \frac{17z^7}{315} + \cdots \quad \text{if } |z| < \tfrac{\pi}{2}$$

(6) $$\exp z = 1 + z + \frac{z^2}{2!} + \frac{z^3}{3!} + \cdots \quad \text{for all } z$$

(7) $$\sinh z = z + \frac{z^3}{3!} + \frac{z^5}{5!} + \frac{z^7}{7!} + \cdots \quad \text{for all } z$$

(8) $$\cosh z = 1 + \frac{z^2}{2!} + \frac{z^4}{4!} + \frac{z^6}{6!} + \cdots \quad \text{for all } z$$

(9) $$\log(1+z) = z - \frac{z^2}{2} + \frac{z^3}{3} - \frac{z^4}{4} + \cdots \quad \text{if } |z| < 1$$

$$\frac{1}{2}\log\frac{1+z}{1-z} = z + \frac{z^3}{3} + \frac{z^5}{5} + \cdots \quad \text{if } |z| < 1$$

(10) $$(1+z)^\alpha = 1 + \alpha z + \frac{\alpha(\alpha-1)z^2}{2!} + \frac{\alpha(\alpha-1)(\alpha-2)z^3}{3!} + \cdots$$

if $|z| < 1$

Taylor Series

One dimensional

$$f(x) = f(x_0) + (x-x_0)f'(x_0) + \tfrac{1}{2!}(x-x_0)^2 f''(x_0) + \cdots$$

Two dimensional

$$f(x,y) = f(x_0,y_0) + (x-x_0)f_x(x_0,y_0) + (y-y_0)f_y(x_0,y_0) + \cdots$$

Remainder term

Lagrange Remainder(Taylor Series):

$$f^{n+1}(c)\frac{(x-a)^{n+1}}{(n+1)!} \text{ for some } c \text{ between } a \text{ and } x.$$

Integral form:

$$\int_a^x \frac{(x-t)^n}{n!} f^{(n+1)}(t)\,dt$$

Generalized factorial function

(1) $\Gamma(x) \equiv \int_0^\infty t^{x-1} e^{-t}\,dt$

(if x is a positive integer n then $\Gamma(n) = (n-1)!$)

(2) $\Gamma(x+1) = x\Gamma(x)$, $\Gamma\left(\frac{1}{2}\right) = \sqrt{\pi}$, $\int_{-\infty}^{\infty} e^{-x^2}\,dx = \sqrt{\pi}$

Vectors

(1) The length of the vector (x, y, z) is $\sqrt{x^2 + y^2 + z^2}$

(2) The scalar (dot) product is

$$\mathbf{a} \cdot \mathbf{b} = a_1 b_1 + a_2 b_2 + a_3 b_3 = |\mathbf{a}||\mathbf{b}| \cos\theta$$

(3) The scalar component of $\mathbf{a}$ in the direction of $\mathbf{b}$ is $\mathbf{a} \cdot \hat{\mathbf{b}}$

The vector projection of $\mathbf{a}$ in the direction of $\mathbf{b}$ is $(\mathbf{a} \cdot \hat{\mathbf{b}})\hat{\mathbf{b}}$

(4) The vector (cross) product is $\mathbf{a} \times \mathbf{b} = \begin{vmatrix} \mathbf{i} & \mathbf{j} & \mathbf{k} \\ a_1 & a_2 & a_3 \\ b_1 & b_2 & b_3 \end{vmatrix}$

(5) The scalar triple product is $\boldsymbol{a} \cdot \boldsymbol{b} \times \boldsymbol{c} = \boldsymbol{b} \cdot \boldsymbol{c} \times \boldsymbol{a} = \boldsymbol{c} \cdot \boldsymbol{a} \times \boldsymbol{b}$

Also $\boldsymbol{a} \cdot \boldsymbol{b} \times \boldsymbol{c} = \boldsymbol{a} \times \boldsymbol{b} \cdot \boldsymbol{c}$

(6) The line through the point (x_0, y_0, z_0) parallel to (a, b, c) is

$$(x, y, z) = (x_0, y_0, z_0) + t(a, b, c)$$

(7) The equation of the plane with normal (a, b, c) is

$$ax + by + cz = d$$

(8) $\boldsymbol{a} \times (\boldsymbol{b} \times \boldsymbol{c}) = (\boldsymbol{a} \cdot \boldsymbol{c})\boldsymbol{b} - (\boldsymbol{a} \cdot \boldsymbol{b})\boldsymbol{c}$

B.2 Numerical Methods

Newton's Method

Newton's method for obtaining numerical approximations to solutions of the equation $f(x) = 0$ is given by

$$x_{n+1} = x_n - \frac{f(x_n)}{f'(x_n)}.$$

Trapezoidal and Simpson's Rules

Numerical approximations to the value of $\int_a^b f(x)\,dx$ can be obtained using either the Trapezoidal or Simpson's rule.

Trapezoidal Rule

$$\int_a^b f(x)\,dx = \frac{1}{2}h(y_0 + 2y_1 + 2y_2 + 2y_3 + \cdots + 2y_{n-1} + y_n)$$

Simpson's Rule

$$\int_a^b f(x)\,dx = \frac{1}{3}h(y_0 + 4y_1 + 2y_2 + 4y_3 + 2y_4 + \cdots + 2y_{2n-2} + 4y_{2n-1} + y_{2n})$$

B.3 Vector Calculus

Gradient

The gradient of a scalar function $f(x, y, z)$ is a vector quantity given by

$$\nabla f = \frac{\partial f}{\partial x}\mathbf{i} + \frac{\partial f}{\partial y}\mathbf{j} + \frac{\partial f}{\partial z}\mathbf{k}.$$

Divergence

The divergence of a vector function $\mathbf{F} = F_1\,\mathbf{i} + F_2\,\mathbf{j} + F_3\,\mathbf{k}$ is a scalar quantity given by

$$\nabla\cdot\mathbf{F} = \frac{\partial F_1}{\partial x} + \frac{\partial F_2}{\partial y} + \frac{\partial F_3}{\partial z}.$$

Curl

The curl of a vector function $\mathbf{F} = F_1\,\mathbf{i} + F_2\,\mathbf{j} + F_3\,\mathbf{k}$ is a vector quantity given by

$$\nabla\times\mathbf{F} = \operatorname{curl}\mathbf{F} = \begin{vmatrix} \mathbf{i} & \mathbf{j} & \mathbf{k} \\ \dfrac{\partial}{\partial x} & \dfrac{\partial}{\partial y} & \dfrac{\partial}{\partial z} \\ F_1 & F_2 & F_3 \end{vmatrix}.$$

Directional Derivative

The directional derivative of f in the direction of the vector $\mathbf{u}$ is given by

$$\frac{df}{du} = \nabla f \cdot \hat{\mathbf{u}}.$$

Chain Rules

1. If $w = w(x, y)$ and both $x = x(t)$ and $y = y(t)$ then

$$\frac{dw}{dt} = \frac{\partial w}{\partial x}\frac{dx}{dt} + \frac{\partial w}{\partial y}\frac{dy}{dt}.$$

2. If $w = w(x, y)$ and both $x = x(u, v)$ and $y = y(u, v)$ then

$$\frac{\partial w}{\partial u} = \frac{\partial w}{\partial x}\frac{\partial x}{\partial u} + \frac{\partial w}{\partial y}\frac{\partial y}{\partial u}$$

and

$$\frac{\partial w}{\partial v} = \frac{\partial w}{\partial x}\frac{\partial x}{\partial v} + \frac{\partial w}{\partial y}\frac{\partial y}{\partial v}.$$

Appendix C

A Collection of Integrals

C.1 An Elementary Set

	Function	Integral
(1)	x^n	$\frac{1}{n+1}x^{n+1}, \quad n \neq -1$
(2)	$\dfrac{1}{x}$	$\log \lvert x \rvert$
(3)	a^x	$\dfrac{1}{\log a} a^x$
(4)	$\sin x$	$-\cos x$
(5)	$\cos x$	$\sin x$
(6)	$\tan x$	$-\log \lvert \cos x \rvert$
(7)	$\cot x$	$\log \lvert \sin x \rvert$
(8)	$\sec^2 x$	$\tan x$
(9)	$\operatorname{cosec}^2 x$	$-\cot x$
(10)	$\sec x \tan x$	$\sec x$
(11)	$\operatorname{cosec} x \cot x$	$-\operatorname{cosec} x$
(12)	$\sec x$	$\dfrac{1}{2} \log \dfrac{1+\sin x}{1-\sin x} = \log \lvert \sec x + \tan x \rvert$
(13)	$\operatorname{cosec} x$	$\dfrac{1}{2} \log \dfrac{1-\cos x}{1+\cos x} = \log \lvert \operatorname{cosec} x - \cot x \rvert$

C.2 An Advanced Set

	Function	**Integral**
(14)	$\arcsin x$	$x \arcsin x + \sqrt{1-x^2}$
(15)	$\arccos x$	$x \arccos x - \sqrt{1-x^2}$
(16)	$\arctan x$	$x \arctan x - \frac{1}{2}\log(1+x^2)$
(17)	$\sinh x$	$\cosh x$
(18)	$\cosh x$	$\sinh x$
(19)	$\tanh x$	$\log\cosh x$
(20)	$\coth x$	$\log\lvert\sinh x\rvert$
(21)	$\operatorname{sech}^2 x$	$\tanh x$
(22)	$\operatorname{cosech}^2 x$	$-\coth x$
(23)	$\operatorname{sech} x \tanh x$	$-\operatorname{sech} x$
(24)	$\operatorname{cosech} x \coth x$	$-\operatorname{cosech} x$
(25)	$\operatorname{sech} x$	$2\arctan e^x$
(26)	$\operatorname{cosech} x$	$\log\left\lvert\tanh\frac{x}{2}\right\rvert = \log\left\lvert\frac{e^x-1}{e^x+1}\right\rvert$
(27)	$\frac{1}{a^2-x^2}$	$\begin{cases} \frac{1}{a}\operatorname{artanh}\frac{x}{a} = \frac{1}{2a}\log\frac{a+x}{a-x}, & \text{if } \lvert x\rvert < a \\ \frac{1}{a}\operatorname{arcoth}\frac{x}{a} = \frac{1}{2a}\log\frac{x+a}{x-a}, & \text{if } \lvert x\rvert > a \end{cases}$
(28)	$\frac{1}{x^2+a^2}$	$\frac{1}{a}\arctan\frac{x}{a}$
(29)	$\frac{1}{\sqrt{a^2-x^2}}$	$\arcsin\frac{x}{a}$
(30)	$\frac{1}{\sqrt{x^2-a^2}}$	$\operatorname{arcosh}\frac{x}{a}$
(31)	$\frac{1}{\sqrt{x^2+a^2}}$	$\operatorname{arsinh}\frac{x}{a}$

C.3 An Advanced Set (continued)

	Function	Integral
(32)	$\sqrt{a^2 - x^2}$	$\frac{1}{2}x\sqrt{a^2 - x^2} + \frac{1}{2}a^2 \arcsin\frac{x}{a}$
(33)	$\sqrt{x^2 - a^2}$	$\frac{1}{2}x\sqrt{x^2 - a^2} - \frac{1}{2}a^2 \operatorname{arcosh}\frac{x}{a}$
(34)	$\sqrt{x^2 + a^2}$	$\frac{1}{2}x\sqrt{x^2 + a^2} + \frac{1}{2}a^2 \operatorname{arsinh}\frac{x}{a}$
(35)	$e^{ax}\sin bx$	$\frac{e^{ax}}{a^2 + b^2}(a\sin bx - b\cos bx)$
(36)	$e^{ax}\cos bx$	$\frac{e^{ax}}{a^2 + b^2}(a\cos bx + b\sin bx)$
(37)	$\sin^m x\cos^n x$	$\begin{cases} \dfrac{\sin^{m+1} x\cos^{n-1} x}{m+n} + \dfrac{n-1}{m+n}\displaystyle\int \sin^m x\cos^{n-2} x\,dx \\ -\dfrac{\sin^{m-1} x\cos^{n+1} x}{m+n} + \dfrac{m-1}{m+n}\displaystyle\int \sin^{m-2} x\cos^n x\,dx \end{cases}$
(38)	$x^n e^x$	$x^n e^x - n\int x^{n-1}e^x\,dx$
(39)	$\tan^n x$	$\frac{1}{n-1}\tan^{n-1} x - \int \tan^{n-2} x\,dx$
(40)	$\sec^n x$	$\frac{1}{n-1}\sec^{n-2} x\tan x + \frac{n-2}{n-1}\int \sec^{n-2} x\,dx$

Appendix D

Answers to Exercises

D.1 Vectors

1. (a) $10\mathbf{i} + 10\mathbf{j} + 10\mathbf{k}$
   (b) 9
   (c) $\frac{4}{9}\mathbf{i} + \frac{4}{9}\mathbf{j} + \frac{7}{9}\mathbf{k}$
   (d) $20\mathbf{i} + 20\mathbf{j} + 35\mathbf{k}$
   (e) $\frac{20}{9}\mathbf{i} + \frac{20}{9}\mathbf{j} + \frac{35}{9}\mathbf{k}$
   (f) $\frac{23}{9}$

2. (a) $-5\mathbf{i} - 5\mathbf{j} + 5\mathbf{k}$
   (b) 3
   (c) $\frac{1}{3}(\mathbf{i} - 2\mathbf{j} - 2\mathbf{k})$
   (d) $-\frac{14}{3}$
   (e) $2\mathbf{i} - 4\mathbf{j} - 4\mathbf{k}$
   (f) $6\mathbf{i} - 12\mathbf{j} - 12\mathbf{k}$

3. (a) i. $7\mathbf{i} + 13\mathbf{k}$
   ii. 0; Yes
   iii. 6
   (b) $\pi - \arccos\left(\frac{1}{\sqrt{35}}\right)$
   (c) $\frac{5}{3}$
   (d) $\mathbf{0}$

4. $-\frac{4}{3}$

5. Angle BAC; $\pi - \arccos\left(\frac{\sqrt{6}}{9}\right)$

6. (a) i. $\mathbf{j}$
   ii. 10, $-\mathbf{i} + 4\,\mathbf{j} + 3\,\mathbf{k}$
   iii. 3, $\frac{1}{3}(2\,\mathbf{i} - \mathbf{j} + 2\,\mathbf{k})$
   iv. 4

7. (a) -3
   (b) $-5\,\mathbf{i} + 5\,\mathbf{j} + 5\,\mathbf{k}$
   (c) 14
   (d) 0
   (e) $\frac{1}{\sqrt{3}}(-\mathbf{i} + \mathbf{j} + \mathbf{k})$

8. (b) 2
   (c) $\frac{2}{3}(-2\,\mathbf{i} + \mathbf{j} + 2\,\mathbf{k})$

9. (a) 5
   (b) $6\,\mathbf{i} + \mathbf{j} - 4\,\mathbf{k}$
   (c) 0
   (d) $\frac{5}{\sqrt{6}}$
   (e) $\frac{1}{2}\sqrt{53}$

10. (a) -15
    (b) $5\,\mathbf{j}$

11. (a) $\pi - \arccos\frac{4}{9}$
    (b) $-3\,\mathbf{i} + 3\,\mathbf{j}$, -9
    (c) $\frac{1}{\sqrt{2}}(-\mathbf{i} + \mathbf{j})$
    (d) 1

12. (a) $\arccos(-\frac{1}{\sqrt{10}})$
    (b) $\sqrt{2}$, $\sqrt{129}$
    (c) 4, $-3\,\mathbf{i}+2\,\mathbf{j}+4\,\mathbf{k}$, 1
    (d) $\frac{1}{3}(-2\,\mathbf{i}+\mathbf{j}+2\,\mathbf{k})$
    (e) $\frac{4}{3}$

13. $\frac{1}{3}F_0$

14. $(3,-4,2) =$
    $\frac{1}{14}(1,2,3)+\frac{1}{14}(41,-58,25)$

15. (a) $x = 2+2t$, $y = 1$, $z = 3-t$
    (b) $(\frac{8}{5},1,\frac{16}{5})$
    (c) $\frac{1}{5}\sqrt{345}$

16. (a) $-x+y+z=-2$
    (b) $\frac{7}{\sqrt{3}}$

17. (a) $x = 1+3t$, $y = -2t$, $z = -\frac{1}{2}-2t$
    (b) $(-\frac{7}{2},3,\frac{5}{2})$
    (c) $\frac{1}{2}\pi - \arccos\frac{1}{\sqrt{51}}$

18. $\frac{1}{2}\sqrt{53}$

19. (a) $x = 2+t$, $y = 5t$, $z = 7t$
    (b) $(4,10,14)$

20. $4x-3y-z=8$

21. (a) $6x+3y+2z=6$
    (b) $\frac{6}{7}$
    (c) $x=6t$, $y=3t$, $z=2t$

22. (a) $\frac{1}{\sqrt{29}}(2\,\mathbf{i}+4\,\mathbf{j}-3\,\mathbf{k})$
    (b) $\lambda = 1$

23. (a) $\arccos\frac{42}{\sqrt{26}\sqrt{69}}$
    (b) $x=-1+t$, $y=-6-5t$, $z=-2t$

24. (a) $\sqrt{1749}$
    (b) $28x+26y+17z+5=0$
    (c) $\frac{5}{\sqrt{1749}}$

25. Lines are skew; distance between lines is $\frac{1}{\sqrt{5}}$

26. (a) $\alpha=-1$; $2x+5y+z=18$
    (b) $\frac{x-1}{2}=\frac{y-2}{-1}=\frac{z-3}{1}$
    (c) $\frac{3}{\sqrt{30}}$

27. (a) $(-1,1,2)$
    (b) $2x+y+3z=5$
    (c) $x=-1+2t$, $y=1+t$, $z=2+3t$

28. (a) Intersect at $(1,1,1)$
    (b) $2x-y+z=2$

29. (a) $\frac{2}{7}$, Yes
    (b) $6\,\mathbf{i}+10\,\mathbf{j}+7\,\mathbf{k}$

30. $(0,0,0)$

31. $-2x+5y-3z=3$; $\frac{3}{\sqrt{38}}$

32. $3x-y+3z+5=0$

33. (a) $\frac{2}{3\sqrt{2}}$
    (b) $\frac{1}{2}\sqrt{14}$
    (c) $-2x+3y-z=3$

35. $(5,6,1)$ and $(-7,3,-2)$

D.2 Complex Numbers

1. (a) 1
   (b) -1
   (c) i
   (d) 0
   (e) $-4+13i$
   (f) $-2+i$
   (g) $-2i$
   (h) -4
   (i) $34-13i$
   (j) $-1-18i$
   (k) $67+69i$
   (l) -4

2. (a) 2
   (b) 3
   (c) -1
   (d) $-1-5i$
   (e) $-\frac{1}{26}$
   (f) -24
   (g) 1

3. (a) $\frac{1}{5}(1-2i)$
   (b) $\frac{1}{13}(3+2i)$
   (c) $2-3i$
   (d) $2i$
   (e) $\frac{36}{25}$
   (f) $\frac{1}{5}(11-2i)$
   (g) $\frac{1}{5}(2-i)$

4. (a) $-7+16i$
   (b) 29
   (c) $\frac{1}{29}(-2-5i)$
   (d) $-\frac{800}{841}(20+21i)$
   (e) $\frac{1}{5}(12-59i)$

5. (a) x^2-y^2
   (b) $3x+4$

6. $-3(6+i)$

7. $-1+2i$, $1+2i$

---

9. (a) $\frac{x}{x^2+y^2} - \frac{y}{x^2+y^2}i$
   (b) $(x^2-y^2+y+1)+i(2xy-x)$
   (c) $\frac{x}{x^2+(y-1)^2} + \frac{1-y}{x^2+(y-1)^2}i$
   (d) $\frac{x^2+y^2-1}{x^2+(y+1)^2} - \frac{2x}{x^2+(y+1)^2}i$

10. (a) $-1-i$
    (b) $\frac{1}{5}(1-2i)$
    (c) $\frac{1}{7}(1-i)$

11. $3+i$, $3-i$, 1

12. (a) $1+i$, $1-i$
    (b) $2-i$, $1-i$

13. (*a*) $\sqrt{13}$, (*b*) 5, (*c*) 2, (*d*) 4, (*e*) $\sqrt{37}$, (*f*) $\frac{1}{\sqrt{10}}$, (*g*) 1, (*h*) $\sqrt{26}$, (*i*) $\sqrt{26}$, (*j*) 10

14. (a) 1.6, (b) 2^{10}

15. (a) $\sqrt{x^2+(y+1)^2}$
    (b) $\sqrt{(x-1)^2+y^2}$
    (c) $\sqrt{(x+2)^2+(y-1)^2}$

16. Distance between z_1 and z_2

18. $|z| = \sqrt{R^2+(\omega L - \frac{1}{\omega C})^2}$;
    $\omega = 1/\sqrt{LC}$

19. Centre at $-1+i$; radius 3

20. (a) $\frac{3}{\sqrt{2}}(1+i)$
    (b) $-1+\sqrt{3}i$
    (c) $0.6946-3.9392i$
    (d) $\frac{3}{2}(\sqrt{3}+i)$
    (e) $2-2\sqrt{3}i$
    (f) $0.5099+2.9563i$
    (g) $-1+\sqrt{3}i$
    (h) $\frac{5}{2}(1-\sqrt{3}i)$
    (i) $-14.6302+2.0855i$

21. (a) $4\,\text{cis}\,\pi/3 = 4e^{i\pi/3}$
    (b) $8\,\text{cis}\,\pi = 8e^{i\pi}$
    (c) $2\,\text{cis}\,\pi/2 = 2e^{i\pi/2}$
    (d) $2\,\text{cis}\,5\pi/6 = 2e^{i5\pi/6}$
    (e) $\sqrt{2}\,\text{cis}\,(-\pi/4) = \sqrt{2}e^{-i\pi/4}$
    (f) $5\,\text{cis}\,0.9273 = 5e^{0.9273i}$
    (g) $5\,\text{cis}\,(-2.498) = 5e^{-2.498i}$

22. (a) $-5.7956-1.5529i$
    (b) $0.5176-1.9319i$
    (c) $0.5176+1.9319i$
    (d) $6i$
    (e) $-0.2588-0.9659i$
    (f) $-1.5-2.5981i$

24. $x = \frac{2}{3}\pi + 2n\pi$

25. (b) $\int e^{2x}\cos 3x\,dx = \frac{1}{13}e^{2x}(2\cos 3x + 3\sin 3x) + c$

26. (a) $8i$
    (b) $2^6(-1-\sqrt{3}i)$
    (c) $2^7(-1+\sqrt{3}i)$
    (d) $-2(\sqrt{3}+i)$
    (e) $2^7\{(-1+\sqrt{3})-(1+\sqrt{3})i\}$
    (f) $2^{-6}(-1+\sqrt{3}i)$

28. (c) $\sin^5\theta = \frac{1}{16}(\sin 5\theta - 5\sin 3\theta + 10\sin\theta)$

29. (a) $\frac{1}{\sqrt{2}}(1+i)$, $\frac{1}{\sqrt{2}}(-1-i)$
    (b) $3i, -3i$
    (c) $\frac{1}{\sqrt{2}}(1+\sqrt{3}i)$, $\frac{1}{\sqrt{2}}(-1-\sqrt{3}i)$

---

30. (a) $1.4917-0.5429i$, $-0.2756+1.563i$, $-1.216-1.021i$
    (b) 1, $\frac{1}{2}(-1+i\sqrt{3})$, $\frac{1}{2}(-1-i\sqrt{3})$
    (c) $1+i$, $\frac{1}{2}\{(-1-\sqrt{3})+i(\sqrt{3}-1)\}$, $\frac{1}{2}\{(\sqrt{3}-1)-i(1+\sqrt{3})\}$

31. (a) $2(1+i\sqrt{3})$, -4, $2(1-i\sqrt{3})$
    (b) $0.3085+2.3311i$, $1.8644-1.4327i$, $-2.1730-0.8983i$
    (c) $-1.6767-5.3676i$, $1.6767+5.3676i$

32. (a) $-2i$, $\sqrt{3}+i$, $-\sqrt{3}+i$
    (b) $2.0987-0.4551i$, $-0.0987+0.4551i$
    (c) $0.3660(-1+i)$, $0.3660(1-i)$, $1.3660(-1+i)$, $1.3660(1-i)$

33. (a) Exterior of circle centre i, radius 2 including boundary

(b) Exterior of circle centre $\frac{1}{2}i$, radius 2 including boundary

(c) Annular region between circles centred at $2i$ and radii 1 and 3 excluding boundaries

(d) $y \geq -1$

(e) $6y - 2x - 2 \geq 0$

(f) Exterior of circle centre $(-\frac{19}{8}, \frac{1}{8})$ radius $\frac{3}{8}\sqrt{10}$

34. Straight line

35. (a) $2 + i, 3i, 2 - i, 4 + 3i, 7i, -2 + 5i$

(b) $\frac{1}{5}(1 - 7i), \frac{2}{3}, -\frac{1}{2}(1 + 7i), \frac{1}{13}(18 - 14i), \frac{8}{5}, \frac{1}{17}(23 + 7i)$

(c) $e, \cos 1 + i \sin 1, e(\cos 1 - i \sin 1), e^2(\cos 1 + i \sin 1),$
$\cos 3 + i \sin 3, e^{-1}(\cos 2 + i \sin 2)$

(d) $1, -i, -2 - 2i, 2 + 11i, -27i, 11 - 2i$

36. (a) Line $u = 0$, line $u = 4$, line $v = 1$, line $v = 5$

(c) Unit circle centre $(0, 0)$, circle centre $(0, 0)$ and radius e^2, positive real u-axis, straight line $v = \tan 2 \cdot u$ in first quadrant

37. $u = e^x \cos y, v = e^x \sin y$

Straight line $v = \tan d \cdot u$

38. Semicircle in upper half plane centre $(0, 0)$ and radius 8 excluding boundary

39. (a) The straight line $u + 2v = 1$

(b) Circle centre $(\frac{1}{2}, \frac{1}{4})$ radius $\frac{1}{4}\sqrt{5}$

(c) Circle centre $(-\frac{1}{3}, \frac{2}{3})$ radius $\frac{1}{3}\sqrt{5}$

D.3 Functions and their Derivatives — Part 1

5. (a) $12x^2 + 14x$; $24x + 14$
   (b) $2x - \frac{4}{x^2} - \frac{1}{2\sqrt{x}}$; $2 + \frac{8}{x^3} + \frac{1}{4x^{3/2}}$
   (c) $4 + 12x + 4\cos x + \frac{3}{x^2} - 12\sin x$; $12 - \frac{6}{x^3} - 12\cos x - 4\sin x$
   (d) $t^2\cos t + 2t\sin t$; $4t\cos t - (t^2 - 2)\sin t$
   (e) $\frac{1}{(1+x)^2}$; $-\frac{2}{(1+x)^3}$
   (f) $3x^2\cos x - x^3\sin x$; $(6x - x^3)\cos x - 6x^2\sin x$
   (g) $5\sin\theta + 5\theta\cos\theta$; $10\cos\theta - 5\theta\sin\theta$
   (h) $\frac{2}{(1-x)^2}$; $\frac{4}{(1-x)^3}$
   (i) $x^2\sec^2 x + 2x\tan x$; $4x\sec^2 x + 2\tan x + 2x^2\sec^2 x\tan x$
   (j) $2\cos 2x$; $-4\sin 2x$

6. (a) $4\cos 4x$
   (b) $-7\sin 7x$
   (c) $3\sec^2 3x$
   (d) $3x^2\cos x^3$
   (e) $-3\sin(3x + 1)$
   (f) $8\cos 2x$
   (g) $6(\cos 3x + \sin 6x)$
   (h) $-2x\sin(1 + x^2)$
   (i) $2\sec^2(2x + 1)$

7. (a) $3(1 + x)^2$
   (b) $-\frac{2x}{\sqrt{1-2x^2}}$
   (c) $2\sin x\cos x = \sin 2x$
   (d) $8x\tan 3x + 12x^2\sec^2 3x$
   (e) $-\frac{12}{(1+4x)^2}$
   (f) $-\frac{1+4x}{(1+2x+4x^2)^{3/2}}$
   (g) $4(1 - 4x^3)(1 + x - x^4)^3$
   (h) $\frac{2x}{(1+3x^2)^{2/3}}$
   (i) $6\sin 3x\cos 3x = 3\sin 6x$
   (j) $6\sin^2 x + 12x\sin x\cos x$
   (k) $\frac{12}{(1-3x)^3}$
   (l) $\frac{1}{2\sqrt{1+x}}$
   (m) $\frac{1}{(1+x)^{1/2}(1-x)^{3/2}}$
   (n) $3(x + \frac{1}{x})^2(1 - \frac{1}{x^2})$
   (o) $-12\cos^2 4x\sin 4x$
   (p) $4\tan 2x\sec^2 2x$

8. (a) $\frac{\cos 2x}{\sqrt{\sin 2x}}$
   (b) $\frac{2+x}{\sqrt{1+4x+x^2}}$
   (c) $8\sin 4x\cos 4x = 4\sin 8x$
   (d) $\tan^2 x + 2x\tan x\sec^2 x$
   (e) $\sec x\tan x$
   (f) $\frac{3}{(1-x)^2}$
   (g) $4\sin(2x - 3)\cos(2x - 3)$
   (h) $-6\sin 3x\cos 3x = -3\sin 6x$
   (i) $12\tan^2 4x\sec^2 4x$
   (j) $\tan(1 + x^2) + 2x^2\sec^2(1 + x^2)$
   (k) $-\frac{1}{(\sin x+\cos x)^2}$
   (l) $\frac{2\cos x}{(1-\sin x)^2}$
   (m) $\frac{1+2x^2}{\sqrt{1+x^2}}$

9. $y = 4x + 3$; $4y + x = 12$

10. (a) $y = 2x$

(b) $-5/3$

11. (b) 2

12. (a) $4\cos\theta, 2\sin\theta$
    (b) $\frac{1}{2}\tan\theta$
    (c) $2\sqrt{3}y = x - 3$
    (d) $\frac{1}{2}\sec^2\theta$
    (f) Ellipse

13. (a) $\sin 2t + 2t\cos 2t$, $\cos 2t - 2t\sin 2t$
    (b) $\frac{\cos 2t - 2t\sin 2t}{\sin 2t + 2t\cos 2t}$
    (c) $4\cos 2t - 4t\sin 2t$, $-4\sin 2t - 4t\cos 2t$
    (d) $-\frac{4+8t^2}{(\sin 2t + 2t\cos 2t)^3}$

14. (a) $\frac{xy^3\sin x - 2xy - y^3\cos x}{3xy^2\cos x + x^2}$
    (b) $-\frac{y^2\sin x + xy^2\cos x + 4xy}{2xy\sin x + 2x^2}$

17. 1.6π cm/min

18. (a) $\frac{1}{50\pi}$ cm/sec
    (b) 1.6 cm^2/sec

19. (a) $\frac{8}{9\pi} \approx 0.28$ m/min
    (b) 0.11 m/min

20. 1065 km/hr

21. Decreasing at 5.59 cm^2/min

22. $(8/5, 4/5)$

24. 5.46 metres from P

25. (b) $\sqrt{\frac{2nC_1}{C_2}} = 200$
    (c) Every 2 weeks

26. $r = 1.23$ metres, $h = 4.93$ metres

27. Cut out squares having sides 3.92 cm

29. Height of rectangular section ≈ 1.23 metres

    Radius of semicircular section ≈ 0.69 metres

30. 55 km/hr

31. Radius of semicircular end is

$$\left(\frac{(\pi+2)M}{\rho\pi^2}\right)^{1/3}.$$

32. $\approx 294^0$

34. 0.75

35. (b) 2.09

37. 1.532

38. (b) 1.15

39. 4% increase

40. 2.5%

42. 1.44% decrease

D.4 Functions and their Derivatives — Part 2

1. (a) 1.0517
   (b) 1.0005
   (c) 1.0000000 to seven places

3. (a) $2xe^{x^2+2}$
   (b) $\cos x\ e^{\sin x}$
   (c) $(1-2x)e^{-2x}$
   (d) $x^2(3+2x)e^{2x}$
   (e) $xe^{\sqrt{x^2+1}}/\sqrt{x^2+1}$
   (f) $e^{3x}(3\sin 2x + 2\cos 2x)$
   (g) $e^{2x}/\sqrt{1+e^{2x}}$
   (h) $\frac{3e^{-3x}}{2(1+e^{-3x})^{3/2}}$
   (i) $4(e^{2x}-2e^{x})(e^{2x}-e^{x})$
   (j) $e^{-2x}(3\sec^2 3x - 2\tan 3x)$
   (k) $e^{2x}(2\cos^2 3x - 6\cos 3x \sin 3x) = 2e^{2x}\cos 3x(\cos 3x - 3\sin 3x)$
   (l) $\frac{-4}{(e^x-e^{-x})^2}$

7. (a) $P_0 = 120,\ k = \frac{1}{3}\log\frac{5}{3}$
   (b) 237
   (c) 20.4 bacteria per hour; 34.1 bacteria per hour

8. (a) i. $(-\infty, \frac{1}{2})$
   ii. $(\frac{1}{2}, \infty)$
   iii. $(1, \infty)$
   iv. $(-\infty, 1)$

13. (a) $e^{2x}(2\sinh 3x + 3\cosh 3x)$
   (b) $x^2(3\cosh 2x + 2x\sinh 2x)$
   (c) $6\cosh 3x \sinh 3x$
   (d) $-2e^{-2x}$
   (e) $6\tanh^2 2x\ \mathrm{sech}^2 2x$
   (f) $\frac{2\cosh x}{(1-\sinh x)^2}$
   (g) $e^{-3x}(2\sinh 2x - 3\cosh 2x)$
   (h) $12\sinh^2 4x \cosh 4x$

17. (a) $\frac{5}{3}, \frac{12}{5}, \frac{3}{5}$
   (b) $\pm 2\sqrt{2},\ \pm\frac{2\sqrt{2}}{3}$
   (c) $\sqrt{10},\ 3/\sqrt{10}$
   (d) $\frac{3}{2\sqrt{2}}, \frac{1}{2\sqrt{2}}$

18. (a) $f^{-1}(x) = \frac{1}{2}(x^3-1)$
   (b) $f^{-1}(x) = \sqrt[3]{3x+2}$
   (c) No inverse
   (d) $f^{-1}(x) = -1 + \sqrt{9+x}$
   (e) $f^{-1}(x) = -1 - \sqrt{9+x}$

21. (a) $-\frac{1}{3}\log\frac{x-1}{2}$
   (b) $2 + e^{x/2}$
   (c) $\sinh x$

22. (a) $x(1 + 2\log x)$
   (b) $\frac{3x^2}{1+x^3}$
   (c) $\frac{14}{1+2x}$
   (d) $-\frac{1}{x}$
   (e) $\frac{3x^2+2\cosh x \sinh x}{x^3+\cosh^2 x}$
   (f) $\frac{6}{x}$
   (g) $\frac{6\log^5 x}{x}$
   (h) $\frac{6(\log x)^5}{x}$
   (i) $\frac{8}{(1+6x)(1-2x)}$
   (j) $\frac{4x}{(1+6x^2)(1+2x^2)}$
   (k) $-2 + \cot x - \frac{2x}{1+2x^2}$
   (l) $\frac{1}{x(1+\log x)^2}$

24. (a) $-2^{-x}\log 2$
   (b) $4^{x^2}\cdot 2x\log 4$
   (c) $x^{\sin x}\left(\frac{\sin x}{x} + \cos x \log x\right)$

(d) $y\left(\frac{6x^2}{1+x^3}+\frac{3\cos x}{1+\sin x}-\frac{8x^3}{1+2x^4}\right)$

(e) $y\left(\frac{2x}{1-x^4}\right)$

(f) $y\Big(-2x+\frac{x}{1+x^2}-\frac{2e^{2x}}{1+e^{2x}}$
$-\frac{2\sin x\cos x}{1+\sin^2 x}\Big)$

25. (a) $-\frac{1}{x\sqrt{x^2-1}}$

(b) $\frac{1}{2\sqrt{x-x^2}}$

(c) $\frac{1}{2x\sqrt{x-1}}$

(d) $-\frac{2}{\sqrt{1-4x^2}}$

(e) $-\frac{2}{x^2+4}$

(f) $e^{-2x}\big(-2\arcsin 4x$
$+\frac{4}{\sqrt{1-16x^2}}\big)$

(g) 0

(h) $-\frac{1}{x^2+1}$

(i) $\arccos x$

(j) $\arctan x$

29. (a) $\log(2+\sqrt{5})$

(b) $\log(2+\sqrt{3})$

(c) $\frac{1}{2}\log 1.5$

31. (a) $2x\operatorname{arsinh} 2x+\frac{2x^2}{\sqrt{1+4x^2}}$

(b) $-\frac{1}{x\sqrt{1-x^2}}$

(c) $\frac{1}{2\sqrt{x^2-x}}$

(d) $e^{-2x}\big(-2\operatorname{arsinh} x^2$
$+\frac{2x}{\sqrt{x^4+1}}\big)$

(e) $\frac{2}{1-x^2}$

(f) $\operatorname{arsinh} x$

(g) $2\operatorname{arcosh} 2x$

D.5 Integration and its Applications

1. (a) $\frac{1}{8}(2x-4)^4+c$
   (b) $\frac{8}{7}x^7-\frac{36}{5}x^5+18x^3-27x+c$
   (c) $\frac{1}{2}\log|2x-3|+c$
   (d) $\frac{1}{2}e^{2x-3}+c$
   (e) $\log(1+x^2)+c$
   (f) $-\frac{1}{3}\cos 3x+c$
   (g) $\frac{1}{2}\sin 2x+c$
   (h) $\frac{1}{5}\tan 5x+c$
2. $12+\frac{8}{n}$; $12-\frac{8}{n}$; 12
3. (a) 8
   (b) 3
4. (a) $\frac{9}{2}$
   (b) $\frac{9}{2}$
   (c) $\frac{8}{3}$
5. $\frac{32}{3}$
6. $\frac{1}{2}$
7. $\frac{37}{12}$
9. (a) $\frac{1}{3}1250\pi$
   (b) $\frac{1}{6}625\pi$
   (c) $\frac{1}{2}375\pi$
   (d) $\frac{1}{3}1375\pi$
10. (a) $\frac{1}{5}288\pi$
    (b) 24π
    (c) 8π
    (d) 40π
    (e) $\frac{1}{5}768\pi$
11. $\frac{1}{21}\pi$
12. $\frac{1}{6}\pi$
13. (a) $\frac{8}{27}(10\sqrt{10}-1)$
    (b) $\frac{123}{32}$
14. (a) $4\pi r^2$
    (b) $\int_0^2 2\pi\cosh x$
    $\sqrt{1+\sinh^2 x}\,dx$
    $= \int_0^2 2\pi\cosh^2 x\,dx.$
15. (a) $\rho\Delta x$
    (b) $\rho x^2\Delta x$
16. (a) $y\Delta x$
    (b) $k\rho x$
    (c) $k\rho x y\Delta x$

    $\frac{5000}{3}\rho k$ Newtons
17. 0.9270
18. (a) 1.4430
    (b) 1.4436
19. 6.79
20. (a) 0.3112
    (b) 0.3102
21. (a) $\frac{27}{768}$
    (b) $\frac{1}{10}\left(\frac{3}{8}\right)^4$
22. 0.8647
23. 1.073
24. $2a\sinh 1$

D.6 Methods of Integration

1. (a) $\frac{1}{12}(3x-5)^4+c$
   (b) $\frac{14}{9}$
   (c) $-\frac{1}{6(3x-5)^2}+c$
   (d) $\frac{1}{3}\log 0.4$
   (e) $-\frac{2}{5}\sqrt{3-5x}+c$
   (f) $-\frac{1}{5}\log|3-5x|+c$

2. (a) $-\frac{1}{3}\cos(3x-2)+c$
   (b) $\frac{1}{3}\tan(3x-2)+c$
   (c) $\frac{1}{3}\cosh(3x-2)+c$
   (d) $\frac{1}{3}\tanh(3x-2)+c$
   (e) $\frac{1}{3}\sin(3x-2)+c$
   (f) $\frac{1}{3}\sinh(3x-2)+c$

3. (a) $\frac{1}{10}(-5e^{-x}+e^{-5x})+c$
   (b) $-\frac{1}{10}(5e^{-x}+e^{-5x})+c$
   (c) $\frac{1}{60}(-15e^{-x}+10e^{-3x}-3e^{-5x})+c$
   (d) $\frac{1}{20}(e^{5x}-5e^{-x}-5e^{x}+e^{-5x})+c$
   (e) $\frac{1}{10}(5\sin x-\sin 5x)+c$
   (f) $-\frac{1}{10}(\cos 5x+5\cos x)+c$
   (g) $\tan x-x+c$

4. (a) $\frac{4}{3}x^3-12\log|x|-\frac{3}{x^3}+c$
   (b) $\frac{1}{4}e^{4x}-6e^{x}-\frac{9}{2}e^{-2x}+c$
   (c) $\frac{27}{7}x^7-\frac{54}{5}x^5+12x^3-8x+c$

5. (a) $\log 9$
   (b) $\frac{2}{3}\sqrt{x^3+9x+1}+c$
   (c) $\frac{1}{3}(2x-x^2)^{\frac{3}{2}}+c$
   (d) $\frac{1}{4}(\arcsin 2x)^2+c$
   (e) $\frac{58}{3}$
   (f) $-\frac{1}{3(3e^x+7)}+c$
   (g) $\frac{1}{4}\log 2$
   (h) $\frac{1}{5}\tan^5 x+c$
   (i) $\frac{2}{3}$
   (j) $2\sqrt{\sin x}+c$
   (k) $\frac{1}{2}\tan x^2+c$

6. (a) $x\arcsin x+\sqrt{1-x^2}+c$
   (b) $\frac{1}{3}x^3\log x-\frac{1}{9}x^3+c$
   (c) $x\log x-x+c$
   (d) π
   (e) $\frac{1}{4}(1-3e^{-2})$
   (f) $-x^2\cos x+2x\sin x+2\cos x+c$
   (g) $-e^{-x}(2x+3)+c$
   (h) $\log x\{\log(\log x)-1\}+c$

(i) $\frac{1}{2}(4\log 4 - 3)$
(j) $-2\sqrt{x}\cos\sqrt{x} + 2\sin\sqrt{x} + c$
(k) $x\cosh x - \sinh x + c$
(l) $\frac{1}{3}x^3 - 2x\cos x + 2\sin x + \frac{1}{2}x - \frac{1}{4}\sin 2x + c$

7. (a) $\frac{1}{16}\pi$
(b) $\frac{1}{5}\sin^5 x - \frac{1}{7}\sin^7 x + c$
(c) $\frac{35}{128}\pi$
(d) $\frac{3}{16}\pi$
(e) $\frac{1}{32}(3\pi + 8)$
(f) $\frac{13}{15} - \frac{1}{4}\pi$
(g) $\tan x + \frac{1}{3}\tan^3 x + c$
(h) $\frac{1}{4}\tan^4 x - \frac{1}{2}\tan^2 x - \log|\cos x| + c$
(i) $\frac{8}{15}$
(j) $24 - 65e^{-1}$
(k) $\frac{3}{4}(1 - 7e^{-2})$

8. (b) i. $\frac{16}{35}$ ii. $\frac{5}{32}\pi$ iii. $\frac{5}{256}\pi$

(c) $$\int_0^{\frac{\pi}{2}} \cos^n x\,dx = \frac{n-1}{n}\int_0^{\frac{\pi}{2}} \cos^{n-2} x\,dx$$

9. (a) $\frac{1}{6}(3x\sqrt{9x^2+1} + \text{arsinh}\,3x) + c$
(b) $\frac{1}{2}\text{arcosh}\,2x + c$
(c) $\frac{1}{2\sqrt{2}}(\arcsin\sqrt{2}x + \sqrt{2}x\sqrt{1-2x^2}) + c$
(d) $\frac{1}{2}\{(x+1)\sqrt{x^2+2x} - \text{arcosh}\,(x+1)\} + c$
(e) $\frac{1}{\sqrt{3}}\text{arcosh}\,\sqrt{3}x + c$
(f) $\frac{9}{2}\text{arsinh}\,\frac{x+1}{3} + \frac{1}{2}(x+1)\sqrt{x^2+2x+10} + c$
(g) $\frac{1}{2}(x+1)\sqrt{x^2+2x-3} - 2\text{arcosh}\,\frac{x+1}{2} + c$

10. (a) $\sqrt{13} - \sqrt{10} + 2(\text{arsinh}\,\frac{2}{3} - \text{arsinh}\,\frac{1}{3})$
(b) $\frac{1}{3}19\sqrt{13} - \frac{1}{3}13\sqrt{10} + 9\log\frac{2+\sqrt{13}}{1+\sqrt{10}}$
(c) $3(\sqrt{2} - \sqrt{5}) + 2\log(\sqrt{2} - 1) - 2\log(\sqrt{5} - 2)$
(d) $\frac{2}{3}\sqrt{5} - \frac{1}{6}\sqrt{2} - \frac{1}{2}\log\frac{2+\sqrt{5}}{1+\sqrt{2}}$

11. (a) $\text{arsinh}\,\frac{x}{2} - \frac{\sqrt{4+x^2}}{x} + c$
(b) $\frac{1}{4}\{\log(\sqrt{x^2+4} - 2) - \log(\sqrt{x^2+4} + 2)\} + c$

(c) $\sqrt{4-x^2} - 2\log\left|\frac{2+\sqrt{4-x^2}}{x}\right| + c$

(d) $\operatorname{arcosh}\frac{x}{2} - \frac{\sqrt{x^2-4}}{x} + c$

12. (a) $1 + 2\log 2$
(b) $\frac{1}{2}x^2 - x + \log|x+1| + c$
(c) $x + \frac{1}{2}\log\left|\frac{x-1}{x+1}\right| + c$
(d) $2 - \log 5$
(e) $\frac{7}{3}\log|x-2| - \frac{1}{3}\log|x+1| + c$
(f) $3\log\left|\frac{x}{x+1}\right| + c$
(g) $\log|x| - \frac{1}{2}\log|x^2+4x+3| + c$
(h) $x - \frac{1}{4}\log|x| + \frac{9}{8}\log|x-2| - \frac{7}{8}\log|x+2| + c$
(i) $-\frac{3}{x-2} - \log|x+2| + \log|x-2| + c$
(j) $-\frac{1}{x-1} + 2\log|x+1| - 2\log|x-1| + c$
(k) $2\log|x-1| - 3\arctan(x+1) + c$
(l) $\log\sqrt{3} + \frac{1}{12}\pi$
(m) $2\sqrt{x} - 3\sqrt[3]{x} + 6\sqrt[6]{x} - 6\log(1+\sqrt[6]{x}) + c$
(n) $\log\left(\frac{3e+3}{2e+4}\right)$
(o) $\log 3 - \log\sqrt{5} + 2\arctan 2$
(p) $\log|x| - \log\sqrt{x^2+2x+5} + c$
(q) $\log(x^2+2x+2) + 3\arctan(x+1) + c$
(r) $\frac{1}{2}\log(x^2+4x+13) + \arctan\frac{x+2}{3} + c$
(s) $-\log(1+e^{-x}) + c$

14. $6\arcsin\sqrt{\frac{x+2}{6}} + \sqrt{(x+2)(4-x)} + c$

15. (a) $\frac{1}{5}\sec^4 x\tan x + \frac{4}{15}\sec^2 x\tan x + \frac{8}{15}\tan x + c$
(b) $\frac{67}{48}\sqrt{2} + \frac{5}{16}\log(1+\sqrt{2})$

16. (b) i. $\frac{2}{\sqrt{3}}\arctan\left(\frac{2\tan\frac{x}{2}+1}{\sqrt{3}}\right) + c$
ii. $\log\left(1+\tan\frac{x}{2}\right) + c$

17. $-\frac{1}{13}e^{-2x}(2\sin 3x + 3\cos 3x) + c$

18. $\frac{1}{2}e^{-\cos 2x} + c$

19. $\frac{1}{10}(5e^x + e^{-5x}) + c$

20. $-\frac{1}{4}e^{-4x} + \frac{1}{3}x^3 + xe^{-2x} + \frac{1}{2}e^{-2x} + c$

21. $\frac{1}{4}x^2 - \frac{1}{4}x + \frac{1}{8}\log|2x+1| + c$

22. $-(2x+5)e^{-x} + c$

23. $x\sinh x - \cosh x + c$

24. $\frac{1}{2}(\log 2 - \frac{3}{8})$

25. $2\sqrt{x}\sin\sqrt{x} + 2\cos\sqrt{x} + c$

26. $\arctan x - \frac{1}{2}\arctan\frac{x}{2} + c$

27. $\frac{5}{6} - \log 2$

28. $2\arctan\sqrt{x} + c$

29. $\frac{1}{12}(e^{3x} - e^{-3x} + 3e^{x} - 3e^{-x}) = \frac{1}{6}(\sinh 3x + 3\sinh x) + c$

30. $\frac{1}{2}(\operatorname{arsinh} e^{x} + e^{x}\sqrt{1+e^{2x}}) + c$

31. $\frac{1}{12}(\sinh 6x - 6x) + c$

32. $7\log 1.5 - 1.5$

33. $5\log|x| - \frac{4}{x-1} - 4\log|x-1| + c$

34. $\frac{1}{4}x^4\log x - \frac{1}{16}x^4 + c$

35. $\frac{1}{64}(24x + 8\sin 4x + \sin 8x) + c$

36. $\log 3 - 1$

37. $x\operatorname{arsinh} x - \sqrt{1+x^2} + c$

38. $\frac{1}{4}\tan 4x - x + c$

39. $x\tan x + \log|\cos x| + c$

40. $2\sqrt{\sin x} - \frac{2}{5}(\sin x)^{\frac{5}{2}} + c$

41. $\frac{1}{20}(2x+1)^{5/2} - \frac{1}{6}(2x+1)^{3/2} + \frac{1}{4}\sqrt{2x+1} + c$

42. $\frac{1}{2}\left(\arctan x + \frac{x}{1+x^2}\right) + c$

43. $\operatorname{arsinh}\frac{x+2}{2} + c$

44. $-\frac{1}{3}e^{-x^3} + c$

45. $-\frac{1}{2}(1+x^2)e^{-x^2} + c$

46. $x\log(1+x^2) - 2x + 2\arctan x + c$

47. $\frac{1}{90}$

48. $2x + \log|x+3| + c$

49. $x + 5\log|x-1| - 4\log|x| + \frac{4}{x} + c$

50. $\frac{1}{16}(\sin 8x + 4\sin 2x) + c$

51. $(\frac{1}{2}x^2 - \frac{1}{4})\arcsin x + \frac{1}{4}x\sqrt{1-x^2} + c$

52. $2\log(1+\sqrt{x}) + c$

53. $-2\sqrt{2+\cos x} + c$

54. $2 - \sqrt{3}$

55. $x + 4\sqrt{x} + 4\log|\sqrt{x}-1| + c$

56. $2\arcsin\frac{x-2}{2} + \frac{1}{2}(x-2)\sqrt{4x-x^2} + c$

57. $\operatorname{arcosh}\frac{x+2}{\sqrt{5}} + c$

58. $\frac{4}{3}(4-\sqrt{8})$

59. $(x^2+1)\sin x + 2x\cos x - 2\sin x + c$

60. $\frac{1}{2}\log(x^2+2x+10) + c$

61. $\frac{1}{2}\log(x^2+2x+10) + \frac{1}{3}\arctan\frac{x+1}{3} + c$

62. $2\sqrt{x}\arctan\sqrt{x} - \log(1+x) + c$

63. $e^x - \log(1+e^x) + c$

64. $\log|x+1| - \log\sqrt{1+x^2} + \arctan x + c$

65. $-2\log|x| + \log|x^2-1| + c$

66. $\frac{1}{34}e^{3x}(3\cos 5x + 5\sin 5x) + c$

67. $\frac{1}{2}\log(x^2+4x+5) - 2\arctan(x+2) + c$

68. $\sqrt{9-x^2} - 3\log\left|\frac{3+\sqrt{9-x^2}}{x}\right| + c$

D.7 Differential Equations

1. (a) $y^2 = x^2 + c$
   (b) $y = ce^{x^2}$
   (c) $y^2 = 2 + c/x^2$
   (d) $y = ce^{10x}$
   (e) $y^2 = 2\log x + 2x + c$

2. (a) $y = 2(1 - e^{-2x})$
   (b) $y = 2/(2 - x)$
   (c) $\arcsin y = \arcsin x + \frac{1}{2}\pi$
   (d) $e^y + \sqrt{2} = 1 + \sqrt{1 + x^2}$

3. $y^2 = x^2 + cx$

4. $y + 2\log(x + y) = x + c$

5. (a) $y = ae^{2x} - \frac{1}{4}(2x + 1)$
   (b) $xy = e^x + c$
   (c) $2y = x^4e^{2x} + cx^4$
   (d) $y = c\sec x - 1$
   (e) $y = ae^{-3x} + \frac{1}{2}e^{-x}$
   (f) $y = \cos x + c\cos^2 x$

6. (a) $y = 3\sqrt{1 - x^2}$
   (b) $y(x - 2) = x - 4$
   (c) $\tan y = x - 1$
   (d) $5xy = 2(x^{\frac{5}{2}} - 1)$

7. (a) $y = a + be^x - x - \frac{1}{2}x^2$
   (b) $y = a + be^{2x} - e^x$
   (c) $\log|(A + y)/(A - y)| = Ax + C$

8. $y^2 = 3x^2 + 1$

9. $100 + t + 100/(1 + \frac{t}{100})^2$ kg

10. 276

11. $v = \frac{g}{k}(1 - e^{-kt}) + Ve^{-kt}$

12. 20.6° C

13. (a) $i = \frac{E_0}{R}(1 - e^{-Rt/L})$
    (b) $i = \frac{E_0}{\omega^2L^2+R^2}(\omega Le^{-Rt/L} + R\sin\omega t - \omega L\cos\omega t)$
    (c) $i = \frac{E_0}{R-aL}(e^{-at} - e^{-Rt/L})$ if $a \neq R/L$

14. (a) $y = ae^{3x} + be^{-2x}$
    (b) $y = ae^{-4x} + be^{2x}$
    (c) $y = ae^{3x} + bxe^{3x}$
    (d) $y = e^{-x}(a\cos 3x + b\sin 3x)$
    (e) $y = e^{-2x}(a\cos\sqrt{5}x + b\sin\sqrt{5}x)$
    (f) $y = (ax + b)e^{-\frac{1}{2}x}$

(g) $y = ae^{3x} + be^{-3x}$
(h) $y = a + be^{9x}$
(i) $y = e^t(a\cos t + b\sin t)$
(j) $y = a\cos 2t + b\sin 2t$

15. (a) $y = -2 + ae^{3x} + be^{-2x}$
(b) $y = -2x + \frac{1}{3} + ae^{3x} + be^{-2x}$
(c) $y = -2e^t + ae^{-t} + be^{2t}$
(d) $y = \frac{4}{3}xe^{2x} + ae^{2x} + be^{-x}$
(e) $y = ae^x + be^{-2x} - 2x^2 - 2x - 3$
(f) $y = a + be^{2t} - \frac{1}{2}t(t+1)$
(g) $y = 12x^2e^{4x} + (ax+b)e^{4x}$
(h) $y = e^x(a\cos 2x + b\sin 2x) + \frac{20}{17}\sin 2x + \frac{80}{17}\cos 2x$
(i) $y = a\sin 2x + b\cos 2x - 5x\cos 2x$

16. (a) $y = ae^{3x} + be^{2x} + 3e^{4x}$
(b) $y = ae^{3x} + be^{2x} - 6xe^{2x}$

17. $e^{3x} + 2e^{-x} + 2xe^{3x}$

18. (a) $y = ae^{2x} + be^{-2x} - e^x - 2x^2 - 1$
(b) $y = ae^{2x} + be^{-x} + \frac{1}{2}e^{-2x} + \frac{2}{3}xe^{2x}$
(c) $y = 1 + e^{-x}(a\cos x + b\sin x) + \frac{1}{5}(\cos 2x - 2\sin 2x)$
(d) $y = (ax+b)e^{-x} + \frac{1}{25}e^x(3\cos x + 4\sin x)$
(e) $y = ae^{2x} + be^x + (x + \frac{5}{6})e^{-x}$
(f) $y = ae^x + be^{-\frac{2}{3}x} + \frac{1}{4}(2x+7)e^{-x}$
(g) $y = a\cos 2x + b\sin 2x + \frac{1}{17}e^x(\sin 2x - 4\cos 2x)$

19. $y = (x + a\sin x + b\cos x)e^{x^2}$

20. $y = a + b/x + \frac{1}{3}(\log x)^3 - (\log x)^2 + 2\log x$

21. $y = \frac{1}{\sqrt{x}}(A\sin x + B\cos x)$

23. (a) $x = \frac{1}{2}\frac{t}{b}e^{-\lambda t}\sin bt$
(b) $x = -e^{-\lambda t}\frac{1}{\omega^2-\lambda^2-b^2}\cos\sqrt{\omega^2-\lambda^2}t + \frac{1}{\omega^2-\lambda^2-b^2}e^{-\lambda t}\cos bt$

25. Steady-state solution is

$$\frac{E_0\omega C^2}{(1-\omega^2 LC)^2 + \omega^2R^2C^2}\left\{\omega R\sin\omega t + \frac{1}{C}(1-\omega^2 LC)\cos\omega t\right\}$$

28. (a) $x = Ae^{(-\mu+\sqrt{\mu^2-m^2})t} + Be^{(-\mu-\sqrt{\mu^2-m^2})t}$
$+ \frac{a}{(m^2-\omega^2)^2+4\mu^2\omega^2}\{(m^2-\omega^2)\sin\omega t - 2\mu\omega\cos\omega t\}$

(b) $x = e^{-\mu t}\left(A\cos\sqrt{m^2-\mu^2}t + B\sin\sqrt{m^2-\mu^2}t\right)$
$+ \frac{a}{(m^2-\omega^2)^2+4\mu^2\omega^2}\{(m^2-\omega^2)\sin\omega t - 2\mu\omega\cos\omega t\}$

(c) $x = (At+B)e^{-\mu t}$
$+ \frac{a}{(m^2-\omega^2)^2+4\mu^2\omega^2}\{(m^2-\omega^2)\sin\omega t - 2\mu\omega\cos\omega t\}$

29. $x = \frac{7}{3}(\cosh\sqrt{3}t - 1)$; $y = \frac{7}{2}\cosh\sqrt{3}t$

D.8 Series

1. Sum =1

2. Sum =1

3. (a) $S_n = \log(n+1)$; series diverges
   (b) $S_n = \frac{1}{2}\{\frac{1}{2} + (-1)^n(\frac{1}{n+2} - \frac{1}{n+1})\}$; series converges with sum $\frac{1}{4}$
   (c) $S_n = 3\{1 - (\frac{3}{4})^n\}$; series converges with sum 3
   (d) $S_n = \frac{6}{5}\{1 - (-\frac{2}{3})^n\}$; series converges with sum 1.2
   (e) $S_n = 0.1 + \frac{4}{90}\{1 - (0.1)^{n-1}\}$; series converges with sum $\frac{13}{90}$
   (f) $S_n = \frac{1}{35}\{1 - (-\frac{2}{5})^n\}$; series converges with sum $\frac{1}{35}$
   (g) $S_n = 1 - (-\frac{1}{2})^n$; series converges with sum 1

4. Series converges with a sum of $\log 2$.

9. (a) Divergent
   (b) Convergent
   (c) Divergent
   (d) Convergent
   (e) Convergent
   (f) Convergent
   (g) Convergent
   (h) Divergent

11. (a) Divergent
    (b) Convergent

12. (a) Convergent
    (b) Divergent

13. (a) Convergent
    (b) Divergent
    (c) Convergent
    (d) Convergent
    (e) Convergent
    (f) Convergent
    (g) Convergent
    (h) Convergent

14. Convergent

15. (a) Convergent
    (b) Divergent
    (c) Convergent
    (d) Convergent
    (e) Convergent
    (f) Divergent
    (g) Convergent

18. (a) Convergent; comparison test
    (b) Convergent geometric series (common ratio $= -\frac{3}{4}$)
    (c) Convergent; alternating series test
    (d) Convergent; ratio test
    (e) Convergent; comparison test
    (f) Divergent; comparison test
    (g) Divergent; comparison test

(h) Convergent; ratio test

(i) Convergent; ratio test

(j) Divergent; $a_n \not\to 0$

(k) Convergent; alternating series test

(l) Divergent; $a_n \not\to 0$

(m) Divergent; ratio test

(n) Convergent; ratio test

19. (a) Convergent; alternating series test

(b) Convergent; ratio test

(c) Convergent; ratio test

(d) Convergent; comparison test

(e) Convergent; absolute convergence theorem

(f) Divergent; $a_n \not\to 0$

(g) Convergent; comparison test

(h) Convergent; ratio test

(i) Convergent; comparison test

(j) Convergent; alternating series test

(k) Convergent; comparison test

(l) Convergent; comparison test

(m) Divergent; comparison test

(n) Divergent; ratio test

(o) Convergent; ratio test

20. (a) $2 + \frac{1}{4}(x-4) - \frac{1}{64}(x-4)^2 + \frac{1}{512}(x-4)^3$; $\sqrt{4.2} \approx 2.0494$;

(b) $\frac{\sqrt{3}}{2} + \frac{1}{2}(x - \frac{\pi}{3}) - \frac{\sqrt{3}}{4}(x - \frac{\pi}{3})^2 - \frac{1}{12}(x - \frac{\pi}{3})^3$; $\sin 58^\circ \approx 0.8480$

(c) $\frac{1}{\sqrt{2}}\{1 - (x - \frac{\pi}{4}) - \frac{1}{2}(x - \frac{\pi}{4})^2 + \frac{1}{6}(x - \frac{\pi}{4})^3\}$; $\cos 44^\circ \approx 0.7193$

24. (a) $-1 - (x+1) - (x+1)^2 - (x+1)^3 - \cdots = -\sum_{n=0}^{\infty}(x+1)^n$

(b) $\frac{\sqrt{3}}{2} + \frac{1}{2}(x - \frac{\pi}{3}) - \frac{\sqrt{3}}{4}(x - \frac{\pi}{3})^2 - \frac{1}{12}(x - \frac{\pi}{3})^3 - \cdots$

(c) $\frac{1}{2} - \frac{\sqrt{3}}{2}(x - \frac{\pi}{3}) - \frac{1}{4}(x - \frac{\pi}{3})^2 + \frac{\sqrt{3}}{12}(x - \frac{\pi}{3})^3 - \cdots$

25. $a_0 = 1$, $a_1 = 2$, $a_2 = 2$

26. (a) $R = 2$

(b) $R = 3$

(c) $R = 1$

(d) $R = \frac{1}{2}$

(e) $R = 3$

(f) $R = 3$

(g) $R = 4$

(h) $R = 2$

(i) $R = 1$

(j) $R = \infty$

(k) $R = 9$ (m) $R = 1$
(l) $R = 1$ (n) $R = 1$

27. (c) $x - \frac{x^3}{3^2} + \frac{x^5}{5^2} - \frac{x^7}{7^2} + \cdots$

29. (a) $\sqrt{1-x} = 1 - \frac{1}{2}x - \frac{1}{8}x^2 - \frac{1}{16}x^3 - \frac{5}{128}x^4 + \cdots; R = 1$
(b) $\sqrt{2-x} = \sqrt{2}(1 - \frac{1}{4}x - \frac{1}{32}x^2 - \frac{1}{128}x^3 - \cdots); R = 2$
(c) $(1+2x)^{-3} = 1 - 6x + 24x^2 - 80x^3 + \cdots; R = \frac{1}{2}$
(d) $(1+x)^{-2} = 1 - 2x + 3x^2 - 4x^3 + \cdots; R = 1$
(e) $(1+3x^3)^{-\frac{1}{3}} = 1 - x^3 + 2x^6 - \frac{14}{3}x^9 + \cdots; R = \frac{1}{\sqrt[3]{3}}$
(f) $\frac{1}{\sqrt{4-x^2}} = \frac{1}{2}(1 + \frac{1}{8}x^2 + \frac{3}{128}x^4 + \frac{5}{1024}x^6 + \cdots); R = 2$
(g) $(1+2x^2)^{-2} = 1 - 4x^2 + 12x^4 - 32x^6 + \cdots; R = \frac{1}{\sqrt{2}}$

30. $-\sum_{n=0}^{\infty}[3^{n+1} + \frac{(-1)^n}{2^{n+1}}]x^n = -\frac{7}{2} - \frac{35}{4}x - \frac{217}{8}x^2 - \cdots;\ R = \frac{1}{3}$

31. $\sum_{n=0}^{\infty} \frac{1}{3}[2^{n+1} + (-1)^n]x^n = 1 + x + 3x^2 + 5x^3 + 11x^4 + \cdots;\ R = \frac{1}{2}$

32. $\log(\frac{1+x}{1-x}) = 2[x + \frac{x^3}{3} + \frac{x^5}{5} + \frac{x^7}{7} + \frac{x^9}{9} + \cdots]$ for $|x| < 1$; $\log 3 \approx 1.0981$ using the first four terms

33. $\arcsin x = x + \frac{1}{6}x^3 + \frac{3}{40}x^5 + \frac{5}{112}x^7 + \cdots$; $\operatorname{arsinh} x = x - \frac{1}{6}x^3 + \frac{3}{40}x^5 - \frac{5}{112}x^7 + \cdots$

34. (a) $1 - \frac{x^2}{2} + \frac{x^4}{2^2 2!} - \frac{x^6}{2^3 3!} + \cdots$
(b) $1 + x - \frac{1}{3}x^3 - \frac{1}{6}x^4 + \cdots$
(c) $\log 2 + \frac{1}{2}x + \frac{1}{8}x^2 - \frac{1}{192}x^4 + \cdots$

37. (a) 1.0280
(b) 0.1026
(c) 0.50387 using the first three terms; truncation error $< 7.6 \times 10^{-7}$
(d) 0.19956 using the first two terms; truncation error $< 5.3 \times 10^{-7}$

38. 0.2032

39. $1 - \frac{1}{2}x - \frac{5}{8}x^2 + \frac{3}{16}x^3 + \cdots; 0.0973$

D.9 Matrices

1. (a) $x_1 = x_2 = 1,\ x_3 = 2$

   (b) $x_1 = -1,\ x_2 = 1,\ x_3 = 3$

2. (a) No solutions

   (b) $x_1 = 2 + \alpha,\ x_2 = -2\alpha,\ x_3 = \alpha$

3. (a)
$$\begin{aligned} x_1 &= -5\alpha - 5\beta \\ x_2 &= -2\alpha + 6\beta \\ x_3 &= -3\alpha + 14\beta \\ x_4 &= 5\alpha \\ x_5 &= 15\beta \end{aligned}$$

   (b) $x_1 = -2\alpha,\ x_2 = \alpha,\ x_3 = 0,\ x_4 = -4\alpha$

4. (a) $\begin{pmatrix} a \\ a \\ a \end{pmatrix}$ (b) $\begin{pmatrix} a \\ b \\ -a-b \end{pmatrix}$ (c) $\begin{pmatrix} 0 \\ 0 \\ 0 \end{pmatrix}$

8. (a) No inverse exists.

   (b) $\begin{bmatrix} 3/11 & -5/11 \\ -1/11 & -2/11 \end{bmatrix}$

   (c) $\begin{bmatrix} -1/2 & 1/2 & 1/2 \\ 1/2 & -1/2 & 1/2 \\ 1/2 & 1/2 & -1/2 \end{bmatrix}$

   (d) $\begin{bmatrix} 9/2 & -3/2 & -1/2 \\ -2 & 1 & 0 \\ 1/2 & -1/2 & 1/2 \end{bmatrix}$

   (e) No inverse exists.

   (f) $\begin{bmatrix} -1/4 & 1/4 & 1/4 & 1/4 \\ 1/4 & -1/4 & 1/4 & 1/4 \\ 1/4 & 1/4 & -1/4 & 1/4 \\ 1/4 & 1/4 & 1/4 & -1/4 \end{bmatrix}$

10. $\begin{bmatrix} -4/5 & -7/10 \\ 1/5 & 3/10 \end{bmatrix}$

14. (a) A represents an anticlockwise rotation of $\pi/3$.
    B represents an anticlockwise rotation of $210°$.
    C represents an anticlockwise rotation of $\pi/4$.

    (b) AB and BA represent anticlockwise rotations through $270°$ so that

    $$AB = BA = \begin{bmatrix} 0 & 1 \\ -1 & 0 \end{bmatrix}.$$

    The matrix C^{-1} is a clockwise rotation through $45°$; that is, an anticlockwise rotation through $315°$ so that

    $$C^{-1} = \begin{bmatrix} 1/\sqrt{2} & 1/\sqrt{2} \\ -1/\sqrt{2} & 1/\sqrt{2} \end{bmatrix}.$$

16. (a) A reflection in the line $y = x$
    (b) A dilation of factor 2 in the x-direction
    (c) A reflection in the y-axis

17. (a) A represents an anticlockwise rotation of $45°$ about the x-axis.
    (b) B represents an anticlockwise rotation of $45°$ about the y-axis.

18. The axis of rotation is the line joining the origin to the point $(0, 1, 1)$; the angle of rotation is $\arccos(1/3)$.

19. The vectors $\begin{pmatrix} 1 \\ 0 \\ -1 \end{pmatrix}$, $\begin{pmatrix} 1 \\ 0 \\ 1 \end{pmatrix}$ and $\begin{pmatrix} 0 \\ 1 \\ 0 \end{pmatrix}$ are eigenvectors of A corresponding to the eigenvalues 0, 2 and 1 respectively.

20. (a) Eigenvalues $1,\ -1,\ 2$ with corresponding eigenvectors

    $$\begin{pmatrix} 1 \\ 0 \\ 0 \end{pmatrix}, \begin{pmatrix} 1 \\ 1 \\ 0 \end{pmatrix} \text{ and } \begin{pmatrix} 5 \\ 1 \\ 1 \end{pmatrix}$$

    respectively.

(b) Eigenvalues 0, 0, 3 with corresponding eigenvectors

$$\begin{pmatrix} 1 \\ -1 \\ 0 \end{pmatrix}, \begin{pmatrix} 1 \\ 0 \\ -1 \end{pmatrix} \text{ and } \begin{pmatrix} 1 \\ 1 \\ 1 \end{pmatrix}$$

respectively.

(c) Eigenvalues -1, $1+\sqrt{2}$, $1-\sqrt{2}$ with corresponding eigenvectors

$$\begin{pmatrix} 0 \\ 1 \\ -1 \end{pmatrix}, \begin{pmatrix} \sqrt{2} \\ 1 \\ 1 \end{pmatrix} \text{ and } \begin{pmatrix} -\sqrt{2} \\ 1 \\ 1 \end{pmatrix}$$

respectively.

(d) Eigenvalues are -1, -1, 8 with the corresponding eigenvectors

$$\begin{pmatrix} 1 \\ 0 \\ -1 \end{pmatrix}, \begin{pmatrix} 1 \\ -2 \\ 0 \end{pmatrix} \text{ and } \begin{pmatrix} 2 \\ 1 \\ 2 \end{pmatrix}$$

respectively.

D.10 Vector-valued Functions

1. (b) $-\frac{1}{2}\pi\,\mathbf{i}+\mathbf{j}$

   (c) $\int_0^\pi \sqrt{1+t^2}\,dt = \frac{1}{2}(\operatorname{arsinh}\pi + \pi\sqrt{1+\pi^2}) \approx 6.1$

2. (b) $-2\,\mathbf{j}+\mathbf{k}$

3. (a) $\frac{1}{\sqrt{26}}(2\,\mathbf{i}+3\sqrt{2}\,\mathbf{j}+2\,\mathbf{k}); \mathbf{r}(s) = (2, 2\sqrt{2}, 1) + s(2, 3\sqrt{2}, 2)$

   (b) $\frac{1}{27}(35\sqrt{35} - 17\sqrt{17})$

4. (b) $\mathbf{r}(s) = (0, \frac{1}{2}\pi, \frac{1}{4}\pi^2) + s(-\frac{1}{2}\pi, 1, \pi)$

   (c) $\int_0^{\frac{1}{2}\pi} \sqrt{1+5t^2}\,dt$

6. (a) $-2\sin 2t\,\mathbf{i} + 2\cos 2t\,\mathbf{j} + 2\sinh t\,\mathbf{k}$; $-4\cos 2t\,\mathbf{i} - 4\sin 2t\,\mathbf{j} + 2\cosh t\,\mathbf{k}$

7. (a) $\mathbf{v} = \cos t\,\mathbf{i} - \sin t\,\mathbf{j} + 3t^2\,\mathbf{k}, \mathbf{a} = -\sin t\,\mathbf{i} - \cos t\,\mathbf{j} + 6t\,\mathbf{k}$

   (b) $t = 0$

8. (a) $-6\sin 2t\,\mathbf{i} + 6\cos 2t\,\mathbf{j} + \mathbf{k}$; $\sqrt{37}$

   (b) $3\sqrt{37}$

   (c) $\frac{1}{\sqrt{37}}(-6\sin 6\,\mathbf{i} + 6\cos 6\,\mathbf{j} + \mathbf{k})$

9. (b) $\mathbf{a} = -4\cos t\,\mathbf{i} - 4\sin t\,\mathbf{j}$

10. (a) $\frac{1}{\sqrt{6}}(2\,\mathbf{i} - \mathbf{j} + \mathbf{k})$

    (b)
    $$\begin{aligned} x &= 1+2s \\ y &= -s \\ z &= 2+s \end{aligned}$$

    (c) $\int_1^4 \sqrt{4u^2+1+\frac{1}{u}}\,du$

11. 45°

D.11 Functions of Several Variables

1. (a) $6x, 2y$
   (b) $2y + 2x; 2x - 3$
   (c) $2x\cos xy - x^2 y \sin xy, -x^3 \sin xy$
   (d) $\frac{2y}{(x+y)^2}, -\frac{2x}{(x+y)^2}$
   (e) $\cos(x+2y), 2\cos(x+2y)$
   (f) $-\frac{y}{(x^2+y^2)}, \frac{x}{(x^2+y^2)}$
   (g) $\frac{y^2}{(x+y)^2}, \frac{x^2}{(x+y)^2}$
   (h) $e^{xy}(\cos x + y\sin x), xe^{xy}\sin x$
   (i) $yx^{y-1}, x^y \log x$

4. $-\frac{2}{r^2}$

6. (a) $-\frac{\sin\theta}{r^2}$
   (b) $-\frac{2\cos\theta}{r^3}$

12. 0

15. $2v^2x + 4uvy$

18. (a) $24\,\mathbf{i} - 8\,\mathbf{j} + 8\,\mathbf{k}$
    (b) $\frac{56}{3}$

19. (a) $4\,\mathbf{i} + 6\,\mathbf{j} - 8\,\mathbf{k}$
    (b) $2x + 3y - 4z + 3 = 0$

20. (a) $-2\,\mathbf{i} + 4\,\mathbf{j} + \mathbf{k}$
    (b) i. $-2x + 4y + z = 11$
        ii. $$\begin{aligned} x &= -1 - 2t \\ y &= 1 + 4t \\ z &= 5 + t \end{aligned}$$

21. (a) $-\frac{4}{3}$ (f decreases at this point in the given direction.)
    (b) $\mathbf{i} + 3\,\mathbf{j}$

22. (a) $-4\,\mathbf{i} + 2\,\mathbf{j} + \mathbf{k}$
    (b) $-4x + 2y + z = 17$

23. (a) $\frac{4}{3}$
    (b) $6\,\mathbf{i} + 4\,\mathbf{j} + 5\,\mathbf{k}$

24. $\frac{28}{9}$

25. $2x + 2y - z = 2$

26. (a) $\mathbf{i} + 3\,\mathbf{j} + \mathbf{k}$
    (b) $\frac{7}{3}$

27. (a) $4\,\mathbf{i} + 8\,\mathbf{j} - 6\,\mathbf{k}$
    (b) $4x + 8y - 6z + 2 = 0$

28. (a) $-3\,\mathbf{j} - 8\,\mathbf{k}$
    (b) 6
    (c) $3\,\mathbf{j} + 8\,\mathbf{k}$

29. (a) $6\,\mathbf{i} + 6\,\mathbf{j} + \mathbf{k}$
    (b) $6x + 6y + z = 25$

30. (a) $2\,\mathbf{i} + 2\,\mathbf{j} + \mathbf{k}$
    (b) $\frac{4}{3}$

32. 7% increase

33. 2% increase

34. 2%

35. 12.15

36. (a) $\frac{1}{12}$

37. $\frac{3}{2}$

38. $-\frac{27}{4}$

39. 8

40. $\frac{40}{3}$

41. $e-1$

42. 8

43. $\frac{1}{12}$

44. (a) $(5,3)$
    (b) $8\sqrt{2}$ in the direction $\mathbf{i}+\mathbf{j}$

45. Least at $(\frac{1}{4},0)$.
    Greatest at $(-1,0)$.

Appendix E

Answers to Quick Tests

E.1 Sample Quick Test

Basic Mathematical Skills

1. a
2. d
3. b
4. b
5. a
6. b
7. b
8. c
9. c
10. b
11. a
12. b
13. c
14. b
15. b
16. b
17. c

E.2 Quick Test Number 1

Basic Mathematical Skills

1. b
2. b
3. b
4. b
5. b

Vectors

1. c
2. a
3. b
4. b
5. c
6. a
7. b
8. c
9. b
10. a
11. b
12. b
13. a
14. a
15. c

Lines and Planes

1. a
2. c
3. b
4. c
5. b
6. d
7. a & c
8. b
9. d
10. c

E.3 Quick Test Number 2

Basic Mathematical Skills

1. c
2. c
3. b
4. a
5. b

Complex Numbers

1. b
2. a
3. c
4. a
5. c
6. b
7. b
8. c
9. a
10. a
11. a & c
12. c
13. b

E.4 Quick Test Number 3

Basic Mathematical Skills

1. a
2. c
3. d

Differentiation

1. b
2. b
3. c
4. c
5. d
6. a
7. a
8. a
9. b
10. b
11. $2x$
12. $5(x^2+1)^4(2x)$
13. $\frac{1}{2}(x^2+1)^{-1/2}(2x)$
14. $-\frac{1}{(x^2+1)^2}(2x)$
15. $-\sin(x^2+1)(2x)$
16. $\cos(x^2+1)(2x)$
17. $\sec^2(x^2+1)(2x)$

E.5 Quick Test Number 4

Functions and their Properties — Part 1

1. b
2. a, c & d
3. b
4. b
5. a
6. b
7. b
8. b
9. a
10. a
11. a
12. c
13. b

Functions and their Properties — Part 2

1. b
2. c
3. b
4. b
5. d
6. c
7. c
8. a – A, b – D, c – C, d – E, e – B
9. c
10. b

Functions and their Derivatives

1. a
2. c
3. b
4. b
5. b
6. d
7. c
8. b
9. d
10. b
11. b
12. c
13. b
14. b

E.6 Quick Test Number 5

Basic Mathematical Skills

1. b
2. b
3. b

Integration

1. c
2. c
3. b
4. b
5. a
6. a
7. a
8. b
9. c

E.7 Quick Test Number 6

Methods of Integration

1. b
2. c
3. c
4. c
5. e
6. b then c
7. a then c
8. b
9. a
10. a
11. b
12. a
13. b
14. a
15. a
16. b
17. c
18. b
19. d
20. a
21. c
22. b
23. a

E.8 Quick Test Number 7

Basic Mathematical Skills

1. a
2. d
3. c
4. c
5. b
6. a

First Order Differential Equations

1. b & d
2. a
3. c
4. b
5. b
6. a
7. a
8. a
9. c

Homogeneous Second Order Differential Equations

1. b
2. a
3. b
4. b
5. b
6. c
7. b
8. c
9. b

General Second Order Differential Equations

1. a
2. b
3. c
4. c
5. d
6. b
7. a

E.9 Quick Test Number 8

Introduction to Infinite Series

1. b
2. a
3. a
4. a
5. a
6. b
7. b
8. b
9. a
10. b
11. b
12. b
13. b
14. b
15. b

Tests for Convergence

1. a
2. c
3. a
4. b
5. a
6. c
7. b
8. a
9. a
10. a
11. b
12. c
13. c

Power Series

1. a
2. c
3. b
4. d
5. a
6. b
7. b
8. c
9. c
10. c
11. d

E.10 Quick Test Number 9

Elementary Row Operations

1. c
2. b
3. b
4. c

Finding an Inverse

1. c
2. a or c
3. a
4. b

Eigenvalues and Eigenvectors

1. c
2. b & c
3. -168
4. a
5. b
6. a
7. b
8. d
9. a
10. c
11. a, b & c
12. $\begin{bmatrix} -3 & 0 \\ 0 & 2 \end{bmatrix}$

E.11 Quick Test Number 10

Vector-valued Functions

1. a
2. b
3. b
4. a
5. a
6. a

E.12 Quick Test Number 11

Surfaces

1. d
2. a
3. b
4. c

Partial Derivatives

1. c
2. b
3. b
4. a
5. d
6. b
7. a

Gradient and Directional Derivative

1. b
2. b
3. b
4. b
5. a
6. a

Normal and Tangent Plane

1. b
2. b
3. b
4. a
5. c
6. c

Double Integrals

1. b
2. Upper triangle bounded by the lines $y = 2x$, $y = 2$ and $x = 0$
3. b
4. b

Appendix F

Examination Papers

Sample Examination Paper

for

Semester 1 Material

Suggested time three hours

Candidates should attempt *ALL* questions in *SECTION A* and any *TWO* (2) questions in *SECTION B.*

SECTION A is allotted 70 marks, and each question in *SECTION B* is allotted 15 marks.

SECTION A

Attempt ALL questions in this section.

A1. Let $\mathbf{a} = \mathbf{i} + 3\mathbf{j} - \mathbf{k}$, $\mathbf{b} = -2\mathbf{i} + \mathbf{j} - 5\mathbf{k}$ and $\mathbf{c} = 3\mathbf{i} - 2\mathbf{j} + 7\mathbf{k}$.

(a) Find

(i) the length of $\mathbf{b}$

(ii) the scalar component of $\mathbf{a}$ in the direction of $\mathbf{b}$

(iii) $\mathbf{b} \times \mathbf{c}$.

(b) Verify that

$$\mathbf{a} \times (\mathbf{b} \times \mathbf{c}) = (\mathbf{a} \cdot \mathbf{c})\mathbf{b} - (\mathbf{a} \cdot \mathbf{b})\mathbf{c}.$$

(Note: You may use the result obtained in part (iii) above.)

(c) (i) Determine the parametric equation of the straight line which passes through the point $(1, 0, 3)$ and is parallel to $\mathbf{b}$.

(ii) Determine the equation of the plane which contains the point $(-1, 4, 3)$ and is perpendicular to $\mathbf{c}$.

$((1 + 1 + 2) + 4 + (2 + 2) = 12$ marks)

A2. Find the derivative y' for each of the following:

(a) $y = \mathrm{e}^{-3x} \sinh 4x$

(b) $y = \operatorname{arcosh} 2x^3$

(c) $y = \log \left(\dfrac{\mathrm{e}^{5x}}{x^3} \right)$

(d) $y = \arctan x + \arctan \left(\dfrac{x+1}{x-1} \right)$.

(Simplify your answers to (c) and (d).)

$(2 + 2 + 3 + 4 = 11$ marks)

A3. If $z = 1+5i$ and $w = \sqrt{3}+i$ determine (without the use of a calculator) the values of

(a) $\operatorname{Im}\left(\dfrac{z}{2i}\right)$

(b) $\dfrac{\overline{w}}{z}$

(c) $|z + iw|$

(d) $|w|$

(e) $|w^5|$

(f) $\arg w$

(g) $\arg w^5$.

$(2+2+2+1+1+2+1 = 11$ marks)

A4. Use the tables in Appendix C, or otherwise, to evaluate:

(a) $\int_0^{\pi} \sin x \cos^2 x \, dx$

(b) $\int_0^{\pi/4} \tan^4 x \, dx$

(c) $\int \sqrt{1-9x^2} \, dx$.

$(3+3+3 = 9$ marks)

A5. Without using a calculator:

(a) Express $z = 1 - \sqrt{3}i$ in polar/exponential form.

(b) Hence determine the square roots of $1 - \sqrt{3}i$.

(c) Express the roots found in (b) in the form $a + bi$ and represent them on the same complex plane (Argand diagram).

$(1+2+3 = 6$ marks)

A6. Determine each of:

(a) $\int \frac{3x-1}{x^2(x-1)} \, dx$

(b) $\int_0^{\sqrt{\pi}} 5x \sin x^2 \, dx$

(c) $\int_0^1 xe^{-4x} \, dx$.

$(5+3+3 = 11$ marks)

A7. A curve has equation

$$x^3 - xy + y^2 = 7.$$

(a) Show that the tangent to the curve is horizontal at the point $(1,3)$.

(b) Determine y'' at the point $(1,3)$.

(3 + 2 = 5 marks)

A8. (a) Sketch the (finite) region R bounded by the lines $y = x$ and $y = x + 1$ between $x = 1$ and $x = 2$.

(b) Set up a definite integral (DO NOT EVALUATE THE INTEGRAL) which gives the volume of the solid obtained by rotating the region R about the line $y = -1$.

(2 + 3 = 5 marks)

SECTION B

Attempt TWO (2) questions in this section.

B1. (a) Find

(i) $\displaystyle\int \frac{x-4}{x^2-4x+13}\,dx$

(ii) $\displaystyle\int e^{2x}\sinh 5x\,dx$.

(b) Sketch the set of points in the complex plane defined by $\{z : |z| \leq |z+3-i|\}$.

(c) The Lost Pyramid (shown at right) has two secret passages represented by the lines

$$\ell_1 : \quad (x,y,z) = (3,0,-5) + t\,(-1,1,4)$$

and

$$\ell_2 : \quad (x,y,z) = (4,1,1) + s\,(3,-1,-2).$$

A fabulous treasure is believed to be located at the point of intersection of the passages. Determine the coordinates of this point.

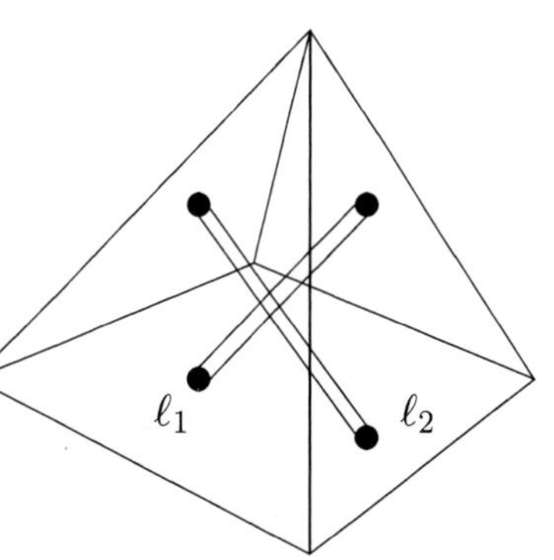

((4 + 3) + 4 + 4 = 15 marks)

B2. (a) If $z = x + iy$ and $f(z) = \dfrac{z-1}{z+1}$,

(i) Write $f(z)$ in the form $f(z) = u(x, y) + i\, v(x, y)$.

(ii) Determine the curve in the complex plane for which $\operatorname{Re} f(z) = 0$.

(b) A rubber band is stretched between two fixed points $P(1, 0, 0)$ and $Q(0, 1, 0)$, and a third point $R(0, 0, \ell)$ which is free to move up or down the z axis (as shown at right; all measurements are in metres).

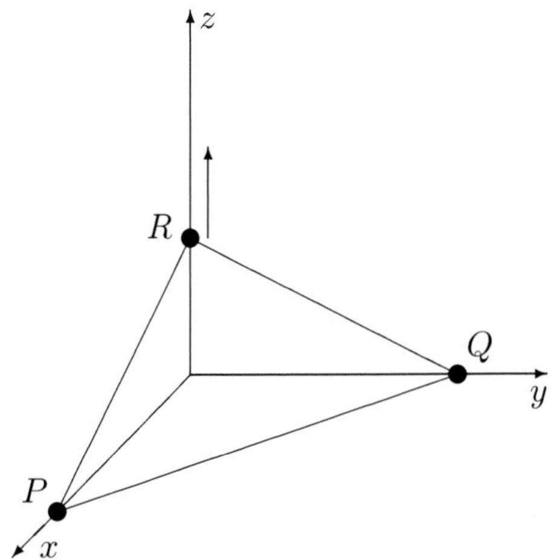

(i) Show that the area of the triangle formed by P, Q and R is

$$A = \frac{1}{2}\sqrt{2\ell^2 + 1}.$$

(Note: You may use the result $A = \frac{1}{2}|\mathbf{a} \times \mathbf{b}|$.)

(ii) If ℓ is steadily increasing at a rate of $0.3\,\mathrm{ms}^{-1}$, how fast is the area A of the triangle increasing when the area is $1.5\,\mathrm{m}^2$.

$((5 + 2) + (4 + 4) = 15$ marks)

B3. (a) Use Newton's method ONCE with an initial guess of $x_0 = 1.5$ to find an approximation to the solution of $x = x^3 - 1$.

(Newton's method for solving $f(x) = 0$: $\quad x_{n+1} = x_n - \dfrac{f(x_n)}{f'(x_n)}$

for $n = 0, 1, 2, \ldots$)

(b) Evaluate $\cosh(\log 7)$ exactly, that is, without the use of a calculator.

(c) A flagpole is to be constructed in the shape of an upside-down Y as shown. The desired height of the pole is 12 m and the desired distance between the legs is 10 m.

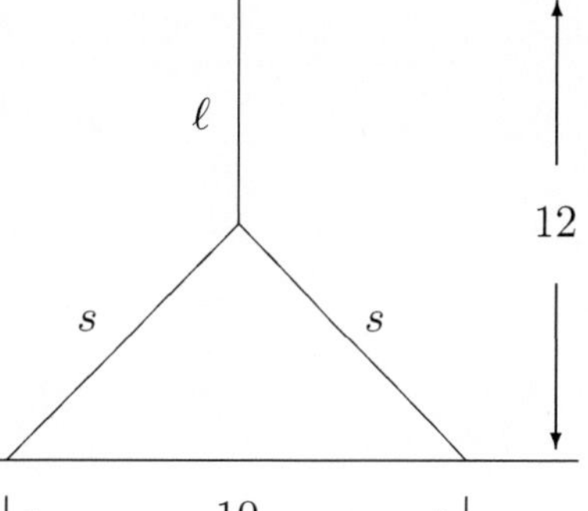

(i) Write down a formula for the length L of material used in terms of s and ℓ.

(ii) Show that $\ell + \sqrt{s^2 - 25} = 12$.

(iii) Explain why s can only take values in the interval $[5, 13]$.

(iv) Hence determine the dimensions s and ℓ which minimize L.

$(4 + 2 + (1 + 2 + 1 + 5) = 15$ marks)

B4. (a) Using logarithmic differentiation, find y' if $y = 3^x e^{x^2}$.

(b) (i) Show that the area A of a circle is related to its circumference C by $A = C^2/(4\pi)$.

(ii) For a given circle, suppose C is measured as $3.0 \pm 0.2\,m$. Use calculus to approximate the maximum possible error in A.

(c) Certain integrals look simple enough, but cannot be evaluated in terms of elementary functions, e.g.

$$I = \int_0^{\pi/4} x \tan x \, dx.$$

(Not even integration by parts will work here!)

(i) Sketch the graph of the function $y = x \tan x$ for $0 \le x \le \pi/4$. Use your sketch to show that $I \le \pi^2/16$. (Hint: $0 \le x \tan x \le \pi/4$ for $0 \le x \le \pi/4$.)

(ii) Use Simpson's rule with four strips to estimate I given

x	0	$\pi/16$	$\pi/8$	$3\pi/16$	$\pi/4$
$x \tan x$	0.000	0.039	0.163	0.394	0.785

(Simpson's rule is:

$$S_n = \frac{h}{3}\{y_0 + y_{2n} + 4(y_1 + y_3 + \cdots) + 2(y_2 + y_4 + \cdots)\}\,.)$$

(4 + (1 + 4) + (2 + 4) = 15 marks)

Sample Examination Paper

for

Semester 2 Material

Suggested time three hours

Candidates should attempt *ALL* questions in *SECTION A* and any *TWO* (2) questions in *SECTION B*.

SECTION A is allotted 70 marks, and each question in *SECTION B* is allotted 18 marks.

SECTION A

Attempt ALL questions in this section.

A1. (a) Determine the general solution of

$$y'' - 5y' + 6y = 0.$$

(b) Using the result from (a) determine the general solutions of

(i) $y'' - 5y' + 6y = 6e^{4x}$

(ii) $y'' - 5y' + 6y = 18x$

(iii) $y'' - 5y' + 6y = 10e^{2x}$.

(2 + (3 + 3 + 5) = 13 marks)

A2. Determine the radius of convergence of the following power series:

(a) $\displaystyle\sum_{n=1}^{\infty} \frac{(3x-1)^n}{n+1}$

(b) $\displaystyle\sum_{n=0}^{\infty} \frac{(-1)^n x^{2n}}{n!}$.

(3 + 3 = 6 marks)

A3. Let $A = \begin{bmatrix} 1 & 2 & 4 \\ 0 & 1 & 0 \\ 1 & 0 & 1 \end{bmatrix}$.

(a) Evaluate AA^T and A^TA. Explain why they differ.

(b) Verify that $\begin{bmatrix} 2 \\ 0 \\ 1 \end{bmatrix}$ and $\begin{bmatrix} -2 \\ 0 \\ 1 \end{bmatrix}$ are eigenvectors of A.

Write down the corresponding eigenvalues.

(4 + 3 = 7 marks)

A4. Find <u>all</u> values of the constant α for which the function $u(x, y) = e^{\alpha y} \cos 2x$ satisfies

Laplace's equation $\dfrac{\partial^2 u}{\partial x^2} + \dfrac{\partial^2 u}{\partial y^2} = 0.$

(4 marks)

A5. Let $f(x, y, z) = 1 + y/x + 4x/z$.

(a) Find ∇f at the point $P(1, 5, 2)$.

(b) Determine the directional derivative of $f(x, y, z)$ at the point P in the direction of
$2\mathbf{i} - 5\mathbf{j} + 4\mathbf{k}$.

(c) In what direction does f <u>increase</u> most rapidly at P?

$(3 + 3 + 1 = 7$ marks)

A6. Evaluate the double integral

$$\iint_R (2y - 3x)\, dx\, dy$$

where R is the (finite) region in the first quadrant bounded by the lines $y = x + 1$, $y = 2x$ and the y axis.

(7 marks)

A7. Determine the solution of the differential equation

$$y' + 3y = 1 + x^2 e^{-3x}$$

which satisfies $y(0) = 1$.

(5 marks)

A8. The equation of a surface S is $z = \sqrt{13 - 3x - 7y}$.

(a) Find a vector normal to S at the point $P(-1, 1, 3)$.

(b) Hence find the equation of the tangent plane to the surface S at P.

$(3 + 2 = 5$ marks)

A9. Using elementary row operations, determine the general solution of the system of linear equations

$$\begin{aligned} 2x_1 + 5x_2 + x_3 - 2x_4 &= 0 \\ x_1 + x_2 - x_3 - x_4 &= 0 \\ 3x_1 + 4x_2 - 2x_3 - 3x_4 &= 0. \end{aligned}$$

(Full marks will not be awarded unless all working is shown.)

(6 marks)

A10. Given

$$e^t = 1 + t + \frac{t^2}{2!} + \frac{t^3}{3!} + \cdots, \qquad \text{valid for all } t$$

and

$$\frac{1}{1-t} = 1 + t + t^2 + t^3 + \cdots, \qquad \text{valid for } |t| < 1$$

(a) Show that

$$\frac{e^{-x}}{1 + x/2} = 1 - \frac{3}{2}x + \frac{5}{4}x^2 - \frac{19}{24}x^3 + \cdots - \cdots.$$

(b) <u>Write down</u> the radius of convergence of the series in (a).

(c) Use the first three terms of the series in (a) to obtain an approximation to the integral

$$\int_0^{0.5} \frac{e^{-x}}{1 + x/2}\, dx$$

and give an upper bound for the error involved.

(4 + 2 + 4 = 10 marks)

SECTION B

Attempt TWO (2) questions in this section.

B1. (a) The position of a moving particle at time t is given by

$$\mathbf{r}(t) = t\,\mathbf{i} + t^2\,\mathbf{j} + te^{3t}\,\mathbf{k}, \qquad t \geq 0$$

(i) Find the velocity $\dot{\mathbf{r}}(t)$ of the particle at $t = 0$.

(ii) Write down an integral (DO NOT EVALUATE THIS INTEGRAL) for the distance travelled between $t = 0$ and $t = 2$.

(b) Show that $u(x,t) = \log\left(\dfrac{x-2t}{x+2t}\right)$ satisfies the wave equation $\dfrac{\partial^2 u}{\partial t^2} = 4\dfrac{\partial^2 u}{\partial x^2}$.

(c) Let $A = \begin{bmatrix} 1 & 4 & 0 \\ 4 & 1 & 3 \\ 0 & 3 & 1 \end{bmatrix}$.

(i) Show that the eigenvalues of A are -4, 1 and 6.

(ii) Find a set of three linearly independent eigenvectors.

((2 + 2) + 5 + (3 + 6) = 18 marks)

B2. (a) Determine the general solutions of

(i) $y'' - y' = 10\sin 2x$

(ii) $y'' + 4y' + 13y = 3e^{-2x}$.

(b) If A_0 dollars are initially invested in a bank account which pays yearly interest at the rate of $x\%$, then after n years the account will contain $A_n = A_0(1+x/100)^n$ dollars. The amount of money in the account will double (i.e. $A_n = 2\,A_0$) when

$$n = \frac{\log 2}{\log(1 + x/100)}.$$

(i) Given the power series

$$\frac{1}{1+t} = 1 - t + t^2 - t^3 + \cdots - \cdots, \qquad \text{valid for } |t| < 1,$$

show by integration that

$$\log(1 + x/100) = \frac{x}{100} - \frac{1}{2}\frac{x^2}{100^2} + \frac{1}{3}\frac{x^3}{100^3} + \cdots - \cdots$$

(ii) State the radius of convergence of the series found in (i).

(iii) Use the first term of the series in (i) to show that the number of years n for the sum of money to double is given approximately by

$$n \approx \frac{100 \log 2}{x} \approx \frac{70}{x}.$$

(This is known as the "Rule of 70.")

$((5 + 5) + (4 + 2 + 2) = 18$ marks)

B3. (a) Determine, with reasons, which of the following series converge:

(i) $\displaystyle\sum_{n=1}^{\infty} \frac{n + \sin n}{4n^3 + 7n + 5}$

(ii) $\displaystyle\sum_{n=1}^{\infty} \frac{3n^2 + 4}{\sqrt{n^5 + 2n^3 + 1}}$

(iii) $\displaystyle\sum_{n=1}^{\infty} \frac{(-1)^{n+1}}{\sqrt{4n + 3}}$.

(b) The depth in metres $d(x, y)$ of water at any point (x, y) of a large bay is given (at low tide) by

$$d(x, y) = 10 - \frac{100}{40 + 5x^2 + 2y^2 + 4xy - 24x - 12y}$$

where x and y are measured in kilometres from a marker buoy. It is known that there is a tiny barely-submerged mud island somewhere in the bay.

(i) Determine the location of the mud island, i.e. find where $d(x, y)$ is a minimum, using calculus.

(ii) What is the depth of water at low tide at the mud island?

$((3 + 3 + 3) + (8 + 1) = 18$ marks)

B4. (a) (i) Sketch the region of integration for the double integral

$$\int_{y=0}^{y=4} \left\{ \int_{x=\sqrt{y}}^{x=2} e^{x^3}\, dx \right\} dy.$$

(ii) Hence evaluate the integral by reversing the order of integration.

(b) A simple electric circuit consists of a battery (producing 12 volts), a resistor (10 ohms) and an inductor (5 henries) connected in series. By one of Kirchhoff's laws, the current $i(t)$ in the circuit satisfies the first order differential equation

$$5\frac{di}{dt} + 10i = E \qquad \text{where} \qquad E = 12.$$

(i) Solve this equation subject to the initial condition $i(0) = 0$.

(ii) Show that at time $t = 0.5\,s$ the current in the circuit is $i = 1.2\,(1 - \mathrm{e}^{-1})$.

(iii) Suppose that the battery is disconnected at $t = 0.5$. The current in the circuit will still satisfy the above differential equation, but with $E = 0$ instead of $E = 12$. Solve this equation for $i(t)$ subject to the initial condition given in (ii).

(iv) Give a sketch of $i(t)$ for $t \geq 0$.

$((2 + 5) + (5 + 1 + 3 + 2) = 18$ marks)

Bibliography

[1] Anton, H. & Herr, A. *Calculus with Analytic Geometry*, 5th ed., New York, John Wiley & Sons, 1995.

[2] Edwards, C.H. & Penney, D.E. *Calculus and Analytic Geometry*, 5th ed., Englewood Cliffs, New Jersey, Prentice-Hall, 1998.

[3] Grossman, S. *Calculus*, 5th ed., Fort Worth, Texas, Saunders College, 1992.

[4] Farrand, S.M. & Poxon, N.J. *Calculus*, Fort Worth, Texas, Harcourt Brace Jovanovich (College Outline Series), 1984.

[5] Fitz-Gerald, G.F. (Ed.), *Tables*, 4th ed., RMIT Lecture Notes in Mathematics, Melbourne, Maths Dept. RMIT, 2003.

[6] Hoffmann, L.D. & Bradley, G.L. *Brief Calculus with Applications*, 7th ed., New York, McGraw-Hill, 2000.

[7] Jordan, D.W. & Smith, P. *Mathematical Techniques*, 3rd Ed., Oxford, OUP, 2002.

[8] Stewart, J. *Calculus*, 5th ed., Belmont, California, Brookes/Cole (ITP), 2003.

[9] Thomas, G.B., Finney, R.L., Weir, M.D. & Giordano, F.R. *Calculus and Analytic Geometry*, Updated 10th ed., Reading, Massachusetts, Addison-Wesley, 2003.

[10] Zill, D.G. *Calculus with Analytical Geometry*, 3rd ed., Boston, PWS-Kent, 1992.

Index